Next Generation Data Communication Technologies:

Emerging Trends

Debashis Saha
Indian Institute of Management - Calcutta, India

Varadharajan Sridhar
Sasken Communication Technologies, India

Information Science REFERENCE

Managing Director: Lindsay Johnston
Senior Editorial Director: Heather Probst
Book Production Manager: Sean Woznicki
Development Manager: Joel Gamon
Development Editor: Hannah Abelbeck
Acquisitions Editor: Erika Gallagher
Typesetters: Lisandro Gonzalez
Print Coordinator: Jamie Snavely
Cover Design: Nick Newcomer

Published in the United States of America by
Information Science Reference (an imprint of IGI Global)
701 E. Chocolate Avenue
Hershey PA 17033
Tel: 717-533-8845
Fax: 717-533-8661
E-mail: cust@igi-global.com
Web site: http://www.igi-global.com

Library of Congress Cataloging-in-Publication Data

Next generation data communication technologies: emerging trends / Debashis Saha and Varadharajan Sridhar, editors.
p. cm.
Includes bibliographical references and index.
Summary: "This book contains case studies, theories, and empirical research aimed to assist individuals and organizations in understanding the critical concepts of data networking and communications"--Provided by publisher.
ISBN 978-1-61350-477-2 (hardcover) -- ISBN 978-1-61350-478-9 (ebook) -- ISBN 978-1-61350-479-6 (print & perpetual access) 1. Data transmission systems. 2. Computer networks. I. Saha, Debashis, 1965- II. Sridhar, Varadharajan.
TK5105.N495 2012
621.39'81--dc23

2011044137

British Cataloguing in Publication Data
A Cataloguing in Publication record for this book is available from the British Library.

Table of Contents

Detailed Table of Contents

The wireless technology market has witnessed a complete paradigm shift as multiple standards and protocols are emerging almost every day. Each and every standard has its limitations and merits, which can be either masked or complemented by some other standards. The demands from the service providers are now sky-high and for the complete commercialization, it is expected that even with scarce network resources all kind of services would be provided, especially in a cost effective manner. This burning issue compels a service provider to roll out some integrated wireless networks to exploit the virtues of each. This chapter formulates the planning problem of an overlay network integrating particularly, 3G, WiMAX, and WLAN. The issue of planning is to establish proper connectivity amongst the three network standards which is unique in its nature. In the proposed planning approach, the authors have endeavored to minimize total cost for vertical handoff generated in the overlay network as well as the cost for wire line connection amongst the various network gateways of the overlay hierarchy. In this work, the authors have focused on the initial planning phase. For validating the novel planning problem, the chapter has taken recourse to simulated annealing (SA) and a well cited meta-heuristic H-II. The authors have also presented comparison of the performances of SA and H-II with a variant of distance based planning (DBP) scheme in this domain.

The success of fourth generation networks (4G) will be driven by intelligent interactivity between access networks and user devices and their ability to meet increasing user needs in term of bandwidth, availability, and especially session continuity. Handling mobility issues over heterogeneous environments is crucial for transparent handoff and efficient utilization of the scarce radio resources with QoS provisioning and is still an open issue. The purpose of this chapter is to develop a novel mobility aware adaptive resource reservation scheme for fourth generation networks that optimizes system performances and guarantees end to end QoS for individual users. Considering the importance of an appropriate use of scarce radio resources, three resource reservations algorithms are proposed respectively for UMTS, WLAN, and WiMax access networks. These algorithms take into consideration both future location of the mobile terminal expressed in probabilistic way and the load conditions of target Access Points (AP)/Base Stations (BS) to determine the amount of radio resources to be reserved for both uplink and downlink. Simulations show that the proposed scheme achieves better results than those of other resource reservation schemes for metrics like bandwidth utilization, handoff latency, and packet loss.

In WSN transport, layer protocol plays a significant role in maintaining the node's energy budget. To find out the dependency of Transport layer on MAC/PHY layer, the authors have extensively tested various transport protocols using IEEE 802.11, IEEE 802.15.4 MAC/PHY protocols for WSN. For IEEE802.11 and IEEE802.15.4 with RTS/CTS ON the TCP variants has shown >80% packet delivery ratio and 5-20% packet loss, while for UDP it is around >63% and 19.54-35.18% respectively. On average 1-3% additional energy is consumed for packet retransmissions in IEEE 802.11 with RTS/CTS OFF whereas significant energy efficiency is observed in IEEE802.15.4 case. For IEEE 802.11 with RTS/CTS ON high throughput, low packet drop rate and increased E-2-E delay is observed, while for IEEE 802.15.4 improved power efficiency and jitter behavior is observed. This has led the foundation for the future development of the cross-layered energy efficient transport protocol for WSN.

A Mobile Ad hoc Network (MANET) is a collection of wireless nodes communicating over multi-hop paths without any infrastructure. Nodes must cooperate to provide necessary network functionalities. The security in routing protocols like Dynamic Source Routing (DSR) can be compromised by a "Black Hole" attack. Here, a malicious node claims to have the shortest path to the destination and attracts all traffic and drops them, leading to performance degradation. The situation becomes worse when two or more nodes cooperate and perform the "Cooperative black hole" attack. This chapter proposes a solution based on probing to identify and prevent such attacks. The proposed solution discovers a secure route between the source and destination by identifying and isolating the attacking nodes. Simulation results show that the protocol provides better security and performance in terms of detection time, packet delivery ratio, and false negative probability in comparison with trust and probe based schemes.

Security is a prime concern in providing communication between vehicles in a Vehicular Ad-Hoc Network (VANET), where they share information among themselves and nearby fixed equipments. Unlike wired networks, the unique characteristics of Vehicular Ad hoc Networks (VANETs) pose a number of non-trivial challenges to security design. This chapter presents a threshold security mechanism with a mobility based clustering for Inter Vehicle Communication (IVC). Nodes that have similar moving pattern are grouped into a cluster. Unlike other clustering algorithms, it takes the moving pattern of the vehicles into consideration along with the driver's intention. The stability of clusters is estimated based on relative mobility of cluster members. A threshold cryptographic scheme is employed on top of the clusters to protect routing information and data traffic. To ensure distributed trust in the clustered environment, the private key (k) is divided into n pieces in such a way that k is easily reconstruct able from any p number of pieces. Even complete knowledge of (p-1) pieces reveals absolutely no information about k.

In this paper, the authors propose a dynamic Public Key Infrastructure (PKI) for vehicular ad hoc networks to distribute the role of the central certification authority (CA) among a set of dynamically elected CAs. The election process is based on a clustering algorithm relying on trust levels and relative mobility. Furthermore, the authors have adapted the Dynamic Demilitarized Zones to protect the elected CAs from malicious nodes and enable them to act as registration authorities (RA). Extensive simulations are conducted to evaluate the performance of the clustering algorithm and investigate the impact of the vehicle speed, the vehicle average arrival rate, and the percentage of confident vehicles on the stability and efficiency of the security infrastructure. The authors demonstrate the percentage of confident nodes has a little impact on these performance metrics and that the minimum number of CAs to cover the entire platoon.

Radio frequency identification technology is becoming ubiquitous, and as an unfortunate side effect, more and more authentication solutions come with more security issues. In former contributions, the authors introduced a solely hash-based secure authentication algorithm that is capable of providing protection against most of the well-known attacks and performs exceptionally well even in very large systems. The authors gave a theoretical analysis of Simple Lightweight Authentication Protocol (SLAP) protocol from security and performance point of view. This chapter gives a detailed examination of small computational capacity systems from the point of view of security. The authors define the model of attacker and the well-known attacks which can be achieved in these kinds of environments. Furthermore, the chapter gives a summary of the significant RFID authentication protocols which are found in literature. The authors present several lightweight authentication protocols and some novel elliptic curve cryptography based methods. Besides, the chapter illustrates the SLAP protocol's performance characteristics with measurements carried out in a simulation environment and compares with the theoretical results. The authors show the effects of numerous attacks and the system's different parameters on the authentication time. Finally, the chapter examines the performance and security characteristics of two other protocols chosen from the literature in order to compare to SLAP algorithm and give proper explanation for the differences between them.

Section 3
Network Data Services and Applications: Improving Interoperability and Performance

Preface

The editors bring you, in this edited volume, a collection of select articles, which were published in the International Journal of Business Data Communications (IJBDCN) in recent past. These research essays provide a snapshot of the emerging communications technologies, related telecom market evolution, associated novel business models, and the resulting research issues. The book is divided into four sections, namely (a) Next Generation Communication Networks: Towards enhanced design, interoperability and performance, (b) Communication Network Security: Improved Architecture and Prevention Mechanisms, (c) Network Data Services and Applications: Improving Interoperability and Performance, and (d) Network Economy: New Business Models, which are briefly introduced later.

Ever since mobile wireless networks have taken the center stage, wired networks have gone into the background just as they have been pushed to the core by the wireless access. It all began with voice in the wireless domain, but now data as usual is surging ahead as the killer service. A recent research study points out that mobile data traffic will grow at a Compound Annual Growth Rate of 92 percent from 2010 to 2015, reaching 6.3 Exabytes per month by 2015. The study also points out "The mobile-only Internet" population will grow 56-fold from 14 million at the end of 2010 to 788 million by the end of 2015 (Cisco, 2010). These trends clearly indicate the possible exponential growth in the use of mobile devices to access Internet and bandwidth intensive applications. Though technologies such as Long Term Evolution (LTE), LTE-Advanced, WiMAX, and mobile WiMAX (IEEE 802.16e) are gaining ground in different geographical areas and being actively adopted by the mobile operators worldwide to provide enhanced user experience, spectrum crunch is evident. In tune with these trends, there are two active research areas in networking. One is Cognitive Radio (CR), which enables flexible and dynamic spectrum usage with cognitive capabilities for opportunistically seizing vacant Radio Frequency spectrum blocks in both licensed and unlicensed bands (Chapin & Lehr, 2007). In a major breakthrough to the proponents of CR technologies, IEEE released 802.22 standards earlier this year that define the unlicensed use of frequencies between TV channels in the Very High Frequencies (VHF) and Ultra High Frequency (UHF) bands. However, there are still evolving research problems on interference management, frequency allocation algorithms, and power management that need to be addressed as early as possible.

Wireless technology is also in the midst of an important stage in its techno-commercial evolution from radios *with behavior fixed in hardware* to radios *with behavior determined by software*. The flexibility demanded by the radio systems can potentially be embedded best in software rather than hardware and hence the development of Software Defined Radios (SDR) (Partridge, 2011). Though the SDRs on their own could not stand commercially, the CR combined with SDRs to provide flexible spectrum management provides an enlarged space for wireless researchers to find ways to allow software radios to use the spectrum appropriately.

Mobile Ad-hoc Networks (MANETs) continue to generate interesting research problems in the area of connectivity, routing, security, and power management. Dynamic topologies, bandwidth constraints, energy-constrained operations, wireless vulnerabilities, and limited physical security are among the characteristics that differentiate mobile ad hoc networks from fixed multihop networks. These networks when connected to high-bandwidth wireless backhaul networks require efficient hand-off mechanisms, and bandwidth shaping rules.

However, the non-proliferation of multi hop mobile ad hoc networks, despite their practical usefulness and need, is mainly due to their poor performances caused by the lack of viable low signaling and high validity routing algorithms. Routing still plays the central and vital role to yield an appropriate and efficient running of these networks, yet it stands out as one of the most crucial actors to attain adequate performances and viable deployments. The crux of routing in such networks lies in its ability to maintain and provide routes that are valid at the instant of their requests otherwise traffic may wander inside the network without ever being able to be delivered to its ultimate destinations. Worse yet, invalid routes amount to over using the wireless communication medium and hence throttling the network from being able to yield good performances.

The proliferation and advancement of wireless communication technologies have revolutionized human lifestyles by providing the best convenience, safety, and flexibility ever in deploying various types of personal communication applications and accessing the Internet services. Car manufacturers have already started equipping cars with the technology that allows drivers and passengers to communicate with each other as well as with the roadside infrastructure, in order to improve the driving experience and make driving safer through Vehicular Ad hoc NETworks (VANETs).

Developments of low power, low cost, embedded wireless devices capable of gathering useful environmental information has led to the developments of Wireless Sensor Networks (WSNs). The WSN technology has gained importance in a range of application scenarios such as military, healthcare, environmental monitoring, and surveillance. In this context, there is a strong need for open protocols to enable the cohabitation of various medium technologies and interfacing facilities between devices.

The field of sensor and mesh networking is reemerging amid unprecedented growth in the scale and diversity of Machine-to-Machine (M2M) networking. M2M enables one device to communicate its status continually or sequentially to another device, often linked to a central management system. People and vehicles can be internetworked in areas without a pre-existing communication infrastructure, or when the use of such infrastructure requires wireless extension. Therefore, such networks are designed to operate in widely varying environments, from military networks (with hundreds of nodes) to low-power sensor networks and other embedded systems.

As more and more devices get networked, need for managing these heterogonous devices on the network is challenging. The Digital Living Network Alliance (DLNA), comprised of leading consumer electronics, computing industry, and mobile device companies, envisions a wired and wireless network of interoperable consumer electronics, personal computers, and mobile devices in the home and on the road, enabling a seamless environment for sharing and growing new digital media and content services. This provides a fertile ground for researchers to work on issues relating to addressability, routing of multimedia content, protocol optimization, radio access network problems, content security, and authorization.

The ubiquitous Power Line Communication (PLC) medium - the electricity distribution network, is being looked at actively for providing high-speed access. While PLC proprietary protocols restrain its usage, collaboration with classic RF solutions is possible using open standards, such as IPv6. Efforts are on to adapt the IEEE 802.15.4 standard over PLC.

Topology control is one of the fundamental research topics in wireless ad hoc and sensor networks. There are varieties of means to perform topology control such as turning on/off network nodes, changing transmission power of network nodes, and building hierarchical network topologies. When designed properly, topology control can improve the operation of wireless ad hoc and sensor networks in terms of energy efficiency, mobility resilience, network capacity increase, interference reduction, et cetera. Nevertheless, many aspects need to be addressed when designing suitable topology control mechanisms for wireless ad hoc and sensor networks. For instance, both the gains by means of topology control and the drawbacks (e.g., relevant overhead on communication & management) from it need to be taken into account. In addition, the realistic networking environments (such as homogeneous & heterogeneous devices, stationary or mobile nodes, QoS requirements, traffic patterns) pose challenges on topology control to work together with other networking tasks. Moreover, topology control has interactions with other layers in the protocol stack such as MAC (Medium Access Control) and routing mechanisms. Hence, comprehensive research efforts on topology control techniques are necessary to meet the various requirements in wireless ad hoc and sensor networks.

Sensor networks and grid computing are being deployed to better manage public utilities such as electricity, water and transport. Smart grids improve the efficiency of electricity grids through active monitoring and reducing reliance on centralized electricity production. Smart city networks monitor traffic, carbon emission, energy usage and help reduce traffic congestion and help earn cities earn carbon credit. Smart M2M communications are a growing area in wireless telecommunications and are expected to be behind 80% of the total carbon savings. Research issues abound in this area relating to optimization of sensor network topologies, performance improvement, energy and power optimization.

Historically, telecom service providers were focused on supplying connectivity based on the concept of Virtual Private Networks (VPNs). Over the years, these services have evolved to facilitate outsourcing of advanced functionalities such as routing and even security (L3 IP VPNs, security as a service, etc.). Soon VPN service providers started to offer dynamic access (through VPNs) to their own skeleton cloud services and also to 3rd party cloud services; e.g., the "VPN gallery service" of Orange Business Services. Recently, advances in distributed systems technology have made it easy for the provisioning of services on an unprecedented scale and with increasing flexibility. At the same time, businesses have started to embrace a model wherein 3rd party services that can be acquired with minimal service provider interaction, replace or complement those that are managed internally. Organizations have only started to grasp the economic implications of this evolution where service and cloud computing enable resource to be employed in utility-based fashion.

With the evolution of wireless Wide Area Networks, the wireless Local Area Networks (LANs) technologies have also evolved from 802.11b/g to 802.11n, offering about 600 Mbps access speeds. Interconnection of LANs with WANs presents opportunities for researchers to explore problems with capacity and access constraints. Seamless Connectivity in such networks entirely depends on efficient optimized handoff mechanisms in order to ensure non -disruptive session continuity and traffic redirection in a seamless and secure transparent manner. The network dynamics due to node mobility, the nature of the wireless channel, interference between different users and collisions adversely impact network performance across all layers of the protocol stack.

In the core network, through IP multimedia sub systems are being deployed currently, research on next generation core network protocols, addressing schemes, routing methods, and dealing with more sophisticated and bandwidth intensive edge networks will require substantial attention in the years to come.

To take care of the burgeoning demand for the Internet bandwidth, scientists are turning to optoelectronic technology – transmitting data using light. However, research question remains on its capacity to handle huge bandwidth demand at low cost in an energy efficient way. As the Internet paves way for people across the globe to engage in intense collaborative activity, need for efficient, scalable, and energy efficient Internet architecture, and technologies are required and research questions in this domain remain abundant.

While these bandwidth intensive technologies are getting deployed, it becomes imperative for these heterogeneous networks to inter-operate seamlessly. On the other, as various multimedia applications with different Quality of Service (QoS) requirements are becoming popular, provisioning these services seamlessly to the end customers is becoming a business challenge for wireless service providers. Research problems in the area of defining, measuring, and monitoring multimedia QoS are of interest in the days to come.

With this brief background on what is happening around us nowadays, the editors introduce the four relevant sections of the book next.

NEXT GENERATION COMMUNICATION NETWORKS: TOWARDS ENHANCED DESIGN, INTEROPERABILITY, AND PERFORMANCE

In the first section on "*Next Generation Communication Networks,*" the editors included research articles that address the challenges in design and performance optimization in the next generation wireless enabled networks.

In their chapter titled "*Planning and Dimensioning of the 3G UMTS Core Networks*", Ouyang and Fallah discuss about Universal Mobile Telecommunications System (UMTS) - one of the third-generation (3G) mobile telecommunications technologies, which is also being developed into a 4G technology. The chapter provides detailed guidelines and algorithms for dimensioning the UMTS core networks to enable any mobile operator's network planning process to be independent from the vendor bias. The authors claim that though the current literature provides many practical tools or theoretical methods to design, plan, and dimension Global System for Mobile Communications (GSM) and UMTS radio networks, they mostly overlook the algorithms for network planning and dimensioning for core networks of GSM, UMTS, and IP Multimedia Subsystem (IMS). This chapter introduces an algorithm for traffic, bandwidth, and throughput dimensioning of the network entities in the UMTS core network. The analysis is based on the traffic and throughput generated or absorbed in the interfaces of the network entities in the UMTS core network. The chapter ends with a case study to verify the algorithms created for UMTS core network. The algorithms developed have been successfully applied in dimensioning a nationwide UMTS network in North Africa and also adopted in an optimization tool by a mobile operator in the United States recently. This chapter is aimed at helping UMTS network operators in dimensioning an optimum network size and building an optimum network structure to deliver an optimum quality of service for users. The dimensioning rules and guidelines provided in the chapter could also help the mobile operators to appropriately size their networks to minimize their Total Cost of Ownership (TCO), which includes capital and operational expenditure.

In the chapter "*Cartography and Stability to Enhance and Self Regulate Proactive Routing in MANETs*" by Abid and Belghith, the authors propose a novel proactive routing protocol for wireless mobile ad hoc networks. The novelty of their proposal stems from the fact that they allow different nodes to adopt different routing period sizes calculated individually, locally, and dynamically. As such, stationary or slowly moving nodes adopt a large period size while fast moving nodes utilize rather a small period size. The size of the routing period depends on the level of the dynamics perceived by each node in an autonomous way where no synchronization is required. Moreover, and to better track node mobility and further mitigate its effect, every node integrates a self- regulating process that keeps calibrating the currently chosen period size. A detailed performance investigation is carried out to demonstrate the betterment brought by their proposal against the de facto OLSR.

The chapter by Paul et al., titled "*Handoff Cost Minimization and Planning of Heterogeneous Integrated Overlay Networks: Meta-Heuristics Based Approach,*" analyzes the problem of hand-off cost minimization in such overlay heterogeneous networks. The authors present an optimization problem that seeks to find an optimal assignment of each WLAN to a specific WiMax base station subject to capacity constraints of the WiMax base station. The problem is modeled as a generalized assignment problem with a non-linear hand-off cost function. A simulated annealing based heuristic approach is used to solve the problem.

The chapter on "*QoS-Aware Scheme for Mobility Management and Adaptive Resource Reservation in 4G Wireless Networks,*" by Trabelsi and Boudriga, tackles one of the major issues in 4G networks: optimizing the handoff procedure to ensure seamless session continuity with minimum latency. They proposed various dynamic and predictive radio resource reservation algorithms in order to enhance the handoff performance in 4G networks. These algorithms concern both of the uplink and downlink and maximize the handoff success probability. A detailed performance study is carried out to demonstrate the efficiency of their proposals.

In general, WSNs consume high energy due to path loss, retransmissions, congestion, and high Bit Error Rates. Hence, optimizing energy consumption in WSNs is the biggest challenge. In WSNs, the transport layer protocol plays a significant role in maintaining the nodes's energy budget. In the chapter on "Dependency of Transport Functions on IEEE 802.11 and IEEE 802.15.4 MAC/PHY Layer Protocols for WSN: A Step towards Cross-Layer Design," Sharif et al. focus on the role of transport agent for ensuring data delivery subject to nodes' energy cost. To find out the dependency of the transport layer on physical and Medium Access Control layer, the authors have extensively tested various transport protocols using IEEE 802.11, 802.15.4 physical, and MAC layers protocols for WSNs. The work paves way for the future development of cross-layer energy efficient transport protocols for WSNs.

This concludes Section 1.

As bandwidth-intensive applications and networks diffuse, the need for effective security management, not only in organizations, but also across national infrastructure, continues to increase. As each country is creating equivalents of the US Department of Homeland Security, the need for providing surveillance, monitoring, and management is increasing.

Besides weak security, the biggest weakness in today's Internet and Intranet is the lack of built-in network management techniques. Among the ideas under consideration are automated ways to reboot systems, self-diagnosing protocols, finer grained data collection, and better event tracking. All of these tools will provide better information about the health and status of networks so that security, availability, and reliability of networks can be improved.

The VANETs come with certain challenges of security and privacy preservation aspects, in which any malicious behavior of users, such as a modification to the disseminated messages, could be fatal to the other users.

COMMUNICATION NETWORK SECURITY: IMPROVED ARCHITECTURE AND PREVENTION MECHANISMS

In the second section of the book, "*Communication Network Security,*" contemporary research articles address the challenges in the design of a proper security architecture and appropriate prevention mechanism.

The chapter by Subathra, et al., "*Secure Route Discovery in DSR against Black Hole Attacks in Mobile Ad Hoc Networks*" addresses one of the important problems of security in Mobile Ad-hoc NETworks (MANETs), which are still an interesting option for the regions where infrastructure deployment is not exactly profitable. The MANETs that use Dynamic Source Routing (DSR) protocol to discover routes between the nodes often suffer from single and cooperative black hole attacks. The chapter discusses a solution for probing the nodes of the network to determine whether they are malicious and discovers a secure route between the source and destination by identifying and isolating such malicious nodes. Simulation results show that the protocol provides better security and performance in comparison with trust based and probe based schemes.

The chapter by Sivagurunathan, et al., titled "*VANET Authentication Using Trust Distribution in a Clustered Environment,*" discusses next generation mobile wireless networks that are infrastructure-less and dynamic. In particular, the chapter addresses the security issue in VANETs. Analysts predict that inter-vehicle networks, when fully deployed across the globe, will be the largest open MANET. VANETs have their own challenges not encountered in usual MANETs due to vehicular mobility that may range from a few kilometers to hundreds of kilometers. To support common daily life applications on VANET requires the creation of a virtual network infrastructure that is scalable, robust, and secure. The authors claim that a stable clustering of nodes is the key to creating this infrastructure, and to this end, present a threshold security mechanism with mobility based clustering for open inter-vehicle communication networks.

The chapter "*A Distributed Secure Architecture for Vehicular Ad Hoc Networks,*" by Gazdar, et al., proposes a novel distributed Public Key Infrastructure for VANETs based on the election of a small subset of trusted vehicles to play the role of the Certification Authorities. The novelty of their proposal stems essentially from the following facts: first an election process based on various metrics including the relative mobility, a constrained cluster size, a *d*-connected cluster, and a security criterion; second, a dynamic demilitarized zone composed of confident nodes within one hop of the elected cluster head to shield it from any malicious node behavior; third, a trust metric to measure the level of confidence of each vehicle within the platoon. Using extensive simulations, they studied the performance of their proposal and investigated the impact of the vehicle speed, the vehicle average arrival rate, and the percentage of confident vehicles on the stability and efficiency of the security infrastructure. In particular, they showed that their proposal is scalable, stable, and efficient, and that only a very small fraction of the nodes need to be confident to attain a very adequate high level of stability and efficiency.

With the extensive usage of Radio Frequency Identification (RFID) in applications such as personal identifications in passports, payment systems in the financial sector, ticketing in transportation sector, supply-chain management in manufacturing and process industries, and asset tracking in the retail sector, vulnerability and security risks of these systems have increased recently. Since the RFID medium can be easily accessed, the messages between communicating nodes can be easily observed, eavesdropped, or forged. Furthermore, customers possessing RFID tags can be tracked, and the transactions can be recorded, leading to violation of their privacy. In their chapter on "*Security Aspects in Radio Frequency Identification Systems,*" Gódor and Imre model the attacker and the well-known attacks in

RFID environment with small computational capacity. They analyze the performance of hash based secure Simple Lightweight Authentication Protocol under various attach scenarios. Using simulation, the authors benchmark the performance of their algorithms and indicate the possibility of embedding this in commercial RFID systems.

This concludes Section 2.

The demand for wireless broadband continues due to exponential growth of Smartphone adoption. Smartphone growth has also fuelled intense mobile platform technologies and associated business models. Apple's iOS, Microsoft's Windows Platform, Google's open source Android, and RIM's Blackberry have sparked interesting research problems in two-sided market theories and applications. These multiple platforms have also resulted in the need to look at the importance of Intellectual Property (IP) and patents.

The availability of multiple platforms and devices has provided users with tremendous choice. This has necessitated the IT infrastructure in organizations to be more flexible in accommodating the diverse needs of the users in terms of device support and management. Research problems in the area of converging IT infrastructure and optimization of organization policies do become important.

With convergence in devices, the network operators are thus are forced to differentiate themselves in the market place by way of Mobile Value Added Services (MVAS). One such MVAS service that provides new business opportunities to both the network operator as well as the content provider is mobile media content service. There are business alliances possible with media firm partnering with wireless service providers to offer content. Financial models using the theory of real options are being used by researchers to analyze firms' responses towards such alliances and investment.

Business communication has undergone a sea change lately, thanks to mobile telepresence that connect virtual meeting rooms to smart phones via wireless broadband networks. Using wireless telecommunications products, organizations create virtual offices that enable employees to work remotely and from home, reducing travel and office space. However, these throw up challenges in the management of remote offices, collaborative tools for enabling efficient knowledge sharing, and effective configuration systems for communication of work artifacts.

While the newer technologies in networking and applications are getting diffused through society, the dire need of applications and solutions for addressing the digital divide continues to increase. Innovative applications that use less energy and memory, and are at the same time capable of scaling up to provide the desired economies, continue to pose interesting research problems.

NETWORK DATA SERVICES AND APPLICATIONS: IMPROVING INTEROPERABILITY AND PERFORMANCE

In the third section, on "*Network Data Services and Applications,*" the editors present a rich set of research papers that address the challenges in improving interoperability and performance in next generation data services and supported applications.

Cabrero, et al. discuss the performance issues of one of the popular services on the Internet, namely the Internet radio, in their chapter on "*Deployment of Live Audio Services on FTTx Networks.*" The authors of this chapter have built a simulation model of an Internet radio service configured with user behaviour and corresponding traffic generation patterns adopted from real cases. The model is validated by comparing simulation results with the data gathered in a real service. The model has been used to evaluate various possible architectures which may be requested by a content provider to deploy an Internet radio service on a cable network. The evaluations show the performance of each of these architectures and the effect they have on the underlying network infrastructure. Moreover, for the least efficient architectures, the most suitable parameters to configure the network infrastructure have been presented in order to improve the behaviour of the service as much as possible. The evaluation of different architectures and the results of the analysis will be useful for service administrators to achieve the optimal configuration for their Internet radio services.

With the commercialization of Internet Protocol Television (IPTV) and other server based video distribution, congestion problems in a multimedia network are becoming common. In a conventional model for multimedia distribution, network resources are shared amongst the users, and hence, as more and more users join the network, network capacity is exhausted, degrading the quality. Alhaisoni, et al., in their chapter on "*Scalable P2P Video Streaming,*" introduce a new methodology using Peer-to-Peer (P2P streaming that can overcome the congestion problems in video distribution over networks. With P2P streaming, as more users want to consume content, the very same users act as distribution relays, contributing to increasing the overall throughput. The authors introduce a layered scalable video coding technique that enables the users to use the portion of the video stream that suits their needs best. The overlay network in which the users' nodes are interconnected to each other provides virtual, transient, and dynamic inter-user communication. The authors explore the effectiveness of a combination of P2P strategies to assess the viability and benefits of scalable video coding over a multi-layered network. The resulting performance shows the promise of this technique over the conventional video distribution mechanisms.

The chapter "*Interoperable IPv6 Sensor Networking over PLC and RF Media,*" by Chauvenet, et al., investigates the interoperability between Power Line Communications (PLC) and Wireless Sensor Networks (WSN). Different applications from smart metering and environment monitoring to home control and energy efficiency are targeted. They proposed an adaptation of the IEEE 802.15.4 standard to the low power, lossy channel and low data rate context of the PLC transceivers using pulse modulation. They focused on the convergence of the IPv6 protocol with the 6LoWPAN adaptation layer and proposed an implementation, called RPL, for the IETF proposed Routing Over Low Power and Lossy (ROLL) networks. The interoperability is then demonstrated using a real test bed integrating PLC and WSN running IEEE 802.15.4/6LoWPAN/IPv6/RPL stacks.

With the availability of Smartphones and the associated location information provided systems such as Global Positioning Systems (GPS), location dependent information services are gaining popularity. By including location as part of users' context information, many VAS can be provided, targeted at mobile users in the areas of geographic, traffic, and logistic areas, apart from providing vital dynamic information for emergency and disaster management services. In such applications, data caching is an important technique to enhance data availability and data access response time. Due to cache size limitations, cache replacement policies are used to find a suitable subset of items for eviction from the cache. Jane, et al., in their chapter on "*A Cache Replacement Policy for Location Dependent Data in Mobile Environments,*" propose a new cache replacement policy for location dependent data. The policy, based on re-entry probability of cache item, area of scope of cache item, and rate of access of the item, takes in to account both spatial and temporal characteristics of cache data items in order to provide better system performance in terms of cache hit ration of mobile clients.

Elcock and Chaudhuri show how to schedule processes in heterogeneous multiprocessor systems using task duplication method in "*Process Scheduling Using Task Duplication in Heterogeneous Distributed Systems.*" Process scheduling in a multiprocessor environment is still a challenging problem, and this classical problem deals with the assignment of tasks to different processors, satisfying a given set of constraints, so that the schedule length or overall task completion time is minimized. In this chapter, the authors have combined the idea of task duplication with the ranking concept used in an earlier algorithm and developed an algorithm that outperforms the earlier algorithm significantly when the conventional metrics, such as average schedule length, speedup, and frequency at which the schedule length is better, were used. Using the metric average running time, the performance of the proposed algorithm, is comparable to the others.

This concludes Section 3.

Internet and the associated WWW have introduced a second economy, normally referred to as the "*Digital Economy" (Arthur, 2011)*. The processes in the physical economy are being entered into the digital economy, where they are speaking to other processes in the digital economy – mainly through the Internet, in a constant conversation among multiple servers and multiple semi intelligent nodes that are updating things, querying, checking off, readjusting, and eventually connecting back with processes and humans in the physical economy.

Along with the digital economy, the emergence of Internet-based social media has made it possible for one person to communicate with hundreds or even thousands of other people in the network. Social media encompasses a wide range of online, word-of-mouth forums including blogs, discussion boards and chat rooms, consumer-to-consumer e-mail, consumer product or service ratings websites and forums, Internet discussion boards and forums, moblogs (sites containing digital audio, images, movies, or photographs), and social networking websites, to name a few (Mangold & Faulds, 2009). An important disruption is that the high-speed access networks and Smartphones enable users to access and use these social media through mobile phones, anytime, anywhere. These many-to-many group forming transactional communication networks and the associated exponential value generation in such networks was characterized in 2001 by Reed (Reed, 2001). The analysis of communication patterns in such social media and peer-to-peer networks will shed a light on their impact on network capacity, topology, routing, and furthermore, the various business models that can be followed by various stakeholders.

The growth of mobile Internet, Smartphones, and the mobile operating platforms has renewed research interest in two and multi-sided markets (Cusumano, 2011). The interaction between mobile network operators, mobile device manufacturers, and mobile application providers enables strong cross-side network effects. The success or failure of a platform, and hence the associated ecosystem, depends on the magnitude of this effect, and hence, pose challenging research problems in the area of network economics (Basole & Karla, 2011).

NETWORK ECONOMY: NEW BUSINESS MODELS

Finally, in the last section, "*Network Economy,*" the editors picked a few research papers that discuss the emerging alternative business models suitable to address upcoming network evolution.

One of the significant entry in to the digital economy is content provisioning. The disruption caused by Apple's iPad, followed by other Tablet manufacturers, have provided easy-to-use, affordable, and compact devices through which digital content can be accessed and consumed. This content provisioning has introduced many new business opportunities to both the network operator as well as the content providers. In their chapter "*Using Real Options Theory to Evaluate Strategic Investment Options for Mobile Content Delivery: A Case Study,*" Liginlal, Khansa, and Chia present a case study of a media company. They analyze two business models: one wherein media firm partners with wireless service providers, and another in which the firm provides media content through strategic alliances with mobile content syndicators. The models are evaluated based on their resource requirements, market share acquisition, revenue generation, and nature, scope, and control of content and bandwidth. Real options analysis is used to value the firm's managerial flexibility in responding to uncertainty in investment. The authors prove how real options analysis can be better than traditional discounted cash flow analysis for long-term decision making, especially for determining the value of technology investments under uncertainty in growing areas such as mobile media markets.

The chapter by Nascimento and Santos, "*Valuation of Alternative Business Models in Information, Communication and Media Markets: Convergence, Ubiquity and Pervasiveness,*" discusses business models for convergent and ubiquitous services. The adoption of the Internet Protocol (IP) platforms has allowed service providers to mix and match telephony, Internet, and broadcast services. It is in the context of the EU Framework Program, called DAIDALOS (Designing Advanced network Interfaces for the Delivery and Administration of Location independent, Optimized personal Services), which is in its second phase of development, that the authors propose a valuation method for investing in pervasive and ubiquitous services. Real options approach is used for the evaluation of various alternatives apart from the widely used net present value analysis. The authors claim that the value of a project to a service provider is highly dependent on the exercise of its inherent options and that operators are likely to adopt a vertical integrated business model in urban areas.

This concludes Section 4 and completes the summary of the book.

All the above chapters lead to the emerging domain of next generation networks – something akin to a hyper-Internet. When Tim Berners-Lee, then a graduate of Oxford University, invented the World Wide Web at CERN circa 1990 as an internet-based hypermedia initiative for global information sharing, little did he think that after 20 years it would become such a hyper-phenomenon. In its early days, the Internet and the associated WWW involved read-only traffic, retrieved from the server to the client terminals. There was no possibility of adding new content or modifying the Internet and WWW content. Evidently, WWW has travelled a long way from what it started with. It has come of age by breaking the digital divide and making the information available for everyone. The move from information to interaction has changed the web from read-only to read-n-write. From a developer's perspective, the Web moved from *html* based web to more sophisticated *Ajax* platform.

Since the Web now is more about the user-developed content, the underlying networks adapt more to users' preferences, likes, and dislikes. As far as the ownership over the resources is concerned, the individual has gained in terms of democracy as she has every right to express freely on the Web. The flow of information has recently changed further drastically with the emergence of online social networks. People are connected with their friends, acquaintances, and relatives through virtual communities and are constantly fed with what they want to share. The scalability of the next generation communication platforms from voice-only or data-only or video-only services to multimedia, value-added services is very much critical to the diffusion of the future Web services and applications.

Debashis Saha
Indian Institute of Management - Calcutta, India

Varadharajan Sridhar
Sasken Communication Technologies, India

REFERENCES

Arthur, W. B. (2011, October). The second economy. *The McKinsey Quarterly*, 1–9.

Basole, R., & Karla, J. (2011). On the evolution of mobile platform ecosystem structure and strategy. *Business & Information Systems Engineering*, *5*, 313–322. doi:10.1007/s12599-011-0174-4

Chapin, J. M., & Lehr, W. H. (2007). Cognitive radios for dynamic spectrum access-the path to market success for dynamic spectrum access technology. *IEEE Communications Magazine*, *45*(5), 96–103. doi:10.1109/MCOM.2007.358855

Cisco. (2010). *Cisco visual networking index: Global mobile data traffic forecast update, 2010–2015.* Retrieved March 15, 2011, from www.cisco.com

Cusumano, M. (2011). The platform leader's dilemma. *Communications of the ACM*, *54*(10), 21–24. doi:10.1145/2001269.2001279

Mangold, W. G., & Faulds, D. (2009). Social media: The new hybrid element of the promotion mix. *Business Horizons*, *52*, 357–365. doi:10.1016/j.bushor.2009.03.002

Partridge, C. (2011). Realizing the future of wireless data communications. *Communications of the ACM*, *54*(9), 62–68. doi:10.1145/1995376.1995395

Reed, D. (2001, February). The law of the pack. *Harvard Business Review*, 23–24.

Section 1
Next Generation Communication Networks:
Towards Enhanced Design, Interoperability and Performance

Chapter 1
Planning and Dimensioning of the 3G UMTS Core Networks

Ye Ouyang
Stevens Institute of Technology, USA

M. Hosein Fallah
Stevens Institute of Technology, USA

ABSTRACT

The current literature provides many practical tools or theoretical methods to design, plan, and dimension Global System for Mobile Communications (GSM) and Universal Mobile Telecommunications System (UMTS) radio networks, but overlooks the algorithms of the network planning and dimensioning for core networks of GSM, UMTS, and IP Multimedia Subsystem (IMS). This chapter introduces an algorithm for traffic, bandwidth, and throughput dimensioning of the network entities in the UMTS core network. The analysis is based on the traffic and throughput generated or absorbed in the interfaces of the network entities in the UMTS core network. Finally a case study is provided to verify the algorithms created for UMTS core network. This chapter is aimed at helping UMTS network operators dimension an optimum network size and build an optimum network structure to deliver an optimum quality of service for users. The algorithms developed in the chapter have been successfully applied in dimensioning a nationwide UMTS network in North Africa and adopted in an optimization tool by a mobile operator in the United States in 2008-09.

INTRODUCTION OF THE UMTS NETWORKS

Rapid changes in mobile telecommunications have always been evolutionary, and the deployment of UMTS to Long Term Evolution (LTE) will be the same. It will be a transition from third generation (3G) to 4G over a period of several years, as is the case still with the transition from 2G to 3G. As a result, mobile operators must find algorithms and rules that will dimension their emerging 3G networks, while addressing their potential 4G deployment requirements and will not require a "forklift" upgrade.

Radio access solutions are a primary concern of the UMTS deployment strategy, as it impacts the

DOI: 10.4018/978-1-61350-477-2.ch001

mobile operators' most valued asset: spectrum. As an equally important part of this equation, the core network will play an essential role in enhancing mobility, service control, efficient use of network resources and a seamless migration from 2G/3G to 4G. Hence, the network evolution calls for a transition to a "flat," all-IP core network with a simplified architecture and open interfaces.

As mobile operators evolve their networks to UMTS or even LTE, they will try to minimize cost and maximize subscriber usage. Therefore, a new problem appears: how to correctly plan and dimension the emerging UMTS Core Networks (CN) with a new flat and all-IP structure to avoid configuring unnecessary network resources and maintaining high Quality of Service (QoS) to subscribers? Meanwhile, the dimensioning algorithms for UMTS CN should be significantly differentiated from the traditional design philosophy for Circuit Switched (CS) and Time Division Multiplexing (TDM) networks such as 2G GSM and CDMA networks.

In order to accurately plan, design, and dimension the UMTS CN, this paper will develop the algorithms of traffic and throughput for the UMTS CN Network Entities (NEs) described in Section 3. The analysis will be based on the live traffic and throughput generated or absorbed in the interfaces of CN NEs. Our approach provides the mobile operators with a capability to assess and plan their capacity requirements independent of any particular vendor product. This vendor neutrality is further discussed later in the paper. A case study is provided to verify the algorithms created for UMTS CN. This paper is aimed at helping UMTS network operators dimension an optimum network size and build an optimum network structure to deliver an optimum quality of service for users.

In addition, the network optimization and expansion is the further effort for the mobile operator after the rolling out of mobile networks. To minimize the CAPEX/OPEX and maintain the QoS of mobile core networks, we propose that the impact of cell cite re-homing on the mobile core should be studied. It is believed that the appropriate cell site re-homing in radio domain, via correct algorithms applied, not only optimizes the radio network but also helps improve the QoS of the core network and minimize the mobile operator's CAPEX/OPEX investment in their core networks.

The rest of the article is organized as follows: Section 2 summarizes the literature in the related area and the challenges in dimensioning core networks. Section 3 introduces the architecture of the UMTS network and in particular the key network entities in UMTS. Section 4 which is the core of the paper discusses the algorithms for traffic and throughput in those interfaces of UMTS CN networks such as Iu-CS, Iu-PS, Nb, Mc, and Mc interface. Section 5 provides two case studies to illustrate application of the algorithms created in Section 4 for Iu-CS and Iu-PS interfaces. Section 6 is the conclusion to the paper.

LITERATURE REVIEW

The current literature provides many practical tools or theoretical methods to design, plan and dimension GSM and UMTS radio networks but overlooks the algorithms for planning and dimensioning of core networks of GSM, UMTS and IMS. No previous literature provides a unified approach to calculate the throughput or traffic of the UMTS core network. Very few studies have addressed the mobile core network planning and dimensioning topic. This is because that the core network in either logical or physical structure is more complicated than the radio access network and the internal throughput or traffic may vary from different vendors' NEs.

Neruda, M. and Bestak, R. (2008) summarizes the evolution path from GSM, UMTS to IMS from the aspect of network entities so that service providers will be able to progressively migrate from GSM to UMTS and IMS. Shalak, R. et al (2004) make a qualitative study of the performance of UMTS core network, in which equipment of multiple vendors of UMTS CN is compared. Harmatos, J. (2002) proposes a model to plan UMTS core network based on the requirements for the radio access network. The model also considers the premise of planning work in cost minimization, which helps mobile operators minimize Capital Expenditure (CAPEX). Because of the complexity of these networks, Harmatos, J. (2002) divided the problem into two parts. First, he finds the location of Media Gateways and a reasonable topology using a linear cost function. In the second part, he uses the real cost function (step function) in order to reduce the cost of the network. Britvic, V. (2004) specifies the strategic steps to plan and deploy the UMTS radio network, the core network, and the access transport network. Previous studies have provided many solutions to plan, dimension, and deploy the UMTS radio networks (Harmatos, J. et al, 1999; Wu, Y. et al, 2003 and 2005; Jamaa, SB. et al, 2004; Maple, C. et al, 2004; Juttner, A. et al, 2005; and Neubauer, T. et al, 2005). Different models and methods have been developed to find the optimal topology of the cells if the basic traffic models and location information to install base stations can be provided by mobile operators. Ouyang, Y and Fallah, M.H. (2009) propose an evolution path and a 'three-layer structure' solution to seamlessly converge the UMTS R4 core network with the legacy GSM core network. The proposed solution is a high level guideline of the mobile core network topology rather than a detailed network dimensioning tutorial. In addition, Kunz, A., et al (2005) have studied the QoS mechanism for GPRS and UMTS PS networks.

Hence, the current literature is relatively mature on dimensioning of the radio networks. The literatures on planning the mobile core networks are limited to high level description for designing core network architecture. This literature gap in the detailed planning and dimensioning of the 3G core networks was the motivation behind our study and the specific focus on estimating the throughput and traffic generated and absorbed in the interfaces in the UMTS core network.

ARCHITECTURE OF UMTS CORE NETWORKS

The core network is the heart of a mobile network. Whether in 2G or 3G phase, the CN plays an essential role in the whole mobile network system to provide such important functions as mobility management, call and session control, switching and routing, charging and billing, and security protection. In the R99 version, the first version of 3G UMTS, the CN portion stays as the same network entity (NE) type and network topology architecture as that in the GSM phase. However, there is a change in R4, the second version of UMTS, which supports a networking mode where the bearer is separated from control. Meanwhile multiple bearer modes such as ATM/IP/TDM are supported by CN. Consequently the Mobile Switching Center (MSC) in GSM/UMTS R99 is split into two NEs: the MSC Server (MSS) and the Media Gateway (MGW).

With a logical division, the CN in UMTS is classified into the circuit switched domain (CS) including such logical NEs as MSC Server, MGW, Visitor Location Register (VLR) integrated in the MSC Server physically, Home Location Register (HLR), Authentication Center (AUC), Equipment Identity Register (EIR) and the packet switched domain (PS) including NEs: Serving GPRS Support Node (SGSN) and Gateway GPRS Support Node (GGSN) (3GPP TS 25.401; 3GPP TS 23.002). Figure 1 displays the topology of UMTS CN with the logical NEs mentioned above.

Figure 1 shows that the UMTS CN consists of these mandatory NEs: MGW, MSC Server, HLR, SGSN, and GGSN. Below is a short description on these NEs.

HLR is responsible for storing, updating, revising or deleting subscriber related information, covering the basic service subscription informa-

Figure 1. Topology of UMTS CN: CS+PS domain

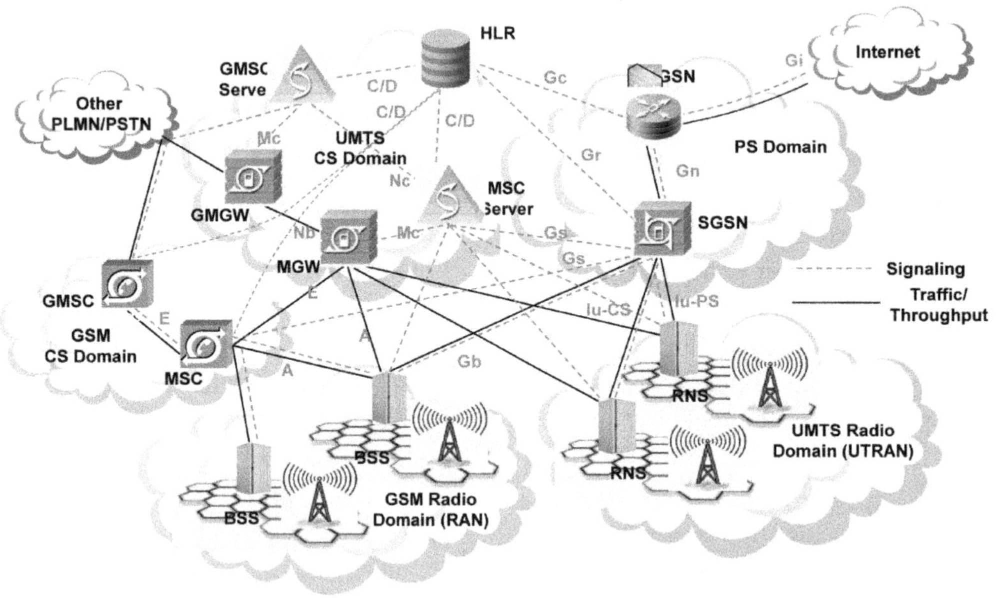

tion, supplementary service subscription information, and location information of subscribers. In addition, it also implements the function of subscriber security management. From the physical connection aspect, HLR provides the D interface to connect with the VLR in the MSC Server, the C interface to connect with the MSC Server or the MSC in the GSM CN, the Gr interface with the SGSN, and the Gc interface with the GGSN. The type of signaling message delivered to and from HLR is the Mobile Application Part (MAP).

As the core NE of the CN in UMTS, the MSC Server is a functional entity that implements mobile call service, mobility management, handover, and other supplementary services. Due to the philosophy of separation of the control function from the bearer function in the UMTS CN, it is actually a controller of the MGW to establish call routes between Mobile Stations (MS) via the Mc interface. The MSC Server also physically integrates with a VLR to hold subscriber's data. The MSC Server provides the Nc interface to connect with its peer MSC Server, the Mc interface with the MGW, the C/D interface with the HLR, the A interface with the 2G Base Station Controller (BSC), and the optional Gs interface with the SGSN. The main types of signaling messaged going through MSC Server are Bear Independent Call Control messages (BICC) or ISDN User Part messages (ISUP) between the MSC Servers, the MAP between the MSC Server and the HLR, and the H.248 between the MSC Server and the MGW.

A MGW in a UMTS implements bearer processing functions between different networks. It implements UMTS voice communication, multimedia service, CS domain data service, and interworking between PSTN and UMTS CN and between GSM CN and UMTS CN. MGW provides Iu-CS interface to connect with the Radio Network Controller (RNC) in the Radio Access Network (RAN), Nb interfaces with its peer MGW, E interfaces with 2G MSC, Mc interfaces with MSC Server, A interface with BSC, and Ai interface with Public Switched Telephone Network (PSTN).

The SGSN is responsible for the delivery of data packets to and from MSs within its serv-

Table 1. Iu-CS UMTS protocol stack

<table>
<tr><th>Radio Network Control Plane</th><th>Transport Network Control Plane</th><th>Circuit Switching Data User Plane</th><th>CS Voice User Plane</th></tr>
<tr><td>MM/SM/CC</td><td></td><td>Application</td><td rowspan="3">AMR Codec</td></tr>
<tr><td rowspan="2">RANAP</td><td></td><td>TAF</td></tr>
<tr><td>ALCAP</td><td>RLP</td></tr>
<tr><td>SCCP</td><td>STC</td><td colspan="2" rowspan="3">Iu UP</td></tr>
<tr><td>MTP3-D</td><td>MTP3-D</td></tr>
<tr><td>SSCF NNI</td><td>SSCF NNI</td></tr>
<tr><td>SSCOP</td><td>SSCOP</td><td colspan="2">AAL2-SAR SSCS</td></tr>
<tr><td>AAL5</td><td>AAL5</td><td colspan="2">AAL2</td></tr>
<tr><td colspan="4">ATM</td></tr>
</table>

ing area. The tasks include packet routing and transfer, mobility management (attach/detach and location management), logical link management, and authentication and charging functions. The interfaces include Iu-Ps interface connecting to RNC, Gn/Gp interface to GGSN, Gr interface to HLR, Gs interface to MSC Server or MSC, Gd interface to Short Message Center (SMC), and Ga interface to Charging Gateway.

GGSN is a gateway between the UMTS PS/GPRS network and the external data networks (e.g. Internet). It performs such functions as routing and data encapsulation between the MS and external data network, security control, network access control and network management. From the UMTS PS/GPRS aspect, the MS selects a GGSN as its routing device between itself and the external network in the activation process of the PDP context in which the Access Point Name (APN) defines the access point to destination data network. From the external data network aspect, the GGSN is a router that can address all the MS IPs in the UMTS PS/GPRS network. The GGSN provides the Gc interface to connect with the HLR, Gn/Gp interfaces with SGSN, Gi interfaces with external data networks, and Ga interfaces with CG.

ALGORITHMS TO DIMENSION UMTS CORE NETWORKS

Since Iu-CS, Iu-PS, Nb, Mc, and Mc interfaces are newly developed in the UMTS CN, this section is focused on the algorithms for these new interfaces. The throughput algorithms for the other interfaces such as A to E and Gb interface, since they have existed in GSM CN, is based on a general rule: multiply the total traffic (Erlang or message size) times the traffic proportion to obtain the traffic distribution for each NE and each link.

Iu-CS Interface

The Iu-CS interface is located between the MGW and the RNC to establish the voice channel and transport the Radio Access Network Application Part (RANAP) signaling message (3GPP TS 25.413). The transmission medium in the Iu-CS interface is the ATM in R4 and is suggested to be replaced by the IP from UMTS R5. The interface Iu-CS consists of user plane based on ATM Adaption Layer 2 (AAL2) and control plane based on AAL5 (3GPP TS 25.401; ITU-T I.363.2). The protocol stack for the Iu-CS interface is shown in Table 1.

Table 2. AAL2 CPS PDU

← Start field →			← AAL2 CPS-PDU payload →	
OSF	SN	P	AAL2 PDU payload	PAD
6 bits	1 bit	1 bit		0-47 bytes
AAL2 CPS PDU				

Table 3. AAL2 CPS PDU payload

← AAL2 Header →				← Information Payload →
CID	LI	UUI	HEC	Information payload
8 bits	6 bits	5 bits	5 bits	1-45/64 bytes
AAL2 PDU Payload				

In the CS voice user plane, the Iu Interface User Plane Protocol (Iu-UP) stands on the top layer and is followed by the AAL2 and the ATM. 3GPP TS 25.415 defines the PDU format for Iu-UP in which we are able to obtain the overhead of Iu-UP frame = Frame Control Part (FCP) + Frame Check Sum Part (FCSP). The typical Iu-UP Packet Data Unit (PDU) formats are Iu-UP PDU type 0, 1 and 14 in which both the FCP and the FCSP occupy 2 bytes respectively. One exception the FCSP is 1 byte for type 1 defined to transfer user data over the Iu UP in support mode for the pre-defined SDU sizes when no payload error detection scheme is necessary over the Iu UP. But this scenario is not usually adopted for the reason that error detection is always needed in transmission. Generally the overhead is obtained Iu-UP frame = FCP + FCSP = 2+2=4 bytes. This value is used for the following calculation.

AAL2 below the layer of the Iu-UP provides bandwidth-efficient transmission of low-rate, short and variable packets in delay sensitive applications. So it is the ideal bearer medium for the circuit switching service of the UMTS. As per ITU-T I.363.2 and ITU-T I.366.2, the AAL2 can be subdivided into two layers: the Common Part Sub-layer (CPS) and the Service Specific Convergence Sub-layer (SSCS). The latter is normally void so only the CPS is considered in this case. The structure of the AAL2 CPS PDU is given in the following illustration. From the PDU structure, we obtain the Start Field=8 bits=1byte=1 Octet; AAL2 Header=8+6+5+5=24 bits=3 bytes=3 Octets. In addition, the ATM cell is 53 bytes and the header of ATM cell is 5 bytes.

The AMR (Adaptive Multi-Rate) codec encodes narrowband (200-3400 Hz) signals at variable bit rates ranging from 4.75 to 12.2 kbps. The mode 7with a codec speed at 12.2kbps for voice signal and use 64 kbps as the codec speed for video call service was adopted in this case. Table 4 summarizes the necessary parameters for Iu-CS interface.

Based on the conditions obtained in Table 5, the functions for the throughput of voice channel in the Iu-CS interface are as follows. Without the Voice Activity Detection (VAD)[1] technique, the maximum throughput of a single channel (unit: bps) in Iu-CS is given by:

Table 4. Overhead of protocols in Iu-CS interface

	Iu-UP Overhead	AAL2 Start Field	AAL2 Header	ATM Header	ATM Cell	AMR Payload (at 12.2 kbps)	G.711 Payload (at 64kbps)
Size (Octets)	4	1	3	5	53	31	40

Table 5. Codec parameters

Codec Type	Codec Speed (kbps)	Payload per Frame (Octets)	Video/Audio Speech Frame (ms)
AMR. Type 7	12.2	31	20
AMR_SID. Type 8	Not a fixed value	6	160
G.711	64	40	5
G.729	8	10	10
G.729_SID	Not a fixed value	2	160

$$TH_{NonVAD} = SP_{AMR}/E_{AAL2} \quad (1)$$

Where SP_{AMR} denotes the codec speed of AMR, obtained from Table 5, E_{AAL2} denotes the efficiency of AAL2 encapsulation. It is given by formula 2 below.

From Table 2, Channel Identification is 8 bits, meaning 2^8=256 CIDs are available. However CID 0 is not used and CID from 2 to7 are reserved, so only from 8 to 255, 248 CIDs are actually provided for AAL2 user:

$$E_{AAL2} = N_{CID} \text{ x } P_{Frame}/(N_{ATMcell} \text{x} S_{ATMcell} \quad (2)$$

where NCID denotes the number of CID, P_{Frame} denotes the different types of payload of frame in Table 5, $N_{ATMcell}$ denotes the number of ATM cells, obtained by formula 4, $S_{ATMcell}$ denotes the size of ATM cell which is 53 octets.

The payload of codec (in octets) can be given by:

$$P_{Codec} = S_{Codec} \text{ x } F_{Speech}/8 \quad (3)$$

where F_{Speech} denotes the speech frame in Table 5, S_{Codec} denotes the codec speed in Table 5.

$$N_{ATMcell} = \left(H_{IuUP} + H_{AAL2} + P_{Codec}\right) \times N_{CID} \,/\, \left(S_{ATMcell} - H_{ATMcell} - SF_{AAL2}\right) \quad (4)$$

where H_{IuUP} denotes the header of IuUP, H_{AAL2} denotes the header of AAL2, P_{Codec} denotes the payload of Codec obtained from formula 3, $H_{ATMcell}$ denotes the header of ATM cell which is 5 octets, SF_{AAL2} denotes the start field of AAL2 obtained from Table 4.

Then substituting the known parameters from Table 2, 3, 4 and 5 into the conditions in Formulas 1, 2, 3, and 4 to obtain $TH_{Non\text{-}VAD}$=16.95kbps.

With the VAD technique, the codec speed of a AMR_ Silence Descriptor (SID) = 1.8kbps, obtains TH_{VAD}=4.5kbps through formula 5 in which E_{AAL2}=0.4 (obtained through Formula 2):

$$TH_{VAD} = SP_{SID} \,/\, E_{AAL2} \quad (5)$$

where SP_{SID} denotes the codec speed of AMR SID, obtained from Table 5.

So the $TH_{Voice\ Channel}$ is given by:

$$TH_{Voicechannel} = TH_{NonVAD}(1\text{-}F_{VAD}) + TH_{VAD} \text{ x } F_{VAD} \quad (6)$$

where F_{VAD} denotes VAD factor: the ratio of silence time in a call to the total time of call.

The effect of VAD directly impacts the throughput caused by voice service. That is, the result of the formula 6. As the controller of VAD function, the mobile operator determines whether to activate VAD or partially activate VAD function. The effect of VAD occurs when the mobile operator activates VAD function in the MGW. In addition, the VAD effect is also impacted by VAD factor which is an estimated parameter identified by the mobile operator. That is, the ratio of silence time in a call to the total time of the call. Thus, if VAD is not activated, the formula 1 is applicable; if VAD is at least partially activated, the formula 6 is applicable.

Similarly the throughput (unit: bps) of single channel for video call service is provided below:

$$TH_{Videochannel} = SP_{Video} / E_{AAL2} \qquad (7)$$

where SP_{video} denotes the codec speed of video call, obtained from Table 5.

In Iu-CS interface, the major throughput is generated by voice service and video call service. At last the total throughput of Iu-CS interface (in bps) is provided by:

$$TH_{IuCS} = N_S \begin{pmatrix} P_{Voice} \times Erl_{VoU\ /\ BH} \times TH_{Voicechannel} \\ +P_{Video} \times Erl_{ViU\ /\ BH} TH_{Videochannel} \end{pmatrix} / F_{\mathrm{Re}\, dudancy} \qquad (8)$$

where N_S denotes the number of 3G subscribers in RNC, P_{Voice} denotes the percentage of subscribers using voice call to total subscribers (normally it is 100%), P_{Video} denotes the video call service penetration rate, $Erl_{VoU/BH}$ denotes the average voice call traffic in Erlang per user per busy hour, $Erl_{ViU/BH}$ denotes the average video call traffic in Erlang per user per busy hour, $F_{Redundancy}$ denotes redundancy factor which prevents the network from traffic overflow (normally set it as 0.7).

With respect to the throughput in MGW, the total throughput obtained via Formula 8 is considered from the subscriber perspective. The traffic (or throughput) which is generated (or caused) by subscribers does have two directions (inbound and outbound). However, in obtaining the total throughput value (such as Iu-CS interface in Formula 8), our thoughts is to accumulate traffic of subscriber level and finally obtain the total traffic generated by the subscribers active in the MGW.

Nb Interface

According to 3GPP TS 29.414 and 3GPP TS 29.415, the protocol stack of the user plane and the control plane of the Nb interface are shown in Table 6 and Table 7.

When an Nb UP layer protocol entity receives an initialization status request from the upper layer, it will start the initialization procedure. Consider the throughput of initialization:

$$TP_{Initial} = S_{Initial} \text{ x } BHCA_{Call} \text{ x } 8 / 3600 \qquad (9)$$

where TP denotes the throughput of an initialization per user, $S_{Initial}$ denotes the size of an initialization message, and $BHCA_{Call}$ represents the average call attempts in busy hour per subscriber Take the IP based fast forward bearer setup as an example, TP=53×1.2×8/3600=0.14bps which only counts for a very small value. Therefore it can be overlooked in calculating the throughput in Nb interface.

As per Table 6, the user data can be transported via three mediums: TDM, ATM or IP, last two of which provides different protocols stacks to achieve the transport process. Table 9 lists sample protocol stacks of the user plane in the Iu interface. Since the protocol stacks are more complicated in the interface Iu, the overhead size shown in Table 9 is larger than in interface Iu-CS in Table 4.

Same with the Iu-CS interface, VAD and non-VAD should be considered individually for the Nb interface. Without VAD, the throughput of a

Table 6. User plane of Nb interface

<table>
<tr><th colspan="2">Transport over IP</th><th>Transport over ATM</th><th>Transport over TDM</th></tr>
<tr><td colspan="3">AMR/G.711</td><td rowspan="4">G.711</td></tr>
<tr><td colspan="3">Nb-UP</td></tr>
<tr><td colspan="2">RTP/RTCP</td><td>AAL2 SAR SSCS</td></tr>
<tr><td colspan="2">IP/UDP</td><td>AAL2</td></tr>
<tr><td>VLAN/MAC</td><td>MPLS/PPP</td><td>ATM</td><td>PCM</td></tr>
</table>

Table 7. Control plane of Nb interface

<table>
<tr><th>Transport by ATM</th><th>Transport over IP</th></tr>
<tr><td rowspan="2">AAL2 Connection Signaling (Q.2630.2)</td><td>IPBCP (Q.1970)</td></tr>
<tr><td>BCTP (Q.1990)</td></tr>
<tr><td>AAL2 Signaling Transport Converter for MTP 3b (Q.2150.1)</td><td>BICC (Q.765.5)</td></tr>
<tr><td>MTP 3b</td><td>M3UA</td></tr>
<tr><td>SSCF-NNI</td><td>SCTP</td></tr>
<tr><td>SSCOP</td><td rowspan="2">IP</td></tr>
<tr><td>AAL5</td></tr>
<tr><td>ATM</td><td>MAC</td></tr>
</table>

Table 8. Overhead of protocols in Nb interface

	Nb-UP	RTP	UDP	TCP	IP	MPLS	PoS	MAC	MAC/VLAN
Size (Octets)	4	12	8	8	20	4	10	34	38

Table 9. Overhead of protocol stacks

Protocol Stack	Overhead (Octets)
NbUP/RTP/UDP/IP/MAC	78=4+12+8+20+34
NbUP/RTP/UDP/IP/VLAN/MAC	82=4+12+8+20+38
NbUP/RTP/UDP/IP/POS	54=4+12+8+20+10
NbUP/RTP/UDP/IP/MPLS/POS	58=4+12+8+20+4+10
RTP/UDP/IP/MAC	74=12+8+20+34
RTP/UDP/IP/VLAN/MAC	78=12+8+20+38
RTP/UDP/IP/POS	50=12+8+20+10
RTP/UDP/IP/MPLS/POS	54=12+8+20+4+10

single channel (in bps) in interface Nb is given by:

$$TH_{NonVADVoice} = (P_{AMR} + O_p) \times 8 / F_{Speech} \quad (10)$$

where P_{AMR} denotes the AMR payload for voice service in Table 5, F_{Speech} denotes the speech frame in Table 5, O_P denotes the overhead of protocol stacks in Table 9.

With VAD technique, we have:

$$TH_{Voicechannel} = TH_{NonVADVoice}(1 - F_{VAD}) + TH_{VADVoice} \times F_{VAD} \quad (11)$$

in which TH_{VAD} is given by formula 12. F_{VAD} is the VAD factor:

$$TH_{VADVoice} = (P_{SID} + O_p) \times 8 / F_{Speech} \quad (12)$$

where P_{SID} is the payload of the AMR SID can be found in Table 5. O_p denotes the overhead of protocol stacks in Table 9. The value depends on which protocol stack group is chosen in transport.

For video call service, obtain the throughput (in bps) in both non-VAD and VAD scenario:

$$TH_{NonVADVideo} = (P_{Video} + O_p) \times 8 / F_{Speech} \quad (13)$$

in which P_{Video} denotes the payload of video service. The value can be obtained from Table 5.

With VAD available:

$$TH_{Videochannel} = TH_{NonVADVideo}(1 - F_{VAD}) + TH_{VADVideo} \times F_{VAD} \quad (14)$$

in which $TH_{VADVideo}$ is equivalent to $TH_{VADVoice}$ in equation 12.

Formula 10, 11 and 12 are based on the application of Transcoder Free Operation (TrFO) technique which enables the voice transported at a speed of AMR 12.2kbps but not G.711 64kbps in CN. However the Tandem Free Operation (TFO) technique, a previous technique used in GSM that transports voice at standard 64 kbps via PCM (Pulse Code Modulation) in CN, may still be applied in UMTS CN. If TFO is fully applied in the UMTS CN:

$$TH_{NonVAD\_TFO} = (P_{TFO} + O_P)\, 8 / F_{Speech} \quad (15)$$

where P_{TFO} denotes the payload based on TFO. The value is different under TFO scenario.

With the VAD technique in the TFO scenario, we have:

$$TH_{VoicechannelTFO} = TH_{NonVADTFO}(1 - F_{VAD}) + TH_{VAD} \times F_{VAD} \quad (16)$$

A more possible scenario is that both TrFO and TFO are adopted by the wireless operator in UMTS CN with a ratio of R_{TrFO} and R_{TFO} which R_{TrFO} + R_{TFO}=1, the overall throughput in Nb interface is provided by:

$$TH_{Nb} = N_S \times \left\{ \begin{array}{l} R_{VO} \times Erl_{VoU/BH} \times \left[\begin{array}{l} TH_{VoCTrFO} \times R_{TrFO} \\ + TH_{VoCTFO} \times (1 - R_{TrFO}) \end{array} \right] \\ + R_{Vi} \times Erl_{ViU/BH} \times TH_{ViC} \end{array} \right\} / F_{Redundancy} \quad (17)$$

where the major throughput in Nb interface is also generated by voice and video call service, N_S denotes the number of 3G subscribers in RNC, R_{Vo} denotes the ratio of subscribers using voice call to total subscribers (normally it is 100%), R_{Vi} denotes the video call service penetration rate, TH_{VoC_TrFO} represents $TH_{Voice\ Channel\_TrFO}$ obtained from equation 11, TH_{VoC_TFO} represents $TH_{Voice\ Channel\_TFO}$ from equation 16, TH_{ViC} denotes TH_{Video} Channel in equation 14, $Erl_{Vo\ User/BH}$ denotes the average voice call traffic in Erlang per user per busy hour, $Erl_{Vi\ User/BH}$ denotes the average video call traffic in Erlang per user per busy hour, Redundancy Factor prevents the network from traffic overflow (normally set at 0.7).

The throughput distribution between MGWs is a tricky issue. The number of trunks or links between the MGWs are dependent on the respective traffic in two distinct MGWs. However, it can be resolved by identifying the traffic distribution ratio in the MGWs. It requires the mobile operator to apply traffic distribution ratio to the result obtained through the formula 17, case by case. Formula 17 helps the mobile operator obtain the throughput for the MGW itself. Then the traffic distribution ratio will determine how to distribute the traffic and throughput generated via the formula 17.

As an illustration of the throughput distribution, the examples below present distinct scenarios of throughput (traffic) distribution.

Assume MGW A connects with a MGW B and a PSTN gateway only. Assume the traffic distribution ratio (No distinction between local and long distance in this scenario) is: Mobile to Mobile 40% and Mobile to PSTN 60%. In this case, the traffic (throughput) going into MGW A will have 40% distributed to MGW B and 60% to PSTN gateway. That is, if 100Mbps throughput obtained based on the calculation, 40Mbps will be assigned to MGW B and 60 Mbps will be distributed to PSTN gateway.

If MGW A connects to MGW B only, it needs to consider whether MGW A and B belong to the same local area or distinct areas (long distance). In the latter case, the throughput is distributed according to the ratio of local calls to long distance calls. Assume local to local calls are 80% while long distance calls are 20%.

If MGWs A and B belong to the same local area, no long distance traffic will be considered. The throughput resources configured between A and B should be able to support the MGW which overtakes the higher local throughput. Say a RNC generates 100Mbps throughput, in which 40% going to MGW A and 60% going to MGW B. In this case at least 60%*100Mbps rather than 40%*100Mbps throughput resource should be configured between MGWs A and B.

If MGWs A and B belong to different local areas, then we need to consider the long distance call ratio. Say RNC A connects to MGW A with 100 Mbps throughput while RNC B connects to MGW B with 200 Mbps. For MGW A with RNC A, 100Mbps*20%=20Mbps will flow out of MGW A and go to MGW B. For MGW B with RNC B, 200Mbps*20%=40Mbps will flow out of MGW B and flow into MGW A. If this is the case, the appropriate throughput resource must be configured to suffice at least 40Mbps (the heavier one) between MGWs A and B.

Mc Interface

The Mc reference point describes the interfaces between the MSS and the MGW. It is full compliance with the H.248 standard (ITU-T, SERIES H). The interface enables the MSC Server to dynamically share the MGW physical node resources. Also it is dynamic sharing of transmission resources between the domains as the MGW controls bearers and manages resources according to the H.248 protocols. The protocol stack in Mc interface is shown in Table 10. As per Table 10, we can infer that the available protocol stack groups with its overhead in Table 11. For pure IP links, H.248/SCTP/IP is preferred and H.248/M3UA/SCTP/IP is optional. For ATM/IP mixed links, H.248/M3UA/SCTP/IP is mandatory and H.248/MTP3b/SSCF/SSCOP is optional.

H.248 message flow transported through Mc interfaces usually includes two message types which are mobile call service and handover service. The Table 12 summaries the H.248 message flow type and its payload for flow going through Mc interface. As per the payload summarized in Table 12, the size of each message flow is given by:

$$S_K = P_K + N_K \times O_{H.248} \tag{18}$$

Table 10. Protocol stack in Mc interface

<table>
<tr><th colspan="3">H.248</th></tr>
<tr><td>M3UA</td><td rowspan="2">SCTP</td><td>MTP 3B</td></tr>
<tr><td>SCTP</td><td>SSCF</td></tr>
<tr><td colspan="2">IP</td><td>SSCOP</td></tr>
<tr><td>VLAN/MAC</td><td>MPLS/PPP</td><td>AAL5</td></tr>
</table>

Table 11. Overhead of protocol stack groups in Mc interface

Protocol stack type	Overhead (Octets)
H.248/M3UA/SCTP/IP/VLAN/MAC	126
H.248/SCTP/IP/VLAN/MAC	86
H.248/M3UA/SCTP/IP/MPLS/PPP	102
H.248/SCTP/IP/MPLS/PPP	62

Table 12. Suggested message flow in Mc interface

Message Flow Type	Notes	Direction	No. of Message	Suggested Message Flow Payload (Octets)
1. Call between 3G (subscriber) and 3G (subscriber)	Internal office (MSS) call	Downlink. MSS to MGW	10	1697
		Uplink. MGW to MSS.	10	1658
2. Call between 3G and 3G	Inter-office call. Mobile Station (MS) originated.	Downlink	11	1969
		Uplink	11	1964
3. Call between 3G and 3G	Inter-office call. MS terminated.	Downlink	11	1915
		Uplink	11	1815
4. Call between 3G and PSTN	MSS as visiting office.	Downlink	11	1959
		Uplink	11	1746
5. Call from 3G to PSTN	MSS as gateway office	Downlink	7	1217
		Uplink	7	881
6. Call from PSTN to 3G	MSS as gateway office	Downlink	9	1465
		Uplink	9	
7. Inter-office handover	Handover into the MSS.	Downlink	7	1505
		Uplink	7	1436
8. Inter-office handover	Handover out of the MSS	Downlink	7	1330
		Uplink	7	1188
9. Internal handover	Handover in the same MSS.	Downlink	5	970
		Uplink	5	831
10. Call failure	Tone in call fails	Downlink	2	303
		Uplink	2	243

in which, S_K denotes the size of each H.248 message flow, K denotes the flow type from 1 to 10 in Table 12, P_K represents the payload of each message flow in Table 12, N_K denotes the number of H.248 messages in each flow in Table 12, $O_{H.248}$ denotes the overhead of H.248 message in Table 11.

It is easy to find that the payload in the downlink direction is heavier than that in the uplink direction, so the payload in the downlink direction is adopted in further calculations. The next step is to obtain the total throughput in the Mc interface via considering two scenarios: the MSC Server (MSS) functions as a visiting office or a gateway office.

With a visiting function for MSS, the message flow going through Mc interfaces include both call message flow and handover message flow. So message type 1, 2, 3, and 4 will be considered for call service. Meanwhile, message type 7, 8, and 9 will be considered for handover service in the Mc interface. At last type 10 in Table 12 will be considered when a call fails to establish. As a result, the Throughput in the Mc interface can be displayed below:

$$TH_{Mc} = N_S \begin{bmatrix} \left(\sum_{K=1}^{4} S_K R_{Call} / 4 + S_{10} R_{Fail} \right) \times BHCA_{Call} \\ + (\sum_{K=7}^{9} S_K / 3) \times BHCA_{Handover} \end{bmatrix} \times 8 / 3600 \quad (19)$$

where N_S denotes the number of 3G subscribers, S_K is obtained from function 18, R_{Call} denotes the ratio of established calls to total calls, R_{Fail} denotes the ratio of failed calls to total calls, $BHCA_{Call}$ represents the average call attempts in busy hour per subscriber, $BHCA_{Handover}$ denotes the average handover times in busy hour per subscriber.

With a sole gateway function for the MSS connection with the PSTN network, the message flow going through the Mc interface includes call service only. So message types 5 and 6 shall be considered for the bandwidth of the Mc interface in a Gateway MSC Server. The Handover function is not implemented in the Mc interface Mc of a Gateway MSS, so the handover portion in equation 19 will be removed for the interface Mc of a Gateway MSS.

Besides the H.248 messages, other messages may also be transported via the Mc interface if an internal Signaling Gateway (SG) is integrated with MGW. The signaling gateway can transport the RANAP message over ATM between RNC and MGW and then over IP between MGW and MSS and the transport Base Station System Application Part (BSSAP) message over TDM between the Base Station Controller (BSC) and the MGW and over the IP between the MGW and the MSS. In addition, it may also transfer the MAP message for HLR or Short Message Center (SMC), ISUP/TUP message for Gateway MSS or CAMEL Application Part (CAP) message for Service Control Point (SCP).

Nc Interface

Nc interface stands between the MSC Servers to implement inter-office call service and handover service. The communication protocol in the Nc interface is Bearer Independent Call Control (BICC), an advanced version evolved from the ISUP protocol, which can be borne in TDM, ATM and IP due to its bearer independent feature. ITU-T Q1901 SERIES Q, ITU-T Q1902.1 SERIES Q, ITU-T Q1902.2 SERIES Q, ITU-T Q1902.3 SERIES Q, ITU-T Q1902.4 SERIES Q, and ITU-T Q1902.4 SERIES Q define two modes to setup the BICC bearer: forward bearer setup which is subdivided into no tunnel case, fast tunnel case and delayed tunnel case; and backward bearer setup which includes no tunnel case only. The two modes have different sizes of message flows, so which mode is selected slightly impacts the throughput in interface Nc. As the preferable mode, the IP based forward bearer setup with fast tunnel case is selected in our case. The protocol stack in the Nc interface is shown in Table 13. The overhead

Table 13. Protocol stack in Nc interface

<table>
<tr><td colspan="4">BICC</td></tr>
<tr><td>M3UA</td><td rowspan="2">SCTP</td><td>MTP 3B</td><td>MTP3</td></tr>
<tr><td>SCTP</td><td>SSCF-NNI</td><td rowspan="3">MTP2</td></tr>
<tr><td colspan="2">IP</td><td>SSCOP</td></tr>
<tr><td>VLAN/MAC</td><td>MPLS/PPP</td><td>AAL5</td></tr>
</table>

Table 14. Suggested payload of BICC message

Message	**Direction**	**Payload (Octets)**
1.IAM	Forward	68
2.APM	Backward	41
3.APM	Forward	184
4.APM	Backward	184
5.COT	Forward	7
6.ACM	Backward	8
7.ANM	Backward	21
8.REL	F or B	12
9.RLC	Reversed to REL	6
Total		531
Number of messages		9

size of protocol stacks in the Nc interface is the same as that in Table 11.

In the case of the forward bearer setup with fast tunnel case, there are 9 messages going through the Nc interface. All messages serve for inter-office call services and inter-office handover services between the MSSs. The message types and suggested payload are summarized in Table 14. Since the direction of steps 8 and 9 in Table 14 are flexible, it is impossible to confirm the payload or message size in each direction. The method to calculate throughput in the Mc interface does not fit the throughput calculation for the Nc interface. An alternative is to calculate the average payload and message size of the message flow in two directions. The formula is:

$$TH_{Nc} = N_S \left\{ \begin{matrix} \left(P_{BICC} + N_{BICC} O_{BICC}\right) / \\ \left[2 \times \left(BHCA_{InterCall} + BHCA_{InterHO}\right)\right] \end{matrix} \right\} \times 8 / 3600 \quad (20)$$

where N_S denotes the number of 3G subscribers, P_{BICC} denotes the total payload of BICC messages which can be obtained from Table 14, N_{BICC} denotes the number of the BICC messages, obtained from Table 14, O_{BICC} denotes the overhead of the BICC message, same as it in Table 11, $BHCA_{\text{Inter-Office Handover}}$ denotes the average inter-office handover times in busy hour per subscriber, $BHCA_{\text{Inter-Office Call}}$ represents the average inter-office call attempts in busy hour per subscriber, it's given by:

$$BHCA_{intercall} = Erlang_{VOU} / {}_{BH} \times P_{interoffice} \times 3600 / T_{call} \quad (21)$$

Table 15. Protocol stack of Iu-PS interface

<table>
<tr><th colspan="2">Radio Network Control Plane</th><th>PS Data User Plane</th></tr>
<tr><td colspan="2">RANAP</td><td rowspan="2">Iu-UP</td></tr>
<tr><td colspan="2">SCCP</td></tr>
<tr><td>MTP3-B</td><td>M3UA</td><td>GTP-U</td></tr>
<tr><td>SSCF-NNI</td><td>SCTP</td><td>UDP</td></tr>
<tr><td>SSCOP</td><td colspan="2">IP</td></tr>
<tr><td colspan="3">AAL5</td></tr>
<tr><td colspan="3">ATM</td></tr>
</table>

Table 16. Suggested header size for Iu-PS interface

User Plane	Header Size (Octets)
Iu-UP	4
GTP-U	12
UDP	8
IP	20
AAL5	3
ATM	5
Total	52

where $Erlang_{VoUser/BH}$ denotes the average voice call traffic in Erlang per user per busy hour, $P_{interoffice}$ denotes the inter-office call rate (percentage), T_{call} denotes the average call time.

The values in Table 14 are based on TrFO fully applied in the network. However, the values based on TFO are very close to those in Table 14. Therefore it may not needed to differentiate the TrFO and TFO scenarios in dimensioning the bandwidth of interface Nc. If the bear is set up by other modes such as the forward bearer setup with delayed tunneling or the backward bearer setup with no tunnel case and so on, formula 20 universally applies for all cases.

Iu-PS Interface

Iu-PS interface, situated between the Radio Network Controller (RNC) and the Serving GPRS support Node (SGSN) and the Iu-CS interface between the RNC and the Media Gateway (MGW) composes the Iu interface. The Iu-PS and the Iu-CS interface define the same protocol stacks of the transport network user plane and the control plane, whereas they have a different transport network user plane. As per ITU-T I.366.2 and 3GPP TS 29.414, the protocol stacks of the Iu-PS interface are shown in Table 15, in which a significant difference is AAL5 rather than AAL2 in Iu-CS interface is adopted in layer 2 of the Iu-PS to transport the data in both control and user plane via IP over ATM. The total throughput in the Iu-PS interface is the sum of the throughput of the user plane and the control plane in the Iu-PS interface. The following paragraphs will introduce the algorithms of the user plane and the control plane of the Iu-PS interface.

Throughput of Iu-PS User Plane

The header sizes of the protocols in the Iu-PS user plane can be easily identified from 3GPP TS 25.401, 3GPP TS 29.060, ITU-T I.363.2, ITU-T I.363.5, IETF RFC 2225, IETF RFC 791, and IETF RFC 761. Table 16 displays the header size of each protocol in the user plane of the Iu-PS interface.

The packets sent via the Iu-PS are carried by ATM. So in order to calculate the throughput in the Iu-PS interface, the first step is to obtain how many ATMs are needed to load the transported packets. The total packet size consists of the sum of average packet size, header of the Iu-UP, header of the GTP-U, header of the UDP, header of the IP and header of the AAL5. The actual size in an ATM cell to load encapsulated packets is 53-header of ATM. As a result, the number of ATM Cells to load the encapsulated packets is given by:

$$N_{ATMCell} = \left(\bar{S}_{Packet} + H_{IuUP} + H_{GTP} + H_{UDP} + H_{IP} + H_{AAL5}\right) / (53 - H_{ATM}) \quad (22)$$

in which S_{Packet} denotes the average packet size which can be obtained from the traffic model provided by the mobile operators, H_{IuUP} denotes the header of Iu-UP packet which is obtained from Table 16, H_{IP} denotes the header of IP packet which is obtained from Table 16, H_{AAL5} denotes the header of AAL5 packet which is obtained from Table 16, H_{ATM} denotes the header of ATM cell which is obtained from Table 16.

In planning a packet switched network, the mobile operators will estimate some important traffic parameters given by a traffic model, such as the average packet size, the estimated number of subscribers, the ratio of attached users in busy hour, and the average throughput per user in one busy hour and the redundancy factor. With these conditions provided by the traffic model, the "pure throughput" value without any overhead can be obtained by:

$$PureThroughput = N_S \times R_{Attach} \times R_{Active\,/\,Attach} \times Th_{User\,/\,S} \times 8 \quad (23)$$

where N_S denotes the number of subscribers with 3G packet switched service subscription, $Th_{User/S}$ denotes the average throughput per user per second (bps), R_{Attach} denotes the ratio of attached users in busy hour, $R_{Active/Attach}$ denotes the rate of attached users who activate PDP in busy hour, 8 denotes the conversion to bits from bytes.

However, in the actual network environment the extra overhead and redundancy will be considered. So based on formula 23, the throughput of user plane of Iu-PS interface is given by:

$$TH_{UPIuPS} = N_S \times R_{Attach} \times R_{Active\,/\,Attach} \times \left(R_{Down} \times Th_{User\,/\,S}\right) \times \left(N_{ATMCell} \times 53 \,/\, \bar{S}_{Packet}\right) \times 8 \,/\, F_{Redundancy} \quad (24)$$

where N_S denotes the number of subscribers with 3G packet switched service subscription, 8 denotes the conversion to bits from bytes, R_{Attach} denotes the ratio of attached users in busy hour, $R_{Active/Attach}$ denotes the rate of attached users who activate PDP in busy hour, S_{Packet} denotes the average packet size which can be obtained from the traffic model provided by the mobile operators, RDown denotes the ratio downstream throughput to down + upstream data throughput, $Th_{User/S}$ denotes the average throughput per user per second (bps), $N_{ATMCell}$ denotes the number of ATM Cells which can be obtained by formula 22, $F_{Redundancy}$ denotes the redundancy factor (normally it is 0.7).

The $\left(N_{ATMCell} \times 53 \,/\, \bar{S}_{Packet}\right)$ portion denotes the proportion of ATM cell sizes to pure packet size. It explains the impact of network overhead on the Iu-PS interface.

The $\left(R_{Down} \times Th_{User\,/\,S}\right)$ portion denotes the data throughput per subscriber in one way direction. It is assumed that downstream is heavier than upstream. If reversed, R_{down} should be changed to R_{up}.

Table 17. Footnotes for formula 25

	N_{IuPSi}	L_{IuPSi}
1	Authentication times per busy hour	Length of messages per authentication
2	Attachment times in busy hour	Length of messages per attachment in Iu-PS
3	Detachment times in busy hour	Length of messages per detachment in Iu-PS
4	Inter SGSN route update times in busy hour	Length of messages per inter SGSN route update
5	Intra SGSN route update times in busy hour	Length of messages per intra SGSN route update
6	Intra SGSN SRNC times in busy hour	Length of messages per intra SGSN SRNC.
7	PDP activation times in busy hour	Length of messages per PDP activation
8	PDP deactivation times in busy hour	Length of messages per PDP deactivation
9	Periodic SGSN route area update times In busy hour	Length of messages per periodical SGSN route update
10	Short message service mobile originated (SMS MO) times in busy hour	Length of messages per SMS service
11	SMS MT times in busy hour	Length of messages per SMS service

Throughput of Iu-PS Control Plane

The control plane of the SGSN provides such four major functions as mobility management, session management, path management and short messages services etc. The primary messages adopted for throughput calculation are categorized by each function.

- Mobility management
    - Authentication message
    - Attach message
    - Intra SGSN routing area update message
    - Inter SGSN routing area update message
    - Service RNC relocation
- Session management
    - Packet data protocol (PDP) activation message
    - PDP deactivation message

The 8 primary messages compose the majority of the throughput in the control plane of the Iu-PS interface. However, Table 17 lists 11 primary types of messages which can be estimated and provided by the mobile operators according to their historical operation data. The first 8 messages are what we introduced above while the last 3 messages, as the optional messages, may also be adopted by mobile operators and applied into formula 25.

The throughput in the control plane of the Iu-PS interface is given by:

$$TH_{CPIuPS} = N_S \ x \ R_{Attach} \ x \ \sum_{i=1}^{11} (N_{IuPSi} \ x \ L_{IuPSi}) \ x \ 8 / 3600 \quad (25)$$

where N_S denotes the number of subscribers with 3G packet switched service subscription, 8 denotes the conversion to bits from bytes, 3600 denotes the conversion to second from busy hour, R_{Attach} denotes the ratio of attached users in busy hour.

The other parameters are explained in Table 17.

The sum of the throughput of those 8 or 11 messages composes the total throughput in the control plane of the Iu-PS interface. Besides the 11 messages discussed, other messages such as P-Temporary Mobile Subscriber Identity (TMSI)

Figure 2. Layout of CS network

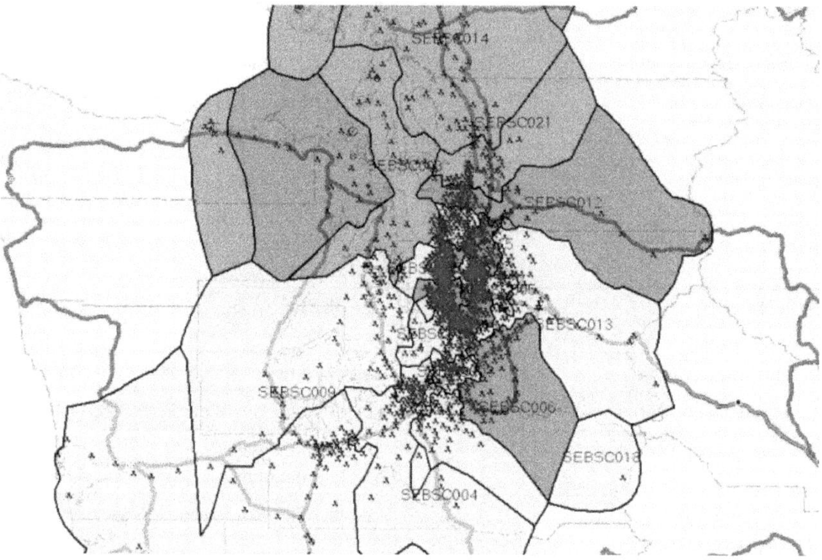

re-allocation message, identification check message, and service request message etc, due to the smaller message size and lower utilization, are not considered in our throughput calculation for the control plane of the Iu-PS interface. If any of these messages are requested by mobile operators and their parameters are difficult to estimate, a redundancy factor can be imposed in formula 25 as a rough calculation.

Total Throughput of Iu-PS Interface

Based on the results from Section 4.51 and 4.52, the total throughput in the Iu-PS interface is the sum of the throughput in the control plane and the user plane of Iu-PS interface. The algorithm is given by:

$$TH_{IuPS} = TH_{UPIuPS} + TH_{CPIuPS} \quad (26)$$

Summary of Section 4

Section 4.1 to 4.5 provides the algorithms of throughput for the new interfaces that exist in UMTS CN. The algorithms for the other interfaces such as A, C, E, Gb, Gs, Gi, Gs and Gc interfaces are still the same as those in GSM/GPRS stage. In the control plane of the Iu-CS and the Mc interface, throughput of the RANAP protocol may also be considered in dimensioning the CN topology. Section 4.1 for the Iu-CS interface and Section 4.3 for the Mc interface only consider the primary factors contributing to the overall throughput. Throughput generated by the RANAP may be accumulated onto the result of formula 8 and 19. In the calculation of throughput for the control plane of the Iu-PS interfaces, only the primary messages that contribute the majority of the throughput for control plane are selected. Considering the throughput from control plane only accounts for a very small portion of the total throughput (less than 1-5%), overlooking the non-primary messages is acceptable.

CASE STUDY

A Case of Circuit Switched (CS) Domain

In Figure 2, a mobile operator intends to roll out a new 3G UMTS CN in the area with slashes

Table 18. Traffic model for circuit switched domain

Parameter	Value	Notes
Network Volume	300,000	3G subscribers
Local 1 Volume	100,000	3G subscribers
Local 2 Volume	100,000	3G subscribers
Local 3 Volume	100,000	3G subscribers
Voice traffic per Sub at BH	0.025	Unit: Erlang.
Video traffic per Sub at BH	0.005	Unit: Erlang.
VAD Factor	0.5	
TrFO rate	100%	
Video Call penetration rate	10%	
Redundancy factor	0.7	Range: 0.7-1
Nb bearer protocol stack	NbUP/RTP/UDP/IP/VLAN/MAC	
Mc bearer protocol stack	H.248/M3UA/SCTP/IP/VLAN/MAC	
Nc bearer protocol stack	BICC/M3UA/SCTP/IP/VLAN/MAC	
Nc bearer setup mode	Forward bearer, fast tunnel.	
$BHCA_{Call}$ per sub	1.5	
$BHCA_{Handover}$ per sub	0.5	Unit: Time/user/BH.
$BHCA_{Inter\text{-}officeHO}$ per sub	0.1	Unit: Time/user/BH.
Inter-office call rate	50%	
Call fail rate (Call fail tone played)	1%	

(represents heavy traffic loading) to replace the legacy GSM systems. Each MGW supports 100,000 3G subscribers in its local area. MSS supports 300,000 3G subscribers. The traffic model is shown in Table 18.

Based on the formulas in section 4, we obtain the results below.

$$TH_{IuCS1} = 100,000 \times \begin{pmatrix} 100\% \times 0.025 \times 17 \times 10^3 \\ +10\% \times 0.005 \times 85 \times 10^3 \end{pmatrix} / 0.7$$

$$= 66.79Mbps = TH_{IuCS2} = TH_{IuCS3}$$

$$TH_{Nb1-2} = 100,000 \times \begin{pmatrix} 100\% \times 0.025 \times 24.775 \times 10^3 \\ +0.1 \times 0.05 \times 99.775 \times 10^3 \end{pmatrix} / 0.7$$

$$= 159.75Mbps = TH_{Nb1-3} = TH_{Nb2-3}$$

$$TH_{Mc1} = 100,000 \times \begin{bmatrix} \left(3207.105 + 5.55\right) \times 1.5 \\ +1033.16 \end{bmatrix}$$

$$\times 8 / 3600 = 1.3004Mbps = TH_{Mc2} = TH_{Mc3}$$

$$TH_{Nc} = 30,000 \times \left[\left(531 + 9 \times 126\right) / 2\right]$$

$$\times \left(0.75 + 0.1\right) \times 8 / 3600 = 471.75Kbps$$

To verify the accuracy of the algorithms in this case, we captured 200 real time throughput values of the three Iu-CS interfaces from the network logs. Compared with the threshold value (66.79Mbps) obtained via the formula for Iu-CS interface, 100% real time throughput value is below the threshold value. We randomly select 50 sample values to plot the blue line in Figure 6. It shows all the throughput values is below the 73.17% of threshold value. This is consistent with the fact of redundancy factor= 0.7.

The throughput values in Figure 4 are real time recorded values and show a randomly selected sample plotted in the figure. The real time values suggest a rough estimation to the interval of the real time throughput.

Threshold 1: 66.79Mbps when $F_{Redundancy}$=0.7.
Threshold 2: 58.44Mbps when $F_{Redundancy}$=0.8.
Threshold 3: 51.95Mbps when $F_{Redundancy}$=0.9.

Figure 3. Topology of CS domain

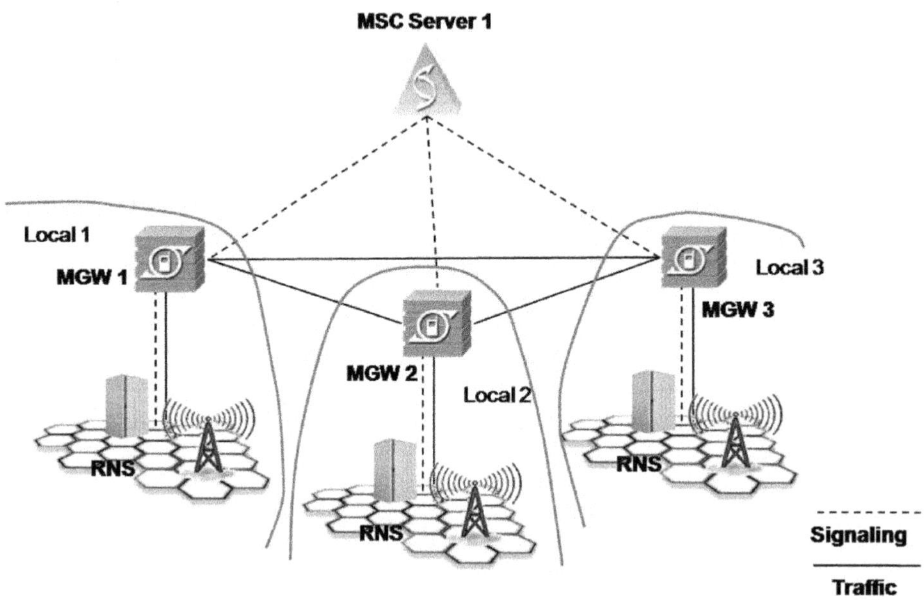

Figure 4. Throughput trial of CS domain

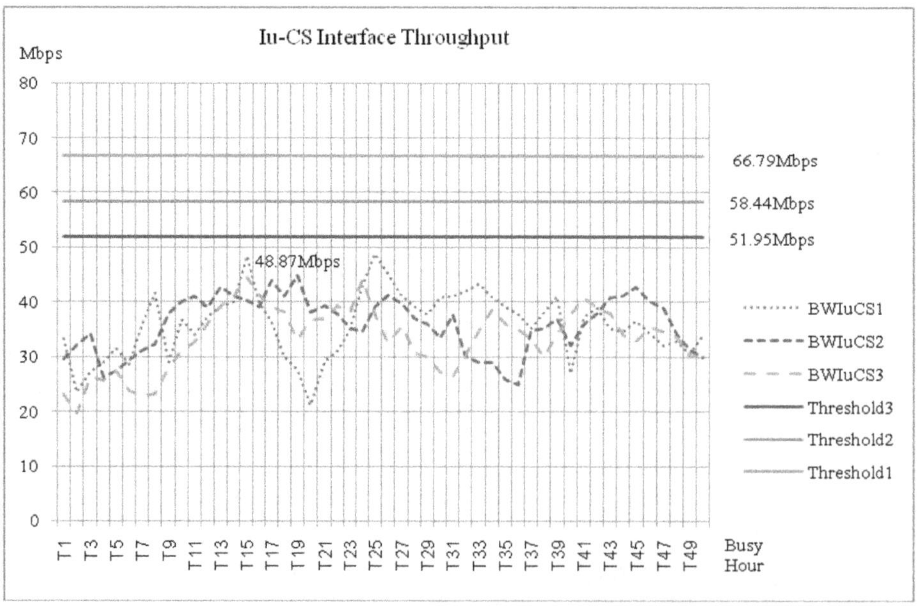

Figure 5. Layout of network coverage

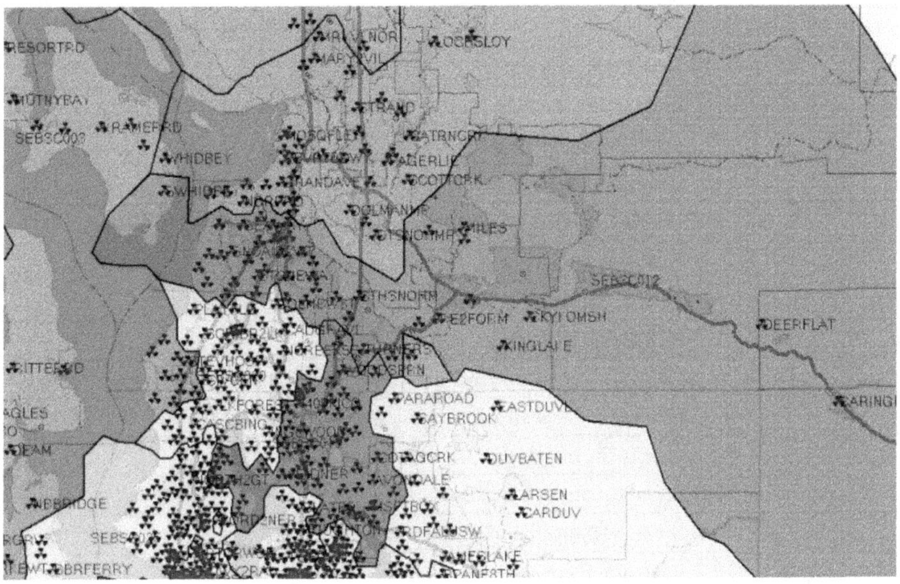

Figure 6. Packet switched network topology

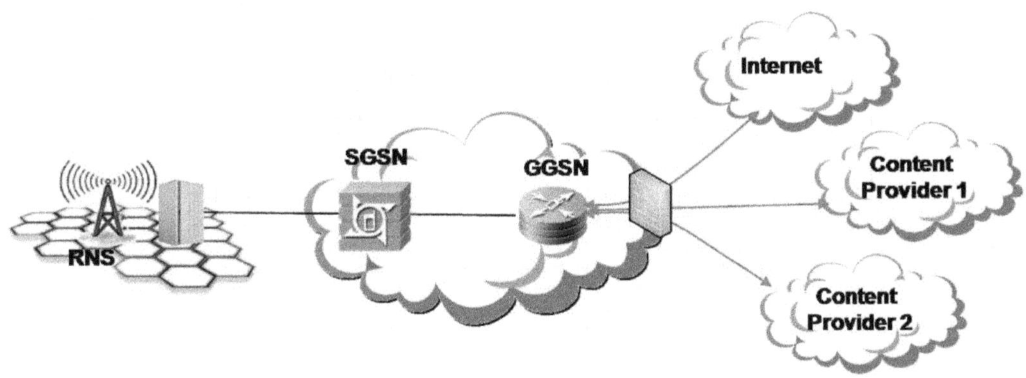

A Case of Packet Switched (PS) Domain

As Figure 5 shows, a mobile operator in North Africa intends to build a new 3G UMTS Packet Switched Network in the slashed area (represents heavy traffic loading) to enhance the data service coverage. The markers in the map represent the cell sites. We randomly select 48 sample values to visualize the throughput in Figure 6.

Table 19 defines the traffic model with required parameters for dimensioning a UMTS PS network. Those parameters shall be pre-estimated and provided by mobile operators. The parameters in Table 19 comprise the traffic model (traffic parameter template) in which most of the parameter values are identified by the mobile operator based on a multivariate analysis of its real time statistical performance data in a long term.

Table 20 provides the suggested length of messages in a certain service in UMTS PS domain. The values may be slightly varied from vendors' products since some fields in the message defined by 3GPP are optional to adopt. In addition, this

Table 19. Traffic model for packet switched domain

Traffic parameters	Value	Description
Network volume	100,000	Subscribers with UMTS PS subscription.
ThUser/bps	600	Average data throughput per user per second Unit:bps.
SPacket	400	Average size of a IP packet.
RAttach	75%	The ratio of attached users to total UMTS PS users.
RActive/Attach	25%	The ratio of activated to attached users
NAttach	0.75	Attachment times per user in busy hour.
NDetach	0.75	Detachment times per user in busy hour
NPDP-Activation	1.5	PDP activation times per user in busy hour.
NPDP-Deactivation	1.5	PDP deactivation times per user in busy hour.
NRoute-intraSGSN	4	Intra SGSN route area update times per user in busy hour
NRoute-interSGSN	0.1	Inter SGSN route area update times per user in busy hour.
NRoute-periodic	0.3	Periodic route area update times
NRoute	4.4	Nroute= NRoute-intraSGSN+interSGSN+periodic
RJoint-Route	18%	Ratio of Joint route area update
RJoint-Location	18%	Ratio of joint location update
NSRNC-IntraSGSN	0.07	Service RNC relocation times per user in busy hour(Intra SGSN)
NSRNC-interSGSN	0.01	Service RNC relocation times per user in busy hour(Inter SGSN)
NSRNC	0.08	NSRNC= NSRNC-Intra + Inter SGSN
NAuth	3.55	Authentication times per user in busy hour
RAuthToHLR	20%	Ratio of authentication that needs to obtain authentication parameters from HLR.
NSMS-MO	0.1	Short messages times per user in busy hour (mobile originated)
NSMS-MT	0.5	Short message times per user in busy hour (mobile terminated)
NSMS	0.6	NSMS=NSMS-MO+MT
RDown-Up	3	Ratio of downstream to upstream data
LdMAP	0.2	Link load of MAP message.

case is used to verify the algorithms, so roaming, intelligent network users and pre-paid users are not considered. All users are assumed to be post paid UMTS PS users.

As per the formula for Iu-PS interface, the throughput for Iu-PS interface is obtained.

The number of ATM cells is obtained by:

$$N_{ATMCell} = \left(\bar{S}_{Packet} + H_{IuUP} + H_{GTP} + H_{UDP} + H_{IP} + H_{AAL5}\right) / (53 - H_{ATM}) = 447 / 48 \approx 10$$

The throughput of user plane in Iu-PS interface is given by:

$$TH_{UPIuPS} = N_S \times R_{Attach} \times \left(Th_{User\ /\ S} \times R_{Down}\right) \times \left(N_{ATMCell} \times 53 / \bar{S}_{Packet}\right) \times 8 / F_{\text{Re dundancy}} = 100,000 \times 75\% \times 25\% \times \left(600 \times 3 / 4\right) \times \left(10 \times 53 / 400\right) \times 8 / 0.7 = 127.77 Mbps$$

The throughput of control plane in Iu-PS interface is provided by:

Table 20. Parameters of PS domain

Parameters	Suggested message length at single direction	Description
LAttach at Iu-PS	336	Length of messages per attachment at Iu-PS interface.
LDetach at Iu-PS	336	Length of messages per detachment at Iu-PS interface.
LPDP-Active at Iu-PS	768	Length of messages per PDP activation at Iu-PS interface.
LPDP-Deactive at Iu-PS	768	Length of messages per PDP deactivation at Iu-PS interface.
LRoute at Iu-PS	144	Length of messages per route area update at Iu-PS interface.
LSRNC at Iu-PS	1152	Length of messages per SRNC relocation at Iu-PS interface.
LAuthen at Iu-PS	192	Length of messages per authentication at Iu-PS interface.
LSMS at Iu-PS	1022	Length of messages per short message at Iu-PS interface.
LPDP-Active at Gn	300	Length of messages per PDP activation at Gn interface.
LPDP-Deactive at Gn	50	Length of messages per PDP deactivation at Gn interface.
LAttach at Gr	294	Length of messages per attachment at Gr interface.
LRoute at Gr	71	Length of messages per route area update at Gr interface.
LAuthen at Gr	259	Length of messages per authentication at Gr interface.
LRoute at Gs	82	Length of messages per route area update at Gs interface.

$$TH_{CPIuPS} = N_S \times R_{Attach} \times \sum_{i=1}^{11}\left(N_{IuPSi} \times L_{IuPSi}\right) \times 8 \,/\, 3600$$
$$= 100,000 \times 75\% \times 4829 \times 8 \,/\, 3600$$
$$= 804833.33bps = 0.80Mbps$$

Total throughput of Iu-PS interface:

$$TH_{IuPS} = TH_{UPIuPS} + TH_{CPIuPS} = 127.77 + 0.80 = 128.57Mbps$$

To verify the precision of the algorithms in this case, we captured 300 real time throughput values of Iu-PS interface from the network logs. Compared with the threshold value (128.57Mbps) obtained via the formula for Iu-PS interface, 100% real time throughput value is below the threshold value. We randomly select 48 sample values to plot the blue line in Figure 6. It shows all the throughput values is below the 68% of threshold value. This is consistent with the fact of redundancy factor= 0.7.

Threshold 1: 128.57Mbps when $F_{Redundancy}$=0.7.
Threshold 2: 112.50Mbps when $F_{Redundancy}$=0.8.
Threshold 3: 99.99Mbps when $F_{Redundancy}$=0.9.

CONCLUSION

The chapter first reviewed the current literatures in planning and designing UMTS networks and analyzed a problem that the mobile operators currently meet in dimensioning and planning UMTS core networks: the past experience of mobile operators in core network dimension and planning is relying on the proposed solutions provided from vendors which dimension the network and calculate the throughput based on the performance of their own products. Mobile operators themselves generally do not have a mature and global approach, which is totally independent from vendors, to neutrally dimension the UMTS core network and estimate the proposals from different vendors.

Also the current literatures introduced many applied methods and tools to plan and design 3G radio networks. However, not much effort has been focused on the evolution of the core network. This paper illustrated the encapsulation, delivery and transport process of packets and messages in UMTS core network. Based on the traffic flow,

Figure 7. The real time throughput in Iu-PS interface

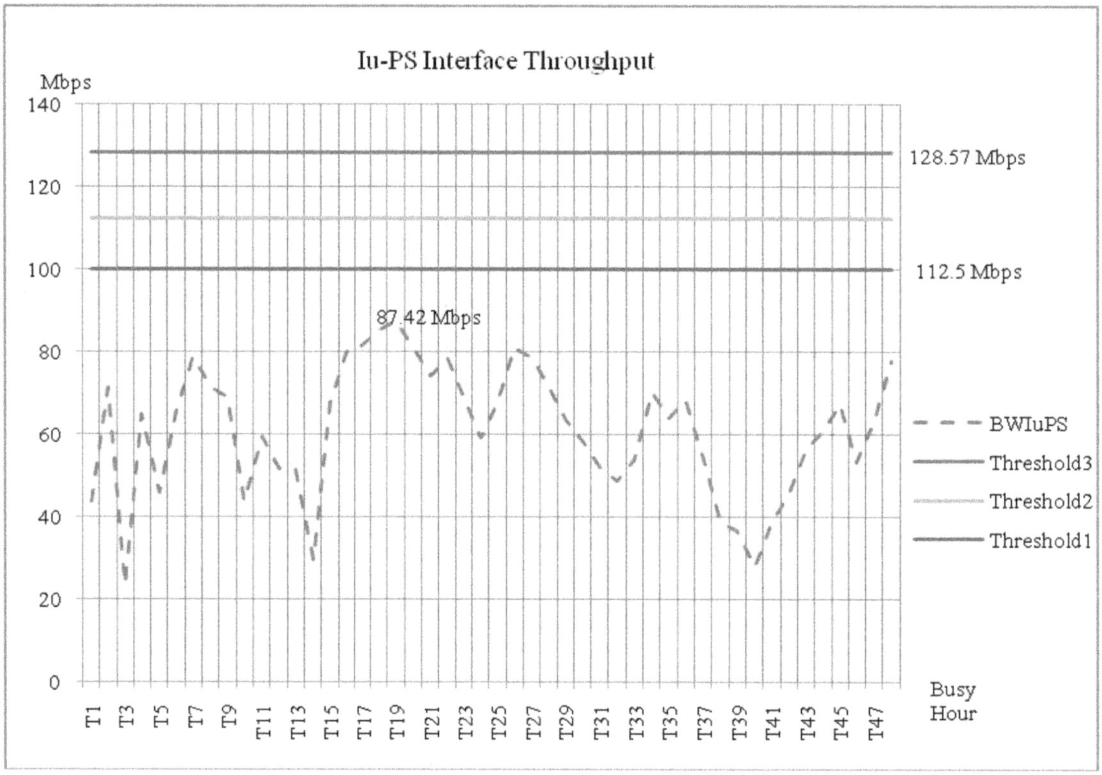

message flow and service process defined by 3rd Generation Partnership Project (3GPP) and the International Telecommunications Union (ITU), the algorithms and formulas to calculate the traffic and throughput of all the interfaces in UMTS core network are provided. Since some parts in the message packet are optional to use by vendors according to 3GPP, the message size, header size and overhead size are suggested values in dimensioning the UMTS CN. The actual values may slightly vary by different vendor's products.

The most significant contribution of this article is to help mobile operators achieve vendor neutrality in network planning. The article provides detailed guidelines and algorithms for dimensioning the UMTS core networks to enable any mobile operator's network planning process to be independent from the vendor bias. The dimensioning rules and guidelines provided in Section 4 could also help the mobile operators to appropriately size their networks to minimize their Total Cost of Ownership (TCO) which includes Capital Expenditure (CAPEX) and Operation Expenditure (OPEX).The importance of vendor neutrality in network planning can be illustrated with the following simple example: Assume the current network traffic is 0.01Erlang/subscriber. If the mobile operator decides to deploy a new UMTS network to support 1 million subscribers, 477 E1 trunks (0.01*10^6/0.7/30=477 E1) would be required. The mobile operator actually needs a budget to cover 477 E1s. Let's say each switching card from a given vendor can support up to 500 Erlang/card and 4E1/card. If selecting this vendor, the mobile operator needs to buy 500 E1s (0.01*10^6/500/4=500 E1). That means the mobile operator will be over-investing in its network. Our model and algorithms enable the operator to estimate its capacity needs totally independent of the vendor products, hence optimizing its investments in capital and operations.

This study needs to be extended further to network dimension and planning towards R6 and on to R8 phase with IP Multimedia Subsystem (IMS) and System Architecture Evolution (SAE) converged with UMTS core network. The evolution from TDM to IP is a lengthy process and requires a systematic and optimal approach. However, confirming the algorithms for UMTS CN is the foundation from which we can extend the research in planning IMS and SAE. The dimensioning work for UMTS network is another step in the evolution of the mobile network. While the deployment of UMTS radio access networks receives considerable attention, the UMTS core network has emerged as a critical element in the delivery of next generation mobile broadband services. As such, the algorithms provided in the paper are and will benefit mobile operators to address the issues in network dimension and plan while positioning them for future technologies.

ACKNOWLEDGMENT

The authors acknowledge the thoughtful suggestions on VAD effect and speech frame by Dr. Kevin Ryan, the helpful editing work of Sharen Glennon, the insightful comments from three anonymous reviewers, and the prompt response from the journal editor on several drafts of this paper.

REFERENCES

3GPP TS 23.002. (2011). *Technical specification group services and systems aspects: Network architecture.*

3GPP TS 25.401. (2011). *Technical specification group radio access network: UTRAN overall description.*

3GPP TS 25.413. (2011). *Technical specification group radio access network: UTRAN Iu interface radio access network application part (RANAP) signaling.*

3GPP TS 25.415. (2011). *Technical specification group radio access network: UTRAN Iu interface user plane protocols.*

3GPP TS 29.060. (2011). *Technical specification group core network, general packet radio service (GPRS): GPRS tunnelling protocol (GTP) across the Gn and Gp interface.*

3GPP TS 29.414. (2011). *Technical specification group core network and terminals: Core network Nb data transport and transport signaling.*

3GPP TS 29.415. (2011). *Technical specification group core network and terminals: Core network Nb interface user plane protocols.* Britvic, V. (2004). Steps in UMTS network design. *Proceedings of the 12th IEEE Mediterranean Electrotechnical Conference,* vol. 2 (pp. 461-464).

Harmatos, J. (2002) Planning of UMTS core networks. *Personal, Indoor and Mobile Radio Communications. The 13th IEEE International Symposium, 2*(15-18), 740–744.

Harmatos, J., Jüttner, A., & Szentesi, Á. (1999). *Cost-based UMTS transport network topology optimisation.* Paper presented at the International Conference on Computer Communications, Tokyo, Japan.

IETF RFC 2225. (1999). *Classical IP and ARP over ATM.*

IETF RFC 761. (1980). *User data protocol.*

IETF RFC 791. (1981). *Internet protocol.*

ITU-T. (2002). *Series H: Audiovisual and multimedia systems, infrastructure of audiovisual services- Communication procedures. H.248.1: Gateway control protocol.*

ITU-T Q1901. (2001). *Series Q: Switching and signalling: Specifications of signaling related to bearer independent call control (BICC), bearer independent call control protocol corrigendum 1.*

ITU-T Q1902.1. (2001). *Series Q: Switching and signalling: Specifications of signaling related to bearer independent call control (BICC), bearer independent call control protocol (capability set 2): Functional description.*

ITU-T Q1902.2. (2001). *Series Q: Switching and signalling: Specifications of signaling related to bearer independent call control (BICC), bearer independent call control protocol (capability set 2): General functions of messages and parameters.*

ITU-T Q1902.3. (2001). *Series Q: Switching and signalling: Specifications of signaling related to bearer independent call control (BICC), bearer independent call control protocol (capability set 2): Formats and codes.*

ITU-T Q1902.4. (2001). *Series Q: Switching and signalling: Specifications of signaling related to bearer independent call control (BICC), bearer independent call control protocol (capability set 2): Basic call procedures.*

ITU-T Q1902.5. (2001). *Series Q: Switching and signalling: Specifications of signaling related to bearer independent call control (BICC), bearer independent call control protocol (capability set 2): Extensions to the application transport mechanism in the context of the bearer independent call control.*

ITU-T I.363.5. (1995). *B-ISDN ATM adaptation layer specification: Type 5 AAL - Series I: Integrated services digital network overall network aspects and functions - Protocol layer requirements.*

ITU-T I.363.2. (1997). *B-ISDN ATM adaptation layer specification: Type 2 AAL Series I: Integrated services digital network - Overall network aspects and functions - Protocol layer requirements.*

ITU-T I.366.2. (2007). *AAL type 2 service specific convergence sub-layer for narrow-band services.*

Jamaa, S. B., Altman, Z., Picard, J. M., & Fourestie, B. (2004). Multi-objective strategies for automatic cell planning of UMTS networks. *59th IEEE Vehicular Technology Conference (VTC), vol. 4* (pp. 2420-2424).

Juttner, A., Orban, A., & Fiala, Z. (2005). Two new algorithms for UMTS access network topology design. *European Journal of Operational Research, 164*, 456–474. doi:10.1016/j.ejor.2003.11.027

Kunz, A., Gonzalez, M., Tcaciuc, S., Ruland, C., & Rapp, V. (2005). Analysis of the QoS requirements under radio access network planning aspects for GPRS/EDGE and UMTS. *Wireless Networks, Communications and Mobile Computing, 2005 International Conference, vol. 1* (pp. 386-391).

Maple, C., Guo, L., & Zhang, J. (2004). Parallel genetic algorithms for third generation mobile network planning. *International Conference on Parallel Computing in Electrical Engineering* (pp. 229–236).

Neruda, M., & Bestak, R. (2008). Evolution of 3GPP core network. *Systems. Signals and Image Processing, IWSSIP, 2008*, 25–28.

Neubauer, T., & Toeltsch, M. (2005). UMTS radio network planning-maximizing return on investment. *Elektrotechnik und Informationstechnik, 122*(3), 108–113.

Ouyang, Y., & Fallah, M. H. (2009). Evolving core networks from GSM to UMTS R4 version. *International Journal of Mobile Network Design and Innovation, 3*(2), 93–102. doi:10.1504/IJMNDI.2009.030841

Ouyang, Y., & Fallah, M. H. (2009). A study of throughput for Iu-CS and Iu-PS interface in UMTS core network. *Performance Computing and Communications Conference (IPCCC), IEEE 28th International* (pp. 349-353).

Ouyang, Y., & Fallah, M. H. (2010). A study of throughput for Nb, Mc and Nc interface in UMTS core network. *Performance Computing and Communications Conference (IPCCC), 28th IEEE International* (pp.371-375).

Shalak, R., Sandrasegaran, K., Agbinya, J., & Subenthiran, S. (2004). UMTS core network planning model and comparison of vendor product performance. *Advanced Communication Technology, the 6th International Conference, vol. 2* (pp. 685- 689).

Wu, Y., & Pierre, S. (2003). Optimization of access network design in 3G networks. *IEEE Canadian Conference on Electrical and Computer Engineering (CCECE)* (pp. 781–784).

Wu, Y., & Pierre, S. (2004). A new hybrid constraint-based approach for 3G network planning. *IEEE Communications Letters*, *8*(5), 277–279. doi:10.1109/LCOMM.2004.827437

Wu, Y., & Pierre, S. (2005). Optimization of 3G mobile network design using a hybrid search strategy. *Journal of Communications and Networks*, *7*(4), 471–477.

ENDNOTES

1. P_{frame} and E_{AAL2} in formula 1 and 2 are distinct in VAD and Non-VAD mode. We have stated that "P_{Frame} denotes the different types of payload of frame in Table 5". Since there are many AMR and Video formats, we did not want to make notations for all parameters. So here we set E_{AAL2} and P_{frame} as a general notation to respectively represent the encapsulation efficiency by AAL2 which includes all types of formats and to represent all types of frame payloads.
2. We have not shown other interfaces besides the Iu interface in the case studies for two reasons: First, because of the page limitation, we selected the most typical interface for the case study. Second, we believe Iu-interface is the most typical interface in the core network from the aspect of traffic estimation. Normally the traffic going through the Iu interface is the total traffic going into the core network side. Then the traffic is distributed into different nodes in the core network. The Iu interface is like the front door of the core network. If the threshold value set for this front door is safe, we may be less worried about the threshold values for other interfaces in the core network.

APPENDIX: GLOSSARY OF TERMS

3GPP: 3rd Generation Partnership Project
AAL2: ATM Adaption Layer 2
ALCAP: Access Link Control Application Part
AMR: Adaptive Multi-Rate
APN: Access Point Name
ATM: Asynchronous Transfer Mode
AUC: Authentication Center
BCTP: Bearing Control Tunneling Protocol
BHCA: Busy Hour Calling Attempt
BICC: Bear Independent Call Control message
BSSAP: Base Station System Application Part
BSC: Base Station Controller
CAP: CAMEL Application Part
CAPEX: Capital Expenditure
CN: Core Network
CPS: Common Part Sub-layer
CS: Circuit Switching Domain
EIR: Equipment Identity Register
FCP: Frame Control Part
FCSP: Frame Check Sum Part
FMC: Fixed Mobile Convergence
GGSN: Gateway GPRS Support Node
GSM: Global System for Mobile Communications
GTP: GPRS Tunneling Protocol
HLR: Home Location Register
IMS: IP Multimedia Sub-system
IPBCP: IP Bearer Control Protocol
ITU: International Telecommunications Union
ISUP: ISDN User Part message
Iu-UP: Iu Interface User Plane Protocol
LTE: Long Term Evolution
MAC: Media Access Control
MAP: Mobile Application Part
MPLS: Multi Protocol Label Switching
MGW: Media Gateway
MS: Mobile Stations
MSC: Mobile Switching Center
MSS: MSC Server
MTP 3: MTP Level 3
NE: network entity
NGN: Next Generation Network
OPEX: Operation Expenditure

PDU: Packet Data Unit
POS: PPP Over SONET/SDH
PPP: Point to Point Protocl
PS: packet switched domain
PSTN: Public Switched Telephone Network
RAN: Radio Access Network
RANAP: Radio Access Network Application Part
RNC: Radio Network Controller
RNS: Radio Network Subsystem
RTP: Real-time Transport Protocol
SAE: System Architecture Evolution
SCP: Service Control Point
SCCP: Signaling Connection Control Part
SCTP: Stream Control Transmission Protocol
SG: Signaling Gateway
SGSN: Serving GPRS Support Node
SID: Silence Descriptor
SMC: Short Message Center
SSCF: Service Specific Coordination Function
SSCOP: Service Specific Connection Orientated Protocol
SSCS: Service Specific Convergence Sub-layer
TCO: Total Cost of Ownership
TFO: Tandem Free Operation
TrFO: Transcoder Free Operation
UDP: User Datagram Protocol
UMTS: Universal Mobile Telecommunications System
VAD: Voice Activity Detection
VLAN: Virtual LAN
VLR: Visitor Location Register
VOIP: Voice over IP

This work was previously published in International Journal of Business Data Communications and Networking, Volume 6, Issue 2, edited by Debashis Saha and Varadharajan Sridhar, pp. 25-51, copyright 2010 by IGI Publishing (an imprint of IGI Global).

Chapter 2
Cartography and Stability to Enhance and Self Regulate Proactive Routing in MANETs

Mohamed Amine Abid
University of Manouba, Tunisia

Abdelfettah Belghith
University of Manouba, Tunisia

ABSTRACT

In this chapter, the authors propose a novel proactive routing protocol where routing advertisement frequency is self regulated locally at and by each node depending on its mobility level. The proposal relays on an asynchronous distributed cartography gathering algorithm and a link stability criterion. The cartography of the network is kept constantly valid in time and for all applied speeds. Each node senses its own dynamics and locally chooses an appropriate routing advertisement period size. As such stationary nodes generate little signaling traffic while fast moving nodes prefer small routing periods to mitigate the effect of their mobility.

INTRODUCTION

The performances of our proposed routing protocol are evaluated and compared to the known Optimized Link State Routing (OLSR) protocol through extensive simulations. We first show that the collected network cartography maintains over time a validity ratio near one hundred percent even for high node speeds. We then show that our proposed routing protocol provides around 97 percent routing validity while that of OLSR can hardly deliver more than 60 percent at moderate to high speeds and workloads. Finally, we show that our proposed protocol provides better throughput than OLSR reaching a 50% increase at moderate to high speeds and workloads and far much less end to end delays.

Routing in multi hop mobile ad hoc networks (MANETs) plays a central and vital role, yet it stands out as one of the most and fundamental issues to attain adequate performances and viable deployments. The crux of routing in such networks

DOI: 10.4018/978-1-61350-477-2.ch002

lies in its ability to maintain and provide valid routes without which traffic may wander inside the network without being able to be delivered to its ultimate destinations. Worse yet, on such invalid routes, the underlying Media Access Control (MAC) protocol retransmits unacknowledged unicast frames for several (by default seven) times which consequently amounts to over using the wireless communication medium and hence yields poor performances. We may stipulate here that the basic reason of the non proliferation of such spontaneous networks, despite their practical usefulness and need, is mainly due to their poor performances caused by the lack of viable low signaling and high validity routing algorithms.

In mobile ad hoc networks, nodes and links can appear and disappear spontaneously as a consequence of several facts such as the behavior of users, the depletion of energy resources, but more inherently and subtly the random mobility of the different nodes. These aspects imply a dynamically and randomly evolving topology in both time and space making the routing function a real difficult and challenging task. As a result, and as the elapsed time since the start of the routing period gets farther, the topological information collected at the beginning of the routing period becomes inaccurate and obsolete leading patently to invalid paths.

Conventional protocols, whether reactive - such as the Dynamic Source Routing (DSR) protocol (Johnson, Maltz and Yih-Chun Hu, 2004) and the Ad-hoc On-demand Distance Vector (AODV) routing protocol (Perkins, Royer and Chakeres, 2003) - or proactive - like OLSR (Clausen and Jacquet, 2003) and the Destination Sequenced Distance Vector (DSDV) routing protocol (Perkins and Bhagwat, 1999) -, use in general the hop-count metric to compute shortest paths towards destinations. However, shortest paths are not always reliable especially in case of dynamic networks. Finding stable routes is rather the main concern for dynamic multi-hop ad hoc networks. Several works already established that choosing routes based on positions, battery level, etc. of the nodes would make selected paths more resilient to topological changes.

Numerous simulation studies were conducted on different scenarios to evaluate the performance of conventional protocols: for both proactive and reactive routing protocols as those conducted by Clausen, Jacquet and Viennot (2002), Novatnack, Greenwald and Arora (2005), or Mbarushimana Amine and Shahrabi (2002). Nevertheless, due to the huge number of relevant and complex events that can take place in mobile wireless ad hoc networks and their impacts on the performance of the protocols, results do not necessarily agree as to which protocol yields satisfying performances and lower control traffic and overhead. It is now understood that proactive protocols in particular, such as OLSR which is the main representative of this family, cannot withstand high mobility in an appropriate manner since the validity of the routes decreases rapidly as we get farther from the start of the routing period (Belghith and Abid, 2009-1, 2009-2).

In this paper, we propose a novel cartography-based protocol that builds the network cartography in an asynchronous distributed fashion, i.e., no synchronization is required among the nodes in the network. This network cartography provides better vision of the network connectivity and allows selecting viable links in a way that improves the routing pertinence as a function of the network dynamics. Yet, the signaling overhead required to build the network cartography and to establish the routing tables consumes less bandwidth than that of the well known and adopted OLSR. Our proposed routing protocol allows different nodes to have different routing period sizes. The size of the routing period adopted by a mobile node is decided locally and dynamically. Conducted simulations brought out the betterments achieved by our routing proposal in terms of a much better throughput, much less end to end delay and less signaling traffic rate than those provided using OLSR.

The paper is organized as follows:

1. In section II, we present some of the relevant research work done in the field.
2. Section III presents a mechanism to collect the cartography of the network in a distributed way, and that can be integrated to common proactive routing protocols.
3. In section IV, we present our routing protocol and describe its functioning.
4. Section V is devoted to the conducted simulations. We compare the results given by our proposed protocol to that of the OLSR (Clausen and Jacquet, 2003) routing protocol and we show the remarkable efficiency achieved by our proposal.
5. Finally, in section VI, we present concluding remarks and some future directions.

RELATED WORK

To withstand the network dynamics, stability-based routing is an approach aiming essentially at choosing routes which are more stable over time and hence more resilient to dynamic changes in the network topology.

In 2004, Chung proposed to classify links based on the mobility behavior of their end point nodes. Links between stationary or very slowly moving nodes are considered as stationary links. Links which exist only for a short period of time are handled as transient links. Newly formed links are also considered to be transient as they are more likely to break down. Routing should favor the use of stationary links whenever this is possible. A ticket-based probing procedure is proposed to find stable routes.

Classification of links is previously adopted in Dube, Rais, Wang and Tripathi (1997) work. Authors used the strength of the received signal from each neighboring node to determine whether the associated link is either weak or strong. Routing is then made through paths maximizing the received signal strength.

Toh (1997) proposed the Associativity-Based Routing (ABR) which uses a new metric called associativity which defines the stability of a link between any given two nodes. ABR considers that the longer the two nodes have being neighbors, the longer they would keep connected. To express its associativity, each node broadcasts periodically a Beacon to indicate its presence. Upon the reception of a Beacon, a counter associated to the generating node is updated. The counter is reset to zero if the associated node is no longer accessible. The optimal route towards a destination is the one maximizing its cumulative associativity metric.

Further stability-based routing techniques can be found in (Beraldi, Querzoni and Baldoni, 2006), (Dubois-Ferriere, Grossglauser and Vetterli, 2003), (Roth and Wicker, 2003), (Yu, Li and Gruber, 2003), (Camp, Boleng and Davies, 2002) and (Hui Zhang and Yu-Ning Dong, 2007).

In 2009, Belghith, Ben Mnaouer and Abid presented the Probabilistic Source (or flow oriented) Routing (PSR) which is a routing protocol using localization information gathered at the start of the routing period, along with a probabilistic technique to compute the most stable paths. This protocol focuses rather on the existence probability of complete routes. The stability of routes may change as time progresses, and used paths selection is made accordingly. The PSR protocol requires, however, synchronization among all nodes in the network.

Along the same line, other probabilistic techniques were proposed to cope with the unpredictable behavior of ad hoc topologies. In (Tseng, Li and Chang, 2003), a probabilistic technique is proposed to estimate the residual lifetime of routes. Routing is then made through the ones with maximum residual lifetime. The lifetime of a route is computed as a function of the existence probability of each link which is derived from the distribution governed by the underlying mobility model.

In contrast to the above protocols, our current routing proposal does not require any synchronization between nodes and defines a proactive routing protocol based on the collected cartography of the network. It allows each node to compute its own routing period and routing period size in a local and autonomic way as a function of the locally perceived network dynamics.

NETWORK CARTOGRAPHY

Network cartography is the set of coordinates of the different nodes within the network area and as such it is much richer than the mere network topology that just gives the connectivity within the network. Our aim here is to design a distributed algorithm that collects the cartography of the network in such a way that it can be integrated in any proactive routing protocol with no additional signaling traffic apart from a very few additional fields in routing announcement messages (i.e., Hello messages).

First of all, we intend to get a correct and valid cartography and therefore we will not tolerate delayed routing information. As such, we propose to distinguish between control (i.e., hello) messages and data packets. Hello messages are to be transmitted as soon as possible before any other awaiting data packets. As such, received or locally generated Hellos are put at the head of the IP sending queue in front of any awaiting data packet.

Secondly, we assume that each node is capable of knowing its own geographical location. Such an assumption can be justified by the recent availability of small and inexpensive low power GPS receiver devices (Giordano and Stojmenovic, 2003). In case where these devices are not available or are not applicable (in-door networks case for instance), many other approaches for inducing relative coordinates still could be used. We can find a large set of GPS-less or restrained techniques for localization (Bulusu, Heidemann and Estrin, 2000; Capkun, Hamdi and Hubaux, 2001; Niculescu and Nath, 2003; Nagpal, Shrobe, and Bachrach, 2003; Saad, Benslimane and Jean-Claude, 2008; Boukerche, de Oliveira, Nakamura and Loureiro, 2007; and Takenaka, Mineno, Tokunaga, Miyauchi and Mizuno, 2007).

In this current work, we assume that nodes are location capable, relaying on GPS or one of the already cited localization techniques. A forwarded hello contains then both of its originating node's position and its forwarding node's position. The network cartography is built by every node through the dissemination of the geographic information integrated in the routing announcements (i.e.; the Hello messages). What we need, for our current purpose, is a collected cartography that reflects adequately and accurately the actual real network cartography when the routing table calculation is needed. To evaluate the correctness of the collected cartography, we compare it against the real current cartography of the network at the instant of the observation. Note that the actual instantaneous cartography of the network can be easily extracted from the simulator but it cannot be known in practice. This is of no concern to us here since all what we are looking for is to be sure that the collected cartography does represent adequately the real actual cartography. For now, we will just define the validity of the Cartography; its evaluation will be differed to when we propose our routing protocol.

Consider a target mobile node *"N"*. When N firstly advertised itself (i.e., sent its own generated Hello), it was at position $(x_0; y_0)$. In the Collected Cartography of a node *"A"* that had already heard N's Hello, a new entry for N was created, showing $(x_0; y_0)$ as N's coordinates. Since node N is mobile, its position varies as a function of time, and consequently it will be at position $(x_t; y_t)$ at instant *"t"* as time progresses. We say that N's position, as indicated by A's current cartography, is valid as long as the distance between the recorded position $(x_0; y_0)$ and the actual current position

$(x_t; y_t)$ is less than a tolerated predefined value denoted by *"d"*. That is:

$$\sqrt{(x_t - x_0)^2 + (y_t - y_0)^2} \le d$$

The validity of the cartography, as perceived by any given node, say node *A*, represents the percentage of nodes having valid positions among all nodes in the network. Parameter *d* is a tuning parameter whose value is relative to the transmission range used, and is in general a very small fraction of this range.

The Proposed Routing Protocol

The proposed routing protocol is an asynchronous self adjustable routing protocol since it is based on an asynchronous distributed cartography algorithm and each node regulates locally and dynamically its own routing period size according to its own dynamics.

The Protocol Description

Given a tolerance value *d*, our aim is to collect a valid cartography. That means, if node *N*'s speed, or more exactly its remoteness from the position where it was at its last announcement, is such that this tolerance value *d* is exceeded (Position of Node *N* may be no longer valid in all other nodes' collected cartographies), *N* has to re-announce itself to enforce the update of its position in the collected cartographies at the other nodes. So, each node has to compute the instant, at which it has to re-announce itself based on its currently observed dynamics, so that it will not exceed the tolerance *d* and keeps the validity of collected cartographies at a high acceptable level at all times. The period size is then locally computed by each node according to its own perceived dynamics. Let *"PSize"* be the period size computed by a given target node. Depending on its currently observed dynamics, *PSize* may vary from the smallest defined updating period denoted by *"Tmin"* to the largest updating period denoted by *"Tmax"*. The target node next announcement is then scheduled at instant *"NextUpdate"*.

Furthermore, the proposed protocol makes each node track its remoteness, from the position where it was at the instant of its last announcement, every *Tmin* seconds (i.e., an observation instant every *Tmin* seconds). This is needed by each node to correctly and appropriately follow its dynamic evolution in time. Let *"Last_Update_t"* be the instant at which the target node has made its last announcement. Let *"Δd"* be the remoteness of the target node at instant (*Last_Update_t+Tmin*). The target node instantaneous remoteness speed is then $S_d = \Delta d/Tmin$.

At this Speed, the cartography tolerance value *d* will be exceeded after (d/S_d) seconds. The target node routing period size is then updated as follows:

- *PSize* = *Tmax* : if $S_d = 0$ m/s
- *PSize* = min(*Tmax*; max(*Tmin*; d / S_d)) : otherwise

This procedure is then repeated each *Tmin* seconds. The target node will consequently re-announce itself at:

NextUpdate = min(*Last_Update_t* + *PSize* ; *NextUpdate*)

As previously pointed out, the minimum period size that can be used is *Tmin*. Each node periodically tracks its remoteness each *Tmin* seconds, and if the tolerance limit is exceeded, an immediate announcement is performed.

As a result, each node will adapt its routing period according to its own sensed velocity level. As such, its position is updated in all other nodes' collected cartographies only when needed. Each node includes its own perceived position in its locally generated Hello message, and then broadcasts it to its neighbors. Upon the reception of this Hello, a node updates its perceived cartography, inserts its current position in the Hello and forwards the message. As such, a forwarded hello contains both its originating node's position as well as its forwarding node's one to avoid inconsistency problems already pointed out in (Abid and Belghith, 2009-3).

The proposed cartography based routing protocol works as follow:

1. It is through Hello messages that a node can maintain its Cartography up to date. The Cartography includes entries to already heard destinations. Each entry includes a sequenceNumber (i.e., the period's number), an entry local start instant (recall that an entry is deleted after a maximum duration period (*"Max_duration_period"*), the current perceived position of the destination node and the destination node address.
2. Each node generates locally a Hello message every computed period (at instant *NextUpdate*). The Hello message contains the following fields:
   - Source address: the address of the node that originated this Hello.
   - Source sequence number: the sequence number that the originator of the Hello message is currently using.
   - Next hop Address: the Next hop to reach the Source Address, this is just the address of the node that sent this Hello.
   - Next hop sequence number: the sequence number that the node forwarding the Hello message is currently using.
   - Source position: the source node coordinates at the Hello generation instant.
   - Next hop position: the sender node current coordinates perceived at the very last self announcement it made.
3. Upon receiving a Hello message, a node first consults the sender's relative fields. If no entry exists in the cartography pointing out the position of the next hop address field then an entry is added. Otherwise if the entry is relative to a previous period (old sequence Number compared to the "Next hop sequence number" field), then the entry is updated accordingly. A Hello is automatically created to advertise this new entry to other nodes in the network. Otherwise, the information is silently ignored.
4. Then, the node checks whether it is the originator of this Hello, in which case it just discards the message. Otherwise, the node tests the usefulness of the received hello: if its Cartography already contains an entry for the "Source Address" (which represents the destination to be reached), the current sequence number of this entry and the sequence number included in the Hello are compared. The entry is updated if it is older. A new entry is automatically created in the Cartography, if the message announces a destination not yet known.
5. If the received message is useful, the node forwards it to its neighbors. It puts its address in the "Next hop Address" field, puts its position in the corresponding field, puts its currently used sequence Number in the "Next hop sequence number", and then broadcasts it to its neighbors after a certain delay as explained below .

   We attempt to make the new information spread in the network as fast as possible, to ensure that all nodes in the network are using the same collected cartography along time. For that, and before forwarding the new heard useful Hello, a node computes the distance *"D"*, separating him from the "Next hop Address" node (the node bringing the new information). The bigger this distance is ($0 \leq D \leq$ Transmission Range $= R$), the fastest the forwarding must be accomplished: we proposed that each node forwards the new information after a delay time:

$$\textbf{\textit{delayTime}} = \frac{1 - \min\left(\boldsymbol{R}, \boldsymbol{D}\right) / \boldsymbol{R}}{2} \quad \text{seconds}$$

New information forwarding is distributed over one half a second, accordingly to the distance between the sender and the forwarder, to lower the chance of collisions.

Once the cartography is collected, a node can compute its routing table using any shortest path algorithm executed on the corresponding cartography. Note here that the cartography collection is made in an asynchronous way since no synchronization is required between nodes. Recall that in ad hoc networks, a wireless link exists whenever its two end point nodes are within the transmission range R. In the proposed routing protocol, however, a node does not re-announce itself unless the tolerance value d is exceeded. Furthermore, if we consider the worst case where two neighbor nodes can only move away linearly, then we must reduce the neighborhood range by twice the value of this tolerance. This will make our routes more stable over time since a node can't move farther more than $2 \times d$ meters. In this work, we suppose that this extreme case has low chance to happen, and as such, we limit ourselves to an acceptable threshold as it will be shown later, by applying the chosen shortest path algorithm to our collected cartography using an effective range between nodes limited to R - d. Next, we investigate the validity of the collected cartography which must be kept at its highest levels at all applied speeds, all loads and all times. The load effect, however, can be overcome thanks to the adopted prioritization IP model.

Simulation Set Up

To ascertain the validity of the cartography as a function of the mobility, traffic load and time, we conducted an extensive set of simulations. We have considered a simulation area of 1000m by 1000m, with 100 mobile nodes using a modified Random Waypoint mobility model (Camp, Boleng and Davies, 2002). The simulation surface was divided into 9 areas, each of which contains from 11 to 12 nodes. The Random-Waypoint mobility is then applied separately inside each area (Figure 1). This is basically to mitigate the known middle concentration effect of the Random WayPoint mobility model.

The Effectiveness of the Restricted Random Waypoint Mobility Model

As proven in several previous works (Bettstetter, Lassila, and Virtamo, 2006 ; Hyytia, Lassila and Virtamo, 2006 ; and Yoon, Liu and Noble, 2003), the Random Waypoint Mobility model suffers from the problem of concentration of the mobile nodes towards the middle of the simulation area. This behavior would result in a reduced average number of hops and a smaller signaling than in actual networks.

Figure 1. Restricted random waypoint: Applied random waypoint inside each area

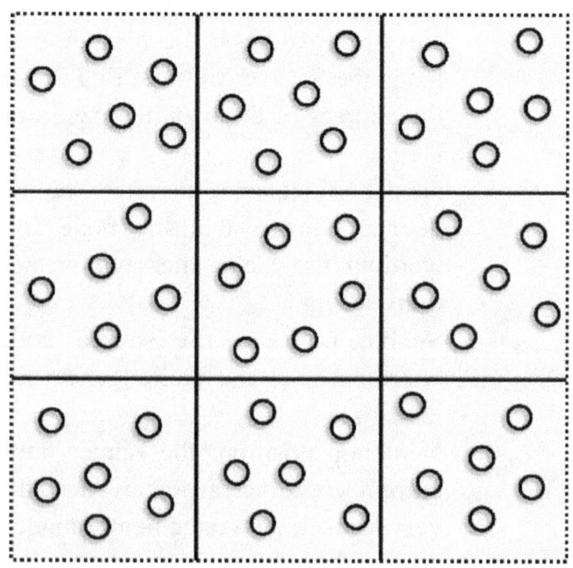

Figure 2. Initial geographical distribution of mobile nodes

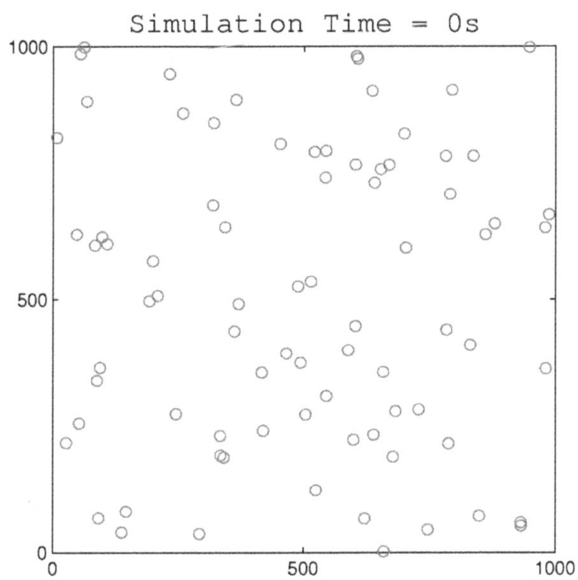

In the quest to mitigate such an undesirable effect, we propose dividing the simulation area into smaller zones, where the RWP mobility model is applied independently. The concentration problem will always be there, but instead inside every sub-area. As such, we will guarantee the presence of nodes at the edges of our area as time progresses, and therefore the average number of hops in the network will be kept enough high during the whole simulation.

Before going any further, we need to better see whether the problem is solved, or at least reduced, using our proposed restricted random waypoint mobility model. It is obvious that this concentration comes out deeper proportionally to the speed of the nodes. Starting the simulation scenarios with the initial distribution of nodes given by Figure 2, Figure 3 highlight the success of our model in

Figure 3. The geographical distribution of the mobile nodes after simulation running time = 1500s

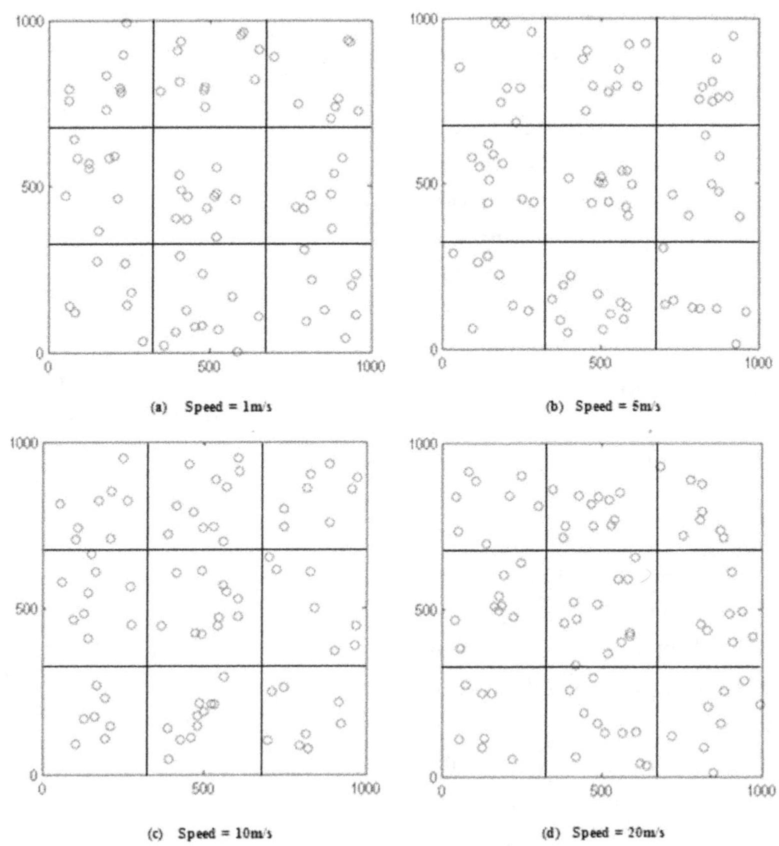

avoiding the concentration problem by showing the nodes' geographical distribution after 1500s of simulation time.

As expected, we observe as the simulation time runs that the nodes using our restricted mobility model occupy the entire simulation area, though some concentration within each sub area.

To better ascertain the odd behavior of the random waypoint mobility model and its impact on simulation results, we propose next to inspect the average length of the shortest paths in the network as a function of time. The statistics are made by a randomly selected mobile node. For all the applied scenarios, we've started by the same initial network with a shortest paths' mean length equal to 3.8 hops. We observe through figure s 4 to 8 that the use of the non restricted random waypoint mobility model leads to smaller average path lengths than when we use our restricted model. The mean length of the shortest paths from the node making the statistics towards all the remaining nodes in the network is smaller when using the non restricted model; and this for all the applied maximum speeds. Notice here that, as pointed out in (Yoon, Liu and Noble, 2003), the hysteresis effect taking place when using high speeds is due to the large simulation time needed by the random waypoint model to reach the permanent regime, as well as some of the initial parameters of this model, such as the use of a non null pause time and a null lower limit of the interval from which the nodes' speed is uniformly picked up.

The Rest of the Simulation Set Up:

We used a transmission range of 250 m, a tolerance ***d*** of 25 meters that is a tolerance equals to one tenth of the used transmission range, a network capacity of 54 Mbps and a maximum MAC retransmission count equals to 3. We used a priority IP module at the network layer to enforce that Hellos are treated before any awaiting data packet. The priority queue maximum size is set

Figure 4. The mean length of the shortest paths as a function of time at a maximum speed of 1m/s

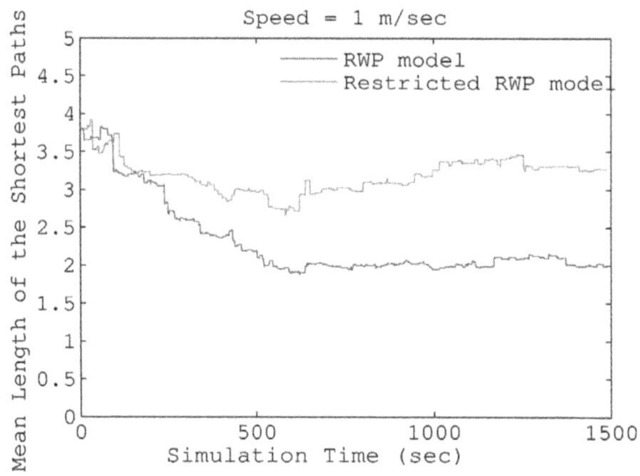

Figure 5. The mean length of the shortest paths as a function of time at a maximum speed of 2m/s

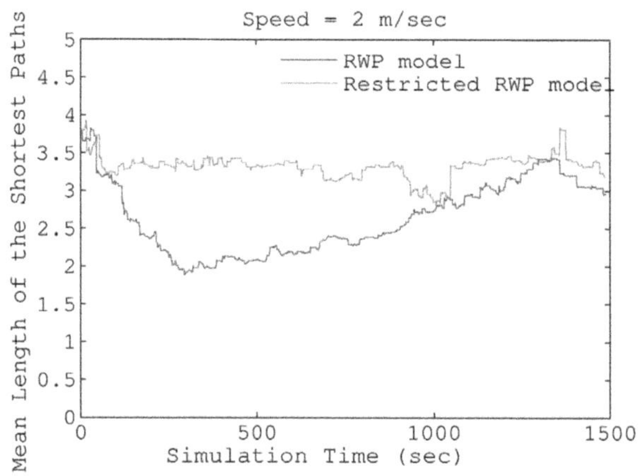

Figure 6. The mean length of the shortest paths as a function of time at a maximum speed of 5m/s

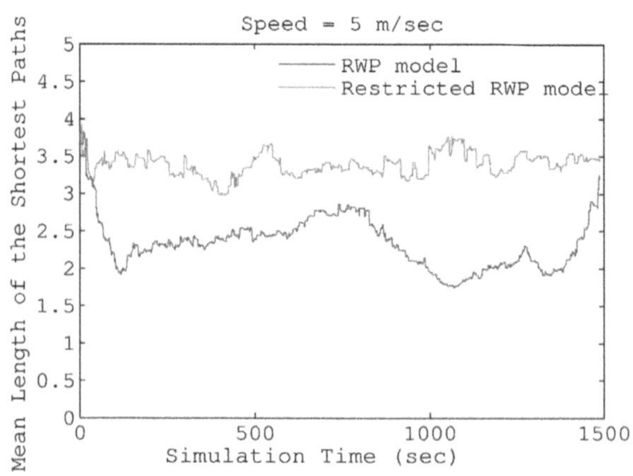

Figure 7. The mean length of the shortest paths as a function of time at a maximum speed of 10m/s

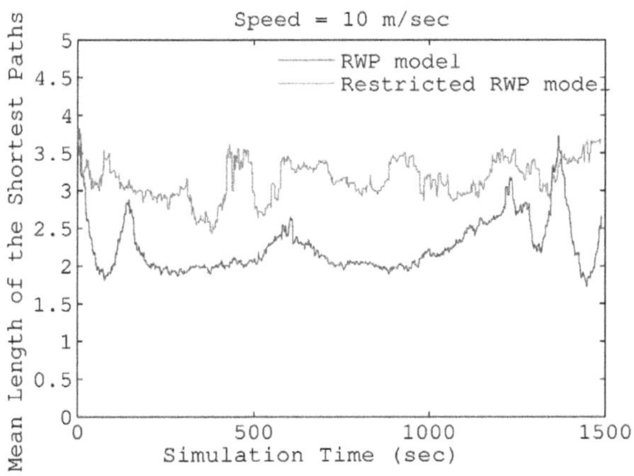

Figure 8. The mean length of the shortest paths as a function of time at a maximum speed of 20m/s

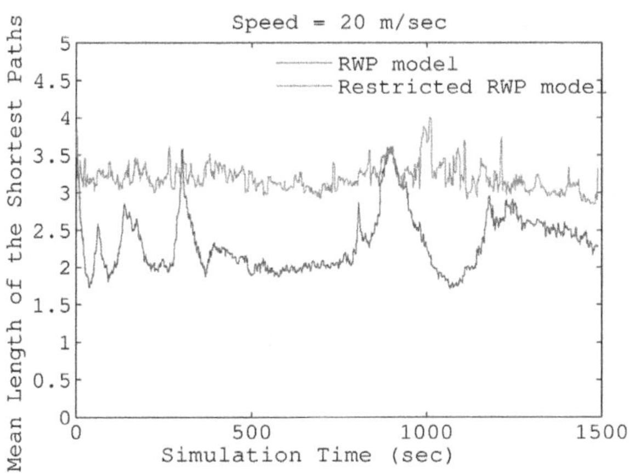

Figure 9. Validity of the cartography as a function of time and speed

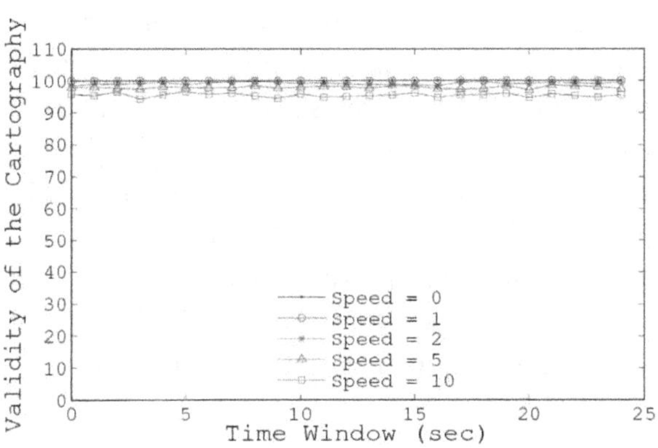

to 100 packets where data packets are admitted only if less than 70 packets are within the queue.

The Maximum updating period size *Tmax* is set to 24 seconds and the minimum, *Tmin*, is fixed to 3 seconds. Finally, the *Max_duration_period*, representing the life time of an entry in the routing table, is set to 30 seconds. All required modifications are ported on the OMNeT++ network simulator.

Simulation Results

Figure 9 portrays the validity of the cartography as a function of time and for different node speeds. Recall that we are using a priority IP handling and therefore the network load has a very little impact on the validity of the cartography. For a null node speed (no mobility) as well as a speed of 1 m/sec, we get a validity of one hundred percent. As the speed gets higher the validity of the cartography declines a little bit but still stays sufficiently high around 97 percent even for a speed of 10 m/s.

SUITABILITY OF THE PROPOSED ROUTING PROTOCOL

To experiment our new stability-based routing proposal, we conducted a set of simulations using the OMNeT++ simulator where we integrated our proposed protocol, using the cartography collecting algorithm and the Priority IP handling. We also integrated OLSR under the same prioritization IP model. Our simulation set up is as defined previously. The OLSR TC period is set to 8 seconds and its Hello period to 2 seconds. The initial topology of 100 nodes is chosen randomly, but in a way to have 11 to 12 nodes in each area (in each area, we place 11 nodes uniformly). We immobilized 10 source nodes at the left edge of the simulation area (randomly distributed over the three sub-areas at the left edge) and similarly 10 destination nodes at the right edge. We also consider ten traffic flows using these fixed source-destination pairs

Figure 10. Validity of the routes as a function of time: our proposal vs OLSR, Speed=0 m/sec

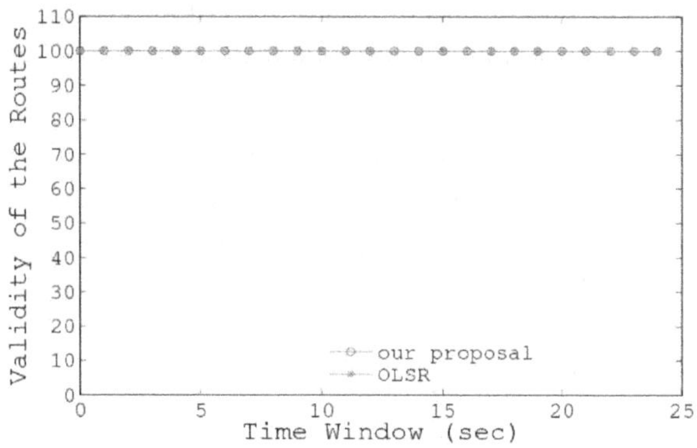

Figure 11. Validity of the routes as a function of time: our proposal vs OLSR, Speed=1 m/sec

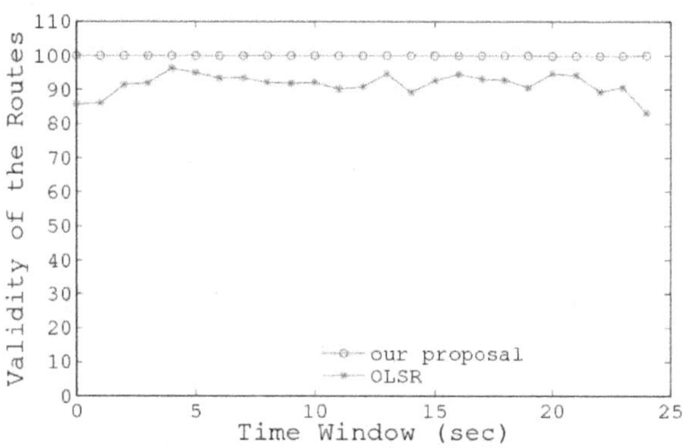

(pairs are chosen randomly). As such, we insured multi-hop routes (4 to 5 hops). Finally, the data packet size used is 200 bytes.

Let us now investigate the correctness or the validity of the routes given by each protocol against the actual real network. The routing period is divided into equally spaced observation points; namely an observation point each second. At each observation time point, we have then three different views of the network: the actual real view provided by the underlying mobility model used in our OMNeT++ simulator (to which we compare the routing protocols), the OLSR routing table and the routing table provided by our

Figure 12. Validity of the routes as a function of time: our proposal vs OLSR, Speed=2 m/sec

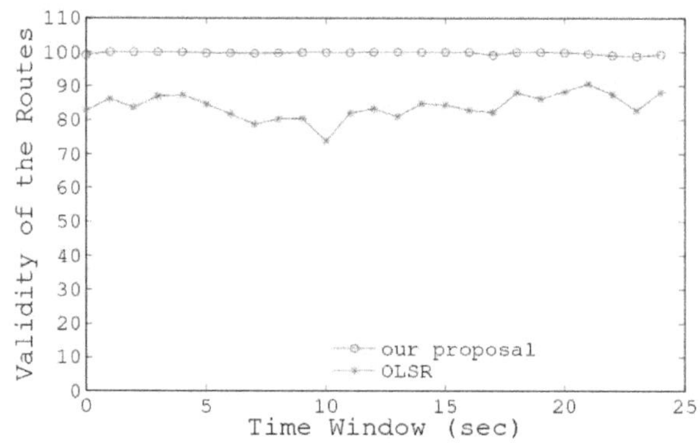

Figure 13. Validity of the routes as a function of time: our proposal vs OLSR, Speed=5 m/sec

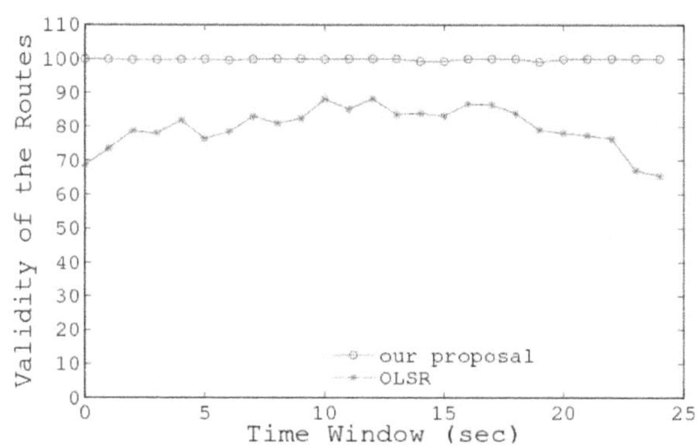

proposed protocol. A route from a source node to a destination node is termed valid if it does exist in the real network regardless whether it is optimal (a shortest route) or not.

To ascertain the efficiency and accuracy of our proposed routing algorithm, we compute the percentage of valid routes from a designated node (node 1) to all the other 99 destinations given by each one of the two protocols. Figures 10, 11, 12, 13, and 14 portray the validity of the routes as a function of time within an observation Time window of 25 seconds, for the two defined protocols with a Null applied data load and respectively

Figure 14. Validity of the routes as a function of time: our proposal vs OLSR, Speed=10 m/sec

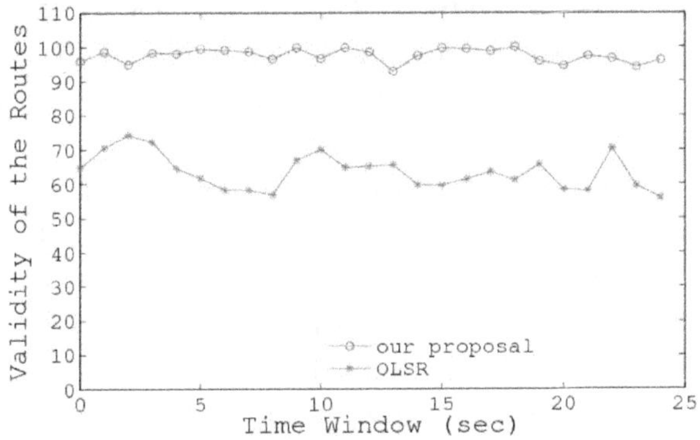

Figure 15. Average number of received packets / flow as a function of the applied load: this proposal vs OLSR, Speed=0 m/sec

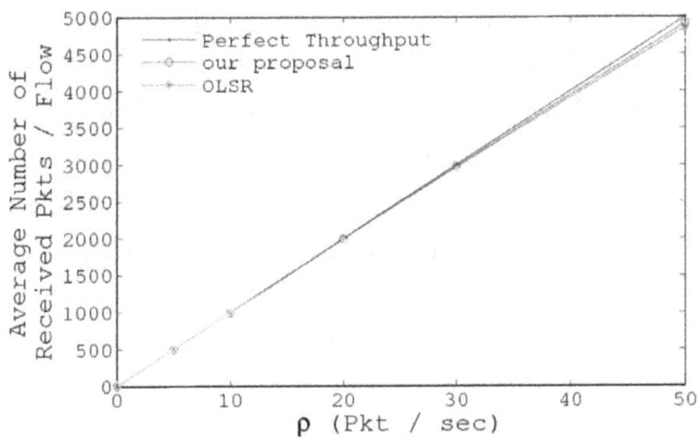

Figure 16. Average number of received packets / flow as a function of the applied load: this proposal vs OLSR, Speed=1 m/sec

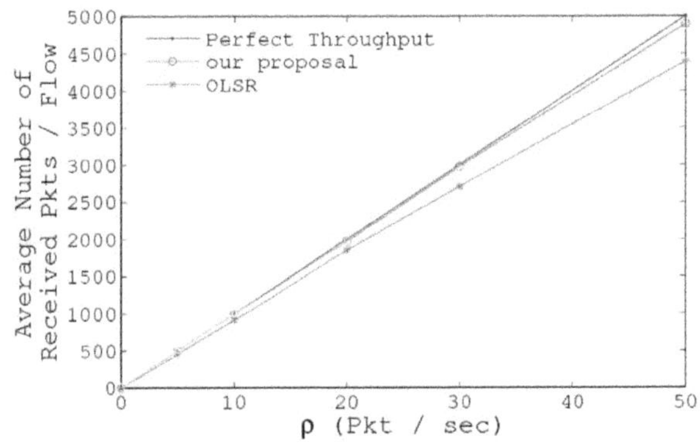

Figure 17. Average number of received packets / flow as a function of the applied load: this proposal vs OLSR, Speed=2 m/sec

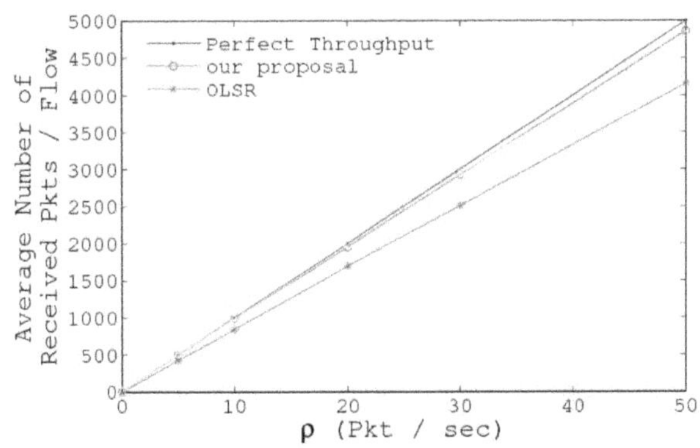

for the speeds 0 m/sec, 1 m/sec, 2 m/sec, 5m/sec and 10 m/sec.

We readily observe that for a null speed the validity provided by both protocols is, as expected, equal to one hundred percent. As node speed gets higher, the OLSR routing protocol starts losing its efficiency by delivering the lowest route validity. More interestingly, we notice that our proposed protocol succeeds in keeping this validity near the 100% over the time even for high mobility levels.

Having such a result, we can expect that our proposed routing protocol is going to completely outperform OLSR. Figures 15, 16, 17, 18 and 19 represent the number of received packets per flow (the network throughput) as a function of the applied data load rate per flow and for different speeds. We clearly observe the sensitivity of OLSR to the network dynamics. As the node speed gets higher, our protocol delivers more throughput than OLSR. While the superiority of our protocol is clear over the entire range of traffic loads, it gets larger as the traffic load gets higher. Indeed, at moderate to high workloads, OLSR is unable to cope with the loss of the routing validity as its routes validity gets deteriorated as shown in the previous Figures (11, 12, 13 and 14) and consequently more and more packets

Figure 18. Average number of received packets / flow as a function of the applied load: this proposal vs OLSR, Speed=5 m/sec

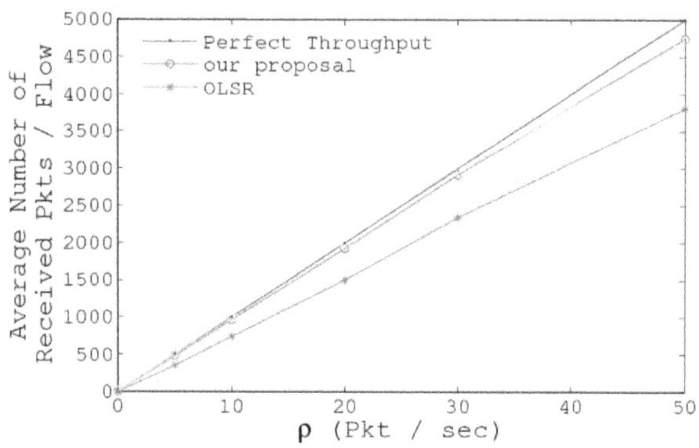

Figure 19. Average number of received packets / flow as a function of the applied load: this proposal vs OLSR, Speed=10 m/sec

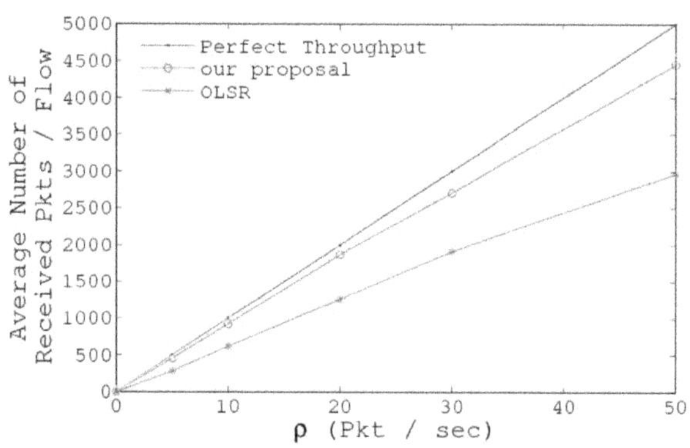

wander through the network wasting resources otherwise used to correctly forward data traffic. For a sustained traffic load of 50 generated data packets per second and a high mobility of 10 m/s (Figure 19), our protocol yields more than 50% additional throughput than that of OLSR.

Let us now, investigate the performance of each protocol in terms of the end to end delay. This metric measures the average end to end delay in seconds spent by the received data packets over all defined flow. Figures 20, 21, 22, 23, and 24 portray this average end to end delay as a function of the traffic load per flow and for different speeds. We clearly observe, except for the Null mobility case, that our routing protocol delivers packets to their ultimate destinations in much less time than OLSR does. As the node speed gets higher, OLSR amounts to very high average end to end delay attaining hundreds of milliseconds (Figure 24) for even moderate traffic loads. Our protocol, at the opposite, sustains both high speeds and workload and maintains a rather very adequate average end to end delay under few tens milliseconds even for a traffic load of 30 packets per second.

When nodes are stationary, OLSR has more chance to slightly outperform our protocol as can

Figure 20. Average delay as a function of the applied load: this proposal vs OLSR, speed=0 m/sec

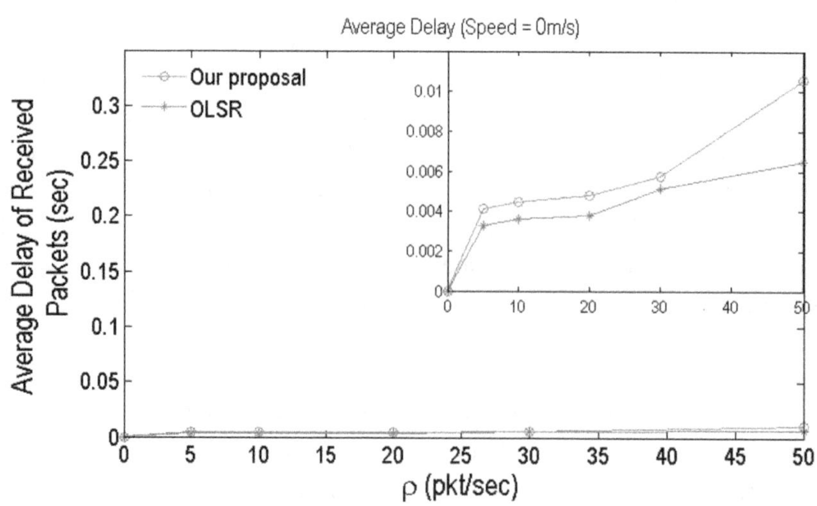

Figure 21. Average delay as a function of the applied load: this proposal vs OLSR, speed=1 m/sec

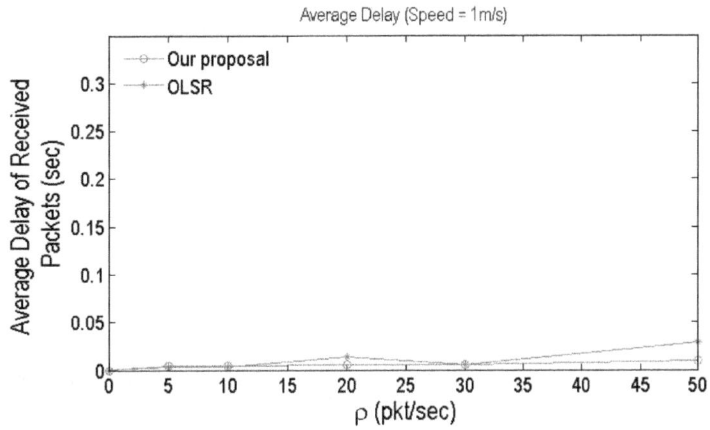

Figure 22. Average delay as a function of the applied load: this proposal vs OLSR, speed=2 m/sec

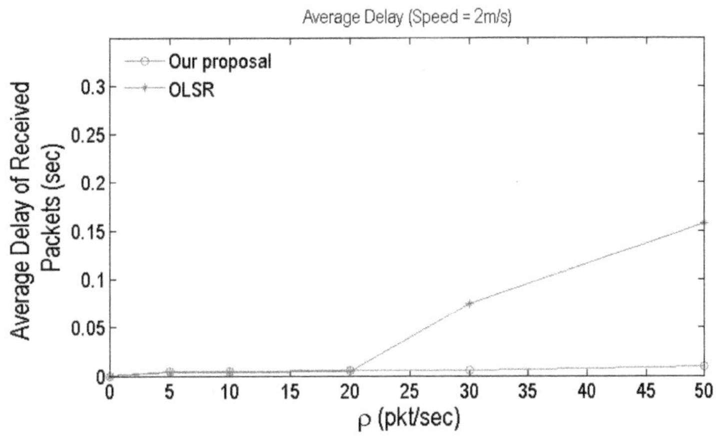

Figure 23. Average delay as a function of the applied load: this proposal vs OLSR, speed=5 m/sec

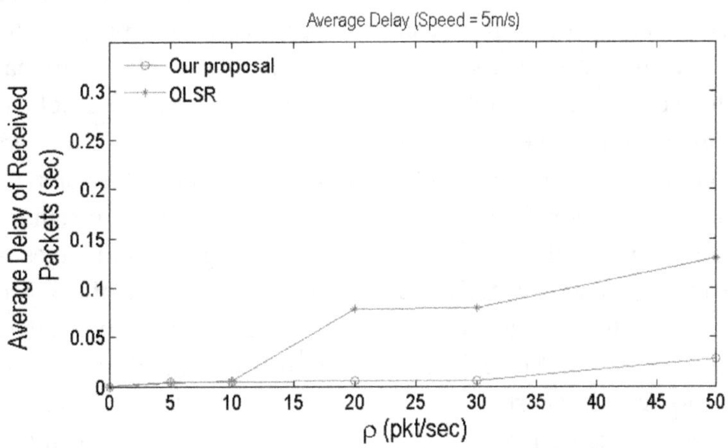

Figure 24. Average delay as a function of the applied load: this proposal vs OLSR, speed=10 m/sec

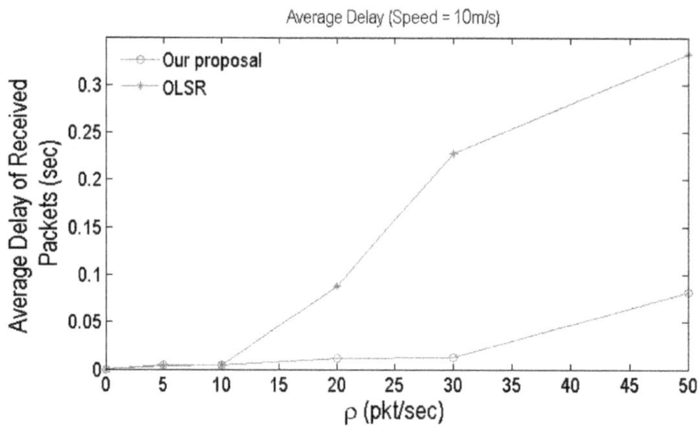

Figure 25. Average length of used routes by received packets: this proposal Vs OLSR, with null speed

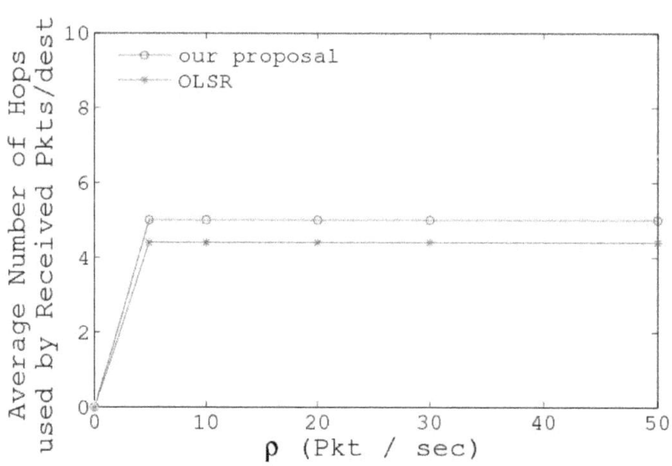

be observed from Figure 20. Indeed our protocol builds slightly longer routes than OLSR as can be seen on Figure 25. Recall that for the sake of stability, we choose to use shorter links (***R*** - ***d***).

As Shown in Figure 25, the average number of hops that data packets have made before reaching their destinations under our protocol is slightly larger than the average length of the routes built by OLSR. The latter provides exactly the shortest routes computed on the real network at the beginning of the routing period. Since, in this case, the network is stationary; the shortest routes remain valid until the end of the period.

Now, we turn to investigate the signaling cost required for our proposal to work and compare it to that of OLSR. Recall that our routing protocol exchanges extended Hello messages to set up and maintain an up to date cartography on which it builds its routing decisions and period sizes. These Hello messages are rather very short compared to the huge TC messages that OLSR builds and distributes among the different MPRs. As such, it is paramount to investigate not only the average number of signaling messages exchanged by each node but instead the average number of signaling bytes per second transmitted by each node within the network. The later provides more insights as

Figure 26. Average number of signaling messages as a function of the network maximum speed: this proposal Vs OLSR, null load

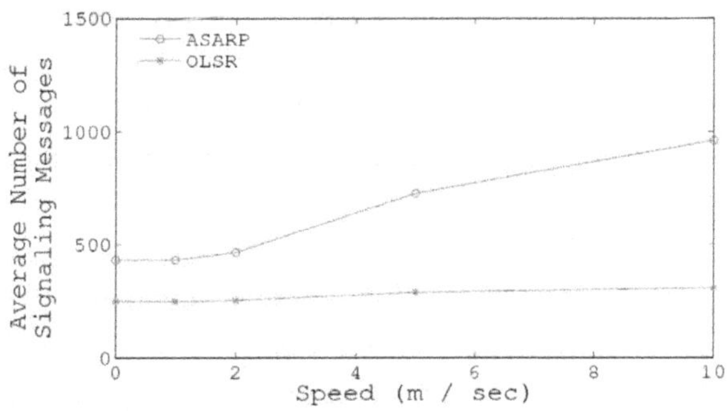

to how much signaling capacity per second is required from the underlying network.

Figure 26 portrays the average number of signaling messages (Hello and TC messages for OLSR and Hello messages for our proposal) transmitted by each node during a simulation time of 100 seconds, as a function of the node speed and for a null traffic load. Recall that the signaling traffic is virtually independent of the traffic load as we have adopted an IP priority mechanism for both implementations of OLSR and our routing protocol. As expected, OLSR transmits less routing messages than our proposed protocol. Furthermore, as the node speed gets higher, our protocol requires more routing messages to be transmitted as the routing periods get smaller. The signaling cost of OLSR, on the other hand, remains virtually insensitive to the node speed as both of its Hello and TC periods are maintained constant.

We may also notice on Figure 26 that the curve relative to OLSR is in fact slightly increasing with the node speed. Indeed, the number of MPR nodes in the network gets slightly higher as the speed increases. Since the network topology changes quickly at higher speeds, new nodes are selected to be MPRs. But some of the older MPRs, despite the fact that they can no more fulfill their status, they persist being MPRs for a period of 3×TC seconds (that is 24 seconds in our case) as specified in the OLSR protocol.

Figure 26 shows that OLSR transmits less routing messages than our proposed protocol, but recall that each MPR accumulates all received signaling messages (Hello and TC messages) and aggregates them into one large TC packet, while our algorithm sends each individual topological change as a single message. This explains the low validity of routes used by OLSR especially when the velocity is high. In OLSR, a collided or lost routing packet leads patently to the loss of a large amount of topological information otherwise required to update routing tables. More interestingly, OLSR utilizes much more capacity to transmit its huge signaling messages to the opposite of our proposed protocol. Figure 27 portrays the average number of signaling bytes transmitted by second by each node which represents the average effective bandwidth consumed by the signaling traffic transmitted by each node. We observe that our proposed protocol outperforms OLSR even at null speed. Indeed, in our proposed protocol, each node adapts its routing period to the perceived degree of its mobility. With a static network, our proposal adopts an updating period of 24 seconds (*Tmax*) for all the nodes, while OLSR maintains a Hello period of 2 seconds and a TC period of 8 seconds regardless of node speeds. This amounts, as portrayed on Figure 27 to almost doubling the signaling traffic for OLSR. As the node speed increases, the signaling overhead gets bigger for

Figure 27. Average number of signaling bytes/sec as a function of the network maximum speed: this proposal Vs OLSR, null load

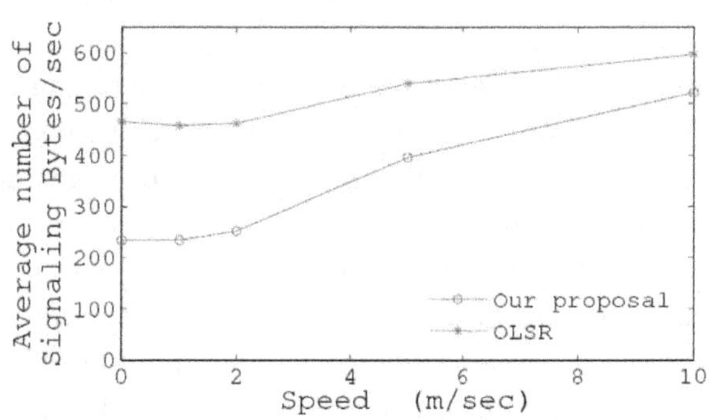

both protocols but the difference between the two protocols gets smaller as our proposal adapts the routing period sizes to the mobility degree. For speeds of 5 m/s, respectively 10 m/s, OLSR requires around 43%, respectively 19%, more signaling traffic than our proposed protocol.

Recall here that our proposed routing protocol generates control traffic proportionally to the sensed node's speed. A node never gets farther than *d* meters from its last announced position without spreading its new position. As a consequence by willingly fixing the node's speed, the tolerance parameter *d* will directly affect the signaling overhead. At a fixed speed, the computed routing period size of a node increases as the tolerance value d increases and vice versa. The other side of the coin is, though, an undesirable increase in the number of hops of the selected routes is obtained when the value of d gets large. This leads also to a high increase of the number of retransmissions needed to deliver data packets to their ultimate destinations. As such, our tolerance parameter becomes a tuning parameter that needs to be further investigated.

CONCLUSION

In this paper, we proposed a new asynchronous-cartography-based-routing protocol that allows each mobile node in the network to calibrate its own routing period size according to its perceived local dynamics. This calibration allows each node to re-announce itself once it exceeds *d* meters from its previously announced position. This tolerance distance represents a real tuning parameter that needs to be appropriately gauged. Increasing *d* leads to reducing the control overhead, but simultaneously increases the number of hops composing the computed routes.

To evaluate our local self regulating routing and compare its performance to that of the seminal OLSR, we conducted extensive simulations. As a first step, we focused on the underling used mobility model: the random waypoint. This particular model suffers from some known problems, mainly the middle concentration problem. We proposed a new restricted random waypoint model to precisely solve this aforementioned problem. We showed through the conducted simulations that the effect of this problem is minimized using our proposed mobility model.

We then moved on to evaluate our proposed routing protocol. We showed in particular, through different scenarios of node speeds and various workloads that our routing proposal outperforms OLSR in terms of a better routing validity, an enhanced throughput, a much lower end to end delay and a lower signaling overhead.

Currently, we are investigating the scalability of our routing protocol and its performance under different mobility and propagation models. We are also investigating the tuning tolerance parameter *d* to best improve the network performances.

REFERENCES

Abid, M. A., & Belghith, A. (2009-3). *Functional enhancements of proactive routing in mobile ad hoc networks*. Third IFIP International Conference on New Technologies, Mobility and Security (IFIP NTMS 2009), Cairo, Egypt, 20-23 December, 2009.

Belghith, A., & Abid, M. A. (2009-1). *Autonomic self tunable proactive routing in mobile ad hoc networks*. The 5th IEEE International Conference on Wireless and Mobile Computing, Networking and Communications (IEEE WiMob'09), Marrakech, Morocco, October 12-14. 2009.

Belghith, A., & Abid, M. A. (2009-2). *Dynamically self adjustable proactive routing protocols for mobile ad hoc networks.* The 34th Annual IEEE Conference on Local Area Networks (IEEE LCN'09), Zurich, Switzerlan, October 20-23.

Belghith, A., Mnaouer, A. B., & Abid, M. A. (2009). *Suitability analysis of probabilistic routing for dynamic ad-hoc networks.* The 5th IEEE International Conference on Wireless and Mobile Computing, Networking and Communications (IEEE WiMob 2009), Marrakech, Morocco, October 12-14, 2009.

Beraldi, R., Querzoni, L., & Baldoni, R. (2006). A hint-based probabilistic protocol for unicast communications in MANETs. *Ad Hoc Networks, 4,* 547-566. Elsevier, 2006.

Bettstetter, C., Lassila, P., & Virtamo, J. (2006). Spatial node distribution of the random waypoint mobility model with applications. *IEEE Transactions on Mobile Computing, 5*(6), 680–694. Retrieved from http://ieeexplore.ieee.org/lpdocs/epic03/ wrapper.htm?arnumber=1624340. doi:10.1109/TMC.2006.86

Boukerche, A., de Oliveira, H. A. B. F., Nakamura, E. F., & Loureiro, A. A. F. (2007). Towards an integrated solution for node localization and data routing in sensor networks. *Proceedings 12th IEEE Symposium on Computers and Communications* (ISCC '07), July 2007.

Camp, T., Boleng, J., & Davies, V. (2002). A survey of mobility models for ad hoc network research. *Wireless Communication and Mobile Computing (WCMC): Special ISSUE on Mobile Ad Hoc Networking: Research. Trends and Applications, 2*(5), 483–502.

Capkun, S., Hamdi, M., & Hubaux, J. P. (2001). GPS-free positioning in mobile ad-hoc networks. *Proceedings of the 34th Hawaii International Conference on System Sciences* (HICSS), 2001.

Chung, W. H. (2004). Probabilistic analysis of routes on mobile ad hoc networks. *IEEE Communications Letters, 8*(8), 506–508. doi:10.1109/ LCOMM.2004.833841

Clausen, T., & Jacquet, P. (2003). *Optimized link state routing protocol (OLSR). Request for Comment 3626.* MANET Working Group.

Clausen, T., Jacquet, P., & Viennot, L. (2002). *Comparative study of routing protocols for mobile ad hoc networks.* The First Annual Mediterranean Ad HOC Networking Workshop, September 2002.

Dube, R., Rais, C. D., Wang, K. Y., & Tripathi, S. K. (1997). Signal stability-based adaptive routing (SSA) for ad hoc mobile networks. *IEEE Personal Communications Magazine, 4*(1), 36–45. doi:10.1109/98.575990

Dubois-Ferriere, H., Grossglauser, M., & Vetterli, M. (2003). Age matters: Efficient route discovery in mobile ad hoc networks using encounter ages. *Proc. of ACM MobiHoc03,* 1-3 Annapolis, MD, USA, 2003.

Giordano, S., & Stojmenovic, I. (2003). *Position-based ad hoc routes in ad hoc networks* (pp. 287–300). Boca Raton, FL: CRC Press, Inc.

Hyytia, E., Lassila, P., & Virtamo, J. (2006). Spatial node distribution of the random waypoint mobility model with applications. *IEEE Transactions on Mobile Computing, 5*(6), 680–694. doi:10.1109/ TMC.2006.86

Johnson, D. B., Maltz, D. A., & Hu, Y.-C. (2004). *The dynamic source routing protocol for mobile ad hoc networks* (DSR). Internet draft, IETF MANET Working Group 2004.

Kleinrock, L. (1975). Queueing systems, volume 1: Theory. In Wiley Interscience (Ed.), *Queueing systems.*

Mbarushimana Amine, C., & Shahrabi, A. (2007). *Comparative study of reactive and proactive routing protools performance in mobile ad hoc networks.* The 21st International Conference on Advanced Information Networking and Applications Workshop, AINA'07, IEEE 2007.

Nagpal, R., Shrobe, H., & Bachrach, J. (2003). Organizing a global coordinate system from local information on an ad hoc sensor network. *Proceedings of the Second International Workshop Information Processing in Sensor Networks* (IPSN), 2003.

Niculescu, D., & Nath, B. (2003). DV based positioning in ad hoc networks. [Springer Netherlands.]. *Telecommunication Systems, 22*(1-4), 267–280. doi:10.1023/A:1023403323460

Novatnack, J., Greenwald, L., & Arora, H. (2005). Evaluating ad hoc routing protocols with respect to quality of service. *Wireless and Mobile Computing Networking and Communications WiMob, 05*, 2005.

Perkins, C. E., & Bhagwat, P. (1999). *Highly dynamic destination-sequenced distance-vector routing (DSDV) for mobile computers*. ACM SIGCOMM'94 Conference on Communications Architectures, Protocols and Applications, 1999.

Perkins, C. E., Royer, E. M., & Chakeres, I. E. (2003). *Ad hoc on demand distance vector routing protocol.* Internet draft, MANET Working Group 2003.

Roth, M., & Wicker, S. (2003). Termite: Emergent ad-hoc networking. *Proceedings of the Second Mediterranean Workshop on Ad-Hoc Networks*, Mehdia, Tunisia, 2003.

Saad, C., Benslimane, A., & Konig, J.-C. (2008). AT-dist: A distributed method for localization with high accuracy in sensor networks. *International Journal Studia Informatica Universalis, Special Issue on Wireless Ad Hoc and Sensor Networks, 6*(1).

Sridhar, K. N., & Chan, M. C. (2005). Stability and hop-count based approach for route computation in MANET. In *Proceedings of ICCCN05*.

Takenaka, T., Mineno, H., Tokunaga, Y., Miyauchi, N., & Mizuno, T. (2007). Performance evaluation of optimized link state routing-based localization. *2007 IEEE International Performance, Computing, and Communications Conference*, (pp. 150-159).

Toh, C.-K. (1997). Associativity-based routing for ad hoc mobile networks. *International Journal on Wireless Personal Communications, 4*(2), 103–139. doi:10.1023/A:1008812928561

Tseng, Y., Li, Y. F., & Chang, Y. (2003). On route life time in multi-hop mobile ad hoc networks. *IEEE TOC, 2*(4).

Yoon, J., Liu, M., & Noble, B. (2003). Random waypoint considered harmful. *IEEE INFOCOM 2003 Twenty-Second Annual Joint Conference of the IEEE Computer and Communications Societies* (pp. 1312-1321). IEEE. Retrieved from http://ieeexplore.ieee.org/lpdocs/epic03/ wrapper.htm?arnumber=1208967

Yu, D., Li, H., & Gruber, I. (2003). Path availability in ad hoc network. *Proceedings of the 10th International Conference on Telecommunications* (ICT03), volume 1, (pp. 383–387).

Yuan, X. (1999). On the extended Bellman-Ford algorithm to solve two-constrained quality of service routing problems. *Proceedings of The Eighth International Conference on Computer Communications and Networks* (IC3N'99), Boston, USA, 1999.

Zhang, H., & Dong, Y.-N. (2007). A novel path stability computation model for wireless ad hoc networks. *IEEE Signal Processing Letters, 14*(12). doi:10.1109/LSP.2007.907994

Zygmunt, J., Haas, M. R., & Pearlman, P. S. (2002). *The interzone routing protocol (IERP) for ad hoc networks*. Internet Draft, work in progress.

This work was previously published in International Journal of Business Data Communications and Networking, Volume 6, Issue 4, edited by Debashis Saha and Varadharajan Sridhar, pp. 21-37 , copyright 2010 by IGI Publishing (an imprint of IGI Global).

Chapter 3
Handoff Cost Minimization and Planning of Heterogeneous Integrated Overlay Networks:
Meta-Heuristics Based Approach

Ayan Paul
Bharat Sanchar Nigam Limited, India

Madhubanti Maitra
Jadavpur University, India

Swarup Mandal
Wipro Technologies, India

Samir Kumar Sadhukhan
Indian Institute of Management Calcutta, India

ABSTRACT

The wireless technology market has witnessed a complete paradigm shift as multiple standards and protocols are emerging almost every day. Each and every standard has its limitations and merits, which can be either masked or complemented by some other standards. The demands from the service providers are now sky-high and for the complete commercialization, it is expected that even with scarce network resources all kind of services would be provided, especially in a cost effective manner. This burning issue compels a service provider to roll out some integrated wireless networks to exploit the virtues of each. This chapter formulates the planning problem of an overlay network integrating particularly, 3G, WiMAX, and WLAN. The issue of planning is to establish proper connectivity amongst the three network standards which is unique in its nature. In the proposed planning approach, the authors have endeavored to minimize total cost for vertical handoff generated in the overlay network as well as the cost for wire line connection amongst the various network gateways of the overlay hierarchy. In this work, the authors have focused on the initial planning phase. For validating the novel planning problem, the chapter has taken recourse to simulated annealing (SA) and a well cited meta-heuristic H-II. The authors have also presented comparison of the performances of SA and H-II with a variant of distance based planning (DBP) scheme in this domain.

DOI: 10.4018/978-1-61350-477-2.ch003

INTRODUCTION

As demand for various multimedia applications with different quality of service (QoS) requirements is growing rapidly, provisioning of seamless wireless service to the end customers is becoming a business challenge for the wireless service providers. Due to technological constraints, it is not possible to meet the varied demands of the customers with a single wireless technology (Stemm & Katz, 1998). This is due to the fact that no wireless technology can provide high data rate with full mobility. Though recent technologies such as long term evolution (LTE) promise to fill this gap, however, it will take some more time for these technologies to have a considerable market presence. Hence in this work, we have considered only those wireless technologies that are readily available in the market. Moreover, presently available technologies also possess different capabilities that suit different application requirements. For example, Wi-Fi, a wireless local area network (WLAN) satisfies requirement of high data rate and limited mobility. On the other hand, universal mobile telecommunication system (UMTS), a third generation (3G) cellular wireless wide area network (WWAN) offers higher mobility at lower data rate. Again, wireless metro area network (WMAN) such as WiMAX provides higher data rate than 3G and at the same time, higher mobility compared to WLAN. The hardware development with multiple technology interfaces (Simić, 2007) has open up the possibility of using this multiple technology infrastructure in a complementary manner.

The costs of services provided to the end user by employing these technologies are not similar. A case in point, WLAN operates in a license-free frequency band, whereas WMAN or 3G operates on expensive licensed frequency band (Zhuang, Gan, Loh & Chua, 2003). Consequently, the cost of service delivery using WLAN becomes lower compared to later ones. Similarly, the deployment cost of WMAN varies with that of 3G. So, it also makes business sense to use multiple technologies as a cost effective measure. This gives rise to the architecture of heterogeneous overlay network where, best features of each technology are kept intact and their weaknesses are mediated by the companion technology. In this scenario, the service provider will also be able to utilize its existing network infrastructure in a better manner.

Cost effective service delivery can be achieved through the combination of an integrated network planning and real-time network resource management policy. Nasser, Hasswa and Hassanein (2006) have addressed the latter issue. To the best of our knowledge, substantial works have not been carried out on integrated network planning by researchers. In this paper, we dwell on the aspects of optimal network planning of heterogeneous network.

This planning may be done in the following scenario:

1. Planning to optimize resource utilization of an overlay network with no existing infrastructure i.e. green field approach.
2. Planning to optimize resource utilization of an overlay network with non integrated existing heterogeneous infrastructure i.e. brown field approach.

In reality, the planning is often done to minimize connectivity cost using popular technique such as distance based planning (DBP) where only distance is considered as a criterion for establishing connectivity amongst the different networks. For example, WLAN is always connected to its nearest WiMAX base station (WiMAX BS). But this kind of planning may lead to inefficient use of other network resources like spectra. In wireless network, one of the usages of spectrum is to support mobility management, namely, location management and handoff management. In heterogeneous network environment, two kinds of handoffs occur namely, horizontal handoff and vertical handoff (Nasser et al., 2006). Horizontal handoff occurs when the user moves from one access point to another while both the access points use the same technology. This is the case when the user moves between two WLAN access points (WLAN AP). Vertical handoff refers to the handoff process of a user among access points that employ different network technologies. Handoff process due to movement of user from WLAN to WiMAX is an instance of the vertical handoff.

Traditionally, service providers have experience in managing the horizontal handoff at the individual network level. Management of vertical handoff is a new challenge faced by them in case of heterogeneous network. During the vertical handoff process, various control messages such as authentication, authorization and accounting (AAA) are generated. These control messages may lead to delay, consumption of precious radio resources. As number of users rises, more vertical handoffs increase the load on the radio resources. One of the most important objectives of network planning is to reduce the load on the radio resource created due to this vertical handoff. So, in case of heterogeneous network, handoff management becomes extremely important for efficient use of network resources and, in turn, for cost effective service delivery.

ORGANIZATION OF THE CHAPTER

The paper is organized as follows. In the next section, we highlight our contribution followed by a section, in which a brief review of related works has been presented. The problem formulation is presented in the fourth section. The fifth section is devoted to the solution methodology. The experimental results have been provided in the sixth section. The last section concludes the work.

Our Contribution

This work, primarily, concerns the handoff management issues, in an integrated all IP network consisting of WLAN, 3G and WiMAX networks. The quest for an optimal overlay network, which could provide enough support for mobility management to an end user, has motivated us to pursue the research in this field. Considering all the constraints posed by each network, the present-day global scenario calls for a hierarchical design, where each mobile terminal can roam around anywhere, anytime. In the process, it should be able to avail the required services demanded from the heterogeneous networks. It is almost mandatory that the end user should enjoy their mobility without bothering about the technological bottlenecks. This consideration, automatically implies that the handoff requests (in this paper, we are working on vertical handoffs, to be precise) by the end users should be catered to in an optimum manner and that should be transparent enough to the end user. Our work attacks these lacunae by planning an integrated overlay network, where handoff requests would be serviced in an efficacious way. For our design problem, we have considered three present network technologies, namely, WLAN, WiMAX and 3G. We have endeavored to lay out a planning problem, where all these technologies are intermixed in such a fashion, that network gateways of each are connected via wire line connection in an optimum manner and enough bandwidth is available to support the handoff requests generated. This work proposes a novel topology where under the umbrella of 3G, WiMAX and WLAN networks have been deployed. We have established the connectivity amongst the three network standards (Figure 1) such that a mobile user can roam around within that heterogeneous yet integrated domain and can enjoy seamless services in an optimum manner. Our planning problem also addresses the issue of minimizing the total cost incurred due to the vertical handoff requests and the amortized cost of cabling. Here, the cost of cabling refers to the non-recurring cost of the network which includes cost of cable laying and also the cost of the transport equipment involved. We have attempted to plan such

Figure 1. Interconnectivity among different networks (WLAN, WiMAX, and 3G). (WLAN access gateway (access router) is connected to WiMAX network gateway (border router), which, in turn, connected to SGSN of 3G network.)

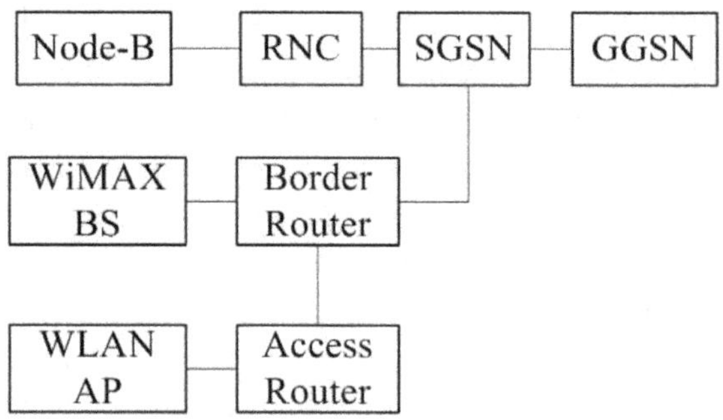

layered network architecture, considering all the design constraints involved, to achieve a robust inter-standard mobility management protocol. We pose a design problem that is applicable for the brown-field scenario. As in case of developed economics, many existing operators have traditionally built up various wireless networks separately. But now, since user device with multiple technology interfaces are available in market, operators may want to integrate these separate infrastructures to optimize the use of their resources. Our work mainly targets this state of affairs.

Finally, to summarize our contribution, we have considered that in our conceived overlay architecture, 3G is overlaid on WiMAX and WiMAX, in turn, is overlaid on WLAN (Figure 2). Presuming this structure, we have tried to reduce the total cost of operation (TCO) by selective assignment of WLAN to WiMAX BS. Here, TCO is defined as the summation of vertical handoff cost and amortized cost of cabling. Assignment refers to the process of interconnecting WLAN access gateway with WiMAX network gateway (Figure 1). Eventually, our planning problem boils down to an assignment problem which can be explained as follows. Suppose the service area consists of n number of WLANs and m number of WiMAX BSs. Theoretically, WLAN may be connected to any WiMAX BS resulting in n^m possible number of solutions. If we consider the distance between WiMAX BS and WLAN, as the only criterion to determine the connectivity, then we can minimize the cost for cabling. However, we may incur a considerable amount of handoff cost, especially due to vertical handoffs. The numbers of vertical handoff will depend upon mobility rate of the users. For example, particularly for highly mobile users, the possibility of occurrence of vertical handoff is pretty high. Therefore, to find out the optimal

Figure 2. Architecture of the proposed heterogeneous overlay network consisting of WLAN, WiMAX and 3G

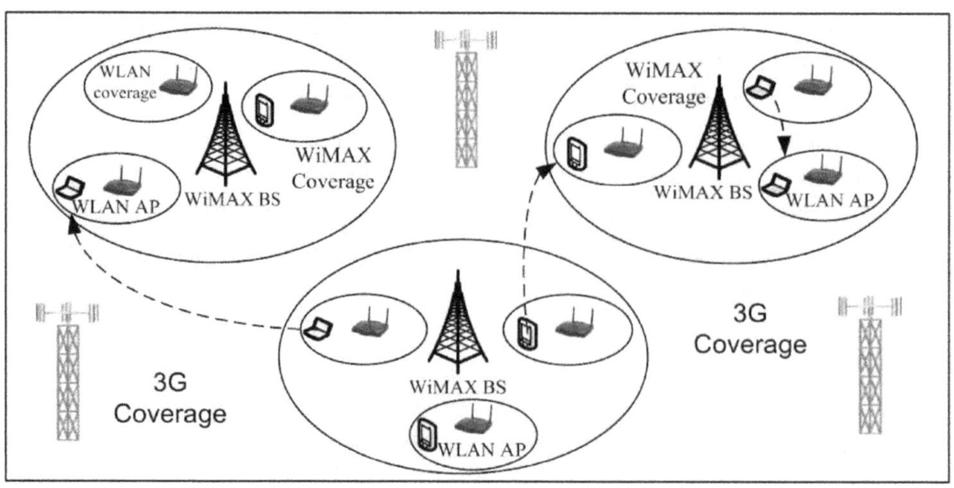

connectivity, we have taken into the consideration the mobility pattern of the users. We have presented a mobility scheme applicable to our scenario. Also, we have formulated the TCO minimization problem as a combinatorial optimization problem. To the best of our knowledge, the proposed formulation of the assignment problem for WLAN-WiMAX-3G overlay network, is the first of its own kind.

We have used meta-heuristic techniques, namely, simulated annealing (SA), H-II (Bhattacharjee, Saha & Mukherjee, 2004) and distance based planning (DBP) to solve the problem.

RELATED WORKS

Similar kind of assignment problems have been typically solved in the literature (Merchant & Sengupta, 1995; Bhattacharjee et al., 2004; Demirkol, Ersoy, Caglayan & Delic, 2004; Mandal, Saha & Mahanti, 2004; Maitra, Mukherjee & Saha, 2002) in the context of cellular networks (2.5G). Merchant and Sengupta (1995) proposed a refined heuristics (called H-I) for solving the NP-hard cell-to-switch (CSA) assignment problem. Later, researchers have proposed several evolutionary techniques (Quintero & Pierre, 2003a; Quintero & Pierre, 2003b) as the solution methodologies for the classical CSA problem. Amongst these evolutionary methods, most popular are Simulated Annealing (Demirkol et al., 2004), Taboo Search (Pierre & Houeto, 2002), Memetic Algorithm (Quintero & Pierre, 2002), several variants of Genetic Algorithm (Din & Tseng, 2002), Ant Colony Optimization (Shyu, Lin & Hsiao, 2006) and so on. However, these works have mostly been carried out for personal communication services networks (PCSN). Nonetheless, the methods, though very computation intensive and time consuming, have their due relevance as solution methodologies for combinatorial optimization problem associated with overlay networks. These methods are most suited for off-line planning phase. Stemm and Katz (1998) have studied the impact of discovery time (the time elapsed before a mobile host can understand whether it moves into or out of a new overlay) upon the handoff latency in an overlay network. Also they have developed a scheme in which a user can enjoy optimal connectivity so that the vertical handoff calls could be maintained. Katz and Brewer (1996), in their work, have presented a new architecture to integrate different wireless technologies that would result in a seamless internetwork application support services. Often vertical handoff management strategies, designed for overlay networks, are some sort of compromise between the customer satisfaction and the overall system performance. To address this tradeoff, a novel vertical handoff strategy has been proposed (Shen & Zeng, 2006) to take care of the system performance as well. The performance comparison has proved the fact that the proposed scheme is superior (so far the blocking probability is concerned) to the most of the schemes cited in the literature. Lo, Lu, Jacobsson, Prasad and Niemegeers (2006), in one of their works, have attempted to integrate wireless personal area network with UMTS. This is an intricate attempt as these two technologies are hard to make inter-work. They have described the complicacies involved and have provided solutions to resolve those. In Lin et al. (2006), challenges posed by wireless broadband networks, WiMAX in particular, have been addressed. For the deployment of a wireless broadband network at a minimum cost, a heuristic based clustering algorithm has been developed. From the simulation results, they have claimed that their algorithm is much suited for the planning of a network. In Chang, Lin and Liang (2006), a vertical handoff management algorithm has been presented, which as per the authors' claim, is capable enough to reduce redundant handoffs and thereby can improve the system performances, namely, network throughput, handoff delay. The algorithm is also successful enough in avoiding connection dropping. The work has considered 3G-WLAN overlaying as well as horizontal handoff between WLANs with MIPv6 support. The numerical results have shown that cost-based handoff (vertical /horizontal) algorithm significantly improves the number of vertical handoffs, decreases connection dropping yet increases throughput compared to some other schemes (based on the same basic premises) developed earlier. Lee, Sriram, Kim, Kim and Golmie (2009) have addressed the issue of vertical handoff performance in case of integrated 3G-WLAN networks. They have developed vertical handoff decision algorithm for connection management and optimal resource allocation for seamless mobility of mobile nodes considering suitable optimization criteria. Authors have claimed that their algorithm maximizes the overall battery lifetime in the same coverage area. In addition, it also aims to equitably distribute the traffic load across available WLAN and 3G networks. Tamea, Vegni, Inzerilli and Cusani (2009) propose a novel vertical handover algorithm based on a probabilistic approach for decision strategy. The proposed technique deals with the assessment of a Wrong Decision Probability (WDP), which assures a trade-off between network performance maximization and mitigation of the ping-pong effect. The recent paper (Shafiee, Attar and Leung, 2011) addresses the problem of optimal vertical handoff (VHO) in a vehicular network setting. A heterogeneous network consisting of a wide-area cellular network interworking with WLAN is considered. Authors have suggested an optimal strategy to carry out VHO in case of lower speed vehicle as well as higher speeds. In an article by Pontes, Silva, Jailton, Rodrigues & Dias (2008), the key issues involved in handoff management in an integrated network comprising IEEE 802.11 WLAN and IEEE 802.16 WMAN have been studied. They have considered a new emerging standard, namely, IEEE 802.21. The standard establishes a framework for "media-independent" handoff (MIH), which helps in handoff management for the integrated environment. Choi and Choi (2007) also consider a heterogeneous network of mobile WiMAX (IEEE 802.16e) and WLAN. Authors propose a VHO decision algorithm that considers impact of service charge due to vertical handover from WLAN to WiMAX. Additionally, they propose a delayed traffic delivery scheme for delay-tolerant applications so that end-users are economically benefitted and also energy consumption is minimized. Recent research trend has been channelized to integrate plethora of technologies to exploit fully the advantages of the wireless technologies. The ultimate goal of these endeavors is to move towards a ubiquitous broadband access irrespective of the standards. Hence, many a works have been reported in the literature, the citation of all is impossible. However, most of the researches are still in their nascent stages and to the best of our knowledge, attempts to plan overlay networks generally uses at most two standards. In our work, we plan the heterogeneous network by employing a three-tier topology, namely, 3G-WLAN-WiMAX. In the next section, we present the analytical formulation of our considerations to plan the integrated network architecture.

PROBLEM FORMULATION

Let us consider that, in the network, there are n WLANs, m WiMAX BSs whose locations are known. Let G ={1,2, ... n} denote the set of WLANs, H= {1,2,... m} denote the set of WiMAX BSs. Throughout this formulation, we use a small letter to denote a member of the set, whereas the set is represented by another capital letter; for example, i $\in$G, j $\in$H. Moreover, we assume WLAN i and WLAN i′ are different WLANs (i $\neq$ i′). Similarly we assume j $\neq$ j′. Let us now define the following notations used in the formulation.

d_{ij}: is 1, if WLAN i is under coverage area of WiMAX BS j, 0, otherwise (this can be pre-computed from the known positions of WLAN and WiMAX BS and their coverage radius).

x_{ij}: is 1, if WLAN i is assigned to WiMAX BS j in network, 0, otherwise.

c_{ij}: is the amortization cost of the wired connectivity between WLAN i and WiMAX BS j.

Cap'_j: is the capacity in packet switching (bits per second) of WiMAX BS j.

Cap_i: is the capacity in packet switching (bits per second) of WLAN i.

H_{WL}^{WM}: is the cumulative handoff cost between WLAN and WiMAX BS per session per unit of time.

$N_{ii'}$: is the number of users moving from WLAN i to WLAN i′ and vice versa.

$H_{ii'}$: is the handoff cost between WLAN i and WLAN i′ involving at least one WiMAX BS per unit of time.

We assume that the handoff cost between WLAN i and WLAN i′ (denoted by $H_{ii'}$) is known. Here, $H_{ii'}$ refers to the cumulative costs of WLAN-WiMAX handoff (H_{WL}^{WM}) and WiMAX-3G handoff that occur when $N_{ii'}$ number of users move from WLAN i to WLAN i′.

Then, the problem can be formulated as:

Minimize (TCO ({ })):

$$(\sum_j \sum_i \sum_{i'} (1 - x_{ij} x_{i'j}) H_{ii'} + \sum_j \sum_i \sum_{i'} x_{ij} x_{i'j} N_{ii'} H_{WL}^{WM}) + \sum_i \sum_j c_{ij} x_{ij}$$

Minimize (TCO({xij})):

$$\forall\ i, i' \in G, j \in H, (i \neq i') \tag{1}$$

Subject to:

$$x_{ij} \leq d_{ij}\ \forall\ i \in G, j \in H \tag{2}$$

$$\sum_{j \in J} x_{ij} \leq 1\ \forall\ i \in G \tag{3}$$

$$\sum_{i \in G} Cap_i x_{ij} \leq Cap'_j\ \ \forall j \in H \tag{4}$$

The constraint (2) implies that the connectivity is only possible under the coverage of WiMAX BS. The constraint (3) indicates that one WLAN can at most be connected to one WiMAX BS. The constraint (4) denotes that the cumulative capacity of WLANs connected to WiMAX BS j should not exceed the capacity of WiMAX BS j. Our objective is to minimize the 3G handoff by connecting two WLANs under one WiMAX BS. The objective function has two components. The first component is the handoff cost. This handoff cost comprises of two costs namely, cost of using 3G network and cost of using WiMAX network. The cost of using 3G network arises when a mobile terminal moves from the coverage of one WiMAX base station to another WiMAX base station with an intermediate support of a 3G network. In a similar fashion, the cost of using WiMAX network arises when a mobile terminal moves from the coverage of one WLAN AP to another WLAN AP with an intermediate support of a WiMAX network. The total handoff cost of the network can be lessened by the following way. If two different WLANs i and i′ can be allocated under same WiMAX BS j, then WiMAX-3G handoff can be avoided. In this case, the total cost of handoff between WLAN i and i′ reduces to only WLAN-WiMAX handoff, which cannot be avoided anyway. Otherwise, the total handoff between WLAN i and i′ remains unchanged. H_{WL}^{WM} is assumed to be same for all pair of WLANs because of the fact that same amount of bandwidth is consumed due to exchange of information required for completion of vertical handoff between WLAN and WiMAX. However, the total cost of WLAN-WiMAX handoff between WLANs connected to the same WiMAX BS varies with $N_{ii'}$. The second part of the objective function denotes the wired connectivity cost among WLANs and WiMAX BSs.

In our formulation, x_{ij} is the decision variable. We need to find out the optimal values of x_{ij} so that the value of the objective function is minimized. The constraint on data traffic is not used in our formulation.

The objective function, formulated above, is not linear and contains product of binary variables. Nonetheless, the nonlinear term XY (i.e., the product of two binary variables; where, X=x_{ij} and Y=$x_{i'j}$) can be converted to linear form using additional variable Z where:

$$Z = XY \tag{5}$$

Here, Z is also a binary variable. We also add the following type of constraints in the above formulation.

$$Z \leq X \tag{6}$$

$$Z \leq Y \tag{7}$$

$$Z \geq X + Y - 1 \tag{8}$$

Thus, the above formulation of the problem can be converted to a 0-1 integer linear programming (ILP) problem. It is obvious from the formulation that an exhaustive enumeration technique for assigning WLAN to WiMAX BS problem requires exploring n^m combinations. It is apparent from the above argument that the problem size and its solution space are pretty large. Hence, we adopt heuristic techniques, which are more suitable to crack such type of ILP problems.

SOLUTION METHODOLOGY

We have formulated the problem (equation 1) as a combinatorial optimization problem and used SA to solve it. We have also used H-II, based on intelligent clustering technique and DBP, based on greedy search technique to solve the same. We describe these techniques in the following paragraphs.

SA Method

The algorithm starts the search process in the combinatorial space with an initial solution (which may be chosen randomly) and iteratively proceeds to improve it by selecting a new solution. The cost of the new solution is calculated. If the cost is reduced then the new solution will be accepted and stored for further exploration. If the cost of the new solution is more than the cost of the best solution stored, then the new solution will be accepted for further exploration with a probability, p, value of which depends on the difference of the cost and stage of the search. This process gets repeated until the termination criterion is met.

Quality of solutions of SA depends on the values of different SA parameters such as λ and T_0. Here, we explain the process of generating initial solution, neighborhood solutions and determining values of above mentioned parameters of SA in the following.

Generation of Initial Solution $S_o \{x_{ij}\}$

To start the algorithm, the decision variable x_{ij} ($\forall$ i$\in$ G, j $\in$ H) is initialized. Here, WLAN i is assigned to its nearest WiMAX BS j when sufficient capacity is available at the WiMAX BS to accommodate the WLAN. Subsequently, the value of corresponding x_{ij} is made 1. Likewise, we assign all the WLANs to their nearest WiMAX BSs. Thus, we get a connectivity solution $S_o \{x_{ij}\}$ as our initial solution (Figure 3).

Figure 3. Generation of new solution (Network consisting of n WLANs and m WiMAX BSs)

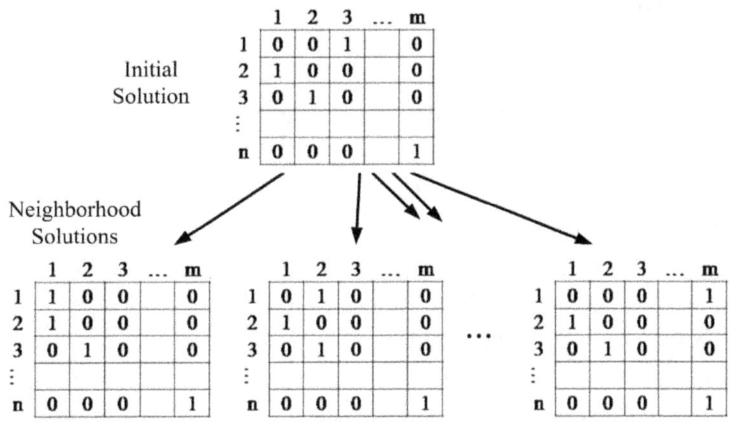

Algorithm 1.

Initialization

1. Generate an initial connectivity configuration $S_o\{x_{ij}\}$. Let the TCO of the initial configuration be TCO (S_o). Let us assume that $B\{x_{ij}\}$ represent the best solution seen during exploration and initially *Set* $B\{x_{ij}\}= S_o\{x_{ij}\}$, $S\{x_{ij}\}= S_o\{x_{ij}\}$, $T=T_0$ and $N=0$.
2. Let the initial temperature be T_0 and cooling factor is λ.

Exploration

Do

While $T>T_{Th}$ and $N<N_{Th}$ //T_{Th} and N_{Th} represent the threshold temperature and maximum number of iterations respectively//

1. Make $S'\{x_{ij}\}$ by modifying $S\{x_{ij}\}$. So, the cost becomes TCO (S').
   a. *If* no new $S'\{x_{ij}\}$ is found, *Go To* Termination Step.
   b. *Set* $N=N+1$
2. Calculate the difference between cost of the solutions
   $\Delta TCO = TCO(S') - TCO(S)$
3. *If* $\Delta TCO < 0$;
   a. *Set* $S\{x_{ij}\}= S'\{x_{ij}\}$
   b. *If* $S\{x_{ij}\}< B\{x_{ij}\}$
      I. $B\{x_{ij}\}= S\{x_{ij}\}$
   c. $T=T^* \lambda$ // λ represents cooling factor//
   Go To Exploration -Step 1

   Else
   a. $p= \exp\left(-\frac{\Delta TCO}{T}\right)$
   b. Generate a random number R between 0 and 1.0.
   c. *If* $R<p$
      I. *Set* $S\{x_{ij}\}= S'\{x_{ij}\}$
      II. $T=T^* \lambda$
      Go To Exploration -Step 1

      Else
      Go To Exploration -Step 1

Termination

Output $B\{x_{ij}\}$ and $TCO(\{x_{ij}\})$

Generation of Neighborhood Solution S′ from Solution S

A new solution is generated by selecting an unassigned WLAN and thereafter assigning it to other WiMAX BSs. In Figure 3, in the initial configuration, WLAN 1 is assigned to WiMAX BS 3. To generate the neighborhood solutions, we have tried to assign WLAN 1 to all other WiMAX BSs taking one at a time, while maintaining the constraints. In that case, the WLAN may be assigned to (m-1) numbers of WiMAX BSs, resulting in (m-1) different configurations. Out of these configurations, we consider a new configuration S′ that satisfies all constraints indicated in equation (2-4) in problem formulation section. If WLAN 1 is moved to all WiMAX BSs and best configuration is found, then we consider the next WLAN for generation of new configuration.

Determination of Values of SA Parameters

Values of different SA parameters need be carefully selected as it affects the performance of the algorithm. To arrive at the optimal values of the parameters, we have run SA algorithm on a sample network consisting of 4 WiMAX BSs and 20 WLANs. Values of the parameters have been varied. For each value, the program is run for 100 instances and the average value is recorded.

Selection of T_0 is important because too high T_0 may take longer time to converge. On the other hand, if the T_0 is kept low, the algorithm may get trapped in local optimum because it may not have enough freedom to explore different solutions. We have selected the value of the threshold temperature as 10 and varied the values of the T_0 from 50 to 500. In Figure 4, improvement of TCO as found by SA over DBP is plotted against T_0 for the sample network stated above. It is evident from the Figure 4 that the improvement in TCO is nominal beyond 400. Hence, we have selected

Figure 4. Percentage improvement in TCO in SA over DBP vis a vis initial temperature (Network consisting of 20 WLANs and 4 WiMAX BSs)

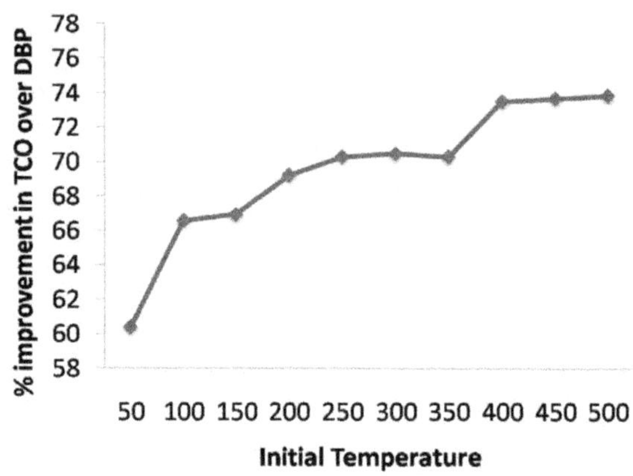

the value of T_0 at 400 as it gives satisfactory result that can be achieved within a reasonable time.

Selection of λ is also vital because low cooling coefficient may decrease the temperature fast, resulting in quick convergence without exploring the search space unduly. Also, higher cooling coefficient makes the search process more time consuming. In Figure 5, we have varied the value of λ and have observed the change in the cost of solution found by SA. If we increase the value of λ beyond 0.9, the improvement in solution cost

Figure 5. Percentage improvement in TCO in SA over DBP vis a vis cooling factor (Network consisting of 20 WLANs and 4 WiMAX BSs)

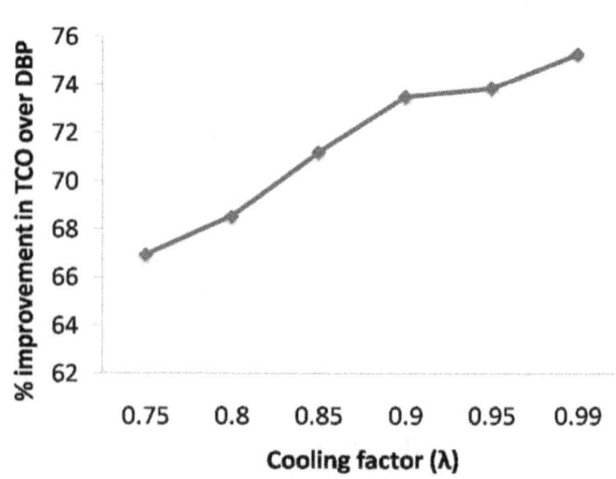

is compromised in terms of computation time. Hence, we have set λ=0.9 for our experiment. So, Judicious selection of λ and T_0 are necessary for efficacious performance of the SA algorithm.

We have observed that a proper choice of the values of T_0 and λ are dependent on the problem size. If we increase the problem size, the solution space increases as well. To attain the near-optimal solution, the search process should explore considerable number of solution points, before convergence, to evade local optima. This calls for higher values of λ. For example, when there are 20 WLANs in the network, we have found good quality solution at the T_0 value of 400 and λ=0.9. But, these values do not ensure similar solution when WLAN count is 32. So, we have found out the proper λ for each network size keeping T_0 fixed at 400. The optimal values for each network size are given in Table 1.

H-II Method

H-II is a meta-heuristic technique based on intelligent clustering. In this method, WLANs that are within a certain distance from WiMAX BSs are assigned to their nearest WiMAX BS, while maintaining the capacity constraint of the WiMAX BS. Rest WLANs are considered one by one by decreasing traffic volume. Among these unassigned WLANs, if any WLAN has more than one feasible WiMAX BSs (feasibility is determined based on distance from WiMAX BS), assignment is made in such a manner where the TCO is minimized. The process continues until all WLANs are properly assigned.

DBP Method

DBP method is a greedy search technique that considers only distance to determine the connectivity between WLANs and WiMAX BSs. DBP tries to assign the WLANs in their nearest WiMAX BSs provided capacity constraint is met.

EXPERIMENTAL RESULTS

In order to evaluate the performances of SA, H-II and DBP, we have implemented these techniques in C++ language and run the simulation in a PC environment on Intel Core™2 Duo platform under Borland C++compiler. In simulation, for a particular size of network we have generated 100 instances. The average of the results found by the above techniques has been recorded. We have found substantial improvement of TCO in both SA and H-II over DBP.

We have generated problem instances programmatically following a set of rules derived from the reality that are described below. For generating a planning instance, we have fixed the WLAN locations a priori in a random manner. Here, we have assumed that the numbers of WiMAX BSs and their locations are fixed. Moreover, the capacities of WLANs and WiMAX BSs are randomly generated with minimum and maximum capacity limits.

Table 1. Optimal values of cooling factor

WLAN Count	Cooling Factor
16	0.87
20	0.9
24	0.93
28	0.96
32	0.99

Algorithm 2. H-II Algorithm

Let set_k denote the set of WLANs assigned to WiMAX BS k

STEP 0: INITIAL ASSIGNMENT

Find the WLAN with B_i that is nearest to WiMAX BS k and initialize set_k to B_i i.e. $set_k^0=\{B_i\}$.

Correspondingly, reduce capacity of WiMAX BS k CAP_k by c_i: $CAP_k^0 = CAP_k - c_i$

STEP ℓ (>0):ITERATION

(ℓ.1) Identify the neighboring WLANs of $set_k^{(\ell-1)}$. Arrange all WLANs in descending order of ρ, where ρ= traffic volume of WLANs.

(ℓ.2) If any WLAN (say the WLAN with B_r) has only one feasible WiMAX BS, allocate the WLAN to that WiMAX BS. Do it for WiMAX BS i.e. $\{set_k^{(\ell-1)}\}$, $k=0,1,2,\ldots, M$

(ℓ.3) If the WLAN with B_r is neighbor to more than one $set_k^{(\ell-1)}$, then assign it to the $set_k^{(\ell-1)}$ for which TCO is minimum provided its capacity can be supported by the WiMAX BS. If there is a tie in TCO, break the tie randomly. If B_r can not be assigned to the nearest WiMAX BS due to capacity limitation, try for the second best WiMAX BS, if any, and so on. If B_r can not be accepted by any WiMAX BS, leave it unassigned.

(ℓ.4) If the WLAN with B_r is assigned to WiMAX BS k, then reduce the traffic handling capacity of set_k by the capacity c_r of B_r.

(ℓ.5) Go to step 2 until the capacity of all the WiMAX BSs have become insufficient or all the WLANs have been assigned.

ENDSTEP: If no WLAN remains unassigned, report SUCCESS
else report FAILURE.

Stop

Algorithm 3. DBP Algorithm

1. Find out feasible WiMAX BSs for the WLANs.
2. Allocate WLANs that have single feasible WiMAX BS.
2.1. Find out WLAN that has single WiMAX BS.
2.2. Allocate the WLAN to its feasible WiMAX BS.
2.3. Reduce the remaining capacity of WiMAX BS by the WLAN capacity c_{WLAN}.
2.4. Go to step 2.1 until all WLANs that have single feasible WiMAX BS are allocated.
3. Allocate WLANs that have multiple feasible WiMAX BSs.
3.1. Find out WLAN having multiple WiMAX BSs.
3.2. Find out the WiMAX BS having minimum distance with the WLAN. If there is a tie break it randomly.
3.3. Allocate the WLAN with the selected WiMAX BS provided the capacity of the WLAN is supported by the remaining capacity of the WiMAX BS. If the WLAN is not assigned due to capacity limitation of WiMAX BS, try for second best WiMAX BS, if any, and so on. If WLAN is not accepted by any of its feasible WiMAX BSs, then leave it unassigned.
3.4. If the WLAN is assigned to WiMAX BS, then reduce the capacity of WiMAX BS by the capacity c_{WLAN} of the WLAN.
3.5. Go to step 3.1 until the capacity of all the WiMAX BSs have become insufficient or all the WLANs have been assigned.

ENDSTEP: If no WLAN remains unassigned, report SUCCESS
else report FAILURE.

Stop

To simulate the movements of the users, several generic models are available in the literature, such as fluid flow model, diffusion model, gravity model and random walk model etc. (Bhattacharjee, Saha & Mukherjee, 2004). Despite the fact, in this work, we present a new mobility model which is close to our problem scenario.

We consider a service area which has many populated places (like airport, restaurants, railway stations, shopping malls etc.) covered by WLANs. In most of the cities, the geographical topology is such that these places are well connected via roads or highways to other key locations. Also, it is assumed that a pattern of mobility of the users will be there where mobile terminals will converge/digress to/from these places. Hence, the mobile users will come under the coverage of the WLAN installed at these sites. Let us assume that a user moves at an average speed of v for a duration of t. To capture the scenario, we have assumed that the direction of user movement is uniformly distributed over $[0,2\pi]$ in the region at regular interval of $(\pi/6)$. In our simulation, we have created a user under the coverage area of any WLAN (WLAN number is randomly generated). Now, we have compelled the user to move with a velocity v for duration of t in the direction d as described above. Likewise, we have repeated the process for multiple users to get a complete mobility profile within range of the area concerned. Thus, we have developed at a mobility pattern which we believe, is closer to the reality. During the movement, if a user enters into the coverage area of any other WLAN, inter-WLAN handoff takes place. Corresponding handoff data serve as an input to our network planning problem.

The results thus obtained by the techniques for the input generated following the above rules are plotted in Figure 6 and Figure 7.

In Figure 6, we have plotted the percentage improvement in TCO. Here, percentage improvement in TCO is given by:

% improvement in TCO over DBP

$$= \frac{TCO(DBP) - TCO(SA / H - II)}{TCO(DBP)} \times 100 \tag{9}$$

From Figure 6, it is found that the reduction of TCO is increasing, while the number of WLANs is varying from 16 to 32. From Figure 7, it is apparent that the TCO reduction is much more effective in case of SA and H-II compared to

Figure 6. Cost reduction in SA vis-à-vis DBP with varied number of WLANs

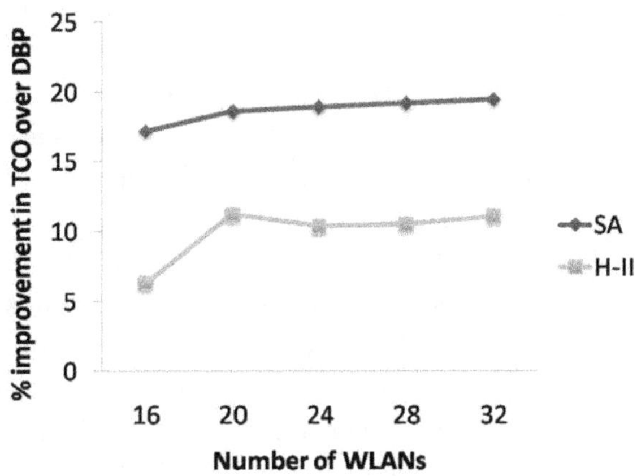

Figure 7. Cost reduction in SA and H-II vis-à-vis DBP with user count in the service area

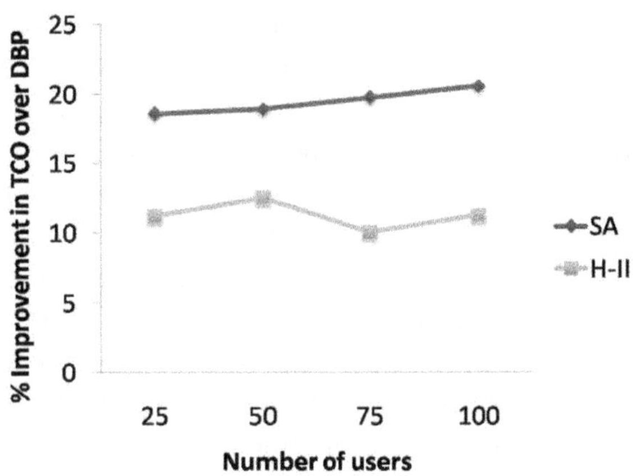

DBP with increase in user numbers from 25,000 to 100,000.

The reason of the outperformance of SA and H-II over DBP can be explained as follows. DBP technique is completely adhoc by nature and takes decision based on distance only, whereas SA and H-II search for optimal solution by taking into account the distance as well as the number of vertical handoffs in the network. When WLAN count increases, the solution space also extends. In that case, SA and H-II find a improved solution with their extensive search methodology. When number of users is increased, instances of vertical handoff also increase. Since vertical handoffs are not considered in case of DBP, H-II and SA outperfom DBP in managing the vertical handoffs and in turn, TCO of the network.

It is also evident from the results that SA outperforms H-II in both the scenarios. The reason is, in case of H-II technique, only boundary WLANs are optimally assigned to WiMAX BSs by considering vertical handoffs as well as distance. On the other hand, SA evaluates every options of assigning all WLANs to reach at the optimal solution. Though this makes the solution space of SA much larger than that of H-II, SA has better chance to find improved solution compared to H-II technique.

CONCLUSION

In this work, we have proposed a heterogeneous overlay network environment consisting of WLAN, WiMAX and 3G networks. As per literature, this kind of integration has not been addressed yet. We have formulated a TCO minimization problem and have solved the problem with our SA based heuristics and H-II. To simulate the problem environment, we have generated the user mobility profile in an unique way to suit our problem. The performances of SA reveals that SA simply outperform H-II and DBP method. Finally, it can be concluded that SA can be used as an efficient heterogeneous network planning tool for minimizing TCO. In this article, though our solution methodology has been popularly used in various studies, but the problem domain is unique. This has opened a new horizon for pursuing research for planning and efficacious resource management for any overlay topology. As a future work, we would propose our own solution methodology, be it traditional, evolutionary methods or any un-endeavored metaheustics of our own.

REFERENCES

Bhattacharjee, P. S., Saha, D., & Mukherjee, A. (2004). An approach for location area planning in a personal communication services network (PCSN). *IEEE Transactions on Wireless Communication, 3*(4).

Chang, B., Lin, S., & Liang, Y. (2006). Minimizing roaming overheads for vertical handoff in heterogeneous wireless mobile networks. *International Conference on Wireless Communications and Mobile Computing* (pp. 957-962).

Choi, Y., & Choi, S. (2007, May). Service charge and energy-aware vertical handoff in integrated IEEE 802.16e/802.11 networks. *26th IEEE International Conference on Computer Communications IEEE INFOCOM 2007* (pp. 589-597).

Demirkol, I., Ersoy, C., Caglayan, M. U., & Delic, H. (2004, May). Location area planning and cell to switch assignment in cellular networks using simulated annealing. *IEEE Transactions on Wireless Communications, 3*(3), 880–890. doi:10.1109/TWC.2004.827767

Din, D. R., & Tseng, S. S. (2002). Genetic algorithm for solving dual-homing cell assignment problem of the two-level wireless ATM network. *Computer Communications, 25*(17), 1536–1547. doi:10.1016/S0140-3664(02)00086-5

Katz, R. H., & Brewer, E. A. (1996, January). The case for wireless overlay networks. *SPIE Multimedia Computing and Networking, 2667*, 77–88.

Lee, S., Sriram, K., Kim, K., Kim, Y. H., & Golmie, N. (2009, February). Vertical handoff decision algorithms for providing optimized performance in heterogeneous wireless networks. *IEEE Transactions on Vehicular Technology, 52*(2), 865–881.

Lin, P., Ngo, H., Qiao, C., Wang, X., Wang, T., & Qian, D. Y. (2006). *Minimum cost wireless broadband overlay network planning.* International Symposium on a World of Wireless, Mobile and Multimedia Networks.

Lo, A., Lu, N. W., Jacobsson, M., Prasad, V., & Niemegeers, I. (2006). *Personal networks: An overlay network of wireless personal area networks and 3G networks*. 3rd Annual International Conference on Mobile and Ubiquitous Systems – Workshops.

Maitra, M., Mukherjee, A., & Saha, D. (2002). *A heuristic-based approach for minimization of inter-switch handoff cost with load balancing for optimal location area planning in a PCSN* (pp. 172–176). IEEE ICPWC.

Mandal, S., Saha, D., & Mahanti, A. (2004). Heuristic search techniques for cell to switch assignment in location area planning for cellular networks. *IEEE International Conference on Communications (ICC)* (pp. 4307-4311).

Merchant, A., & Sengupta, B. (1995). Assignment of cells to switches in PCS networks. *IEEE/ACM Transactions on Networking, 3*(5), 521–526. doi:10.1109/90.469954

Nasser, N., Hasswa, A., & Hassanein, H. (2006, October). Handovers in fourth generation heterogeneous networks. *IEEE Communications Magazine, 44*(10), 96–103. doi:10.1109/MCOM.2006.1710420

Pierre, S., & Houeto, F. (2002). Assigning cells to switches in cellular mobile networks using taboo search. *IEEE Transaction on Systems, Man and Cybernetics. Part B, 32*(3), 351–356.

Pontes, A. B., Silva, D. D. P., Jailton, J. J., Rodrigues, O. J., & Dias, K. L. (2008, October). Handover management in integrated WLAN and mobile WiMAX networks. *IEEE Wireless Communications, 15*(5), 86–95. doi:10.1109/MWC.2008.4653137

Quintero, A., & Pierre, S. (2002, November). A memetic algorithm for assigning cells to switches in cellular mobile networks. *IEEE Communications Letters, 6*(11), 484–486. doi:10.1109/LCOMM.2002.805515

Quintero, A., & Pierre, S. (2003a, September). Assigning cells to switches in cellular mobile networks: A comparative study. *Computer Communications, 26*(9), 950–960. doi:10.1016/S0140-3664(02)00224-4

Quintero, A., & Pierre, S. (2003b, February). Evolutionary approach to optimize the assignment of cells to switches in personal communications networks. *Computer Communications, 9*(2), 927–938. doi:10.1016/S0140-3664(02)00238-4

Shafiee, K., Attar, A., & Leung, V. C. M. (2011, March). Optimal distributed vertical handoff strategies in vehicular heterogeneous networks. *IEEE Journal on Selected Areas in Communications, 29*(3), 534–544. doi:10.1109/JSAC.2011.110304

Shen, W., & Zeng, Q. (2006). *A novel decision strategy of vertical handoff in overlay wireless networks*. Fifth IEEE International Symposium on Network Computing and Applications (NCA'06).

Shyu, S. J., Lin, B. M. T., & Hsiao, T. (2006). Ant colony optimization for the cell assignment problem in PCS networks. *Computers & Operations Research, 33*(6), 1713–1723. doi:10.1016/j.cor.2004.11.026

Simić, I. S. (2007, June). Evolution of mobile base station architectures. *Microwave Review, 13*(1).

Stemm, M., & Katz, R. H. (1998). Vertical handoffs in wireless overlay networks. *Mobile Networks and Applications, 3*(4), 335–350. doi:10.1023/A:1019197320544

Tamea, G., Vegni, A. M., Inzerilli, T., & Cusani, R. (2009, September). A probability based vertical handover approach to prevent ping-pong Effect. *6th International Symposium on Wireless Communication Systems,* 2009 (pp. 181-185).

Zhuang, W., Gan, Y., Loh, K., & Chua, K. (2003, November). Policy-based QoS management architecture in an integrated UMTS and WLAN environment. *IEEE Communications Magazine, 41*(11), 118–125. doi:10.1109/MCOM.2003.1244931

This work was previously published in International Journal of Business Data Communications and Networking, Volume 6, Issue 1, edited by Debashis Saha and Varadharajan Sridhar, pp. 1-16 , copyright 2010 by IGI Publishing (an imprint of IGI Global).

Chapter 4
QoS-Aware Scheme for Mobility Management and Adaptive Resource Reservation in 4G Wireless Networks

Sihem Trabelsi
High School of Communications (Sup'Com), Tunisia

Noureddine Boudriga
High School of Communications (Sup'Com), Tunisia

ABSTRACT

The success of fourth generation networks (4G) will be driven by intelligent interactivity between access networks and user devices and their ability to meet increasing user needs in term of bandwidth, availability, and especially session continuity.

Handling mobility issues over heterogeneous environments is crucial for transparent handoff and efficient utilization of the scarce radio resources with QoS provisioning and is still an open issue.

The purpose of this chapter is to develop a novel mobility aware adaptive resource reservation scheme for fourth generation networks that optimizes system performances and guarantees end to end QoS for individual users. Considering the importance of an appropriate use of scarce radio resources, three resource reservations algorithms are proposed respectively for UMTS, WLAN, and WiMax access networks. These algorithms take into consideration both future location of the mobile terminal expressed in probabilistic way and the load conditions of target Access Points (AP)/Base Stations (BS) to determine the amount of radio resources to be reserved for both uplink and downlink.

Simulations show that the proposed scheme achieves better results than those of other resource reservation schemes for metrics like bandwidth utilization, handoff latency, and packet loss.

DOI: 10.4018/978-1-61350-477-2.ch004

INTRODUCTION

The future-focused networks are designed to satisfy the growing users' need for bandwidth-hungry applications and rich media services. These networks are expected to integrate advanced built-in QoS functionalities and to meet great challenges related namely to high capacity, seamless handoff, intelligent resource reservation, real time service delivery and low latency.

These IP based networks, called 4G networks, are capable of supporting and handling the technological heterogeneity through the integration of various access networks ranging from cellular networks such as (UMTS, Universal Mobile Terrestrial Service) to wireless access networks such as (WLANs, Wireless Local Access Networks and Ad'hoc networks) within a scalable and innovative platform.

Enhancing the capability of these networks to deliver better services for these new applications through the provision of end-to-end QoS is becoming a key challenge and a condition of the success of the fourth generation of mobile wireless networks.

QoS guaranties are particularly required for real time and streaming multimedia applications and services in networks where network capacity is limited. This requirement arises specially with the need for traffic management and when congestion, jitter and latency are likely to affect services availability. But, because of the high mobility constraints and the lack of coordination between management entities of different portions of the heterogeneous networks, the fulfillment of end-to-end requirements have no guarantees. This why guarantying seamless mobility and QoS for end-to-end delivery of services over heterogeneous networks, where more than one management entity exists is still an open issue.

Promising and successful future 4G frameworks will overcome the aforementioned shortcomings and provide personalized user services which demands higher understanding of user's expectations.

Therefore, key features of 4G will be, besides QoS consistency, the implementation of user personalization, deploying heterogeneity in terms of network, terminal and services, evolutionary design of devices and personalized content on mobile devices.

Several research initiatives have been proposed in the literature to address QoS optimization through the enhancement of resource reservation mechanisms and the mobility management across heterogeneous access networks. Some of these works were interested in proposing general frameworks for QoS optimization either by estimating user requirements and peak time of services utilization to save the bandwidth/cost factors (Abbasi, A.A., Iqbal, J. and Malik, A. N. (2010)) or by designing a multiobjective optimization framework which provides a clear and comprehensive description of different options and solution to achieve an optimal network planning (Gordejuela-Sanchez, F., Juttner, A. and Zhang, J. (2009)).

Other works have addressed the QoS aspects in heterogeneous networks from an architectural point of view. Aiash, M., Mapp, G. and Lasebae, A. (2011) proposed a QoS signaling mechanism to be implemented by the Y-Comm architecture as a potential 4G framework. The proposed mechanism is designed to handle heterogeneity in the sense that it introduces new functional modules and interfaces to be run on different network entities. Providing end to end QoS over heterogeneous networks through an extensive 4G architecture is approached by Iancu, B., Dadarlat, V. and Peculea, A. (2010). This architecture which uses an hierarchical approach to organize link bandwidth is based on agreed SLAs and SLSs between autonomous management entities and is characterized by the creation of a new abstraction layer with complex functions called convergence layer and availing the bandwidth organization. All these contributions were innovative in term of potential scenarios and architectures for the 4G networks but their main shortage is that they were merely descriptive and did not provide analysis of the benefits provided schemes in term of end user QoS parameters.

Networking within heterogeneous environments requires the availability of multi-interface terminals in order to take advantage of the network diversity. Moreover these terminals should be QoS support enabled so as to preserve the overall QoS network level. This issue is investigated by Andi, W.C., Yeo, C.K. and Lee, B.S. (2010) who propose an environment-aware interface management framework for QoS support in multi-interface wireless terminals facilitated by the link layer. The proposed framework is based on the Media Independent Handover (MIH) protocol and utilizes link layer metrics to observe the network condition and further exploit them to aid the handoff decision. Besides being environmentally aware, the framework also incorporates policies to support user's preferences but does not address the impact of these policies on network performances.

Another group of related works addressed, more in details the QoS problems. They focused on handoff issues and more particularly mobility management and on reducing handoff latency and packet loss. So far, different mechanisms dealing with handoff management have been proposed. These mechanisms need to ensure that mobile devices continue to receive ongoing communications without any noticeable disruption during handoff events among the heterogeneous networks (Magagula, L.A., Chan, H. A., and Falowo, O.E. (2011)).

The most recent approaches have studied the implementation of IEEE 802.21 for handoff management whose scope is to optimize and facilitate handoff in heterogeneous networks. The contribution proposed by Chang, L.H., Wang, C.C., and Lee, T.H. (2010) presented a handoff mobility management framework by using the IEEE 802.21 protocol to solve some handoff issues, such as latency and packet loss, in homogeneous and heterogeneous networks. The same issues (reducing handoff latency and packet loss) were investigated by Kim, T.W. and Kim, H.G. (2011), Yu, M.J., Choi, S.G., Kim, H.S., and Cho, K.S. (2011) and Magagula, L.A., Falowo, O.E., and Chan, H.A. (2010) as they all discuss how to guarantee seamless handoff performances using different approaches.

In the first contribution, an experimental testbed for reducing the handoff latency for streaming packets was proposed and a comparison with the handoff latency for IP packets was carried out. In their contribution, M.J., Choi, S.G., Kim, H.S., and Cho, K.S. (2011), propose an improved mobility management scheme for fast handoff between WiMAX and WLAN networks, which simplifies re-authentication and IP configuration process and supports low handoff latency by using general AAA key and Mobility Agent Identifier.

The third work addresses the employment of a handoff coordinator between the networks involved in the handoff process. The proposed handoff coordinator utilizes a network-based mobility management protocol to efficiently handle vertical handoffs in a localized heterogeneous wireless environment and to reduce latency and packet loss during the handoff.

While each of these three contributions presented a solution for a particular issue for the enhancement of handoff performances, their main shortage is that did not address and study the effect of the combination of mobility management and resource reservation on end user QoS parameters as we did.

In order to further enhance vertical handoff performances, several resource reservation mechanisms have been proposed Such as the contribution of Todini et al., (2006) where authors describe a packet scheduling algorithm and three resource allocation algorithms, characterized by various degrees of optimality and by different computational complexities. This contribution mainly focuses on OFDMA/TDD system and does not address other types of access networks.

The contribution presented by Samaan, N., Benmammar, B., Krief, F. & Karmouch, A. (2005) describes a novel approach for an advanced reservation protocol, to provide seamless real-time services to mobile users in wireless integrated services networks. Two contributions are presented; the first is the utilization of knowledge about user preferences, goals, and analyzed spatial conceptual maps to predict the user's future location and the second is a predictive advanced resources reservation protocol suitable for mobile environments. The main limitation of this work is that it does not take into account load conditions of access networks where the resource reservation must be done.

Huang, C.J., Chuang, Y.T., Guan, C.T., Yang, D.X. & Chen, Y.J. (2007) present a self-adaptive bandwidth reservation scheme, which adopts a probabilistic mobility prediction model to estimate the bandwidth required in neighboring cells, and to reduce the forced termination probability of multimedia handoffs in 4G mobile WiMAX networks. 4G mobile WiMAX advantages are exploited via cross layer design by using bandwidth reservation for different class traffic. However, authors only addressed resource reservation for WiMAX networks as part of a 4G environment.

Ghosh, D. and C. Lott (2007) describe uplink - downlink imbalance in realistic TDD and FDD wireless networks, its potential causes and its impact on traffic overhead and channel performance, soft handoff, adaptive server selection and sector load control scheme. Authors define imbalance metrics and a model of receive antenna imbalance in their simulation studies of system robustness using cdma2000 1xEV-DO as an example. In particular, they discuss the effect of soft-handoff on system robustness, and the robustness of computed load control, in the presence of uplink-downlink partially-correlated time-varying shadow fading which models link imbalance in real networks. However, authors did not quantify and compare the performance degradation across various technologies in the presence of link imbalance neither the physics of the phenomena leading to link imbalance.

Su, G. M., Han, Z., Wu, M., and Ray Liu, K. J. (2007), propose a network aware and source aware video streaming framework to support interactive multiuser communications within single-cell and multicell IEEE 802.11 networks. This framework integrates cross-layer error protection mechanism and performs dynamic resource allocation across multiple users and explores the diversity of content complexity exhibited by different video sequences and the heterogeneity of uplink and downlink channel conditions experienced by different users. Unfortunately, this paper only addresses the resource reservation issue for WLAN networks.

Although the above-mentioned papers address resource reservation and handoff issues in heterogeneous networks, none of them deals with resource reservation in various access networks. Further, to the best of our knowledge, there is no previous work addressing dynamic radio resource reservation for vertical handoff optimization that takes into account both future mobile terminal location and load conditions of candidate access networks.

In this paper, we aim to propose a novel handoff management scheme which handles resource reservation for uplink and downlink for control channels taking into account future mobile terminal location and load conditions of access networks. Our contributions are three-fold:

- First, we propose a mobility model to estimate future mobile terminals' locations using an extrapolation technique and we then make use of an error estimation model,
- Second, we propose a novel resource management scheme and a new approach for handling asymmetry between uplink and downlink in our resource management algorithms,
- Third, we propose a scheme to evaluate handoff latency and packet loss and carry

out simulations in order to prove that our solution outperforms other solutions.

The remainder of this paper is organized as following: in the first section, the mobility pattern of the mobile terminal is modeled. In Section 2, computations related to the predictive location probabilities are detailed. In section 3, the resource reservation procedure and vertical handoff algorithms are presented. Simulation results are analyzed in section 4 and conclusions are provided in section 5.

MOBILITY MODELING

Movement Extrapolation

In this section we describe how terminal mobility is modeled using an extrapolation technique. When a handoff trigger is detected, the predicted location is used to determine the probable location of the mobile terminal just before the handoff and to reserve an appropriate amount of radio resources in candidate access networks in advance.

We suppose that the mobile terminal crosses its path following a sequence of linear segments and traverses each of them at a constant speed ν. At the end of a given segment, the mobile terminal can stop in order to alternate the phase and the speed at which it moves.

Therefore, according to the described mobility pattern, at some random times $0 \leq T_1 \leq T_2 \leq ... \leq T_n$, a new segment is traversed. Let n denotes the total number of linear segments that could be traversed by the mobile terminal. Hereafter, we denote by "mobility step" the movement made by the mobile terminal at the beginning of each period of time $T_j, j \in \{1, n\}$.

At the beginning of each mobility step, the mobile terminal extrapolates its next coordinates and at the end of the same mobility step, the TDOA (Time Difference of Arrival) localization technique is used in order to determine the mobile terminal's real coordinates. The accuracy of the extrapolation method is determined making use of an error estimation method. This way, at each epoch of time, the mobile terminal has a prior knowledge about the localization that could be reached before the end of the mobility step it is going through. This information is particularly important when a handoff is about to occur as it allows the mobile terminal to have an adjusted value of future coordinates without having to wait for new TDOA coordinates. These values will be used in order to associate to each available access network a probability that the mobile terminal moves into this access network and reserve resources in each of the corresponding APs/BSs.

At epoch T_j, the mobile terminal is located at (x_j, y_j). The system extrapolates the mobile terminal's movement using its current location and direction and expects its next location $(\hat{x}_{j+1}, \hat{y}_{j+1})$

This new location is determined using the Lagrange extrapolation method. This method provides accurate and relatively simple previsions based on polynomial functions. This choice is driven by the need for a low complexity, accurate extrapolation formula. A high complexity scheme is incompatible with resources available at a mobile terminal (limited computation capacity, low energy, etc.). This function is a linear combination of Lagrange basis polynomial.

According to this method, given a set of $k+1$ data points $(x_0, y_0)...(x_k, y_k)$, for each $x_{k+1} = x_k + \Delta x$, we compute $\hat{y}_{k+1}$ as follows:

$$\hat{y}_{k+1} = \sum_{j=0}^{k} y_j l_j(x) \tag{1}$$

where $l_j(x) = \prod_{i=0, i \neq j}^{k} \frac{x - x_i}{x_j - x_i}$

Prediction Error

At the end of the jth mobility step, the mobile terminal reaches its destination $\left(x_{j+1}, y_{j+1}\right)$ which is determined using the TDOA triangulation method. The estimated mobile terminal position using the Lagrange extrapolation method $\left(\hat{x}_{j+1}, \hat{y}_{j+1}\right)$ does not necessarily produce the correct geolocation due to discrepancy between real field and the predicted one. The difference between the two fields depends on the accuracy of the extrapolation and therefore, prediction error estimation must be carried out.

The prediction error estimation is performed using the Mean Square Error (MSE) method. Accordingly, the error expression at the jth mobility step is given by the following expression:

$$e_j = \frac{\sum_{k=1}^{j}\left[y_k - \hat{y}_k\right]^2}{j} \tag{2}$$

All the operations described above are performed for each mobility step; therefore, during its course of movement, the mobile terminal calculates the MSE according to the recursion given by equation (3). This equation is obtained expanding equation (2):

$$e_{j+1} = \frac{j}{j+1} e_j + \frac{\left[y_{j+1} - \hat{y}_{j+1}\right]^2}{j+1} \tag{3}$$

Once the MSE at the iteration $j+1$ is computed, we need to determine the accuracy coefficient $C_{acc} \in \left[0,1\right]$ which reflects the precision of the extrapolation technique. The accuracy coefficient value is computed as follows:

$$C_{acc} = 1 - \frac{e_{j+1}}{\overline{\overline{e_{j+1}}}} \tag{4}$$

Where $\overline{e_{j+1}}$ is given by the following equation:

$$\overline{e_{j+1}} = \frac{\sum_{k=1}^{j+1}\left(y_k - \overline{y}_{j+1}\right)^2}{j+1} \tag{5}$$

$\overline{y}$ is the mean of the predicted values of y_k until the iteration $j+1$ given by:

$$\overline{y} = \frac{\sum_{k=1}^{j+1}\hat{y}_k}{j+1} \tag{6}$$

Then C_{acc} can be written as:

$$C_{acc} = 1 - \frac{\sum_{k=1}^{j+1}\left(y_k - \hat{y}_k\right)^2}{\sum_{k=1}^{j+1}\left(y_k - \overline{y}\right)^2} \tag{7}$$

The operations described above are repeated by both the network and the mobile terminal until the latter detects a handoff trigger. Our goal is to make the mobile terminal have a mobility history cache which allows it to predict further moves using the accuracy coefficient particularly when a handoff is about to occur and this cache is cleared immediately after the handoff in order save storage resources. This information is used for prediction of cells or access networks toward which the mobile terminal is going to move.

PREDICTIVE LOCATION SCHEME

System Architecture Design

In order to facilitate the handling of access network heterogeneity, both network and mobile terminal implement the IEEE 802.21 standard. They exchange mobile and/or, network relevant

Table 1. Handoff triggers

Trigger	Description
RSS triggers	RSS_1: first RSS (Radio Signal Strength) level which triggers handoff preparation
	RSS_2: second RSS level which triggers handoff execution
IEEE 802.21 triggers	The message "Link Going Down" for handoff execution
	The message "Link detected" for handoff preparation

information including localization information through the IS (Information Service). In our case, we suppose that this service has sufficient knowledge about the network topology and the relevant information is transmitted to the mobile terminal in a periodic way.

The Handoff preparation, which is the computation of predictive location probabilities, and the handoff execution, are triggered by events presented in Table 1.

Our system architecture is illustrated in Figure 1. This architecture is made up of three collocated access networks namely an IEEE 802.11 WLAN (Wireless Local area Network), a UMTS-TDD (Universal Mobile Telecommunication System - Time Division Duplexing) access network and a WiMAX (Worldwide Interoperability for Microwave Access) network.

In order to approach real working conditions, we make use of a mobility model that presents geographic restrictions. More particularly, we use a model where the mobile terminal's motion is bounded to freeways or local streets in the urban area. This model gives mobile terminals the freedom to change the lane in the road they are moving in. Therefore, a real city map is used. We will use the notations given in Table 2 to describe our model.

Figure 1. System architecture

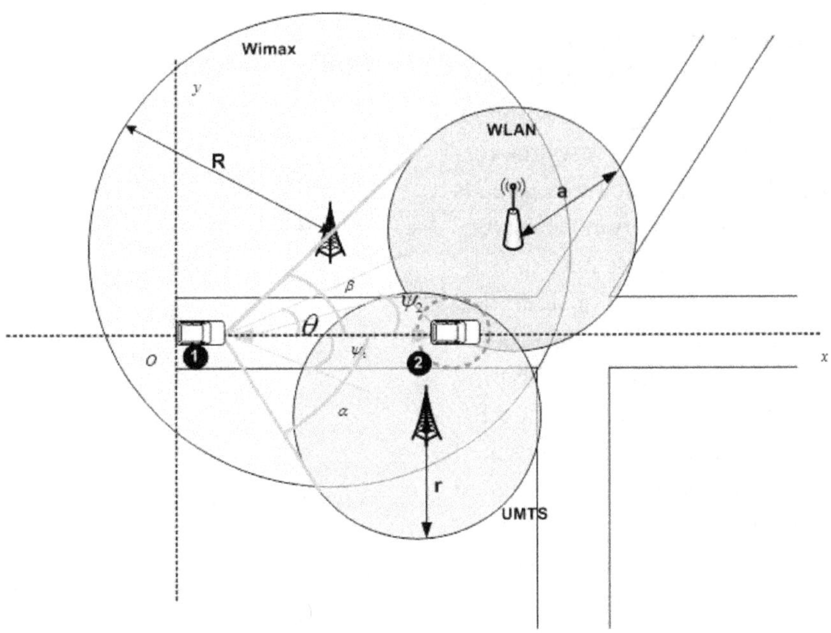

Table 2. Notations and definitions

Notation	Definition
R	The radius of the WIMAX cell
r	The radius of the UMTS cell
a	The radius of the WLAN cell
α	The angle between the horizontal X axis and the border of the UMTS cell
β	The angle between the horizontal X axis and the border of the WLAN cell
ψ_1	The angle between the horizontal X axis and the line passing through the point of intersection of the border of the UMTS cell and the street and the origin of the coordinate system
ψ_2	The angle between the horizontal X axis and the line passing through the point of intersection of the border of the WLAN cell and the street and the origin of the coordinate system

Overlapping Coverage Areas Crossing Probabilities

In next generation networks, providing QoS guarantees with efficient bandwidth utilization is becoming a key issue. During a handoff, a mobile terminal is transferred from one base station to another during an ongoing call. Just before the handoff occurs, an appropriate amount of bandwidth should be allocated in the new cell in order to minimize service disruption.

A successful handoff is performed if the tradeoff between efficient bandwidth utilization and QoS provision is solved. Otherwise, the handoff attempt is blocked and the ongoing call is dropped. To this end, our idea is to develop a resource reservation scheme which aims at reducing overheads among target APs/BSs and at reserving resources in an efficient manner in 4G access networks.

Two factors that are taken into consideration in developing our resource reservation scheme namely the probability of moving to a neighboring cell and the load condition of all neighboring APs/BSs. Hereafter, we will present the details related to computing the motion probabilities associated with each available access network.

When the mobile terminal detects one of the handoff preparation triggers presented in Table 1, we assume that it will mostly hold the same direction. Nevertheless, the mobile terminal can slightly change its initial direction θ. Let τ be the duration of the mobile terminal motion starting from the beginning of the mobility step $j+1$. The Probability Density Function (PDF) of the mobile terminal direction of motion can be assumed to be Gaussian, i.e., the Mobile Terminal's direction of motion can be represented by a Gaussian distribution whose PDF is given by the following formula:

$$P(\theta) = \frac{1}{\sigma\sqrt{2\pi}} \exp\left(-\frac{(\theta-\mu)^2}{2\sigma^2}\right) \tag{8}$$

For simplicity reasons, we have chosen to adopt a standard Gaussian distribution, i.e with $\mu = 0$ and $\sigma = 1$. Then (8) can be written as follows:

$$P_{0,1}(\theta) = \frac{1}{c\sqrt{2\pi}} \exp\left(-\frac{\theta^2}{2}\right), -\pi \le \theta \le \pi \tag{9}$$

Where $c = 2\Phi(\pi) - 1$.

According to the network architecture illustrated by Figure 1, let P_1 be the probability that the mobile terminal moves into the zone where the WiMAX access network overlaps with the UMTS access network. This probability is equal to the relation of the UMTS coverage zone to the WiMAX coverage zone and to the street area bounded by obstacles and sidewalks; multiplied by the coefficient accuracy of the mobility prediction pattern. This probability can be expressed as:

$$P_1 = C_{acc} \times \int P_{0,1}\left(\theta\right) d\theta \qquad (10)$$

which can be written as:

$$P_1 = C_{acc} \times \frac{1}{c\sqrt{2\pi}} \int_{\psi_1}^{\arcsin\left(\frac{rv\tau}{\cos(\psi_1)}\right)-\alpha} \exp\left(-\frac{\theta^2}{2}\right) d\theta \qquad (11)$$

After expansion, it can be written as:

$$P_1 = \frac{C_{acc}}{c}\left(\Phi\left(\arcsin\left(\frac{rv\tau}{\cos(\psi_1)}\right) - \alpha\right) - \Phi\left(\psi_1\right)\right) \qquad (12)$$

Where:

$$\Phi\left(x\right) = \frac{1}{2}\left[1 + erf\left(\frac{x}{\sqrt{2}}\right)\right]$$

and

$$erf(x) = \frac{2}{\sqrt{\pi}} \int_0^x \exp(-t^2) dt \qquad (13)$$

Then, equation P_1 can be written as:

$$P_1 = \frac{C_{acc}}{c}\left(1 + \frac{1}{2}\left(erf\left(\frac{\arcsin\left(\frac{rv\tau}{\cos(\psi_1)}\right) - \alpha}{\sqrt{2}}\right) + erf\left(\frac{\psi_1}{\sqrt{2}}\right)\right)\right) \qquad (14)$$

Let P_2 be the probability of the mobile terminal moves to the zone where a WiMAX and a WLAN access network coexist. This probability can be expressed as:

$$P_2 = \frac{C_{acc}}{c}\left(1 + \frac{1}{2}\left(erf\left(\frac{\beta - \arcsin\left(\frac{av\tau}{\cos\psi_2}\right)}{\sqrt{2}}\right) + erf\left(\frac{\psi_2}{\sqrt{2}}\right)\right)\right) \qquad (15)$$

P_3 is the probability that the terminal moves into the zone where the three access networks coexist simultaneously. Considering that the events whose probabilities are respectively P_1 and P_2 are independent, P_3 can be written as $P_3 = C_{acc} \times P_1 \times P_2$.

THE DYNAMIC RESOURCE RESERVATION ALGORITHMS

Minimizing call dropping probability during a handoff directly relies on maximizing handoff success probability. This can be achieved by providing as many available channels as handoff requests and therefore by reserving sufficient bandwidth to control channels conveying handoff requests.

The resource reservation scheme we propose in this paper focuses on dynamically increasing the proportion of control channels into the frame and reserving radio resources for them only when a handoff is about to occur and then releasing them when the handoff is completed according to the algorithms presented hereafter. Control channels convey handoff requests from the mobile terminal and the response from the network. So, the frame composition is changed in such a way that these channels are prioritized and handoff attempts have more chances to be successful. Depending on data structure of each access network, more time slots are reserved for control channels. The resource reservation scheme is initiated by the network.

We will now proceed to presenting the details of the Dynamic Resource Reservation Algorithms (DRRA) for each access network.

DRRA for UMTS Networks

Reservation is made in a UMTS-TDD base station in order to enhance handoff performances by reserving a given amount of radio resources for uplink/downlink control channels which will be conveyed to the network. The resource reservation algorithm aims to dynamically adjust the frame composition and reserve a fraction of time slots for control channels such as Random Access CHannel, Uplink Shared Channel, and Downlink Shared Channel. This algorithm takes into consideration location probabilities determined in the previous section and the frame structure. The uplink/downlink load factors are determined as follows:

Uplink Load Factor

The uplink load factor is determined by the following equation (Smith, C. & Collins, D. (2007)):

$$F_{UMTS}^{k,u} = \left(1+i\right)\sum_{j=1}^{N} \frac{1}{1+\left(\dfrac{C}{a_j \times R_j \times \left(E_b / N_0\right)_j}\right)} \tag{16}$$

where:

- $i = \dfrac{nearby\,cell\ interference}{local\ cell\ interference}$,
- N: The number of users in the cell,
- C: The chip rate,
- E_b / N_0 : The signal energy per bit divided by noise spectral,
- a_j: The activity factor,
- R_j: The user data rate,

Downlink Load Factor

The downlink load factor is determined by the following equation (Smith, C. & Collins, D. (2007)):

$$F_{UMTS}^{k,d} = \left(1-\gamma+i\right)\sum_{j=1}^{N}\left(\frac{E_b}{N_0}\right)_j \left[\frac{a_j \times R_j}{C}\right] \tag{17}$$

where γ denotes the average value of orthogonality and i denotes the ratio of other cell to own base station received interference by use.

The global load factor associated to cell k is given by (Smith, C. & Collins, D. (2007)):

$$F_{UMTS}^{k,g} = \varpi F_{UMTS}^{k,u} + \left(1-\varpi\right) F_{UMTS}^{k,d} \tag{18}$$

Where ϖ is the weight of the uplink traffic in the whole traffic of cell k.

Resource Reservation Scheme

For dynamic resource reservation, we propose to reserve N_{UMTS}^{k} time slots for control channels at each handoff attempt according to the following scheme:

$$\begin{cases} N_{UMTS}^{k} = X_{umts}^{k} - \left[X_{umts}^{k}\right] TS, \ \ if\ X_{umts}^{k} \geq TS-1, \\ N_{UMTS}^{k} = \left\lfloor X_{umts}^{k} \right\rfloor, \ \ otherwise. \end{cases}$$

where:

$$X_{umts}^{k} = \frac{P_{UMTS}^{k}}{F_{UMTS}^{k,g}} \tag{19}$$

and:

- $\left\lfloor X_{umts}^{k} \right\rfloor$ denotes the floor of X_{umts}^{k},

- TS denotes the number of time slots in the UMTS cell,
- $\lfloor X^k_{umts} \rfloor$ denotes the quotient of the Euclidean division of X^k_{umts} by TS .

The DRRA we developed for UMTS networks is presented in Table 3:

DRRA for WLANs

The DRRA for the WLAN focuses on reserving resources for control frames both in the downlink and uplink. This will help shortening contention periods and avoiding collision problems related to these specific frames especially with the CSMA/CA technique.

The resource reservation schemes for WLAN aim to efficiently manage the contention windows in such a way that the periods where mobile terminals are listening to the communication channel are reduced. During the period where a mobile terminal requests a handoff to the network, $DIFS$ (Distributed Inter Frame Space) interval is slotted in a determined number of intervals, say N^k_{WLAN}, which depends on parameters such as the location probabilities, the load factor of APs, etc. Similarly to UMTS networks, for WLANs, an indicator reflecting the access point's utilization is defined: "The busyness ratio". The busyness ratio for a given AP is given by the following expression according to Zhai, H., Chen, X., & Fang, Y. (2005):

$$B_k = 1 - \frac{p_k \varsigma}{p_k \sigma + p_s T_{succ} + p_c T_{col}} \quad (20)$$

Where ς is the length of an empty backoff time slot, p_k the probability that the observed backoff time slot is idle, p_s the probability that there is one successful transmission, p_c the collision probability and

$$T_{succ} = \overline{data} + ack + sifs + difs \quad (21)$$

$$T_{col} = \overline{data^*} + ack\_timeout + difs \quad (22)$$

Where $\overline{data}$ and $\overline{data^*}$ denote the average length, in seconds, for the successful transmission and collision of the data packets respectively.

Related to AP k, N^k_{WLAN} is determined using the following formula:

$$N^k_{WLAN} = X^k_{wlan}, \text{ where } X^k_{wlan} = \frac{P^k_{WLAN}}{B^k_{WLAN}} \quad (23)$$

The length of the contention interval $DIFS$ when a control frame must be transmitted (Probe Request, Probe Response, Authentication, Re-association Request, Re-association Response) is then given by the following equation:

$$\begin{cases} DIFS = K^k_{WLAN}, & if\ N^k_{WLAN} \geq 1, \\ DIFS = K^k_{WLAN} - \lfloor K^k_{WLAN} \rfloor DIFS_{ini} & else. \end{cases} \quad (24)$$

where:

$$K^k_{WLAN} = \frac{DIFS_{ini}}{N^k_{WLAN}} \quad (25)$$

And $DIFS_{ini}$ denotes the initial value of $DIFS$ before the execution of the DRRA, $\lfloor K^k_{WLAN} \rfloor$ denotes the quotient of the Euclidean division of K^k_{WLAN} by $DIFS_{ini}$.

The DRRA we developed for the WLANs is presented in Table 4.

Table 3. DRRA for UMTS

1. While being connected to network j ($j \neq$ UMTS)
2. if *handoff* preparation *trigger received* then
3. if n > 1 and an UMTS network is available then
4. Compute locations probabilities for each UMTS cell P^{k}_{UMTS}
5. Compute load factors for each UMTS cell $F^{k,g}_{UMTS}$
6. Compute the amount of reserved resources for each cell N^{k}_{UMTS}
7. Reserve N^{k}_{UMTS} Time Slots for control channels in cell k
8. if handoff fails then
9. while $N^{k}_{UMTS} < TS - 1$
10. Increment N^{k}_{UMTS} and
11. Increment control channel periodicity
12. else
13. Release reserved resources
14. endif
15. else
16. If *handoff execution trigger received* then
17. Perform horizontal handoff
18. else
19. Stay connected to network j
20. endif
endif

DRRA for WiMAX Networks

For the WiMAX network, we suppose that modulation and coding technique used is the OFDMA (Orthogonal Frequency-Division Multiple Access), and we chose to use the TDD mode because of its advantages.

The algorithm will focus on dynamically reserving more symbols for control channels. The amount of reserved resources will essentially depend on the handoff probability associated to the WiMAX access network and on the load of the target BS. The DRRA for the WiMAX network is presented in Table 5.

Table 4. DRRA for WLAN

1. While being connected to network j ($j \neq$ WLAN)
2. **if** *handoff preparation trigger received* **then**
3. **if** $n > 1$ and a WLAN network is available **then**
4. Compute location probabilities for each WLAN AP P^k_{WLAN}
5. Compute the busyness ratios for each WLAN AP B^k_{WLAN}
6. Estimate the amount of reserved resources for each WLAN AP N^k_{WLAN}
7. **if** handoff fails **then**
8. **while** $DIFS < DIFS_{ini}$ **then**
9. Increment N^k_{WLAN}
10. **else**
11. Release reserved resources
12. **endif**
13. **else**
14. **if** *handoff execution trigger received* **then**
15. Perform horizontal handoff
16. **else**
17. Stay connected to network j
18. **endif**
endif

ASYMMETRY OPTIMISATION

A substantial amount of research work has investigated downlink and uplink asymmetry in wireless access networks and has associated it with resource reservation issues. For example, Yun, M., Kim, Zhou, Arora, Gomes, & Choi, (2009), developed a new uplink resource management scheme in heterogeneous networking environments that supports multiple radio access technologies. Giupponi, Agust, Pérez-Romero & Sallent, (2006) presented a joint radio resource management strategy for multi-radio access technologies based on fuzzy neural methodology. They showed and discussed, the way the proposed framework deals with the multiservice allocation in heterogeneous scenarios. Authors also presented an approach based on predicting future joint radio resource management decisions and on reserving radio resources for handoff calls. Results in term of radio access technologies assignment percentage are discussed for the uplink.

Table 5. DRRA for WiMAX

1. While being connected to access network j ($j \neq$ WiMAX)
2. **if** *handoff preparation trigger received* **then**
3. **if** n > 1 and a WiMAX network is available **then**
4. Compute location probabilities for each WiMAX cell P^{k}_{wimax}
5. Compute load factors for each WiMAX BS F^{k}_{wimax}
6. Compute the amount of reserved resources for each WiMAX cell N^{k}_{wimax}
7. **for** $i = 1,...,N$ **do**
8. Reserve $N^{k,i}_{wimax}$ Time Slots for control channels in cell k
9. **if** handoff fails **then**
10. Increment $N^{k,i}_{wimax}$
11. **else**
12. Release reserved resources
13. **endif**
14. **else**
15. **If** *handoff execution trigger received* **then**
16. Perform horizontal handoff
17. **else**
18. Stay connected to network j
19. **endif**
20. endif

To address resource reservation issues for both uplink and downlink differently in an heterogeneous networking environment, we propose a reservation scheme which manages traffic asymmetry by assigning a different amount of radio resources to uplink and downlink according to traffic load of cells that could be chosen for the next handoff. The proposed scheme takes into consideration, in addition to traffic asymmetry characteristics for different access networks, the aforementioned parameters namely, the predictive location probabilities and the individual traffic load per target cell. Accordingly, adjacent cells may allocate radio resources independently according to their traffic load and to the predicted future location of the mobile terminal. Let us now describe our proposed scheme for resource reservation for both uplink and downlink in order to enhance overall efficacy and improve handoff performance. In particular, the objective is to find the best tradeoff between downlink and uplink control channel allocation that maximizes the handoff success probability.

TDD Uplink Resource Reservation

The resource reservation strategy described in the previous section allocates a certain amount of radio resources for control channels without differentiation between uplink and downlink. But in reality, asymmetry between uplink and downlink must be taken into account and properly handled in order to increase system performances and handoff success probability. Optimal utilization of radio resources by reserving them, taking into account different uplink/downlink characteristics, allows a fair scheduling and balanced system capacity utilization.

Our idea is to propose a novel uplink/downlink reservation scheme based on previous ideas and on the request type issued by the mobile terminal or by the network. Dealing with handoff, we assume that the mobile terminal is essentially under one of these states: requesting a handoff or receiving a handoff response. Let T_{ho_pr} denote the time when the handoff preparation begins, T_{ho_suc} denote the time when the handoff succeeds and T_{ho_cp} the time when the handoff procedure finishes. Then, the handoff request occurs between T_{ho_pr} and T_{ho_suc}, the handoff response is generated between T_{ho_suc} and T_{ho_cp}.

Depending on the state of the mobile terminal, more radio resources will be required either for the uplink or for the downlink. Obviously, more radio resources are needed when the mobile terminal is requesting a handoff, i.e., when a new channel is granted in the cell chosen for the next handoff. So, more control channels are needed during $\left[T_{ho\_pr}, T_{ho\_suc}\right]$.

For TDD networks, the system utilization in term of control channels is the sum of the downlink and the uplink utilization relating to control channels. We define the utilization ratio of cell k for access network j, $\Delta\rho_j^k$ as:

$$\Delta\rho_j^k = \frac{N_j^{k,u}}{N_j^{k,d}} \qquad (26)$$

where $N_j^{k,u}$ and $N_j^{k,d}$ denote the amount of reserved resources for the uplink and the downlink, respectively. One can note that:

$$N_j^k = N_j^{k,u} + N_j^{k,d} \qquad (27)$$

where N_j^k denotes radio resources reserved according to the algorithms given in Table 4 and Table 5, and 6. According to equations (26) and (27) and to the resource reservation procedure described in the previous section, the amount of resources reserved for the uplink in TDD systems is given by the following equation:

$$N_j^{k,u} = N_j^k \frac{\Delta\rho_j^k}{\Delta\rho_j^k + 1} \qquad (28)$$

In order to keep up with allocation constraints defined above, we must have $\Delta\rho_j^k > 1$. The initial value of $\Delta\rho_j^k$ is set to one, i.e. $\Delta\rho_j^k = 1$. We define a discrete time scale t that increments by one whenever the handoff attempt fails. Accordingly, $\Delta\rho_j^k$ is incremented by one until reaching the time T_{ho_suc}. This amount of radio resources is being reserved in all cells detecting the mobile terminal signal since the detection of the first RSS trigger till the success of the handoff request.

TDD Downlink Resource Reservation

Downlink resources are likely to be more needed when the network side will transmit a handoff response. At $t = T_{ho\_suc}$, the value of $\Delta\rho_j^k$ is reset to one. The amount of resource reserved in the downlink is given by:

$$N_j^{k,d} = N_j^k \left(\frac{1}{\Delta\rho_j^k + 1} \right) \quad (29)$$

Starting from T_{ho_suc}, until T_{ho_cp}, t is incremented and $\Delta\rho_j^k$ accordingly until the channel allocation procedure is completed.

Uplink Resource Reservation in WLAN

According to previous sections, prioritizing a control channel is achieved by a novel cooperative channel prioritization algorithm. We proposed three different algorithms for the purpose of improving spectrum efficiency.

In the same context, we propose a scheduling technique, specific to IEEE 802.11 networks, where the uplink utilization is being enhanced through a technique which reduces collisions. This technique proposes to further reduce the contention window either for the downlink or for the uplink according to the request being transmitted by the network or by the mobile terminal. The number of required time intervals for the uplink is given by:

$$N_{wlan}^{up,k} = \frac{DIFS}{\Delta\rho N_{wlan}^k} \quad (30)$$

where $\Delta\rho = \frac{uplink\ utilization}{downlink\ utilization}$ is normalized system utilization defined as the fraction of time the channel is used to transmit on the uplink, divided by the fraction of time the channel is used to transmit on the downlink.

Downlink Resource Reservation in WLAN

For the downlink, the proportion of resources given to uplink is given as following:

$$N_{wlan}^{k,d} = N_{wlan}^{k,j} - \frac{DIFS}{\Delta\rho N_{wlan}^{k,j}} \quad (31)$$

As mentioned before, the system architecture implements the IEEE 802.21 standard. This standard relies on a set of handoff enabling functions among which the MIHF (Media Independent Handover Function) which provides three types of services, namely MIES (MI Event Service), MICS (MI Command Service), and MIIS (MI Information Service). Each service is responsible of a set of specific commands. MIH primitives optimize handoff latency and suggest the method of link layer trigger. In order to trigger the dynamic adjustment of the frame composition, we propose to add a new primitive called "MIH-FRAME-ADJUST" to the MIH set of primitives. This primitive allows AP/BS to dynamically adjust uplink and downlink frame structure / or to shorten contention window in WLANs according to the aforementioned algorithms.

In order to start off the scheduling scheme for the uplink, the system broadcasts a new message to the concerned AP/BS called "MIH-UPLINK-ADJUST". For the downlink, at T_{ho_suc} the system broadcasts a primitive called "MIH-DOWNLINK-ADJUST".

HANDOFF LATENCY AND PACKET LOSS ESTIMATION

In heterogeneous networks, most significant metrics for evaluating handoff performances are latency and packet loss. Since data packet could be lost during the handoff period, this may not meet the QoS requirements for real time applications. This finding is particularly important in heterogeneous and multi-hop networks as mobility management protocols may be different across the various access networks. This is why we propose a framework to evaluate the handoff latency and the packet loss for the studied architecture. Simulations which will be carried out using this framework will demonstrate the performance of our proposed mobility and resource management scheme.

Handoff Latency

As describe earlier, our heterogeneous architecture comprises three access networks: a WLAN access network, a UMTS access and a WiMAX network. The Handoff latency is defined as the time interval during which a mobile terminal cannot receive and transmit any packet when switching from an access network to another (Yu, M.J., Choi, S.G., Kim, H.S., and Cho, K.S. (2011)). In general, the total handoff latency is divided into four components (George, A., Rashad, S., Kumar, A. and Kantardzic, M. (2008)): 1) the time taken by the mobile terminal to detect that it lost connectivity to the AP/BS denoted T_{dt}, 2) the time to discover a relay gateway that can offer connectivity to the new AP/BS denoted T_{dg}, 3) the transmission delay in sending control packets denoted T_{tx}, 4) the propagation delay denoted T_{prop}.

Therefore the total handoff latency can be written as the sum of all these components $\left(T_{handoff} = T_{dt} + T_{dg} + T_{tx} + T_{prop}\right)$.

For our framework and with the use the IEEE 802.21 MIH services, network entities, are informed about the relevant parameters necessary in handoff decision making prior to the handoff process. This is why, some hypothesis should be made:

- Due to the handoff triggers listed in Table 1, the mobile terminal is able to detects that its connectivity is going down, so T_{dt} is equal to zero.
- Since the next network to select after the handoff is determined in advance, T_{dg} can also be assimilated to zero.

So, the previous equation can be written as:

$$T_{handoff} = T_{tx} + T_{prop} \tag{32}$$

Where:

$$T_{prop} = T_{prop-AP/BS} + T_{prop-IP} \tag{33}$$

Under normal traffic conditions, an AP/BS does not have the capacity to process all received data/control packet at the same time. This is why queuing theory is generally used to model the AP/BS behavior. Let us assume that packets arrival and service at a particular AP/BS follows a Poisson distribution and that they are processed and sent randomly. These packets are divided into different classes ranging from 1 to K having different priorities. Then, the transmission delay at a particular AP/BS is modeled by an M/G/1 queue characterized by λ_k (the packet arrival rate at the AP/BS for the class of traffic k) and by μ_k the service rate for the class of traffic k).

According to the Pollaczek-Khinchin mean formula, the total latency at a given transmission

T_{tx} which is the time taken to process and transmit packets can be written as:

$$T_{tx} = E\left[S_k\right] + \frac{\lambda E\left[S_k^2\right]}{2\left(1 - \rho_k\right)} \tag{34}$$

Where:

$E\left[S_k\right]$ is the individual service time for the class of traffic k.

$E\left[S_k^2\right]$ is the second moment of S_k for the class of traffic k.

$\rho_k = \frac{\lambda_k}{\mu_k}$ is the transmission node load for the class of traffic k.

Using the previous notations, (34) can be written as:

$$T_{tx} = \frac{\rho_k}{\lambda_k}\left(1 + \frac{\rho_k}{2\left(1 - \rho_k\right)}\right) \tag{35}$$

In the third section we proposed a new scheme which prioritizes control channels in comparison with other type of traffics. Therefore the class reserved to control traffic has the most important priority. Furthermore, the target AP/BS for a handoff has, according to the DDRA algorithms, the lowest load. Besides, control packets have the highest priority which is proportional to the amount of reserved resources.

Referring to the contribution of George, A., Rashad, S., Kumar, A. & Kantardzic, M. (2008), we can make the following assumptions:

$T_{prop-AP/BS}$ is estimated as $1\mu s$.

$T_{prop-IP}$ is assumed as $100\mu s$ as an upper bound.

Packet Loss

As the mobile terminal loses connection for the handoff period ($T_{handoff}$), some data packets may be lost, which degrades definitely the QoS in the access network. So, we propose a mobile terminal driven buffering mechanism in order to avoid packets loss when a user moves across heterogeneous networks. The scheme prevents packet loss by proactively buffering packets that will be lost in a particular AP/BS. Buffers are implemented at each access node and packets start getting buffered as soon as a handoff preparation trigger is detected.

Depending on buffer size and buffering period, this alternative allows minimizing the amount of packet lost and enhances network performances which might be required to meet particular QoS requirements.

Let η be the buffer size of a particular network node and, D the average throughput for the studied access networks, PS the packet size, t_{pck} the packet interval. When a handoff preparation trigger is detected, packets buffering will start. Packets are lost only if:

$$\eta \leq \frac{D \times \left(T_{handoff} - t_{pck}\right)}{PS} \tag{36}$$

It can be observed from this inequality that he higher the handoff latency is, the higher the rate of packet loss will be. The number of packets lost can be estimated as follows:

$$N_{loss} = \eta - \frac{D \times \left(T_{handoff} - t_{pck}\right)}{PS} \tag{37}$$

From this equation, it is reflected that the value of packet loss is affected by several parameters such as the buffer size, the handoff latency, the packet interval but the most important thing is that this value is directly related to the performances of our DRRA algorithms through the load of the system and the location of the mobile terminal.

SIMULATION AND ANALYSIS

The model presented in the previous sections is used to evaluate the proposed dynamic resource reservation schemes. In this section, we present and discuss the simulation results for the topology of Figure 1. To the best of our knowledge, there is no previous work that considered both the predictive location probabilities and the load factors for heterogeneous networks.

The simulations focus on both predictive location probabilities and network utilization relating to the amount of reserved resources.

Predictive Location Probabilities

As shown earlier, the location probabilities associated to the three access networks of our network topology depend on the speed of the mobile terminal. In order to study the effect of the mobile terminal speed on the predictive location probability, we consider different values for the mobile terminal speed, the behavior of the predictive location probabilities related to each access network (UMTS, WLAN and WiMAX) is then computed. The system parameters are given in Table 7.

Figure 2 shows the relationship between the predictive location probabilities and the speed of the mobile terminal. We suppose for these simulations that mobile terminal's speed varies from 1 to 10m/s.

This figure shows that for a particular zone, the predictive location probability increases as mobile terminal's speed increases. This implies that, for a mobile terminal moving with high speed, the handoff should be initiated earlier as compared to a slowly moving mobile terminal despite the access network technology.

Dynamic Resource Reservation

In this section, we will focus on the simulation of the dynamic resource reservation schemes proposed in previous sections. Simulations will consider parameters presented in Table 7 and the system architecture of Figure 1. The percentage range for both load factors and location probabilities is 0-100% for the UMTS and WLAN access networks.

The simulation results for the UMTS access networks are presented in Figure 3 where we plot the overall reserved resources for a UMTS candidate BS to a handoff process versus the predictive location probabilities and load factors of this BS.

It can be seen that a large amount of radio resources is reserved for the UMTS target BS with high location probabilities and low load factor values. However, when the BS is overloaded and the location probabilities are low, the amount of reserved resources does not exceed 10% according to the algorithm presented in Table 3. For example, when the load factor is equal to 5%, and the predictive location probability associated to this UMTS BS equal to 95%, up to 80% of resources of this BS can be reserved for control channels in term of control times slots compared to data time slots.

Table 7. Simulation parameters

Parameter	**R**	**r**	**a**	α	β	ψ_1	ψ_2
Value	2000m	500m	100m	30°	20°	25°	60°

Figure 2. Predictive location probabilities

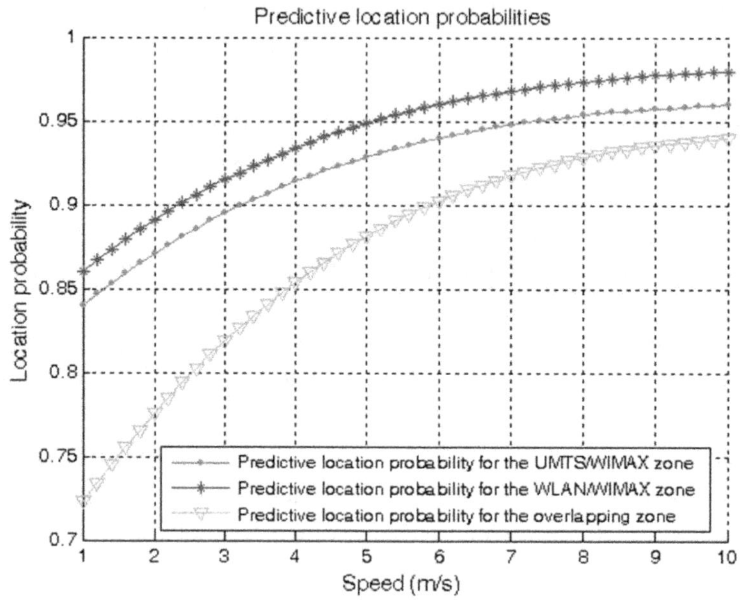

Figure 3. Reserved resources for a UMTS BS

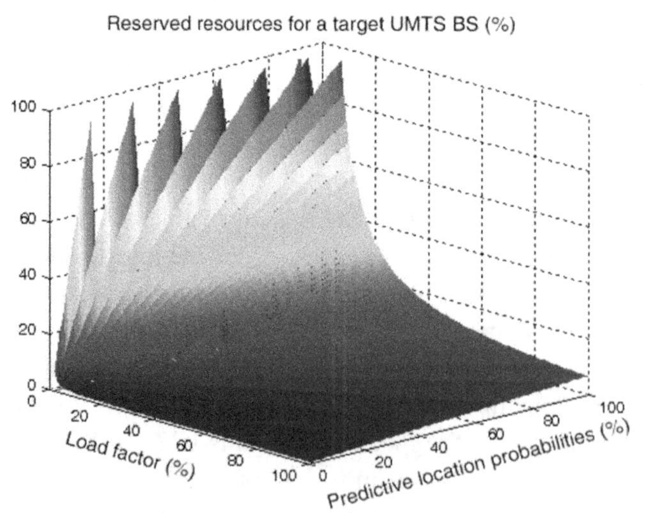

Similarly, when the load factor increases from 5% to 80% and the location probability decreases from 95% to 30%, only 2% of the UMTS frame time slots are reserved for both uplink and downlink control channels for a handing off mobile terminal. In this case, since the target BS is overloaded, data traffic should not be disrupted and therefore resource reservation should be as low as possible.

Figure 4 shows the performance of the DRRA for a WLAN access network. As seen in this figure, the initial *DIFS* interval is shortened as the busyness ratio decreases, and as the predictive location probability increases. For example, the proposed scheme reduces the length of the $DIFS$ interval by 45% when the load factor is equal to 40% and the location probability associated to the target WLAN AP equal to 65%. This means that $DIFS$ interval is slotted making use of a high factor in such a way that a handing off mobile terminal sends its control frames after a space which is further reduced compared with the case when it sends data. Indeed, with this newly proposed scheme, a handoff request has more chances to be successful.

As shown in Figure 3 and Figure 4, the proposed scheme may contribute effectively in improving the overall handoff performance (through the reduction of handoff latencies) while efficiently managing the network utilization.

We will now simulate the utilization of uplink and for a given future location of the mobile terminal. We study the effect of both the overall load and the number of iterations in the DRRA on the uplink utilization based on a fixed value of the probability associated to the UMTS network (84%). For TDD networks, namely UMTS and WiMAX, uplink and downlink resource reservation takes into account the type of transmitted request either by the mobile terminal or by the network and reserves accordingly an adequate amount of bandwidth.

Uplink Utilization

Figure 5 shows the radio network resources reserved for the uplink in TDD networks for different values of N_j^k and $\Delta\rho$. Figure 5 represents the evolution of the distribution of radio resources in the TDD frame with emphasis on the uplink utilization for different values of link sharing. The curve increases with the different scenarios then become stable nearly around 50%. For the lowest values of the individual amount of resources allocated to a particular mobile station (ie, for the highest values of network load and consequently the highest values of mobile terminals), the values of the uplink utilization are also relatively low. As expected, the resources reserved for the uplink, i.e., the amount of resources reserved for control

Figure 4. Reserved resources for a WLAN AP

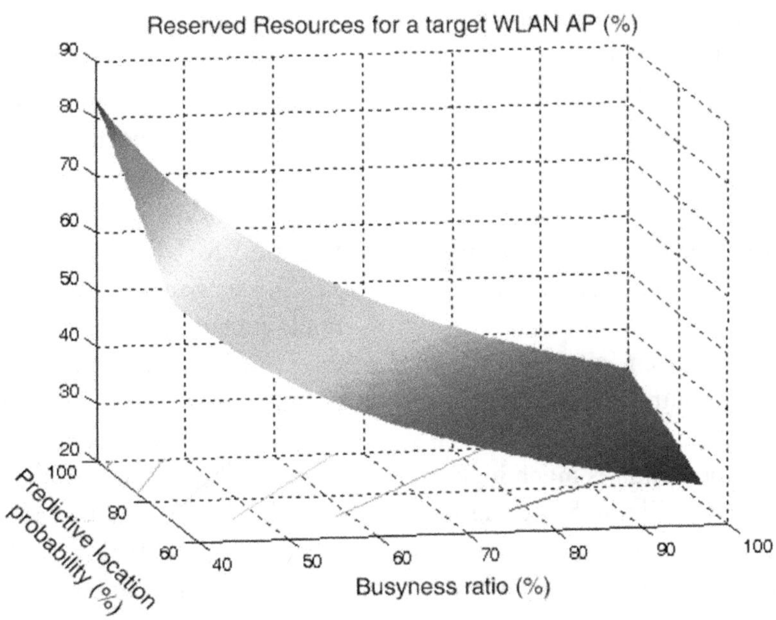

Figure 5. Uplink utilization for TDD networks

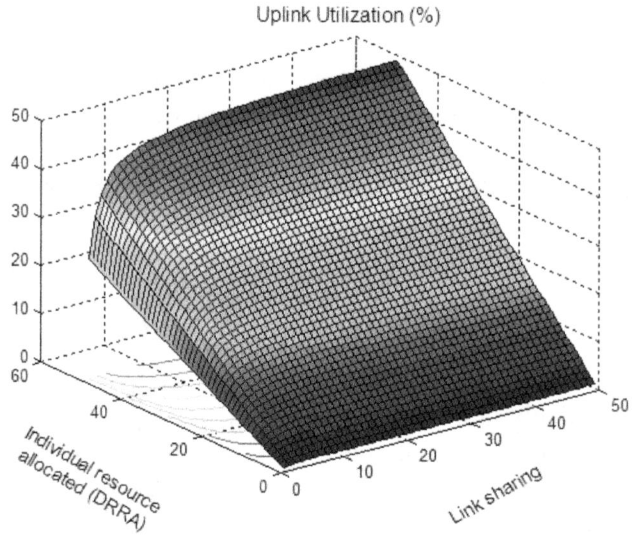

channels in the uplink increase as the utilization ratio increases and as the overall load of the network decreases. The reserved resources increase continuously for different values of the link sharing with a slight advantage for the values that are less than10.

Downlink Utilization

The system utilization of analytic and simulation results for the downlink are plotted in Figure 6 for different values of $\Delta\rho$ and N_j^k respectively. The downlink utilization becomes larger as the amount of individual resources allocated using DRRA increases. However, as the value of utilization factor increases, the downlink utilization tends towards a constant value approximately around 1%.

This is because the system allocates some radio resources for the control messages going from the network to the mobile terminal even though the priority of the uplink is much higher than the downlink one. For Figure 5 and Figure 6 illustrating both uplink and downlink utilization in TDD systems, one can note that significant variations in the amount of links utilizations for both uplink and downlink are recorded for <10. This is because of the frame size limits which should be taken into consideration.

WLAN Uplink Utilization

Figure 7 shows the simulation results of the optimization technique according to the changes of N_j^k and $\Delta\rho$. As the value of utilization ratio increases, the uplink utilization for transmitting handoff requests increases also. Similarly, as the amount of individual radio resources allocated to a particular mobile terminal increases and when the link sharing is near to zero, the downlink utilization for IEEE802.11 networks increases too. This is because the amount of decrement, which is caused by frame retransmissions, gets larger when the value of $\Delta\rho$ increases and because the load under these conditions is maximum, so uplink utilization for control channel is also very important. Thus, more frames are transmitted in the uplink which increases the utilization and decreases the collision probability. For these simulations, the value of $DIFS$ is set to 50μs

Figure 6. Downlink utilization for TDD networks

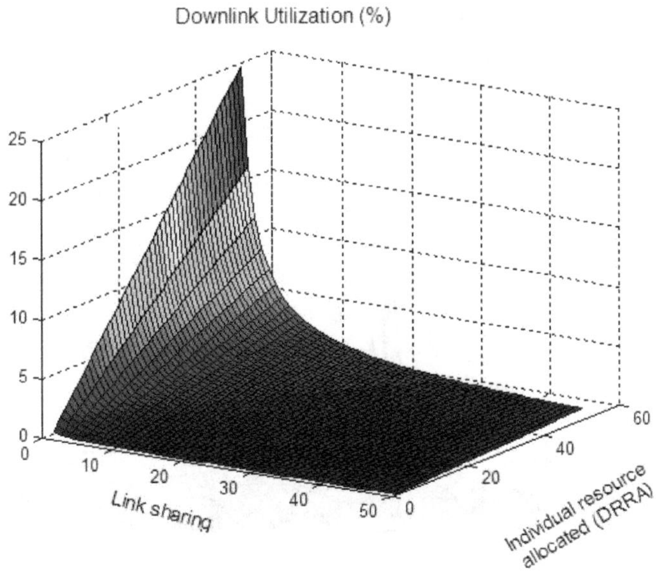

and the value of the probability associated to a WLAN coverage zone to 73%.

WLAN Downlink Utilization

Figure 8 shows the simulation results of dynamic parameters update for both N_j^k and $\Delta\rho$. As the value of increments of the utilization ratio increases, the value of the downlink utilization ratio decreases also. It is also shown that the value of $\Delta\rho$ should increase as the downlink traffic load increases to achieve the ideal utilization ratio. The decrease of utilization contributes in the decrease of the collision and hence the enhancement of overall performances.

For Figure 7 and Figure 8 illustrating both uplink and downlink utilization for IEEE 802.21 systems, one can note that there are two remarkable things for both simulation cases. The first observation is that the two shapes show an important variation in the values of the link utilization ratios, for a value of link sharing smaller than 10. The second observation is that for a value of the link sharing greater that 10 the values of utilization ratios obtained and become stable. This is directly related to the frame size constraints.

Handoff Latency

In order to evaluate and measure the performances of our mobility management and resource reservation scheme and to compare them with those of other schemes, we will now simulate the end user performance parameters which are the handoff latency and the packet loss. We refer to our network topology which is made up of three heterogeneous access networks. The goal is prove that our solution with the proposed DRRA algorithms outperforms other solutions. For the simulations related to the handoff latency, we consider an heterogeneous networking environment where the packet size is fixed at 512 bytes and where packet are generated by mobile terminals at a rate of 5 packets/second.

We performed simulations with different scenarios depending on the load at a particular AP/BS, i.e., on the number of mobile terminal generating this total load. The maximum number of mobile terminals is fixed at 25, each of them

Figure 7. Uplink utilization for IEEE 802.11 networks

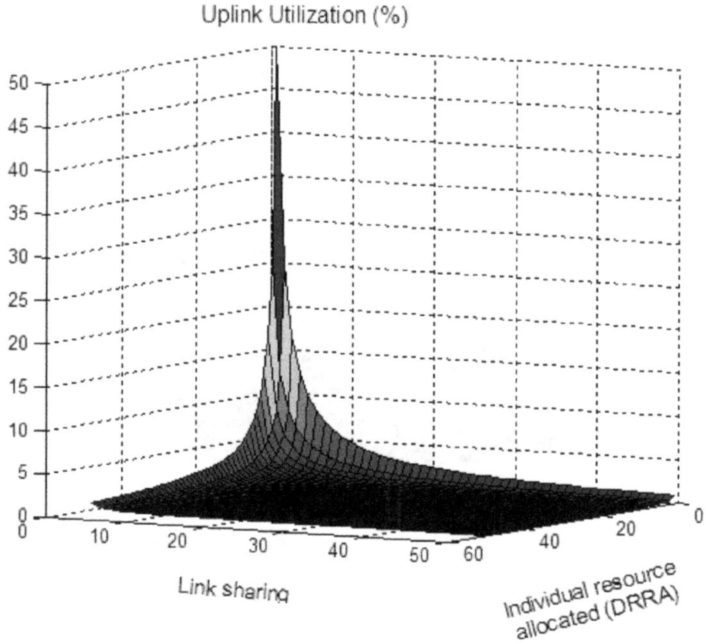

Figure 8. Downlink utilization for IEEE 802.11 networks

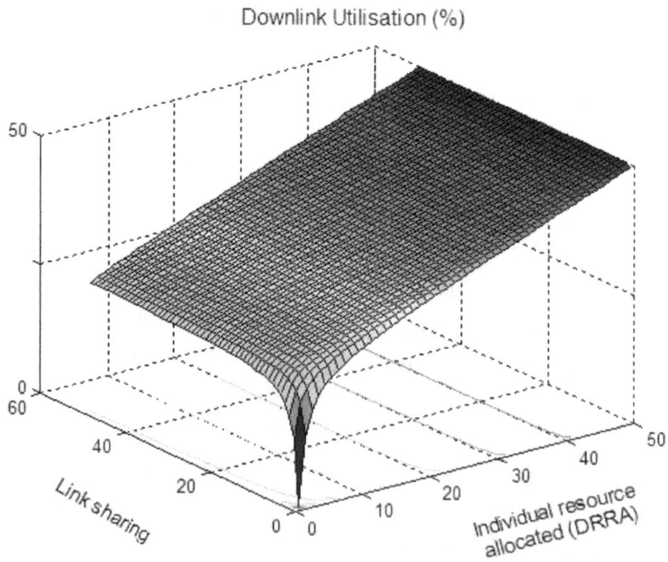

generates a load fixed at 3%. Results obtained from the simulations are depicted in the Figure 9. It can be observed that the results obtained using our solution performs better that those of PMIPv6 with HC, PMIPv6 and FMIPv6 in term of handoff latency. The simulations results for these three protocols investigated by Magagula, L.A., Chan, H.A., and Falowo, O.E. (2010) show that handoff

Figure 9. Handoff latency

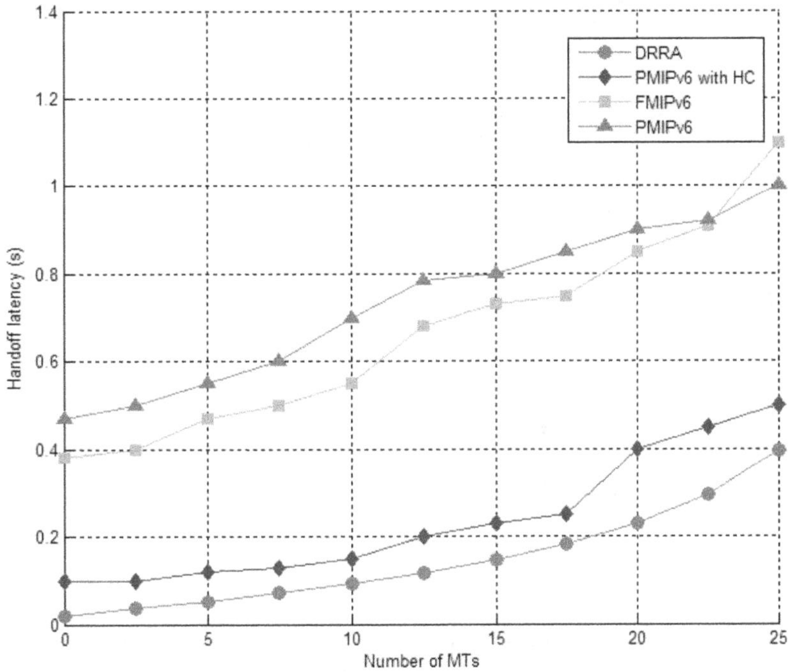

latencies achieved by all the enhancement made to the MIPv6 protocol either with Proxy Mobile IPv6 (PMIPv6) which is a network based network management protocol, or with the Fast Handover for MIPv6 (FMIPv6) which is a host based mobility management protocol, or with the PMIPv6 optimized with a Handover Coordinator (HC) which is a network based mobility management protocol are higher than 0.1 s. For all mobility management scenarios, the handoff latency increases with the number of mobile terminals.

But using our approach, the lowest values of handoff latencies are obtained and the maximum value does not exceed 0,4 s for the highest load or the maximum value of mobile terminals. This is because in our approach the AP/BS selected for a handoff process has the lowest load and is asked to adjust its scheduling scheme in order to prioritize control channel and among other handoff requests in comparison to other types of traffic.

Moreover, our scheme is mobility aware which means that the resource reservation algorithms reserve resources only in AP/BS selected for a handoff process which makes them available for other handoff requests and this contributes in reducing packet processing time and handoff delays.

Packet Loss

Simulation results for packet loss are presented in Figure 10. This figure depicts the impact of the number simultaneous handing off mobile terminals on the average value of packet loss. As expected, results obtained using our approach are better than those achieved with PMIPv6, FMPIv6 and PMIPv6 with HC. With the maximum value of load and with a bandwidth of 100 Mbps, the packet loss does not exceed 5 packets with our approach and can reach 47 with the PMIPv6 for example.

Having these performances gives our approach the ability to satisfy real-time services requirements since the packets are delayed for lower

Figure 10. Packet loss

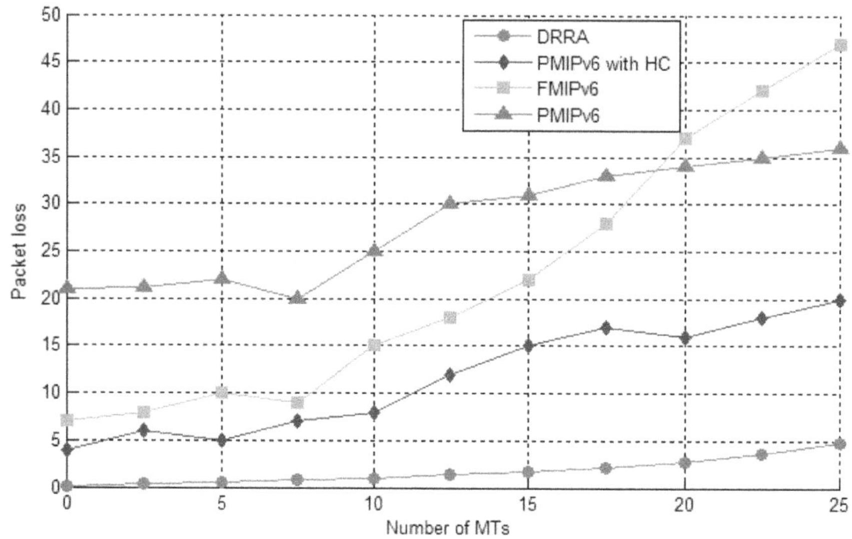

periods. Because the increase of the number of mobile terminals in the network results in more important load at the target AP/BS, the packet loss increases for all the studied schemes. But our solution provides the lowest values.

The enhancement made in comparison with the best results of other solution (achieved by PMIP with HC solution) and with a buffer size as high as 300 packets is up to 25%.

Performance Comparison of the Proposed DRRA to Other Resource Reservation Methods

Although much work has been done on resource reservation in heterogeneous networks, most of it addressed resource reservation for data channels in a particular access network without special interest to the control channels. So, in order to compare the performances of the proposed DRRA, three alternative approaches for resource reservation namely for TDD-OFDAM, WiMAX and WLAN access networks are considered.

The methods considered for comparison purposes are all dealing with resource reservation but each one adopts a different method in a particular access network.

The first approach proposed by Kou and Zhen (2009), deals with Dynamic Uplink/Downlink resource allocation for TDD OFDMA access networks based on Basic Capacity-Based Admission Control (BCBAC). The main objective of this work is to adjust resource assignment according to asymmetry load in order to maximize the utilization of network resource considering cellular network and wireless mesh network. Simulation results, which took into consideration parameters like the standard deviation and the service arrival rate, have shown that when the uplink-load/downlink-load ratio is more adjacent, the system utilization is higher. This approach studies the overall system utilization and the utilization for the uplink and the downlink separately. It can be observed that the system utilization is very close to 1 when the service arrival rate is greater than 4 request/second.

Huang, Chuang, Guan, Yang and Chen (2007), proposed a scheme for self adaptive bandwidth reservation scheme for WiMAX networks based on a probabilistic mobility prediction model to estimate the bandwidth required in neighboring cells. The performance of this approach is measured in terms of the forced termination probability for the handoffs, the call blocking probability for the new connections and bandwidth. Simulation results related to the bandwidth utilization showed that up to 100% of bandwidth is utilized with the mobility prediction based bandwidth reservation scheme when the new arrival rate is equal to 1 request/second.

Sung, Byung, and Yuguang (2005) showed that with their method to control the utilization ratio of uplink and downlink traffic and to enhance the system utilization of the IEEE 802.11 access networks, which is the sum of downlink and uplink utilization can reach 90% when the system utilization which is the sum of the downlink utilization and the uplink utilization is equal to 10. They claim that the utilization sharing between the uplink and the downlink can be controlled by the network operator or by the offered traffic load. The efficiency of the proposed system under saturation load condition has been demonstrated by Markov analysis and computer simulation.

The simulation results comparing our approach to these three approaches for resource reservations in heterogeneous access networks are presented in the following table.

Table 8 shows that our solution outperforms the other methods previously used for resource reservation due to the following considerations:

- Our method deals with more access network technologies than the other approaches,
- Our method reserves radio resources only for control channel and can consequently save these resources. Indeed control channels are fewer than data channels and the amount resource reserved for these channels will be necessarily less than the amount of resource reserved for data channels,
- Our method takes into account more parameters than the other methods. All the parameters considered allow having an optimized resource reservation approach. In fact, considering the network topology allows to predict future movements, then, considering the predictive location probability allows to perform resource reservation only in probable target access networks and considering the traffic load,

Table 8. Comparison between our proposed approach and the other methods

	Type of the reservation	Technology of access networks	Reservation for data/control channels	Links concerned by reservation	Parameters involved in resource reservtation	Bandwidth utilization
Kou and Zhen (2009)	Dynamic	TDD-OFDMA	Data channels	Uplink + Downlink Uplink Downlink	Traffic load Probability of downlink	Up to 100%
Huang, Chuang, Guan, Yang and Chen (2007)	Mobility prediction based	WiMAX	Data channels	Uplink + Downlink	Probability of motion bandwidth for the neighboring cells	Up to 90%
Kim, S.W, Kim, B.S. & Fang. Y (2005)	Dynamic	IEEE 802.11	Data channels	Uplink + Downlink	Traffic load	Up to 100%
Our approach	Dynamic Mobility prediction based	TDD-OFDMA WiMAX IEEE 802.11	Control channels	Uplink + Downlink Uplink Downlink	Traffic load Network topology Predictive location probabilities	Up to 90% for TDD And up to 80% for IEEE 802.11

allows resource reservation only in under loaded AP/BS.

CONCLUSION

In this paper we have investigated the resource reservation problem for an integrated 4G architecture based on predictive location probabilities and load conditions associated to available access networks. We have developed a dynamic resource reservation algorithm for each of these access networks.

Simulation results have shown that the developed algorithms for radio resource reservation efficiently comply with load conditions of the BS/AP where resource reservation must be carried out. This means that traffic exchange disruption is minimized for mobile terminal sending and receiving normal data and that handoff requests are prioritized for handing off mobile terminals. In other words, we have shown that this procedure contributes effectively to reducing handoff disruptions since more radio resources are reserved for control channels in most probable access networks to be selected in the near future.

REFERENCES

Abbasi, A. A., Iqbal, J., & Malik, A. N. (2010). *A novel QoS optimization architecture for 4G networks.* International Conference on Computer Electrical and Systems Science and Engineering.

Achieng, R., & Chan, H. A. (2006). *A queuing theoretic scheme for a QoS-Enabled 4G wireless access*. Retrieved from http://utsescholarship.lib.uts.edu.au

Andi, W. C., Yeo, C. K., & Lee, B. S. (2010). Environment-aware QoS framework for multi-interface terminal. *Journal of Computer Communications*, *33*(9), 1049–1055. doi:10.1016/j.comcom.2010.01.010

Chang, L. H., Wang, C. C., & Lee, T. H. (2010). A handover mechanism using IEEE 802.21 in heterogeneous wireless networks. *Communications in Computer and Information Science*, *77*, 95–108. doi:10.1007/978-3-642-13405-0_10

Endrayanto, A. I., Van Der Berg, H., & Boucherie, R. J. (2005). An analytical model for CDMA downlink rate optimization taking into account uplink coverage restrictions. *International Journal on Performance Evaluation*, *59*(2-3), 225–246. doi:10.1016/j.peva.2004.07.004

Gelabert, X., Pérez-Romero, J., Sallent, O., Agustí, R., & Casadevall, F. (2008). *Radio resource management in heterogeneous networks*. Heterogeneous Wireless Access Networks - Architectures and Protocols.

George, A., Rashad, S., Kumar, A., & Kantardzic, M. (2008). *A predictive handoff scheme for multi-hop heterogeneous wireless networks*. 3rd International ICST Conference on Communication System Software and Middleware. doi:10.1109/COMSWA.2008.4554472

Ghosh, D., & Lott, C. (2007). Uplink-downlink imbalance in wireless cellular networks. *IEEE International Conference on Communications,* (pp. 4275 – 428). doi: 10.1109/ICC.2007.705

Giupponi, L., Agust, R., Pérez-Romero, J., & Sallent, O. (2006). A framework for JRRM with resource reservation and multiservice provisioning in heterogeneous networks. *Mobile Networks and Applications*, *11*(6), 825–846. doi:10.1007/s11036-006-0052-3

Gordejuela-Sanchez, F., Juttner, A., & Zhang, J. (2009). A multiobjective optimization framework for IEEE 802.16e network design and performance Analysis. *IEEE Journal on Selected Areas in Communications*, *2*(27), 202–216. doi:10.1109/JSAC.2009.090211

Huang, C. J., Chuang, Y. T., Guan, C. T., Yang, D. X., & Chen, Y. J. (2007). A self-adaptive bandwidth reservation scheme for 4G mobile WiMAX. *IEEE International Workshop on Cross Layer Design*, (pp. 55–59).

Iancu, B., Dadarlat, V., & Peculea, A. (2010). A novel end-to-end QoS framework over heterogeneous networks - An architectural approach. *In ICAI'10 Proceedings of the 11th WSEAS International Conference on Automation & Information, 51*(2).

Junqi, W., Zhengbing, H., Jiuming, Y., & Jun, S. (2007). The degradation of bandwidth in QoS scheme for 4G mobile networks. *IEEE International Conference on Wireless Communications, Networking and Mobile Computing*, (pp. 2020 – 2024).

Karthikeyan, N. K., & Narayanasamy, P. (2008). A novel mobility based bandwidth reduction algorithm in cellular mobile networks. *International Journal of Computer Science and Network Security*, *8*(5), 62–69.

Kim, S. W., Kim, B. S., & Fang, Y. (2005). Downlink and uplink resource allocation in IEEE 802.11 Wireless LANs. *IEEE Transactions on Vehicular Technology*, *54*, 320–327. doi:10.1109/TVT.2004.838887

Kim, T. W., & Kim, H. G. (2011). A study on the delay of streaming packets during IP handover between heterogeneous networks. *13th International Conference on Advanced Communication Technology*, (pp. 1275–1278).

Kou, M., & Zhen, Y. (2009). *Dynamic uplink/downlink resource allocation for TDD OFDMA access network*. International Conference on Communications and Mobile Computing. doi 10.1109/CMC.2009.191

Lee, T., Seo, S., Lee, U., & Song, J. (2009). A new soft handover mechanism using DCHs in 3GPP HSDPA networks. *Journal of Networks*, *4*(3), 184–191. doi:10.4304/jnw.4.3.184-191

Magagula, L. A., Chan, H. A., & Falowo, O. E. (2010). Achieving seamless mobility through handover coordination in a network-based localized mobility managed heterogeneous environment. *21st Annual IEEE International Symposium on Personal, Indoor and Mobile Radio Communications*, (pp. 2505–2510). doi:10.1109/PIMRC.2010.5671745

Magagula, L. A., Chan, H. A., & Falowo, O. E. (2011). (in press). Handover approaches for seamless mobility management in next generation wireless networks. *Journal of Wireless Communication and Mobile Computing*. doi:10.1002/wcm.1074

Magagula, L. A., Falowo, O. E., & Chan, H. A. (2010). Enhancing PMIPv6 for better handover performance among heterogeneous wireless networks in a micromobility domain. *EURASIP Journal on Wireless Communications and Networking, 2010*. doi:10.1155/2010/274935

Magagula, L. A., Falowo, O. E., & Chan, H. A. (2010). PMIPv6-HC: Handover mechanism for reducing handover delay and packet loss in NGWN. In. *Proceedings of GLOBECOM, 2010*, 1–5. doi:10.1109/GLOCOM.2010.5684108

Niyato, D., & Hossain, E. (2005). Call admission control for QoS provisioning in 4G wireless networks: Issues and approaches. *IEEE Network*, *19*(5), 5–11. doi:10.1109/MNET.2005.1509946

Ramesh, B. H. S., Gowrishankar, G., & Satyanarayana, P. S. (2010). Call admission control for next generation wireless networks using higher order Markov model. *Computer and Information Science Journal*, *3*(1), 192–196.

Salhani, M. (2008). *Modélisation et Simulation des Réseaux Mobiles de 4ème Génération Université de Toulouse* (Doctoral dissertation, Toulouse University)

Samaan, N., Benmammar, B., Krief, F., & Karmouch, A. (2005). *Prediction-based advanced resource reservation in mobile environments* (pp. 1411–1414). Electrical and Computer Engineering.

Smith, C., & Collins, D. (2007). *3G wireless networks* (2nd ed.).

Su, G. M., Han, Z., Wu, M., & Ray Liu, K. J. (2007). Joint uplink and downlink optimization for real-time multiuser video streaming over WLANs. *IEEE Journal of Selected Topics in Signal Processing, 1*(2), 280–294. doi:10.1109/JSTSP.2007.901518

Todini, A., Baiocchi, A., Merliny, S., Messinaz, D., Morettix, M., & Scanavino, B. … Veronesiy, D. (2006). Cross-layer design of packet scheduling and resource allocation algorithms for 4G cellular systems. *Personal Multimedia Communications,* (pp. 100–105).

Yu, M. J., Choi, S. G., Kim, H. S., & Cho, K. S. (2011). An improved scheme for reducing handover latency in heterogeneous networks. *13th International Conference on Advanced Communication Technology,* (pp. 1534–1539).

Yun, M., Kim, T., Zhou, Y., Arora, A., Gomes, J., & Choi, H. A. (2009). Uplink resource management design in multi-access wireless networks. In Liu, B. (Eds.), *Wireless algorithms, systems, and applications* (pp. 499–508). doi:10.1007/978-3-642-03417-6_49

Zhai, H., Chen, X., & Fang, Y. (2005). How well can the IEEE 802.11 wireless LAN support quality of service? *Wireless Communications, 4*(6), 825–846. doi10.1.1.60.9241

This work was previously published in International Journal of Business Data Communications and Networking, Volume 6, Issue 4, edited by Debashis Saha and Varadharajan Sridhar, pp. 64-85 , copyright 2010 by IGI Publishing (an imprint of IGI Global).

Chapter 5
Dependency of Transport Functions on IEEE802.11 and IEEE802.15.4 MAC/PHY Layer Protocols for WSN:
A Step towards Cross-layer Design

Atif Sharif
Curtin University of Technology, Australia

Vidyasagar M. Potdar
Curtin University of Technology, Australia

A. J. D. Rathnayaka
Curtin University of Technology, Australia

ABSTRACT

In WSN transport, layer protocol plays a significant role in maintaining the node's energy budget. To find out the dependency of Transport layer on MAC/PHY layer, the authors have extensively tested various transport protocols using IEEE 802.11, IEEE 802.15.4 MAC/PHY protocols for WSN. For IEEE802.11 and IEEE802.15.4 with RTS/CTS ON the TCP variants has shown >80% packet delivery ratio and 5-20% packet loss, while for UDP it is around >63% and 19.54-35.18% respectively. On average 1-3% additional energy is consumed for packet retransmissions in IEEE 802.11 with RTS/CTS OFF whereas significant energy efficiency is observed in IEEE802.15.4 case. For IEEE 802.11 with RTS/CTS ON high throughput, low packet drop rate and increased E-2-E delay is observed, while for IEEE 802.15.4 improved power efficiency and jitter behavior is observed. This has led the foundation for the future development of the cross-layered energy efficient transport protocol for WSN.

DOI: 10.4018/978-1-61350-477-2.ch005

INTRODUCTION

The idea of low power, low cost, embedded wireless devices capable of gathering useful environmental information has led the development of Wireless Sensor Networks (WSNs) (Akyildiz et al., 2007; Potdar et al., 2009; Sharif et al., 2009). Today WSN technology has gained fundamental importance in a range of application scenarios like military, modern healthcare, environmental monitoring, surveillance etc. Fundamentally energy is considered as the biggest threat to WSN node design including both the hardware and the communication protocol design. WSN pays high energy cost in data retransmissions due to path loss, congestion and high bit error rate (BER) caused by interference or collisions at the receiving node. Either case will have a direct impact on the overall network throughput (Sharif et al., 2010), E-2-E packet latency and of course the longevity of the WSN. Reliable and high data transfer rate are the main concerns in event-critical applications e.g. Wireless Multimedia Sensor Network (WMSN). So facilitating congestion control and E-2-E reliability mechanism will be the primary objective of any transport layer protocol design.

In WSN, the transport layer protocol functionality is comprise of a congestion control component (Sharif et al., 2010) that handles minimum number of data packets (either scheduled or prioritized) in order to ensure the minimum latency and energy conservation (achieved by minimum retransmissions) under congested network conditions. Also the transport function must ensure the event and query reliability. In the WSN network dealing with multimedia information, (i.e., WMSN) the effect of congestion is even more pronounced. The nodes near the sink or those that are experiencing large amount of data above their handling capacity (bottleneck nodes) experience congested paths during data transmission that requires high data rate. This condition would result in performance degradation (due to collisions and retransmissions) and rapidly depletes the node's energy. Thus the transport protocol should effectively detect the congestion and try to mitigate it immediately in order to avoid excessive collisions and retransmissions.

Second important functionality of the transport layer protocol is the data reliability feature (Sharif et al., 2010), which ensures how efficiently the transport protocol detects the data packet loss and notifies it to the nearby storage node for data packet retrieval (to epitome the correct information from the received data) in order to ensure the Quality of Service (QoS) objective of the application is not compromised e.g. event-critical applications (Sharif et al., 2009). Currently the WSN supports packet level and application level reliability. For packet level the transport protocol guarantees the reliable transmission of each and every data packet (packet level reliability) while certain applications (like Military) demand proportionally reliable transmission of all data packets (application reliability).

The Transmission Control Protocol (TCP) and User Datagram Protocol (UDP) schemes of the existing wireless network are not well suited for WSN because of having strict E-2-E reliable aspect, unfairness for multi-hop environment (as in TCP) and connectionless with no congestion control feature (as in UDP). New proposals using multi-path transmission have been introduced such as Congestion Detection and Avoidance (CODA) (Wan et al., 2003) and Multi-flow Real-time Transport Protocol (MRTP) (Mao et al., 2003). CODA has considerable delay since it decides on the basis of the status of the intermediate nodes and MRTP does not consider energy efficiency in WMSN. Some other transport protocols have been designed recently for scalar WSNs with the aim of decreasing energy consumption, providing reliability, and controlling congestion. These protocols do not fulfill major factors of QoS, such as high bandwidth and real-time communication, which are required for multimedia communication in sensor networks, e.g., Reliable Multi-Segment Transport protocol (RMST) (Stann & Heideman,

2003), Reliable Bursty Convergecast (RBC) (Zhang et al., 2005), Garuda (Park et al. 2004), Pump Slowly Fetch Quickly (PSFQ) (Wan et al., 2002), Event-to-Sink Reliable Transport (ESRT) (Sankarasubramaniam et al., 2003), Distributed TCP (DTC) (Dunkels et al., 2004) and Sensor TCP (STCP) (lyer et al., 2005) do not support real-time communication while providing reliability. Also, Fusion (Hull et al., 2004) is another transport protocol that is not compatible with the limited energy sources of video sensor nodes. Also existing work in transport protocol development targets either upstream (source-to-sink) (Park et al., 2004; Wan et al., 2002) or downstream (sink-to-source) reliability (Stann & Heideman, 2003; He et al., 2003; Sankarasubramaniam et al., 2003).

In this paper we primarily focus on the role of transport agent for ensuring data delivery while keeping in the constraints of energy cost. We will test this by observing the behaviors of standard transport agents like UDP and TCP including its variants like Tahoe, Reno, NewReno, Vegas, SACK, Fack etc over various MAC/PHY layer protocols in wireless domain using AODV (Perkins et al., 2003) as a routing agent. This would help us in understanding the dependence of transport layer functionality over MAC/PHY layer. In the next section we will discuss the functionalities of few of above mentioned transport layer protocols.

The rest of the paper is organized as follows. After the introduction to WSN and the transport protocol, the related work is covered in the second section. Problem definition is introduced in the third section followed by the fourth section where we have describe the simulation setup used for observing the transport protocol behavior.

In the fifth section we outline the simulation results and then we conclude the paper in last section.

RELATED WORK

In this section we focus on the existing transport protocols used for WSN. We begin the discussion with TCP, UDP and then extend it to various other WSN transport protocols like CODA (Wan et al., 2003), ESRT (Sankarasubramaniam et al., 2003), RMST (Stann & Heideman, 2003), PSFQ (Wan et al., 2002), RBC (Zhang et al., 2005), GARUDA (Park et al., 2004), DTC (Dunkels et al., 2004), and STCP (lyer et al., 2005).

TCP (connection oriented by nature) assumes congestion as the main cause of packet drop. It assures strict E-2-E reliability at very high energy cost, which is not acceptable in WSN because of its strict energy constraints. UDP on the other hand, being connectionless by nature, offers significant throughput as compared to TCP, but the packet drop rate is significantly higher during congestion. Therefore the basic inference is that UDP offers extremely high throughput but minimum reliability, whereas TCP offers extremely high reliability but very low throughput. Our literature study concludes that the existing research in the field of WSN transport protocol development exists in between these two extremes, which we now discuss in detail.

CODA (i.e., Congestion Detection and Avoidance) protocol (Wan et al., 2003) lacks reliability but provides excellent congestion control in the forward direction. This control is achieved by explicit notification of the congested scenario to the child nodes. It utilizes wireless channel load and nodes buffer occupancy for congestion detection and uses open-loop Hop-by-Hop (H-b-H) back pressure and closed loop End-to-End (E-2-E) multi-source regulation for rate adjustment.

Other than CODA, there are a number of other WSN transport protocols published in the literature like ESRT (Sankarasubramaniam et al., 2003), RMST (Stann & Heideman, 2003), PSFQ (Wan et al., 2002), RBC (Zhang et al., 2005), GARUDA (Park et al. 2004) and DTC (Dunkels et al., 2004). These protocols only aim to provide reliability but completely lack congestion control feature. We now briefly explain this set of protocols.

- **Event-to-Sink Reliable Transport (ESRT):** (Sankarasubramaniam et al., 2003) aims to provide Event Based (EB) reliability support by employing implicit loss detection and notification and E-2-E loss recovery mechanism. It also provides passive congestion control in the Forward (Fwd) direction by incorporating Implicit (Imp) congestion notification feature.
- **Reliable Multi-Segment Transport protocol (RMST):** (Stann & Heideman, 2003) offers Packet Based (PB) reliability support, H-b-H loss recovery in the forward direction by incorporating "Not Acknowledge" (NACK) based loss detection and notification control. It uses timing information for data loss detection and once data loss is detected it sends a NACK packet to the source node or storage node, which then retransmits the corresponding data packet upon receiving the NACK.
- **Pump Slowly Fetch Quickly (PSFQ):** (Wan et al., 2002) aims to address the reliability support only in the reverse (Rev) direction from the sink node to the source nodes. It uses NACK for data loss detection and is facilitated with rapid H-b-H loss recovery mechanism. However it is unable to detect the loss of a single packet when packets are transmitted in bulk. It also requires more buffer storage (being H-b-H nature) at the intermediate nodes for data recovery.
- **Reliable Bursty Convergecast protocol (RBC):** (Zhang et al., 2005) offers E-2-E packet based reliability support in the forward direction. The level of reliability provided by RBC is comparatively higher than RMST and PSFQ because it incorporates implicit H-b-H loss detection, loss notification and loss recovery mechanism at MAC level. It uses NACK/ACK (Acknowledge) as a means for data loss notification and Implicit Acknowledgement (IACK) for loss notification at MAC level only.
- **GARUDA:** (Park et al., 2004) provides reliability in the reverse direction, i.e., from the sink node to the sensor nodes. It uses core nodes located at 3-hop distance apart for temporary sensed data storage, which could be retrieved in the event of loss. It uses NACK for data loss detection and two-tier two-stage loss recovery mechanism.
- **Distributed TCP cashing protocol (DTC):** (Dunkels et al., 2004) aims to provide total packet driven reliability in both directions. It uses ACK and Selective ACK (SACK) for data loss detection. In DTC sink node has loss recovery control and uses H-b-H mechanism for possible loss recovery.
- **Sensor TCP (STCP):** (lyer et al., 2005) is the only protocol so far that offers both congestion control and data reliability. It uses open-loop H-b-H rate control mechanism with implicit congestion notification. It uses packet processing and inter-arrival time for congestion detection.

It offers both packet and event driven reliability and uses ACK and NACK to facilitate E-2-E reliability objective.

In this section we reviewed the existing transport protocol schemes for WSN. In the next section we will explain the problem definition for designing the transport layer protocol for WSN.

PROBLEM DEFINITION

In WSN considerable amount of energy is consumed for communicating each packet i.e., per packet communication. This high energy cost is due to many factors including packet collisions at the receiving node due to transmissions from other nodes having the common destination, crosstalk: resulting in high bit error rate (BER), congestion, route failure etc. All these factors contribute to high energy cost, increased E-2-E packet delay and lower throughput of the system.

To cut down the overall networks energy budget and likewise to enhance the effective throughput of the system by minimizing the E-2-E packet delay it is the responsibility of the WSN transport protocol that takes all these effects into account. Indeed the WSN transport protocol actually can get most of the information from the MAC/PHY layer that help in maintaining a balance between system throughputs, data packet loss and energy consumption. So the purpose of the simulation is to evaluate the existing standard transport protocols (UDP and TCP variants) for WSN. This simulation would help us in better understanding the above said performance limiting features while simultaneously help us in identifying the dependence of transport layer protocol on the underlying MAC/PHY layer protocol and would also allow us to possibly look for new dimensions in transport protocol design for WSN.

SIMULATION SETUP AND PERFORMANCE METRICS

Network Plan

In this section we will describe the network plan which will be used in extensive functionality testing of UDP, TCP and its variant like Tahoe, Reno, NewReno, Vegas, SACK, and Fack etc. for WSN. The network topology (homogeneous by nature) we have used is shown in Figure 1 and is comprised of 9 wireless sensor nodes placed in open/free space environment. Node 0, 1, 2 and 3 are considered as source nodes having TCP/TCP variants or UDP as transport agents and File Transfer Protocol (FTP) or Constant Bit Rate (CBR) as traffic sources.

The packet size is considered to be 1KB in each case and in CBR the data rate is set to 400Kbps for effective utilization of the bandwidth. The Node 8 acts as a sink while rest of the nodes act simply as relay nodes (only involved in routing). The motive of this test setup is to evaluate the dependence of transport protocol on MAC/PHY layer by measuring the network's throughput, packet loss ratio, E-2-E packet delay and energy

Figure 1. The network topology used for evaluation

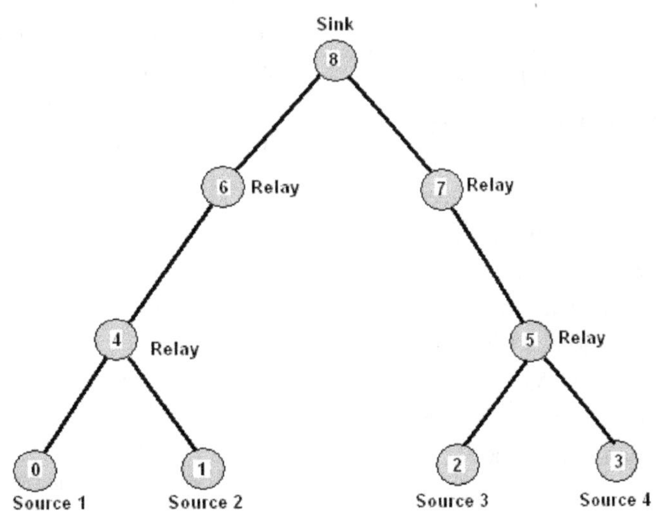

behavior of nodes by employing the node's energy model.

Radio Propagation Model (RPM)

The RPM is generally used to predict received packet power, which is entirely decided by the path loss (L), the loss covered by the distance between the transmitter (*tx*) and receiver (*rx*). If the received packet power is above a certain threshold limit then it is considered as successfully received packet else the packet is considered to be erroneous and dropped. We used open space scenario for our simulation. In open space environment the receiving node, at distance 'd_{tx-rx}' from the transmitting node, will receive incoming transmitted electromagnetic signal from both direct path and ground reflections. And for our network we will use the TwoRayGround as radio propagation model. This model considers both the ground reflections and direct path. For this case the received signal power *P* at the receiver (*rx*) is given by P_{rx} (Layuan et al., 2007; Rappaport, 1995).

$$P_{rx} = \frac{P_{tx} \cdot G_{tx} \cdot G_{rx} \cdot h_{tx}^{\ 2} \cdot h_{rx}^{\ 2}}{d_{tx-rx}^{\ 4} \cdot L} \tag{1}$$

where G_{tx}, G_{rx} is the cumulative gain of the transmitting and receiving antenna, h_{tx} and h_{rx} are the effective heights of the transmitting and receiving antennas, d_{tx-rx} is the distance between transmitter and receiver node and *L* is called the path loss.

Before finding the signal to noise ratio (*SNR*) threshold at the receiver end (as we mentioned above), we first define the interference region, the region linked to the receiver having radii dint, in which the receiving node may be interfered with another transmission (interfering transmission power P_{int}) from a hidden node at a distance dint from it. So the SNR threshold (*SNR_Thre*) at the receiver (for successful reception) is given by the following equation (Layuan et al., 2007)

$$SNR_{rec} = \frac{P_{rx}}{P_{\text{int}}} = \left(\frac{d_{\text{int}}}{d_{tx-rx}}\right)^4 \geq SNR\_Thres \tag{2}$$

So to avoid the packet collision at the MAC level due to hidden node (also called interfering node) the d_{int} (Layuan et al., 2007) should be

$$d_{\text{int}} \geq \sqrt[4]{\text{SNR\_Thres}} \ * d_{tx-rx} \tag{3}$$

By proper selection of the d_{int}, hidden nodes appear out of the receiving node's interference region and this minimizes the effect of packet collisions at the receiving nodes due to transmissions from hidden nodes. So in this case the receiving node is insensitive to the hidden node's communication and this would result in a collision free communication. So it is an important point to consider dint while designing the network.

MAC Protocol

We used IEEE802.11 and IEEE802.15.4 as MAC/PHY layer standards. Both protocols use Carrier Sense Multiple Access with Collision Avoidance (CSMA/CA) for carrier sensing and collision avoidance and operates in 2.4GHz ISM band. In IEEE 802.11 we also exploit the use of Request to Send frame/Clear to Send frame (RTS/CTS) for avoiding the hidden node problem in ad-hoc network situation.

Normally we take *SNR_Thre* = 10 for quality reception, so from Eq.3 we will get

$$d_{\text{int}} \geq 1.778 * d_{tx-rx} \tag{4}$$

$$d_{tx-rx} \leq 0.5624 * d_{\text{int}} \tag{5}$$

Although by making carrier sensing region having radius $d_{cs} > d_{int}$ we can avoid collision since the interfering node actually have to wait for the carrier for its transmission, which is currently being used by the main transmitting node. However this substantially degrades the overall system throughput. Similarly by setting $d_{cs} = d_{int}$ avoids the use of RTS and CTS. So we have to compromise in between the two thresholds.

Energy Model

To test the energy efficiency of the transport layer protocols for WSN using IEEE802.11 and IEEE802.15.4 as MAC/PHY layer standards and Ad hoc AODV as routing agent we employ the energy model in our simulations. Since every node consumes some power during transmission including retransmissions, reception and idle state, this model helps in keeping track of the energy graph of individual node. The purpose is to check the effectiveness of the transport protocol from the energy consumption point of view as because of packet collisions or poor SNR (due to bad channel quality) more retransmissions occur from a particular node which results in rapid depletion of node's energy budget.

Performance Metrics

The performance of various transport protocols is evaluated against the following metrics

1. Average E-2-E packet delivery ratio: It is defined as the ratio of data packets send by all the potential sources to the data packets received by the sink.
2. Average E-2-E packet delay: It is defined as the total time a packet would take from the source to sink (E-2-E). This includes all the possible delays as a result of queuing, MAC layer retransmissions, propagation delays and transfer time.
3. Percentage Average data loss: It is defined as ratio of the data loss (difference of send and receive data) to the send data multiplied by 100. The main contributors of data loss are collisions at the receiving end in the presence of blind nodes, congestion and Link failure due to node's energy depletion. Here we also take into account the effect of collisions due to blind nodes.
4. Average Power consumed: It is defined as the average power, in Watts, consumed by the source, relay and sink nodes during the data communication over wireless channel including the possible retransmissions at the PHY/MAC level.
5. Jitter: In WSN, as in traditional networking subjects, jitter is defined as the time variation or fluctuations in inter-arrival packet time caused by network congestion, route change or timing drifts. Using Equations 1-5 we will now design and simulate a wireless network comprised of 9 nodes and then test the functionality of UDP and TCP variants like Tahoe, Reno, NewReno, Vegas, SACK, and Fack against the performance metrics as outlined above.

SIMULATION RESULTS

In this simulation we have used Network Simulator NS-2 and we consider that the entire multi-hop network is comprised of 9 nodes as an ad-hoc network for the case of IEEE 802.11 and multi hop WPAN for the case of IEEE 802.15.4 (Sink (node 8) act as a PAN coordinator). For these wireless setups we performed extensive analysis of total throughput, E-2-E packet delay and the average power consumption by the source, relay and sink nodes. In this simulation setup we focused more on results from practical perspective and for this we have used the following parameters listed in Table 1. The values for RX Threshold and CS Threshold are calculated by using the idea presented in Equations 1- 5.

Also in our case we are actually considering the amount of power consumed by nodes during their communication, idle and sleep modes by enabling the Energy Model parameter. Also we

Table 1. Wireless sensor network parameters

Parameter	Values
Frequency	2.472e+9
Propagation model	TwoRayGround
Transport Protocols	UDP, TCP-SACK TAHOE, VEGAS, RENO, NEW-RENO, FACK
MAC	IEEE802.11 and IEEE802.15.4
RX Threshold IEEE802.11 (node separation 250m) IEEE802.15.4 (node separation 50m)	1.296e-10W 3.73066e-9W
Routing agent	AODV
Ifqlen (Queue length at MAC level)	200 packets
Energy Model	Yes
CP Threshold	10
CS Threshold IEEE802.11 IEEE802.15.4	1.1664e-10W 3.357594e-9W
Node initial power	100W
Node idle power consumption	712e-6W
Node rx power consumption	35.28e-3W
Node tx power consumption	31.32e-3W
Node sleep power consumption	0.001W

are considering the case of multimedia application with a packet size of 1KB and sending rate is 400Kbps (for each source) for UDP where as for TCP it grows based on its congestion window (*cwnd*). The simulation is performed 5 times and the results mentioned here are averaged for each particular case in order to ensure the reliability in the results obtained.

IEEE 802.11

For the case of IEEE 802.11 we performed the simulation for two cases

- RTS/CTS ON and
- RTS/CTS OFF

Average E-2-E Packet delivery Ratio Vs Reliability

By having RTS/CTS ON we eliminated packet drop caused by packet collision at the receiver side due to transmission from other child nodes and/or interfering or blind node. We can clearly observe the impact of having RTS/CTS ON from Figure 2, Figure 3, and Figure 4 where it is obvious that with RTS/CTS ON the data loss rate (due to collisions) reduces significantly and throughput increases, which results in power efficiency by eliminating the need for packet retransmission. at the MAC/PHY level. Also among various transport protocols UDP has showed its poor reliability and average packet delivery ratio (being connectionless by nature). Although its average data sending rate is quite high (approximately > 190000 bytes for average RTS/CTS ON and OFF cases) but still its packet delivery ratio is poor and is around 29.72% and 35.18% for RTS/CTS OFF

Figure 2. Average data send/received (Bytes) using IEEE 802.11 with RTS/CTS OFF/ON

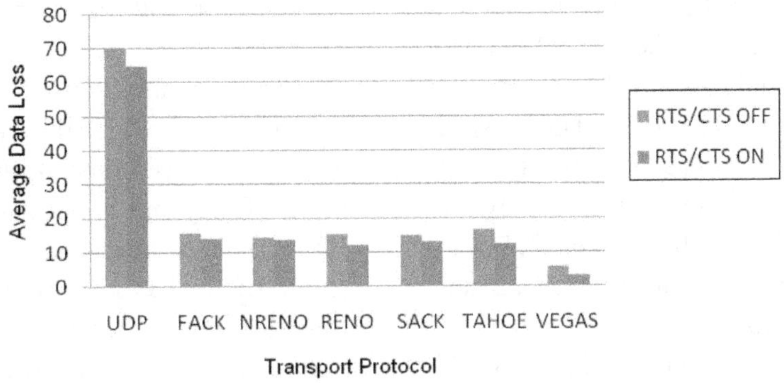

Figure 3. Data reliability analysis using IEEE 802.11 with RTS/CTS OFF/ON

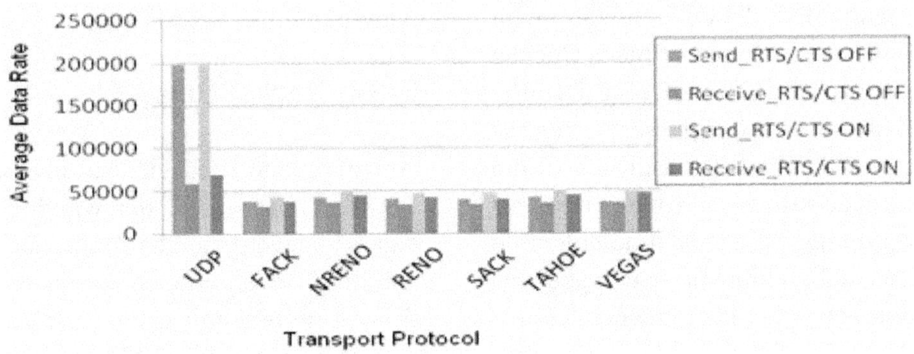

Figure 4. Average packet delivery ratio comparison using IEEE 802.11 with RTS/CTS OFF/ON

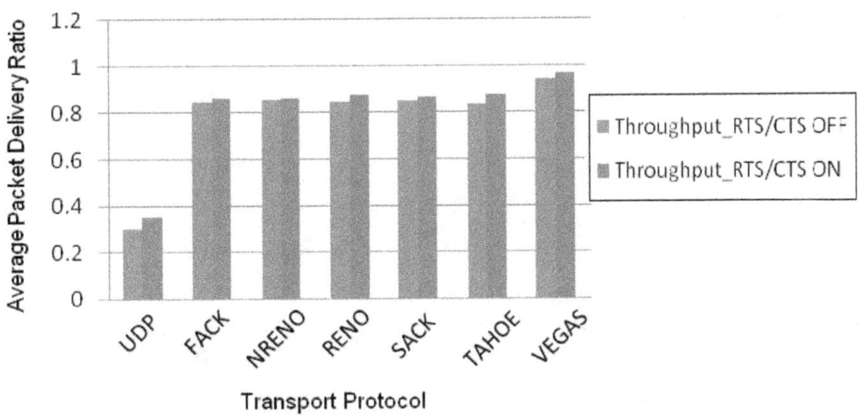

and RTS/CTS ON cases which is significantly lower than TCP variants case (where the average data rate (in bytes) is in between 37KB to 45KB and average packet delivery ratio is around 86% which is higher than UDP).

Similarly for the data reliability in case of TCP variants, only TCP-Vegas (which basically uses

TCP congestion control algorithm with an emphasis on packet delay rather than on packet loss for detecting the congestion) showed better result by having 5.6623% and 3.040613% data loss for IEEE802.11 with RTS/CTS OFF and ON respectively while others have 12-16% for the two cases. But with respect to the available bandwidth of 2Mbps (effectively 1.7 Mbps (Li et al., 2001)) the bandwidth utilization factor of TCP variants is poor as compared to UDP.

The channel access mechanism using RTS/CTS has a great impact on the average packet delivery ratio and helps in reducing the data packet loss due to blind nodes. If we compare the two cases (IEEE802.11 with RTS/CTS OFF and ON) with respect to the received number of bytes (for the two cases due to packet collision, RTS/CTS OFF, at MAC level) about 5.46% of the sent bytes has been lost in case of UDP while for the rest of the cases it is averaged around 5.5Kbytes (3-20%) which is also quite significant and is not tolerable for event driven and multimedia sensor applications where data reliability is of great concern.

Figure 5, Figure 6 and Figure 7 depict the average power consumed by the UDP (if we compare the entire power consumed to total delivered data packets) for both cases is high as compared to TCP variants.

Also from the Figure 5, Figure 6, and Figure 7 it would be obvious that with RTS/CTS ON the average energy consumption of the respective nodes (Relay and Sink nodes) drops. Because of RTS/CTS ON the collisions at receiving side due to blind nodes has been dropped down as compared to RTS/CTS OFF case where nodes on the average consume 1-3% higher energy for retransmitting the lost packet information due to collision. And this 1-3% difference is quite significant especially when we talk about the longevity of such networks.

E-2-E Packet Delay

As we are concerned with multimedia traffic so we can see from Figure 8. that with high data rate the data is queued at the MAC level due to congestion and MAC level retransmissions and this queued delay is more dominant in node's average E-2-E packet delay behavior. Also with the inclusion of RTS/CTS ON the sources actually have to wait

Figure 5. Average power (watts) consumed by source nodes using IEEE 802.11 with RTS/CTS OFF/ON

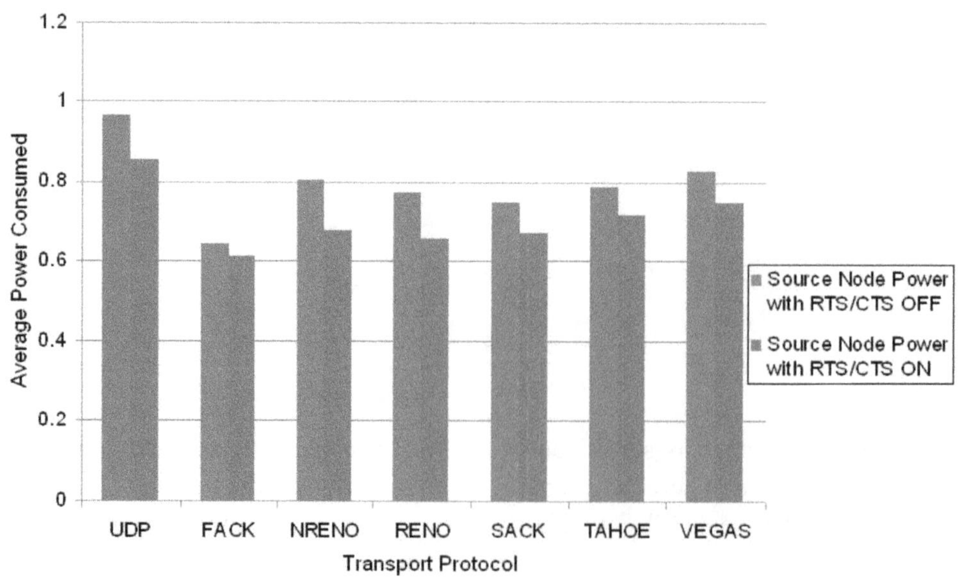

Figure 6. Average power (watts) consumed by relay nodes using IEEE 802.11 with RTS/CTS OFF/ON

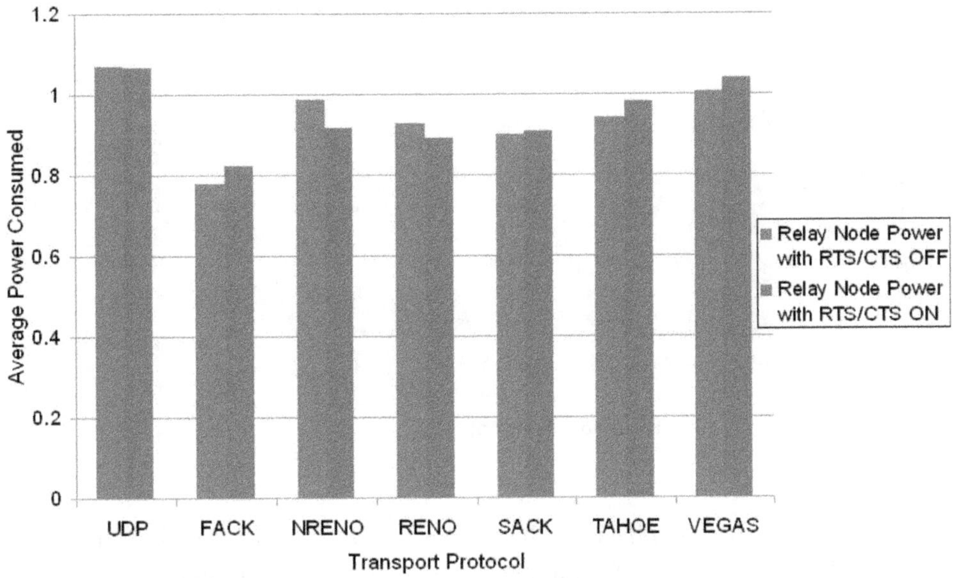

Figure 7. Average power (watts) consumed by sink node using IEEE 802.11 with RTS/CTS OFF/ON

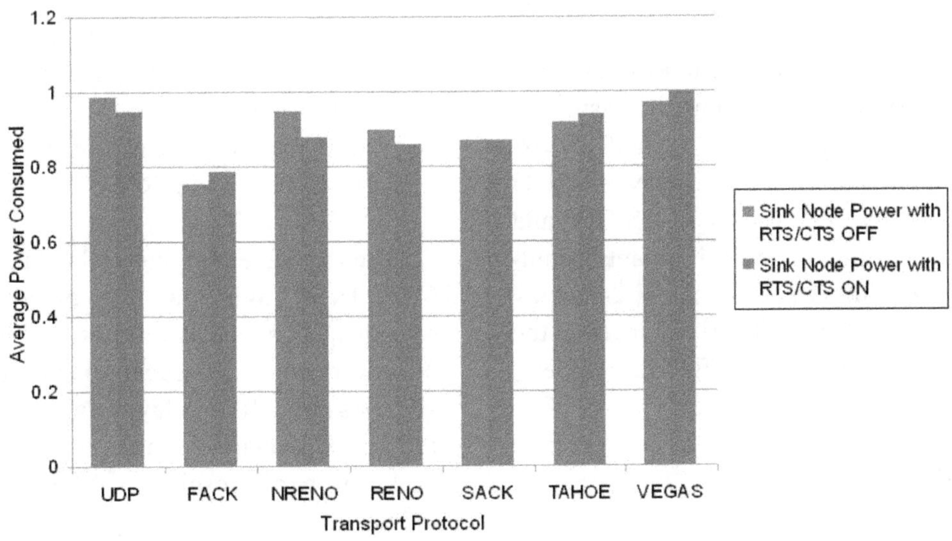

for the carrier sense before starting the transmission. And for that reason we have seen that with RTS/CTS ON the average E-2-E packet delay has increased. UDP amongst all showed poor results in both the cases because in case of UDP the CBR traffic source kept sending data packets of size 1KB at a rate of 400Kbps.

For the case of TCP variants they showed an average E-2-E delay of more than 120 msec for either case. Among all only TCP Vegas showed better results by exhibiting an average E-2-E delay of 128.64 msec (RTS/CTS OFF) and 148.1 msec (RTS/CTS ON). TCP Vegas showed this improved behavior because it uses the TCP con-

Figure 8. Average E-2-E packet delay (sec) analysis using IEEE 802.11 with RTS/CTS OFF/ON

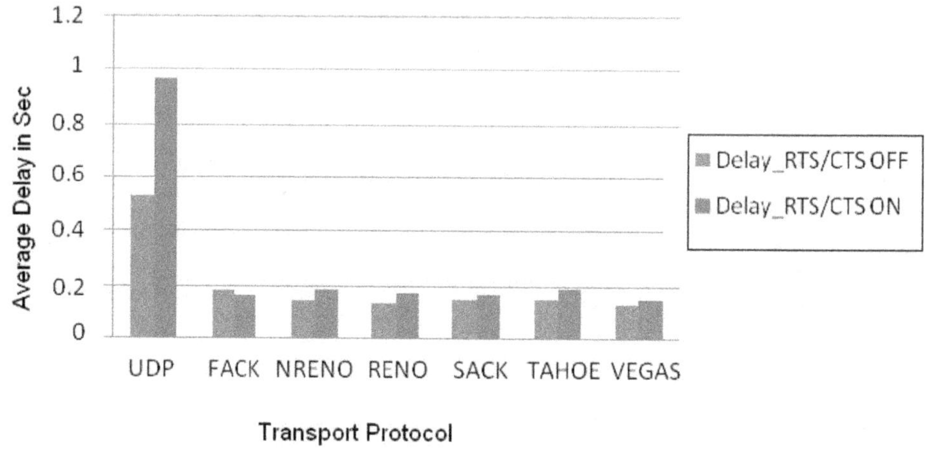

gestion control algorithm that emphasizes the packet delays rather than the packet loss for source rate readjustment. It detects the congestion based on increasing Round Trip Time (RTT) and would cause a linear increase or decrease in the cwnd (when packet loss occurs) during the next RTT against the set limits of traffic thresholds. TCP FACK and TCP Tahoe on the other hand showed worse behavior among the TCP variants by exhibiting an average E-2-E packet delay of 180.85 msec and 190.365 msec for RTS/CTS OFF and ON respectively. In summary TCP variants including the UDP has showed an average E-2-E delay of more than 120msec and it still did not meet the objectives of multimedia traffic (for voice ≤ 120msec).

Jitter Analysis

In this section we are going to present the jitter analysis that we performed for UDP, TCP and its variants like FACK, NewReno, Reno, SACK, TAHOE and Vegas that uses IEEE 802.11 as underlying MAC protocol for two possible scenarios (with and without RTS/CTS control information exchange) and is shown in Figure 9. The figures that are referred here are shown in Appendix A.

- *UDP:* In case of UDP the average value of jitter experienced by its data traffic with and without RTS/CTS ON is approximately 0.2871 sec and 0.5427 sec respectively. From Figure 12.a and 12.b it is obvious that in case of UDP that uses IEEE802.11 with RTS/CTS ON the jitter has sharp spikes with a max of 3.6 sec as compared to 2.55 sec in case of UDP that uses IEEE802.11 with RTS/CTS OFF.

The large average jitter value in case of RTS/CTS ON is obvious as it involves the channel sensing mechanism as an additional timing offset prior to queuing, propagation and other unforeseen possible offset delays (e.g., caused by route failure). This channel sensing, though reduces the packet drop rate, results in substantial increase in the packet queuing delay (as the information is sent only once after the successful reception of CTS against RTS has happened also called 'channel reservation') that would increase the average packet jitter value.

TCP RENO: The jitter analysis for TCP Reno is shown in the Figure 20 and Figure 21 for the two possible cases. From the Figure 20 and Figure 21 it is obvious that the TCP Reno exhibits an average jitter of 0.1050 sec and 0.1384 sec

Figure 9. Average jitter (sec) analysis using IEEE 802.11 with RTS/CTS OFF/ON

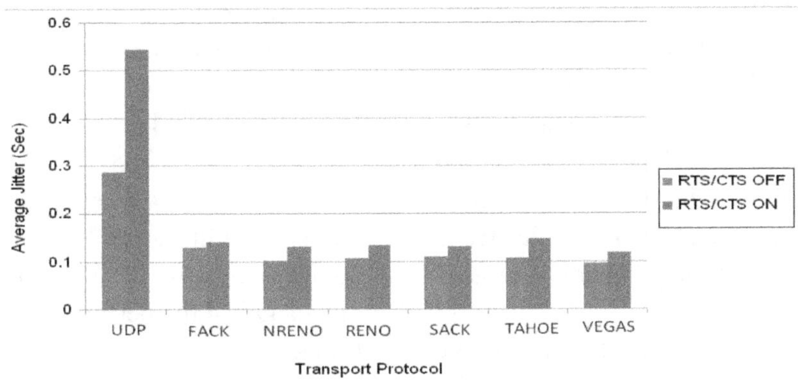

for IEEE802.11 with RTS/CTS OFF and ON respectively. Jitter relates the packet inter-arrival time with the difference in the consecutive packet sequence number and is seriously being affected by the packet drop e.g., TCP Reno, which detects congestion upon packet drop. It will lose its self clocking capability once multiple segments drop resulting in timeouts and serious performance degradations. It controls the cwnd for with and without congestion cases as follows:

$$cwnd = cwnd + \frac{1}{cwnd} \quad (6)$$

$$cwnd = cwnd - \frac{1}{2}cwnd \quad (7)$$

and in-terms of packet loss (p) as

$$cwnd = \frac{1.22}{p^{0.5}} \quad (8)$$

This packet/segments drop causes its congestion control algorithm to reduce the source sending rate in other words increase the latency between the two consecutive packets thereby increasing the jitter.

- *TCP NewReno:*Figure 22 and Figure 23 shows the jitter analysis for TCP NewReno. It exhibits an average jitter of 0.102 sec and 0.1325 sec for IEEE802.11 with and without RTS/CTS ON. As compared to TCP Reno it exhibits shorter jitter for the two cases as the TCP NewReno handles multiple segment loss case without entering into slow start phase and also it supports 'fast retransmission' recovery.
- *TCP Tahoe:* As shown in the Figure 24 and Figure 25 the TCP Tahoe offers an average jitter of 0.1069 sec and 0.1461 sec for IEEE802.11 with and without RTS/CTS ON. As compared to previous TCP variants the present rise in jitter value for either case is due to the channel access mechanism (RTS/CTS sensing or channel sensing as already explained above) and also due to slow start mechanism, where the cwnd starts with 1 and grows exponentially. Once congestion is detected or first packet drop occurs the cwnd drop down to half of the previous value $\frac{cwnd}{2}$.
- *TCP SACK:* TCP SACK jitter analysis is shown in the Figure26 and Figure27 for IEEE802.11 with and without RTS/CTS ON. It supports the data reliability issues and exhibits an average jitter of 0.1091

sec and 0.13 sec for the two cases. The response is better than TCP Tahoe, Reno, UDP and TCP NewReno and slightly poor for TCP NewReno without RTS/CTS OFF case as it lags the congestion control feature, which is quite vital for maintaining the average jitter value to some threshold for applications where jitter significance is quite vital e.g., multimedia applications.

Another reason for this decrease in the average jitter value is that it uses Selective ACK (SACK) so only the selected packets (which are dropped) will be requested for retransmission not the whole block of data as in case of for example TCP Reno.

- *TCP FACK:* The TCP FACK (Forward ACK) jitter behavior is shown in Figure 28 and Figure 29 for IEEE802.11 with and without RTS/CTS ON. It exhibits an average jitter of 0.1289 sec and 0.1388 sec respectively. It uses SACK and congestion control features and supports over damping algorithm to offset the slow start over shoots and ramp down algorithm for smoothening in transmission, but we still did not observe a significant improvement in jitter behavior for the two possible scenarios of WSN (as what we were expecting initially).
- *TCP Vegas:* The TCP Vegas jitter performance graphs are shows in Figure 30 and Figure 31 respectively for IEEE802.11 with and without RTS/CTS ON. TCP Vegas shows better jitter behavior as compared to the rest variant for IEEE802.11 case by exhibiting an average jitter value of 0.0944 sec and 0.1170 sec. The reason for this improvement is quite obvious as its congestion control algorithm is based on packet delay rather than on packet loss for rate change. It detects congestion based on the increase in RTT and will linearly increase or decrease in the cwnd (adjust *cwnd* size based on the difference between expected and actual RTT) during the next RTT against the set traffic threshold limits.

IEEE 802.15.4

For IEEE 802.15.4 we evaluated the transport protocols for multi-hop WPAN targeting multimedia applications with RTS/CTS ON. The evaluation is made against the performance metrics discussed previously.

Average E-2-E Packet Delivery Ratio vs. Reliability

In this section we will compare average data rate, average packet delivery ratio and average percentage data lost for UDP and TCP variants that use IEEE802.15.4 as underlying MAC/PHY layer protocol with RTS/CTS ON. IEEE802.15.4 uses Binary Phase Shift Keying (BPSK), Offset Quadrature. Phase Shift Keying (O-QPSK) and Amplitude Shift Keying (ASK) as coding technique while Direct-Sequence Spread Spectrum (DSSS) and Parallel Sequence. Spread Spectrum (PSSS) as underlying PHY layer modulation scheme opposed to IEEE802.11 that uses BPSK, QPSK, 16-QAM (Quadrature Amplitude Modulation) and 64-QAM as coding technique while OFDM (Orthogonal frequency-division multiplexing) as modulation scheme.

From Figure 10, Figure 11 and Figure 12, it is obvious that TCP Tahoe and TCP Reno performed well as compared to UDP and other TCP variants by having lowest data loss i.e. 18.878% and 19.1011% or having high average packet delivery ratio. TCP FACK, TCP SACK and UDP on the other hand showed worse behavior by exhibiting an average data loss of 20.175%, 19.8845% and 19.543% respectively. For the case of UDP using IEEE802.15.4 with RTS/CTS ON the data loss of 19.54% is far better compared to the IEEE802.11 case where the percentage data loss was 65-70% and this efficiency is because of improved chan-

Figure 10. Average data send/received (bytes) using IEEE802.15.4

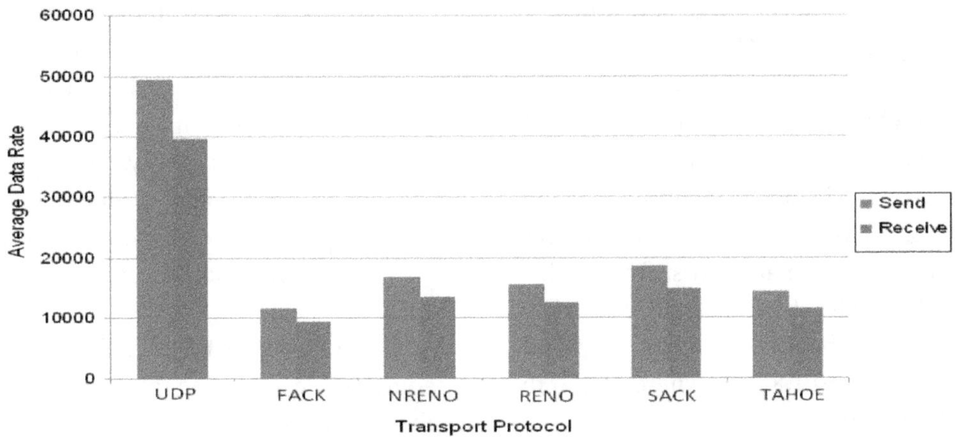

Figure 11. Average packet delivery ratio comparison IEEE802.15.4

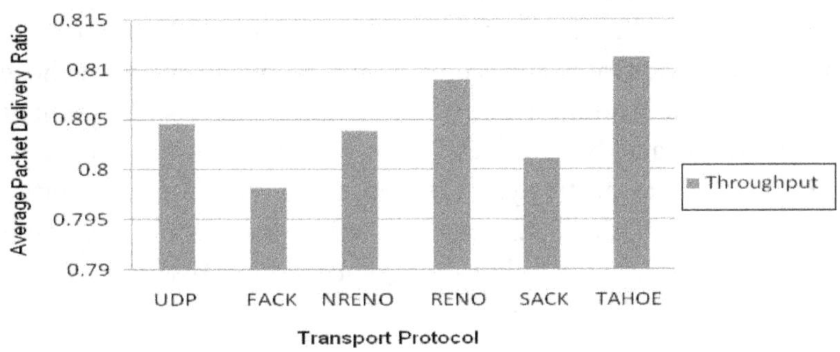

Figure 12. Data reliability analysis using IEEE 802.15.4

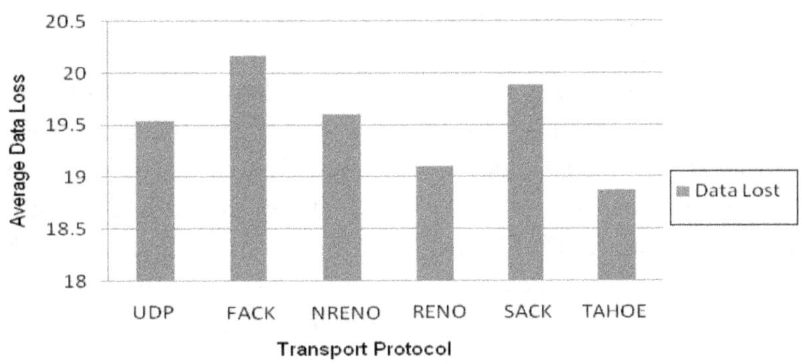

nel control feature of IEEE802.15.4 protocol standard. Figure 11 shows the percentage of packet delivery ratio for UDP and TCP variants that uses IEEE802.15.4 as underlying MAC/PHY protocol with RTS/CTS ON. From the Figure 11 we observe that the average percentage packet delivery ratio of UDP and TCP variants is in the range of 79.825% to 81.125% with maximum of 81.125% for TCP Tahoe and worst for 79.825% for TCP FACK. Also for the UDP over IEEE 802.15.4, the throughput result seems promising when compared to IEEE802.11where it is in the range of 29% to 35%.

Figure 13, Figure 14 and Figure 15 show the average power consumed by the Source, Relay and Sink nodes that uses IEEE802.15.4 as underlying MAC/PHY protocol. As we can see from Figure 13, Figure 14 and Figure 15 the average power consumed by the Source, Relay and Sink nodes in WPAN configuration of IEEE 802.15.4 is less than half power unit (0.5 Watt) as compared to IEEE802.11 case where the average power consumed by Source, Relay and Sink nodes ranges in between 0.7 to 1 Watt respectively.

This power efficient behavior of IEEE802.15.4 as compared to IEEE802.11 is due to the following reasons:

1. The low power consumption is entirely based on how low the node's operating duty cycle is and is consistent with the low peak currents.
2. IEEE802.15.4 allows the use of high data rate but low symbol rate at the PHY level, where the peak currents track low symbol rate rather than high data rate, thus results in power efficiency. But using this allows the use of multilevel signaling technique which results in sensitivity loss.
3. This loss of sensitivity is being offset by the use of orthogonal signal coding technique with bandwidth tradeoff.
4. The use of beacon enabled mode in IEEE802.15.4 (is to support the communication of control or beacon packets between transmitter and receiver) for node synchronization as opposed to beaconless mode i.e., IEEE802.11, where the nodes keep listening to channel which thereby depletes their energy quickly. This synchronization allows the nodes to sleep between coordinated transmissions thus resulting in power efficiency and longevity of the WSN.

Although in IEEE802.15.4 configuration the source node's sending rate is set to 250Kbps (less than the data rate of 2 Mbps set by IEEE802.11) but still the ratio of successfully received data rate for IEEE 802.15.4 is better as compared to IEEE 802.11, and this ratio clearly explains why the nodes having IEEE802.15.4 as underlying MAC/PHY protocol consumed less power when compared to nodes using IEEE 802.11 as underlying MAC/PHY protocol because it reduces the power consumed during the retransmission of the lost data packets (data reliability).

In short the node's power efficiency (or at more general level the power efficiency of the overall WSN) entirely depends on the efficiency of an underlying MAC/PHY protocol.

E-2-E Packet Delay

Average E-2-E delay comparison for UDP and TCP variants using IEEE802.15.4 as underlying MAC/PHY protocol is shown in the Figure 16. Channel sensing (RTS/CTS) and queuing delay (due to multimedia nature of the data) are the main causes of high E-2-E delay. From Figure 16 we can see that except SACK, NRENO and TAHOE rest all have an E-2-E packet delay of more than 200msec which is again not in accordance to QoS specifications for wireless multimedia and event driven applications. As compared to IEEE802.11the transport protocols using IEEE802.15.4 has showed better performance by exhibiting average E-2-E packet delay of 441.65 msec for UDP and lowest of 74.76 msec for TCP SACK.

Figure 13. Average power (watts) consumed by source nodes using IEEE 802.15.4

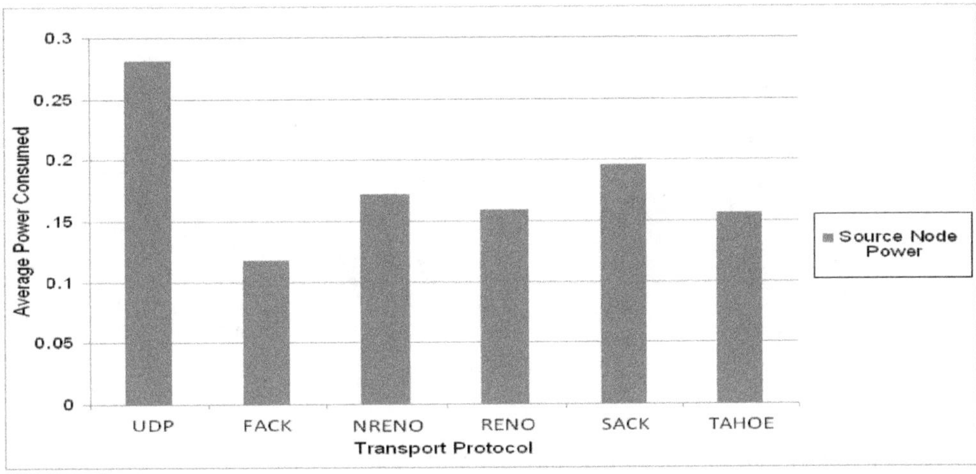

Figure 14. Average power (watts) consumed by relay nodes using IEEE 802.15.4

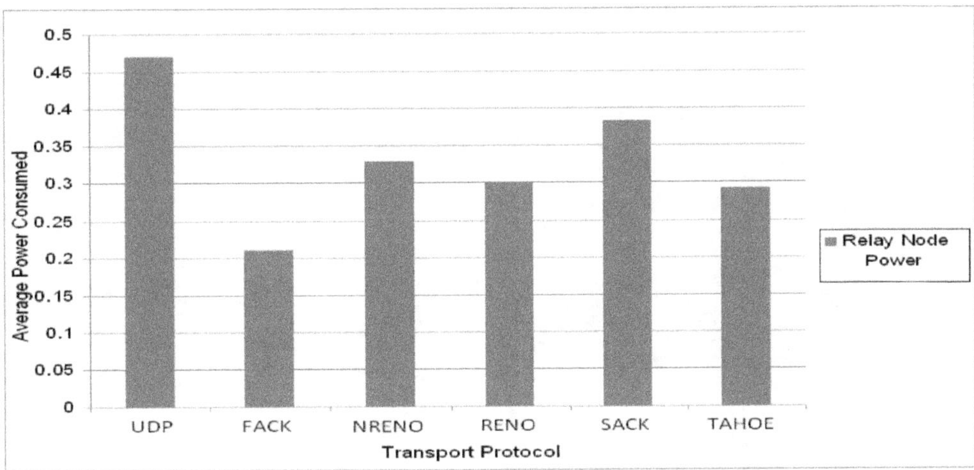

Figure 15. Average power (watts) consumed by sink nodes using IEEE 802.15.4

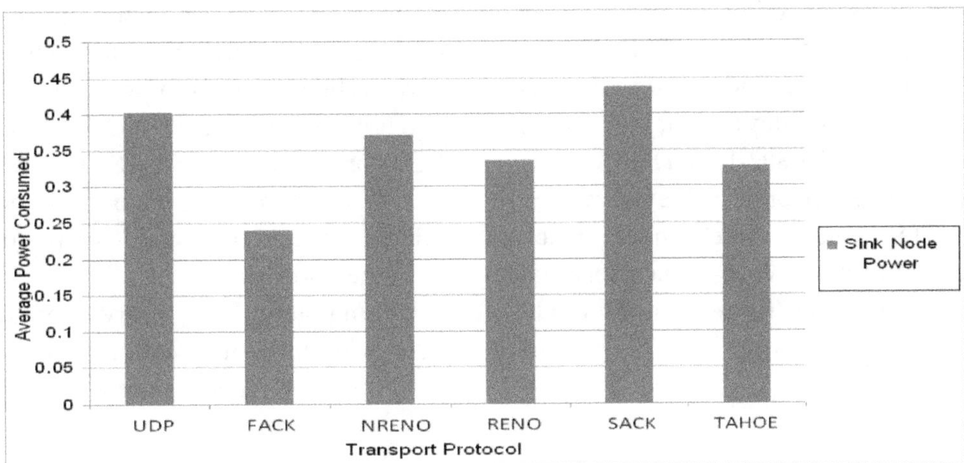

Figure 16. Average E-2-E Packet Delay (Sec) Analysis using IEEE802.15.4

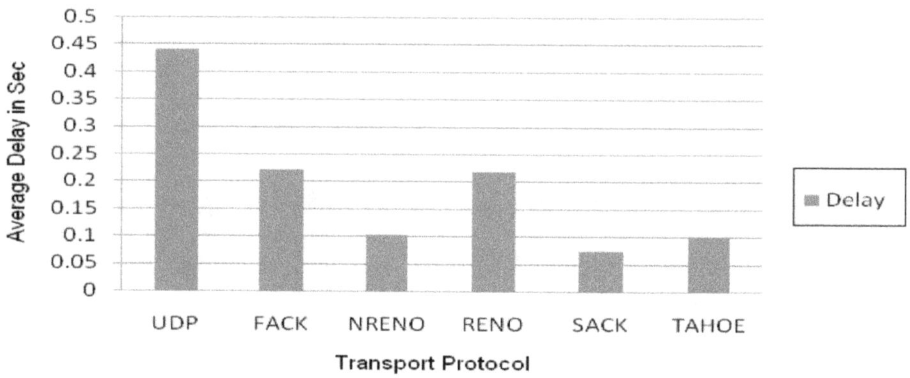

Figure 17. Average Jitter (Sec) Analysis using IEEE 802.15.4

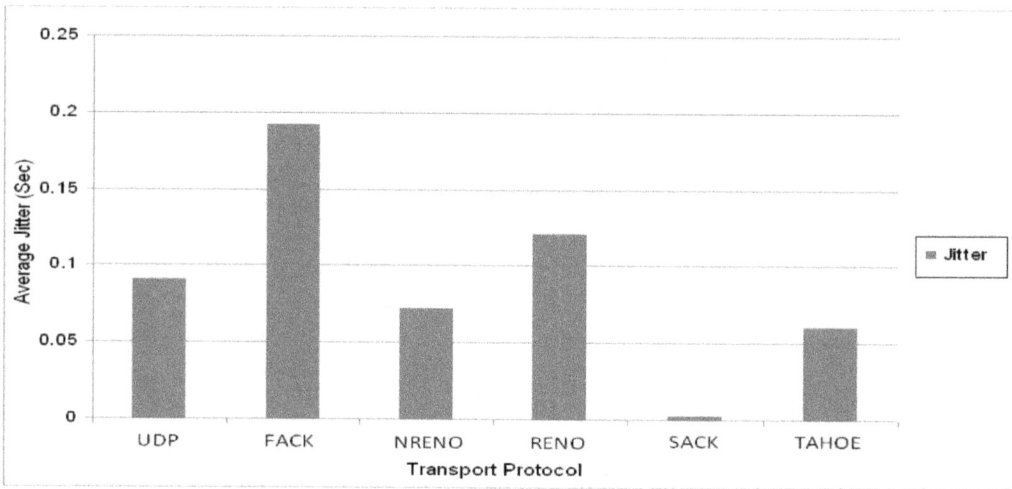

The pivotal reason for this efficiency (besides the protocol congestion handling algorithms as we have already discussed in 5.1.2 and 5.1.3) in average E-2-E packet delay is that IEEE802.15.4 standard also support time and rate sensitive applications because of its ability to operate in TDMA access mode. This IEEE802.15.4 TDMA mode of operation is being supported by the Guaranteed Time Slot (GTS) feature of the standard which helps in preventing interferences from other sensing nodes with in WSN (Krishnamurthy et al., 2008).

Jitter Analysis

In this section we are going to present the jitter analysis that we performed for UDP, TCP and its variants like FACK, NewReno, Reno, SACK, and TAHOE that uses IEEE 802.15.4 as underlying MAC protocol for RTS/CTS ON. The results are shown in Figure 17. A majority of the reasoning related to the protocol architectural role has been discussed previously.

From Figure 17 we observe that except FACK every other protocol performs well using IEEE802.15.4 as underlying MAC/PHY protocol.

Using IEEE802.15.4 in general exhibits low jitter for the case of UDP and TCP variants for the reason that IEEE802.15.4 standard supports quick response (from sleep to active mode) in 15msec or less and therefore the average latency can be very low and hence the devices can be very responsive. Another reason is that IEEE802.15.4 MAC layer is fundamentally the same as that of IEEE802.11 (CSMA/CA system) together with some optional time slot (beacon frames, such frames are used by the coordinator to describe the channel access mechanism to other nodes) structure and security functionality.

From the graph we have drawn the following conclusions:

A. UDP: jitter is less than 0.1 sec compared to 0.5427 sec for IEEE802.11, an 83.27% improvement in case of UDP that uses IEEE802.15.4.
B. TCP FACK: that uses IEEE802.15.4 exhibits an average jitter of 0.1921 sec as compared to 0.1388 sec for IEEE802.11 case (38.4% worst behavior than TCP FACK that uses IEEE802.11).
C. TCP NewReno: exhibits an average jitter of 0.0119 sec compared to 0.13 sec for IEEE802.11 case a substantial improvement of 90.85% which is quite significant.
D. TCP Reno: exhibits an average jitter of 0.1207 sec compared to 0.1325 sec for IEEE802.11 and improvement of 8.91% as compared to TCP Reno with IEEE802.11.
E. TCP SACK: exhibits an average jitter of 0.0023 sec compared to 0.13 sec for IEEE802.11 and improvement of 98.23% as compared to TCP SACK with IEEE802.11.
F. TCP Tahoe: exhibits an average jitter of 0.0603 sec compared to 0.1461 sec for IEEE802.11 and improvement of 58.73% as compared to TCP Tahoe with IEEE802.11.

DISCUSSION AND CONCLUSION

In this paper we tried to map the impact of MAC/PHY layer protocol on transport protocol functionality, for which we took the case of multihop wireless ad-hoc and WPAN for IEEE802.11 and IEEE802.15.4. We actually imposed every possible physical constraint to these networks like the sensing range, interference range, bandwidth etc. We also investigated the power consumption of independent nodes in these networks just to graph the power efficiency of these transport protocols. Following insightful conclusions have been drawn from the above simulations:

IEEE 802.11

Following are the conclusions drawn for the case of IEEE802.11

1. With RTS/CTS ON average throughput is increased
2. With RTS/CTS ON average packet drop rate is reduced.
3. With RTS/CTS ON average E-2-E packet delay is increased.
4. With RTS/CTS ON average 1-3% energy efficiency is achieved.
5. With RTS/CTS ON the average Jitter is increased.

IEEE 802.15.4

Following are the conclusions drawn for the case of IEEE802.15.4

1. UDP performed well having high throughput and lower average data loss as compared to IEEE802.11 with RTS/CTS ON/OFF.
2. Approximately on an average 2-4% more data loss is observed in case of TCP variants when compared to IEEE802.11.

3. For UDP average E-2-E packet delay dropped to a value approximately half to the case of IEEE802.11 with RTS/CTS ON.
4. Similar improved average E-2-E packet delay behavior was also observed in case of TCP variants.
5. Significant reduction in average power consumption was observed as compared to IEEE802.11.
6. Improved Jitter behavior was observed when compared to IEEE802.11 case.

We have seen from the above results that there seems to be a direct relationship between the PHY/MAC layer and Transport Layer protocols for getting high throughput and minimal E-2-E packet delay. For achieving these objectives we have to take following things into account

1. While network planning the transmitting and receiving node's distances should be calculated according to the Equations (1-5) to minimize the packet retransmissions as a result of collisions at the receiving node due to hidden node's transmission. Here packet retransmission not only introduces the delay but also increase the average per packet power required for successful communication.
2. To overcome the effects of hidden node the RTS/CTS should be used at the MAC/PHY level.
3. Another factor, although we did not mention or evaluated above, is the idea of how long the packet stays in MAC buffer (buffer delay) and the actual packet processing time (the time the MAC layer receives packet from the network layer to the time it successfully transmit the last bit of the packet). This helps in combating against E-2-E packet delay challenge by the use of estimation theory for packet processing time at intermediate node. We planned to use this piece of information including some other lower layer information. e.g., channel condition monitoring, channel occupancy etc in building our cross-layer communication model for WSN transport protocol.
4. Also from the results it is obvious that a considerable amount of data was lost at MAC/PHY layer, which then reduced the reliability of the transport protocol. The data loss mainly occurred either due to congestion and collision. Since we addressed the issue of hidden node (means no packet collision at the receiving end) the only thing left was the loss due to congestion. Ideally the protocol should control the rate of data flow and if some data has been lost there should be a rapid data recovery mechanism. So a need of such data recovery mechanism is deemed necessary.
5. The transport layer protocol should also have a clear picture of the wireless channel quality (packet error rate, PER) so that it is aware of the reason of data loss. Again in this case it will be guided by the MAC/PHY layer.

In summary for achieving high throughput and data reliability for a transport protocol we have to relate the functionalities of the two layers (cross-layering). Merging the functionalities of the two layers enables the transport layer protocol to handle the data more efficiently and delivers it to the destination node more reliably with minimal delay. Also in case of data packet loss which results either due to collisions, crosstalk or path failures due to node's energy depletion etc. would invoke fast data recovery mechanism for retrieving the lost data packet with in marginal time limits without affecting the application's QoS. Based on this point we can predict our future research direction. Our next objective is to implement a transport protocol for wireless multimedia applications that guarantee high throughput by having effective congestion control mechanism while simultaneously providing high data reliability by ensuring rapid loss recoveries by introducing the concept of distributed memory storage in WSN.

ACKNOWLEDGMENT

This paper is an extension of the paper we submitted for the 7th International Conference on Advances in Mobile Computing & Multimedia (MoMM2009), Kuala Lumpur, Malaysia, 14-16 December, 2009.

REFERENCES

Akyildiz, I. F., Melodia, T., & Chowdhury, K. R. (2007). A Survey on wireless multimedia sensor networks. *Computer Networks, 51*(4), 921–960. doi:10.1016/j.comnet.2006.10.002

Brakmo, L., & Peterson, L. (1995). TCP Vegas: End to End Congestion Avoidance on a Global Internet. *IEEE Journal on Selected Areas in Communications, 13*(8), 1465–1480. doi:10.1109/49.464716

Dunkels, A., Voigt, T., Ritter, H., & Alonso, J. (2004, June). Distributed TCP Caching for Wireless Sensor Networks. In *Proceedings of the 3rd Annual Mediterranean Ad Hoc Networking Workshop*, Turkey.

He, T., Stankovic, J. A., Lu, C., & Abdelzaher, T. (2003, May). SPEED: A Stateless Protocol for Real-Time Communication in Sensor Networks. In *Proceedings of 23rd International Conference on Distributed Computing Systems (ICDCS)*, USA (p. 46).

Hull, B., Jamieson, K., & Balakrishnan, H. (2004, November). Mitigating Congestion in Wireless Sensor Networks. In *Proceedings of the 2nd ACM Conference on Embedded Networked Sensor Systems (ACM SenSys)*, USA (pp. 134-147).

IEEE. *802.15 WPANTM Task Group*. (n.d.). Retrieved August 2008, from http://ieee802.org/15/index.html

IEEE. *P802.11, Main General Info Page*. (n.d.). Retrieved August 2008, from www.ieee802.org/11/main.html

Iyer, Y. G., Gandham, S., & Venkatesan, S. (2005, October). STCP: a Generic Transport Layer Protocol for Wireless Sensor Networks. In *Proceedings of the 14th IEEE International Conference on Computer Communications and Networks (ICCCN)*, USA (pp. 449-454).

Krishnamurthy, V., Fowler, K., & Sazonov, E. (2008). The Effect of Time Synchronization of Wireless Sensors on the Modal Analysis of Structures. *International Journal of Smart Materials Systems and Structures, 17*(11), 1–13.

Layuan, L., Chunlin, L., & Peiyan, Y. (2007). Performance evaluation and simulations of routing protocols in ad hoc networks. *International Journal of Computer Communications, 30*(8), 1890–1898. doi:10.1016/j.comcom.2007.02.015

Li, J., Blake, C., DeCouto, D. S. J., & Lee, H. I. (2001, July). Capacity of Ad Hoc wireless networks. In *Proceedings of the 7th International Conference on Mobile Computing and Networking (SIGMOBILE)*, Italy (pp. 61-69).

Mao, S., Bushmitch, D., Narayanan, S., & Panwar, S. S. (2003, October). MRTP: a Multiflow Real time Transport Protocol for ad hoc networks. In *Proceedings of the IEEE 58th Vehicular Technology Conference (VTC)*, USA (Vol. 4, pp. 2629-2634).

Park, S. J., Vedantham, R., Sivakumar, R., & Akyildiz, I. F. (2004, May). A scalable approach for reliable downstream data delivery in wireless sensor networks. In *Proceedings of the 5th ACM International Symposium on Mobile Ad Hoc Networking and Computing (ACM Mobihoc)*, Japan (pp. 78-89).

Perkins, C. E., Royer, E. B., & Das, S. (2003). *Ad hoc On-Demand Distance Vector (AODV) Routing (IETF RFC 3561)*.

Potdar, V., Sharif, A., & Chang, E. (2009, May). Wireless Sensor Networks: A Survey. In *Proceedings of the 23rd International Conference on Advanced Information Networking and Applications Workshops (WAINA)*, UK (pp. 636-641).

Rappaport, T. S. (1995). *Wireless Communications: Principles and Practice*. Upper Saddle River, NJ: Prentice Hall.

Sankarasubramaniam, Y., Akan, O. B., & Akyildiz, I. F. (2003, June). ESRT: Event-to-Sink Reliable Transport in Wireless Sensor Networks. In *Proceedings of the 4th ACM International Symposium on Mobile Ad Hoc Networking and Computing (ACM Mobihoc)*, USA (pp. 177-188).

Sharif, A., Ahmad, R. F., & Potdar, V. (2010, April). Adaptive Channel Coding and Modulation Scheme Selection for Achieving High Throughput in Wireless Networks. In *Proceedings of the 6th International Symposium on Frontiers of Information Systems and Network Applications (FINA 2010)*, WA.

Sharif, A., Potdar, V., & Chang, E. (2009, May). Wireless Multimedia Sensor Network Technology: A Survey. In *Proceedings of the 7th IEEE International Conference on Industrial Informatics (INDIN)*, UK (pp. 606-613).

Sharif, A., Potdar, V., & Rathnayaka, A. J. D. (2009, December). Performance Evaluation of different Transport layer protocols on the IEEE 802.11 and IEEE 802.15.4 MAC/PHY layers for WSN. In *Proceedings of the 7th International Conference on Advances in Mobile Computing & Multimedia (MoMM2009)*, Malaysia (pp. 300-306).

Sharif, A., Potdar, V., & Rathnayaka, A. J. D. (2010, April). Prioritizing Information for Achieving QoS Control in WSN. In *Proceedings of the International Conference on Advanced Information Networking and Applications (AINA)*, WA.

Stann, F., & Heideman, J. (2003, October). RMST: Reliable Data Transport in Sensor Networks. In *Proceedings of the 1st IEEE International Workshop on Sensor Network Protocols and Applications (SNPA)*, USA (pp. 102-113).

TCP. (n.d.). Retrieved August 2008, from http://www.ibiblio.org/pub/ docs/rfc/rfc793.txt

The Network Simulator — NS-2. (n.d.). Retrieved July 2008, from http://www.isi.edu/ nsnam/ns/index.html

UDP. (n.d.). Retrieved August 2008, from http://tools.ietf.org/html/rfc768

Wan, C. Y., Campbell, A. T., & Krishnamurthy, L. (2002, September). PSFQ: A Reliable Transport Protocol for Wireless Sensor Networks. In *Proceedings of ACM International Workshop on WSN and Applications (WSNA)*, USA (pp. 1-11).

Wan, C. Y., Eisenman, S. B., & Campbell, A. T. (2003, October). CODA: Congestion detection and avoidance in sensor networks. In *Proceedings of the 1st ACM Conference on Embedded Networked Sensor Systems (ACM SenSys)*, USA (pp. 266-279).

Zhang, H., Arora, A., Choi, Y. R., & Gouda, M. G. (2005, May). Reliable Bursty Convergecast in Wireless Sensor Networks. In *Proceedings of the 6th ACM International Symposium on Mobile Ad Hoc Networking and Computing (ACM Mobihoc)*, USA (pp. 266-276).

ENDNOTES

1. For more details about the simulation results see Appendix A.
2. By counting the duplicate ACK's the algorithm computes the outstanding network data thereby inflates *cwnd* on receiving each duplicate ACK, causing the data to be transmitted as *cwnd* becomes large enough. The algorithm allows one halved of the *cwnd* of the new data to be transmitted following a fast retransmission.
3. For more details about the simulation results see Appendix A.

APPENDIX A: JITTER PLOTS FOR VARIOUS TRANSPORT PROTOCOLS USING IEEE 802.11 AND IEEE 802.15.4 MAC PROTOCOLS

Figure 18. Jitter analysis for UDP with IEEE802.11RTS/CTS OFF

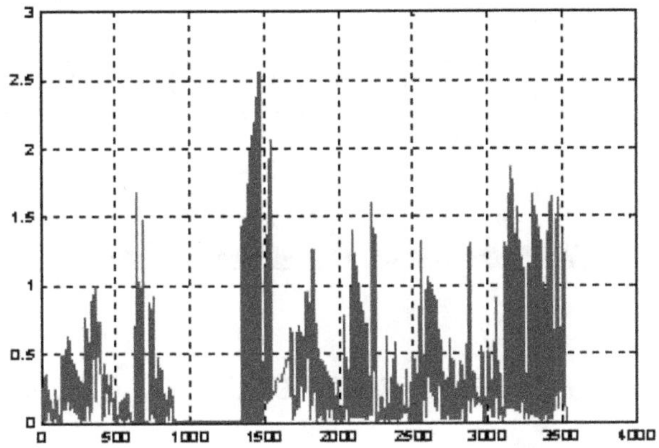

Figure 19. Jitter analysis for UDP with IEEE802.11RTS/CTS ON

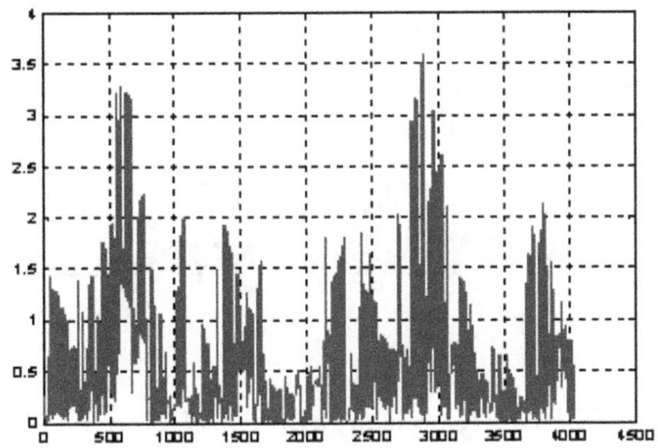

Figure 20. Jitter analysis for TCP FACK with IEEE802.11RTS/CTS OFF

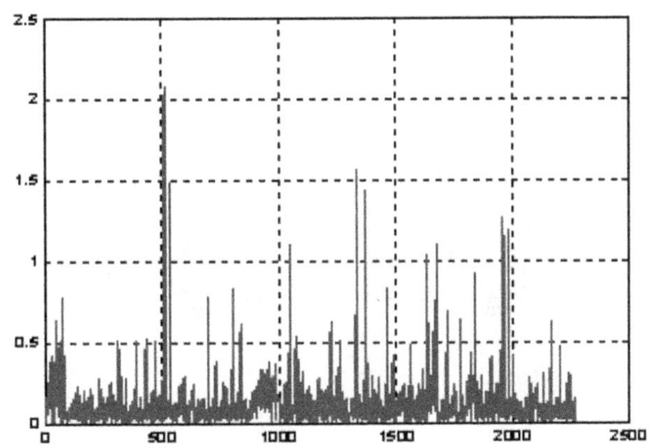

Figure 21. Jitter analysis for TCP FACK with IEEE802.11RTS/CTS ON

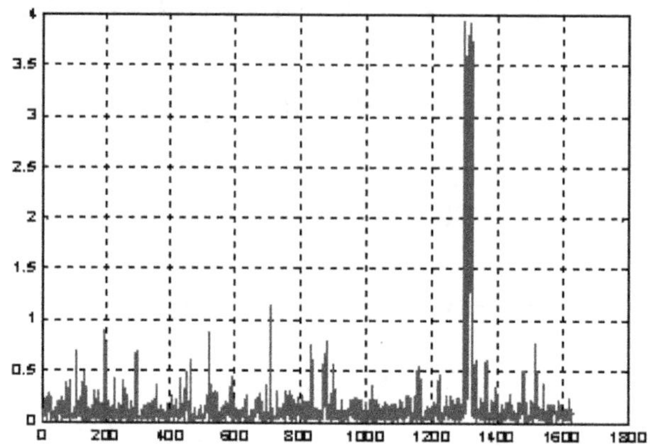

Figure 22 Jitter analysis for TCP NewReno with IEEE802.11RTS/CTS OFF

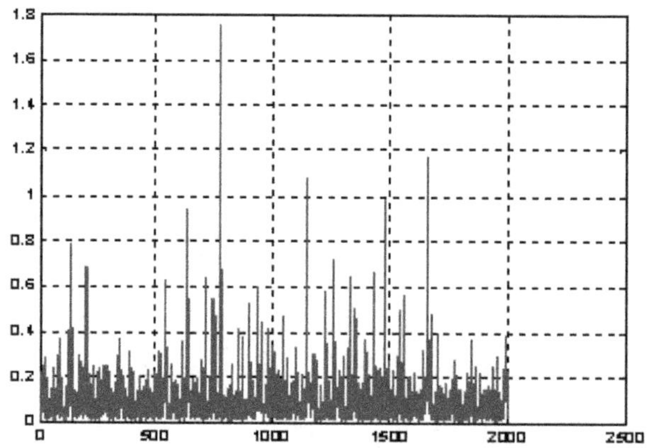

Figure 23. Jitter analysis for TCP NewReno with IEEE802.11RTS/CTS ON

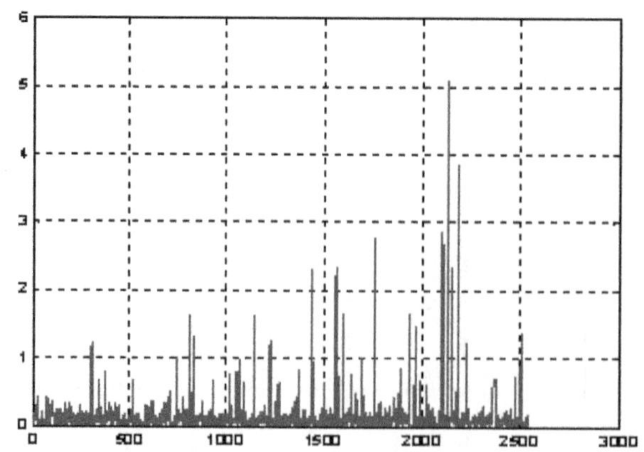

Figure 24. Jitter analysis for TCP Reno with IEEE802.11RTS/CTS OFF

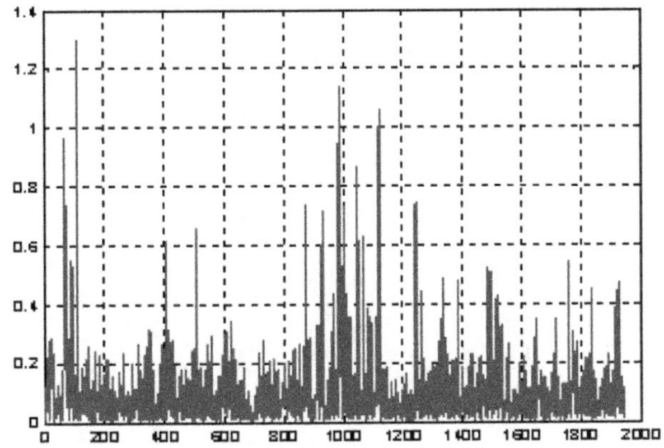

Figure 25. Jitter analysis for TCP Reno with IEEE802.11RTS/CTS ON

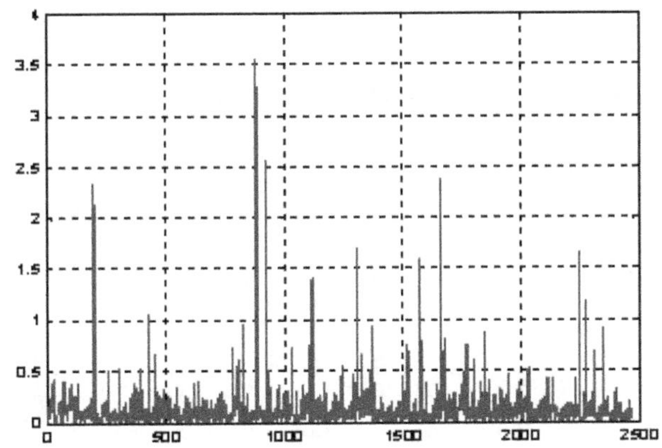

Figure 26. Jitter analysis for TCP SACK with IEEE802.11RTS/CTS OFF

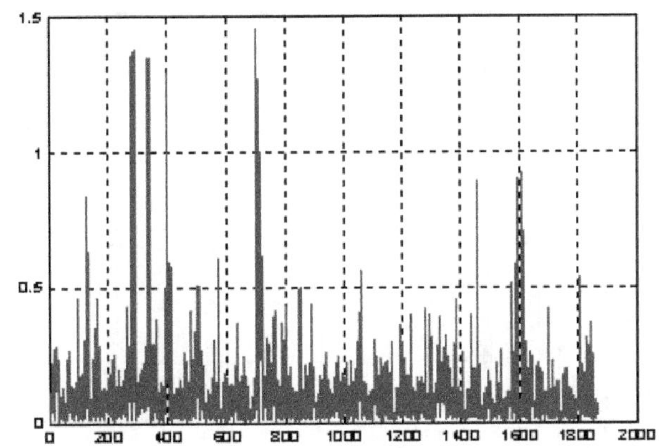

Figure 27. Jitter analysis for TCP SACK with IEEE802.11RTS/CTS ON

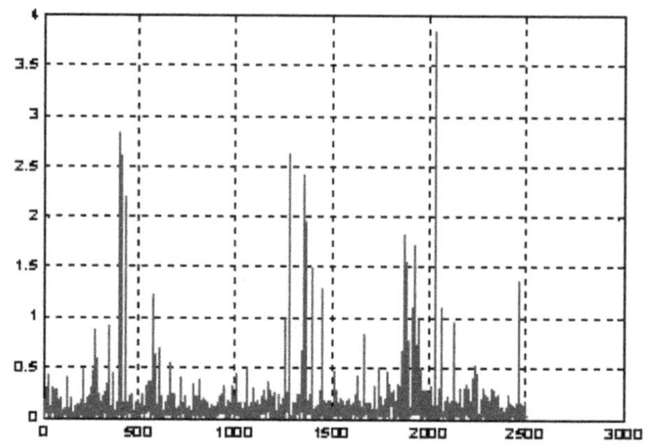

Figure 28. Jitter analysis for TCP TAHOE with IEEE802.11RTS/CTS OFF

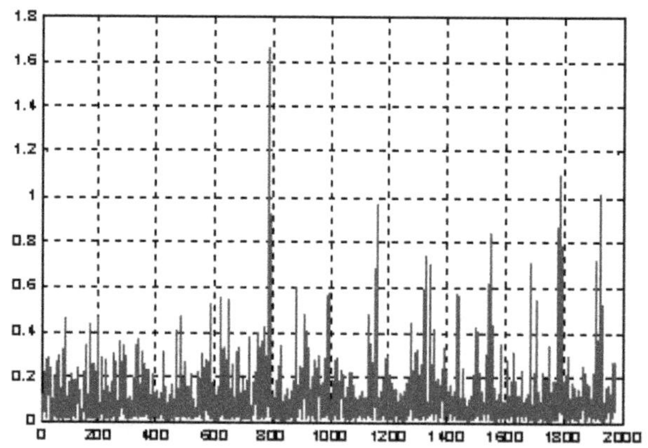

Figure 29. Jitter analysis for TCP TAHOE with IEEE802.11RTS/CTS ON

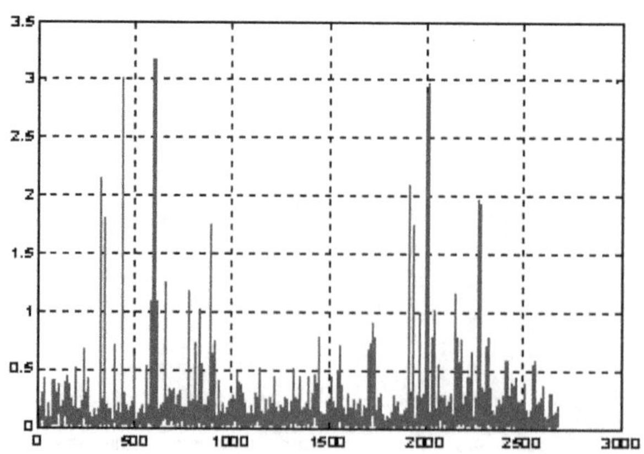

Figure 30 Jitter analysis for TCP Vegas with IEEE802.11 with RTS/CTS OFF

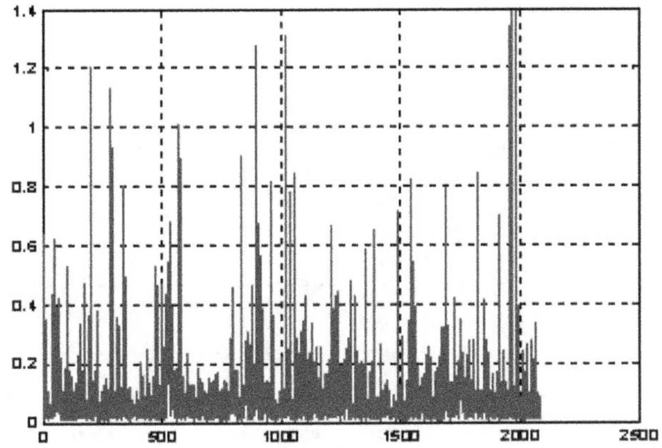

Figure 31. Jitter analysis for TCP Vegas with IEEE802.11 with RTS/CTS ON

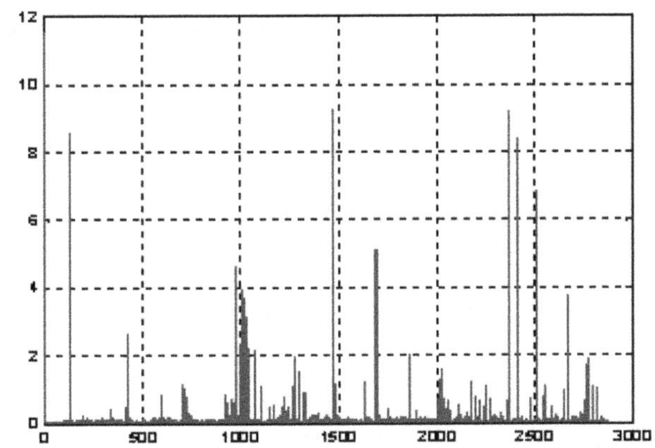

Figure 32. Jitter analysis for UDP with IEEE802.15.4

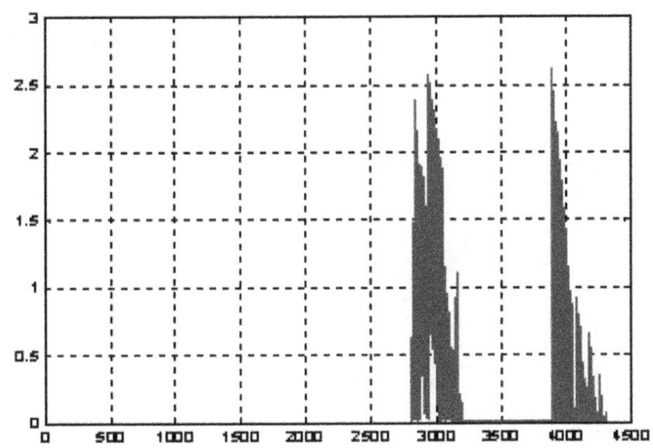

Figure 33. Jitter analysis for TCP FACK with IEEE802.15.4

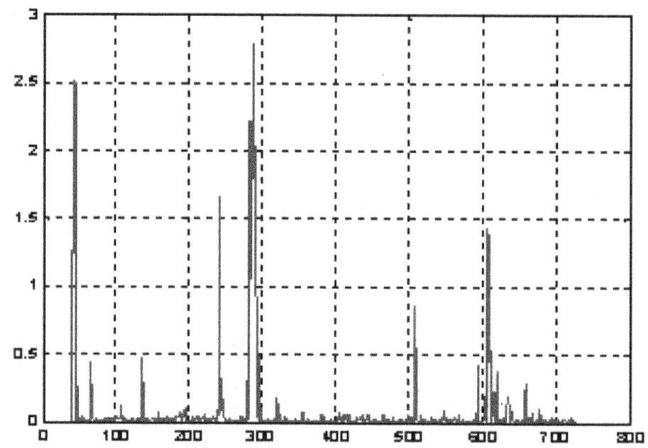

Figure 34. Jitter analysis for TCP NewReno with IEEE802.15.4

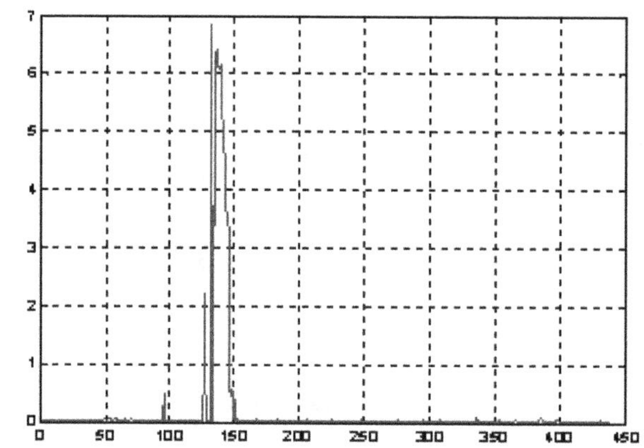

Figure 35. Jitter analysis for TCP NewReno with IEEE802.15.4

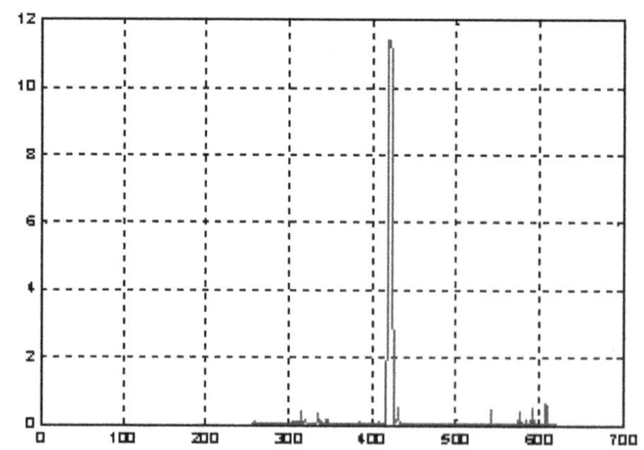

Figure 36. Jitter analysis for TCP NewReno with IEEE802.15.4

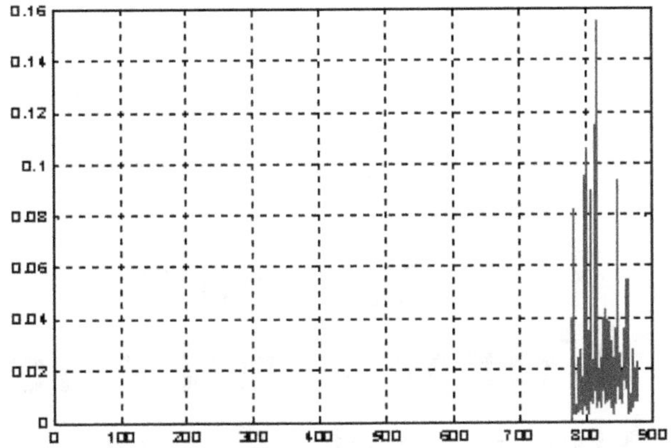

Figure 37. Jitter analysis for TCP NewReno with IEEE802.15.4

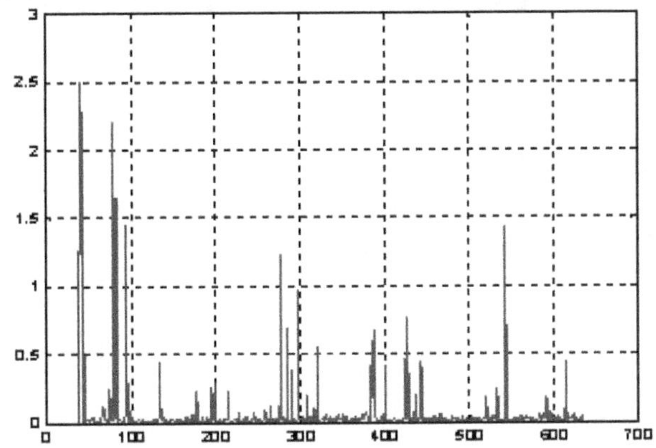

This work was previously published in International Journal of Business Data Communications and Networking, Volume 6, Issue 3, edited by Debashis Saha and Varadharajan Sridhar, pp. 1-30 , copyright 2010 by IGI Publishing (an imprint of IGI Global).

Section 2
Communication Network Security:
Improved Architecture and Prevention Mechanisms

Chapter 6
Secure Route Discovery in DSR against Black Hole Attacks in Mobile Ad Hoc Networks

P. Subathra
Thiagarajar College of Engineering, India

S. Sivagurunathan
Gandhigram Rural Institute – Deemed University, India

ABSTRACT

A Mobile Ad hoc Network (MANET) is a collection of wireless nodes communicating over multi-hop paths without any infrastructure. Nodes must cooperate to provide necessary network functionalities. The security in routing protocols like Dynamic Source Routing (DSR) can be compromised by a "Black Hole" attack. Here, a malicious node claims to have the shortest path to the destination and attracts all traffic and drops them, leading to performance degradation. The situation becomes worse when two or more nodes cooperate and perform the "Cooperative black hole" attack. This chapter proposes a solution based on probing to identify and prevent such attacks. The proposed solution discovers a secure route between the source and destination by identifying and isolating the attacking nodes. Simulation results show that the protocol provides better security and performance in terms of detection time, packet delivery ratio, and false negative probability in comparison with trust and probe based schemes.

INTRODUCTION

Wireless networks can be implemented either as an infrastructure based networks or an infrastructure less networks. The infrastructure based networks use fixed base stations, which are responsible for coordinating communication between the mobile hosts (nodes). The Mobile Ad hoc Network (MANET) falls under the class of infrastructure less networks (Sergio Marti et al., 2000).

Ad hoc network is a collection of nodes that do not rely on a predefined infrastructure to keep the network connected. So the functioning of ad hoc network is dependent on the trust and cooperation amongst the nodes. Nodes help each other in conveying information about the topology of the

DOI: 10.4018/978-1-61350-477-2.ch006

network and share the responsibility of managing the network. Hence in addition to acting as hosts, each mobile node does the function of routing and relaying messages for other mobile nodes (Deng et al., 2002; Milanovic et al., 2004). All network activities, such as discovering the topology and delivering data packets, have to be executed by the nodes themselves collectively.

The MANETs are categorized into two types: closed MANET and open MANET (Miranda & Rodrigues, 2002). In a closed MANET, all mobile nodes cooperate with each other towards a common goal, such as emergency search/rescue or military and law enforcement operations. In an open MANET, different mobile nodes with different goals share their resources in order to ensure global connectivity. However, some resources are consumed quickly as the nodes participate in the network functions. For instance, battery power is considered to be the most important in a mobile environment. An individual mobile node may attempt to benefit from other nodes, but refuse to share its own resources. Such nodes are called selfish /misbehaving nodes and their behavior is termed as selfishness/misbehavior (Miranda & Rodrigues, 2002). One of the major sources of energy consumption in the mobile nodes of MANETs is wireless transmission (Rerney & Nilsson, 2001). A selfish node may refuse to forward data packets for other nodes in order to conserve its own energy.

The misbehavior of a node can be due to any one of the following reasons: A node may try to conserve its energy by not forwarding data packets for other nodes from the network; node may be overloaded, broken, compromised or congested in addition to intentionally being selfish/malicious (Yang et al., 2006 & Hubaux et al., 2001). Misbehavior can be divided into two categories (Yang et al., 2006): routing misbehavior (failure to behave in accordance with a routing protocol) and packet forwarding misbehavior (failure to forward the packets). This work proposes an algorithm that enables packet forwarding misbehavior detection.

A single black hole attack (Papadimitratos & Haas, 2002) is a specific type of attack, where a malicious node injects false route replies to the route requests it receives by advertising itself as having the shortest path to a destination. These fake replies can be fabricated to divert network traffic through the malicious nodes for eavesdropping, or simply to attack all traffic to it in order to perform a Denial of Service (DoS) attack by dropping the received packets. In a cooperative black hole attack two or more nodes cooperate amongst themselves and when a packet is forwarded to any of these nodes they collude with each other and drop it. Since the nodes cooperate with each other they do not reveal the identity of the node that actually drops the packet to the external world and thus the trust information given by these would be a fake one. This makes the identification of such attacks more difficult.

This article provides a novel approach to detect both the single and cooperating black hole nodes and to remove them from further routing path to improve the network's performance. The rest of the article presents the state of the art, the model and the assumptions related to this work along with the proposed algorithm. The evaluation of the algorithm is also presented.

RELATED WORK

The literature review carried out in this area revealed that the problem of packet dropping attack is basically handled by five different types of techniques as shown in Figure 1.

Authentication Based Techniques

Authenticating the nodes of a network may be the first and foremost preventive measure against attacks on MANETs. Cryptographic based authen-

Figure 1. Taxonomy of techniques used to detect/ prevent packet dropping attacks in MANETs.

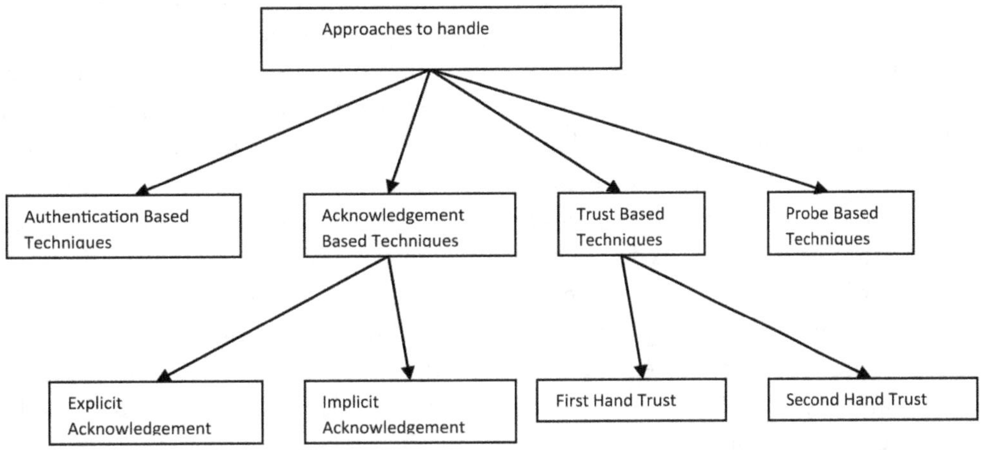

tication schemes have been extensively used in the field of MANETs to provide security. Some of it and other such techniques which try to protect the MANET from packet dropping attacks are:

A threshold cryptography based (Zhou & Hass, 1999) approach is used on all types of routing techniques to provide security to the MANET. Here, only the authenticated nodes are allowed to take part in the functionalities of the network, thus preventing malicious nodes to enter inside the network. In this approach, a (n, t+1) scheme using a certification authority to authenticate nodes is used. But the main drawback is that, it demands preconfiguration of servers and nodes. It is also not always possible to have the availability of (t+1) servers for nodes seeking authentication. Moreover, this technique cannot address a situation in which a node authenticated as a good node, but can become compromised and act maliciously later. To address the issue of the compromised nodes the Ariadne (Hu et al., 2002) was proposed. It is a modification to the existing Dynamic Source Routing protocol (DSR) to protect against compromised nodes. Here an extremely efficient symmetric cryptography has been used. The initiator can verify each transitional node on the path to the destination present in the ROUTE REPLY message. Three different schemes such as TESLA (Timed Efficient Stream Loss Tolerant Authentication), Digital signature and pair wise shared secret key were used for node authentication. The limitations are that the authentication requires excessive preconfiguration. Moreover, TESLA requires clock synchronization which is an unrealistic constraint upon the ad hoc networks.

An authentication mechanism based on the hash function, the Message Authentication Code (MAC), and the Pseudo Random Function (PRF) (Luo et al., 2008) on top of AODV protocol was proposed to identify single black hole nodes in the absence of clock synchronization. But here too there is a difficulty of unlimited message authentication and also there is also a chance for the malicious nodes to forge the hash key. This technique does not address the situation under which multiple nodes cooperate together to act as cooperative black hole nodes. So as an enhancement to this, Zhao and Zhou (2009) proposed a two-authentication mechanism, based on the hash function, the Message Authentication Code (MAC) and the Pseudo Random Function (PRF) to provide fast message verification and group identification which helps in identifying multiple black holes cooperating with each other and also discovers a safe routing. But this involves heavy processing which deprives the nosed of the battery power, which is a valuable resource in a MANET.

Acknowledgment Based Techniques

Acknowledgement based methods trust and uses their immediate neighbor's acknowledgement to know the forwarding nature of the nodes which are ahead of it in the routing path. This is done either through explicit or implicit acknowledgment.

A watch dog and path rater (Sergio Marti et al., 2000) based approach was proposed on a DSR algorithm to identify and exclude malicious nodes in MANET. Here each node acts as a watch dog to its next hop neighbor to check its forwarding nature using an implicit acknowledgment technique. The path rater gives a rating to each node depending on its forwarding (goodness) nature so that paths with good rating could be selected for further routing. But the main limitation of this method is that the watch dog cannot detect the malicious nodes in the presence of ambiguous collisions, receiver collisions, limited transmission power, false misbehavior, collision and partial dropping. Moreover the path rater cannot reconvert a misbehaving node into a trusted one if that node starts behaving properly after sometime. As an enhancement to this technique (Patcha & Mishra, 2003) collaborative security architecture was proposed preventing black hole attack on an Ad hoc On- demand Distance Vector (AODV) protocol in MANETs. They maintained a list of ordinary, trusted and watch dog nodes. Watch dogs are selected from the list of trusted nodes. Each transmitting node should send a message to its nearby watch dog to monitor its next hop neighbor. Colluding attack is also addressed by this scheme. The main limitations are that they address only static or very low mobility node scenario and moreover introduction of new packet types increase the network overhead significantly.

An explicit acknowledgment based approach (Liu et al., 2007) was proposed for the detection of routing misbehavior on a DSR protocol in MANETs. The basic idea of this approach is to send Two-hop acknowledgment packets in the opposite direction of the routing path to monitor the next hop node's forwarding nature. The drawbacks are that it increases overhead and cannot handle cooperative attacks. There is also no punishment scheme for the misbehaving node.

Though the acknowledgement based schemes are quite interesting, the explicit acknowledgement techniques increase the traffic heavily and the implicit acknowledgement techniques cannot detect malicious nodes in the presence of ambiguous collisions, receiver collisions, limited transmission power, false misbehavior, collision and partial dropping. Hence these techniques are not effective in detecting cooperative black hole nodes.

Trust Based Techniques

In trust based schemes, all the nodes in the MANET keep track of all other nodes and rate them according to their goodness depending on whether it forwards the packets given to it or not. The trust messages maintained by a node may either be a first hand trust or a second hand trust. The first hand trusts are those that are directly observed and calculated by the node which maintains them. The second hand trusts are those values which are received from the node's neighboring trusted nodes. Most of the algorithms used in MANET use a combined approach where they use both the first hand and second hand trusts. Various approaches have been done based on these trust based schemes to protect the MANET from black hole attacks. Some of them are:

A Distributed Trust model based on a decentralized approach of trust management was proposed by Rahman and Hailes (1997) to protect the network where all entries or nodes act either as a recommender or a requestor. The trust values are computed by means of the recommended trust value of the target and its recommenders. But this method does not provide measures for safeguarding against false malicious recommendations about other nodes. A CONFIDANT (Cooperation Of Nodes: Fairness In Dynamic Ad-hoc NeTworks) protocol (Buchegger & Le Boudec, 2002) was proposed later which was a reputation based system where the nodes within a MANET rate each other based on observed behavior. Nodes that are deemed to be misbehaving are placed on a blacklist and are consequently isolated. This scheme too does not protect against false accusations and hence it is susceptible to blackmailing. Here each and every node has to maintain a trust level for all other nodes in the MANET which would consume much of the memory space of these nodes. Moreover this scheme does not address the cooperative black hole attack.

A trust based scheme (Bhalaji et al., 2008) was proposed later to increase security of DSR routing protocol in MANET by avoiding packet dropping attacks. Like the other trust based mechanisms this too suffers from the extra load of monitoring and maintaining the trust worthiness of all other nodes available in the MANET. Later two of the authors who contributed to this work made an enhancement to their earlier work by proposing a Grudger protocol (Bhalaji & Shanmugam, 2009) which simulates the behavior of birds grooming parasites off each others head. They use three different categories of nodes: Unknown, Known and Companion. Depending on the association between neighbors in the MANET, nodes select a secure routing path and thus eliminate the occurrence of black hole nodes in the routing path. This proposal too does not address the cooperative black hole attack directly.

To address the cooperative black hole attack, the "Fidelity Table" based approach (Tamilselvan & Sankaranarayanan, 2008) was the next development in this area. Here every participating node will be assigned a fidelity level that acts as a measure of reliability of that node. If this drops to zero, the node is considered to be a Black hole and is eliminated. The main drawback is that the nodes are assumed to be already authenticated. Moreover it addresses only cooperative black holes and even in the presence of a single black hole which is very common in MANETs the black hole node's next hop neighbor is also accused to be black hole. There is no means to identify or recognize a black hole node if it has started to behave properly.

Basically all the trust based schemes need to perform a continuous monitoring of the nodes in the MANET environment. A new node that enters this environment is not treated as a good node even though it is really one. Moreover the new nodes will have to participate in the routing process and forward packets to gain the trust of other nodes. Since the route selection algorithm of MANET is a dynamic one, there are always possibilities for some of these new nodes to be left out from being selected in the routing path. So these left out nodes will have little chance to prove their trustworthiness to the other nodes. In the presence of cooperative black hole nodes each and every node is treated as a separate single black hole nodes and the cooperating nature of it is never identified. These features of the trust based schemes make them less attractive for implementation in the presence of cooperative black hole attacks.

Probe Based Techniques

The probe based techniques are light weight techniques compared to the above mentioned techniques. But some of them use a hybrid approach were a trust table is maintained by all the nodes and in case of any deviations from the normal routing process, probing techniques are used to detect the malicious nodes and to eliminate them from the network. This avoids malicious node from further affecting the MANET.

A hybrid approach which used both the trust based and probe based method was proposed (Awerbuch et al., 2002) as an on-demand secure routing protocol, where all nodes maintain a weight list consisting of reliability metrics of the nodes within the network. This list is used in the route discovery phase to avoid faulty paths. Faulty links are given decreased rating and are avoided consequently. Probing techniques are useful in the identification of faults caused by non-malicious acts. The main drawback is that, the probing messages are distinguishable from data packets, so the malicious nodes may behave better during probing and notoriously under normal data transmissions. Their scheme is a hybrid approach which uses both the trust and probing based techniques which involves heavy processing for the individual nodes.

To overcome this and to simplify the processing a pure probe based scheme (Deng et al., 2002) was proposed later. This method addresses a single black hole attack, which performs a routing misbehavior in AODV protocol. If a node is suspected to be a black hole, then its next hop neighbor is checked whether it has a route to the destination, if it does not then that node is a black hole node. This technique fails to address cooperative black hole attack. If a routing path contains cooperative black hole nodes, then this probing technique identifies only the last node of this misbehaving list to be a (single) black hole node and the others escape with the help from its cooperating next hop neighbor. This provides way for the remaining malicious nodes to be included in the routing paths later on. Thus the performance of the MANET is degraded. A variation of the probing scheme was also proposed (Just & Evangelos, 2003) to identify malicious packet dropping in MANET. Probing is done from the farthest node to the nearest node and identifies the first node that acknowledges. Now probing is done through an alternate path to the next hop neighbor of the node which gave the acknowledgment in the earlier probing. But in this probing path the node ID of the acknowledging node is also appended. If still the same node acknowledges, then it is a black hole node. The implementation is done on a DSR protocol. This article too does not address the cooperative black hole attack.

An attempt was made by Ramasamy et al. (2007) to prevent a cooperative black hole attacks in MANET. Their method was based on each node maintaining a DRI (Data Routing Information) entry for all known nodes. Each node sends its DRI entry for the next hop and its immediate next hop nodes. Further examination of the DRI tables in the consequent path (next hops) to identify cooperative black holes is also done. But these are only proposed and implementation results are not available. Their proposal was later implemented and validated by Weerasinghe and Fu (Weerasinghe & Fu, 2008). The simulation results show that maintenance of the DRI table in each and every node increases the processing power which is an issue not yet addressed by them. Table 1 gives the summary of the related works. From the literature review the following conclusions were made:

The authentication based schemes were very effective in preventing black hole attacks since they used strong cryptographic techniques. But most of them required pre configuration and clock synchronization which may not always be possible in a MANET. The acknowledgement based schemes were able to identify compromised nodes which act as black holes. But the introduction of new packet types increased the overhead. Cooperative nodes still remain undetected.

In trust based schemes, routes are selected based on the trust level of each of the nodes in the routing path. Thus nodes which do not forward properly will not be included in any of the routing paths. So any member of single black hole nodes can be detected. These techniques cannot withstand false accusations and false recommendations. The nodes need to keep track of the forwarding nature of all other nodes in the MANET to have a secure routing guaranteed. The probe based schemes are light weight techniques which try to identify single black holes. Some of them use a hybrid approach by maintaining a table of reliability for all nodes in the MANET, and route

Table 1. Summary of related work done in providing security to MANET against packet dropping attacks

SNo	Authors	Features of the Scheme	Limitations
Authentication Based Schemes			
1	Zhou et al. (1999)	Threshold Cryptography.	Preconfiguration required.
2	Perrig et al. (2002)	Symmetric Cryptography.	Preconfiguration and Clock synchronization required.
3	Junhai Luo et al. (2008)	Message Authentication Code (MAC) and Pseudo Random Function (PRF)	Unlimited authentication required. Hash key can be forged.
4	Zhao et al. (2009)	Two-authentication (MAC & PRF)	Unlimited message authentication required.
Acknowledgment Based Schemes			
5	Sergio Marti et al. (2000)	Watch dog and Path rater.	Black holes undetected under ambiguous collision.
6	Patcha et al. (2003)	Ordinary, trusted and watch dog nodes.	New packet types increase overhead. Little or no movement nodes alone addressed.
7	Lin et al. (2007)	Two hop acknowledgement scheme.	Cooperative black hole nodes go undetected.
Trust Based Schemes			
8	Rahman et al. (2007)	Decentralized Distributed Trust.	False recommendations not handled.
9	Buchegger et al. (2002)	CONFIDANT protocol	False accusations not handled.
10	Tamilselvan et al. (2008)	Fidelity table.	Single black holes falsely accused as cooperative black holes.
11	Bhalaji et al. (2008)	Trust level maintenance.	Continuous monitoring and maintenance of trust table required.
12	Bhalaji et al. (2009)	Grudger Protocol.	Continuous monitoring and maintenance of trust table required.
Probe Based Schemes			
13	Awerbuch et al. (2002)	Weights based on reliability metric. Probing used.	Probing messages are distinguishable from data packets. Processing overhead is high.
14	Deng et al. (2002)	Next hop neighbor checked for reliability.	Cooperative black holes not detected.
15	Just et al. (2003)	Probes through alternate path.	Cooperative black hole nodes are not detected.
16	Ramasamy et al. (2007)	DRI table and Probing.	No simulation results available.
17	Weerasinghe et al. (2008)	Implemented the DRI table and probing scheme of Ramasamy et al (2007).	Processing overhead is high.

selection is based on the reliability metric. Once a malfunctioning is identified in this route, probing is used to identify the malicious node. But the processing overhead is much high. Some other simple probe based techniques are used to identify only single black holes and not cooperative ones.

Though some of the above mentioned techniques are able to identify multiple black holes, they are not able to identify the cooperating nature of these nodes which would be of use while providing security to the MANET. So there is a need for a technique which would really be able to identify two or more nodes that cooperate and drop packets within their links.

DEFINITIONS AND ASSUMPTIONS

Node State

The states of the node are defined as follows: A node is GOOD if it responds to probe messages for itself and forwards other probe messages for others. A benign link failure may also be detected as BAD behavior if it is not cleared by other mechanisms. (e.g. Route error in DSR). A node is considered DOWN if:

1. It is a neighbor node to the probing node and it does not respond to probe messages.
2. It is not a neighbor node and it does not respond to probe messages through all the known paths.
3. A node is considered to be in the UNKNOWN state if on all known paths from the probing node to that node, there is at least one node in BAD or DOWN state.

Assumptions

Probe Messages are Indistinguishable from Normal Packets

The main limitation of the probing technique is that it can be easily defeated if probe message can be distinguished from normal data packets. For example a malicious node may forward probe messages, but drop all the other data packets thereby avoiding detection. This assumption can be realized using end-to-end encryption of IP (Internet Protocol) payload by pair wise shared keys. Since a malicious node can understand only the IP header, it does not have the information of upper layer protocols such as TCP/UDP (Transmission Control Protocol/User Datagram Protocol) and their port numbers. By implementing probing technique over the network layer (e.g. Based UDP), an adversary will not be able to distinguish a probe message from that of a data packet. (e.g. HTTP (Hyper Text Transfer Protocol) or SMTP (Simple Mail Transfer Protocol) packet). Some other assumptions are:

1. Piggybacking a probe message on a normal data packet which requires acknowledgment, such as TCP SYN (Transmission Control Protocol Synchronization). The disadvantage is that such data packets may not be available during the time of probing.
2. An adversary cannot modify the forwarding software of the compromised router. Thus, the adversary can only make decisions based on IP addresses, and so cannot distinguish a probe message from that of a normal data packet.

Multi-hop Source Routing Protocols

The probing technique assumes a multi-hop source routing protocol since a probing node needs to specify the source route which a probe message takes to reach the destination. This assumption is practical since some routing protocols, such as DSR (Johnson et al., 2002) are multi-hop source routing protocols.

Bidirectional Communication Links

Let us assume that all communication links are bi-directional. This assumption is practical in some wireless networks, such as IEEE 802.11 (ANSI/IEEE standards, 1999) where all links have bidirectional connectivity for link layer acknowledgments to work.

THE DISTRIBUTED PROBING SCHEME

In order to monitor the behavior of mobile nodes by the probing technique, a decision on which node should to be probed and how far to be probed should be taken. Given a network with n nodes, there are several interesting possibilities (Johnson et al., 2002).

1. There is only one probing node and it probes all the other nodes.
2. There are *k* probing nodes ($1<k<n$) and each probes over a distance of r ($1<r<$infinity).
3. There are *n* probing nodes and each probe over a distance of infinity.

The last approach is preferred in a MANET since it can detect selective packet dropping (ie, based on IP addresses) and also a node need not rely upon the information from other nodes to detect malicious nodes. But generation of significant network overhead is an issue here. This network overhead can be reduced if probe messages piggyback normal data packets. To simplify the problem, the probing technique is divided into three distinct phases:

1. Probing path selection phase.
2. Modified probing phase.
3. Diagnosis phase.

Probing Path Selection Phase

The probing paths are selected solely from the routing cache maintained by mobile nodes. There are usually many redundant paths in the routing cache. Although probing over each of them may allow for validating all the known paths, it will also produce significant network overhead. The ideal strategy shall select a minimum number of paths but allow for monitoring the forwarding behavior of as many nodes in the routing cache as possible (Just et al., 2004). The probing path selection algorithm returns a set of paths with the following properties.

1. For any two paths P_i and P_j, $P_i \not\subset P_j$. Since probing over a path can always disclose the forwarding behavior of those nodes in any of its subsets, any path which is a subset of another path will be eliminated.
2. When two probing paths have a single node in common, Mike Just et al (Mike Just et al., 2004) used the following procedure to limit the number of nodes being probed. That is, for any two paths P_i and P_j, if the second farthest node in Pi is an intermediate node of P_j, the farthest node of P_i will b e removed. For example, given two paths $P_1 = A \rightarrow B \rightarrow C \rightarrow D$ and $P_2 = A \rightarrow E \rightarrow F \rightarrow C \rightarrow G \rightarrow H$, node D will be removed from P_1. With D in P_1, A can monitor the forwarding function of C. Since such monitoring can be achieved by probing over P_2, there is no need to keep D in P_1. C will still be kept in P_1, since A needs to monitor B by sending a probe message to C. But in this paper let us not make such exclusions since there may be a possibility of B and C to be cooperative black holes and it may go undetected.
3. The length of any path (in terms of number of hops) is greater than one. Since the basic requirement is to monitor the forwarding function of mobile nodes, probing over a one hop path offers no information. A probe message is sent to a neighbor node only when its subsequent node does not respond to the probe message and the probing node needs to know if the neighbor node is BAD or has moved out of its transmission range.

The Modified Probing Phase

From the observations, two ways of probing are identified. One way is to probe from the farthest node to the nearest. The other way is to probe from the nearest to the farthest. Each has its own advantages and disadvantages. Probing from far to near is better if the probing path is GOOD since it takes only one probe message and proves the goodness of all the intermediate nodes. But it may take more probe messages if a BAD node is located near the probing node. This method can be applied to a network where the majority of the nodes are assumed to be GOOD most of the time.

The advantage of probing from near to far is that it generates smaller number of probing messages to detect a BAD node located near the probing node. Another advantage is the prior knowledge of the states of all the intermediate nodes along the path to the probed node except its immediate predecessor node. The disadvantage is that an intelligent attacker may be able to avoid detection by forwarding all packets (including probe messages destined to the downstream nodes) for a certain period of time immediately after receiving a probe message for itself. A received probe message therefore serves as a signature to an attacker that a diagnosis process is ongoing and it would start to behave normally for a short period of time. In this article, the algorithm for the first method i.e., probing from the farthest to the nearest which is stronger than the alternative approaches in detecting malicious nodes has been proposed.

To probe a path, the probing node sends a probe message to the farthest node. If an acknowledgment message is received within a certain period of time, all the intermediate nodes are shown to be GOOD. Otherwise, a probe message is sent to the second farthest node. This process is repeated until one node (a neighbor node) is probed and is not responsive. In the latter case, the understanding is that the neighbor node in the probed path is either DOWN or has moved out to another location. Since the neighbor node is not responsive, nothing can be done to monitor the rest of the nodes in the path. Therefore, probing over this path is stopped. If an intermediate node is responsive but a node subsequent to it is not, then one of the following is possible:

1. The intermediate node failed to forward the probe message to the next node.
2. The link between the two nodes is broken by location change
3. The unresponsive node is not capable of responding to the probe message.

The diagnosis algorithm will then be called to decide which one is the case.

The Diagnosis Phase

With the response that is received for the probed message, the diagnosis phase decides whether the malfunctioning is due to the presence of a single black hole attack or a cooperative black hole attack or simply a link failure. The flow chart that explains the modified probing and the diagnosis for the detection of single black hole, cooperative black hole and link failure is shown in Figure 2. For better understanding the flow chart is drawn by taking a routing path: Vs, V_1, V_2, V_3,…..,V_i, V_{i+1}, V_{i+2},…, V_{i+n}, V_d and an alternate path: V_s, V_{alt1}, V_{alt2}, V_{alt3}, V_{i+2}.

Consider Figure 3 for the following diagnosis algorithm. When V_s finds some malfunctioning in the path (V_s,V_{i-3},V_{i-2},V_{i-1},V_i,V_{i+1},V_{i+2},V_D), V_s probes this path from the farthest node to the nearest node ie, first it probes V_{i+2}, V_{i+1} and then V_i. After these probing, V_s detects that node V_i is responsive but the subsequent nodes V_{i+1}, V_{i+2} are all unresponsive, then any one of the following might have occurred.

1. Link V_i to V_{i+1} could have been broken
2. V_i could be a black hole/malicious node
3. There is possibility for V_i to be a part of cooperative black hole where it may cooperate with any of the nodes that is before V_i in this probing path say, V_{i-1},V_{i-2},etc.

The probing node first searches the routing cache for another path to V_{i+1} or any other node beyond V_{i+1} up to the node that lies just before the destination in the routing path.

Case 1: If an alternate path to V_{i+1} exists as shown in Figure 4, (ie, V_s,V_{alt1},V_{alt2}, V_{alt3},V_{i+1}), now append the sub path from V_i to the node just next to the Source node V_s, i.e., probing node. Now the new path would be V_s,V_{alt1},V_{alt2}, V_{alt3},V_{i+1}, V_i, V_{i-1},V_{i-2}, V_{i-3}.

Now V_s sends a new probing message through this new path, probing from the farthest to the

Figure 2. Flow chart for the proposed algorithm

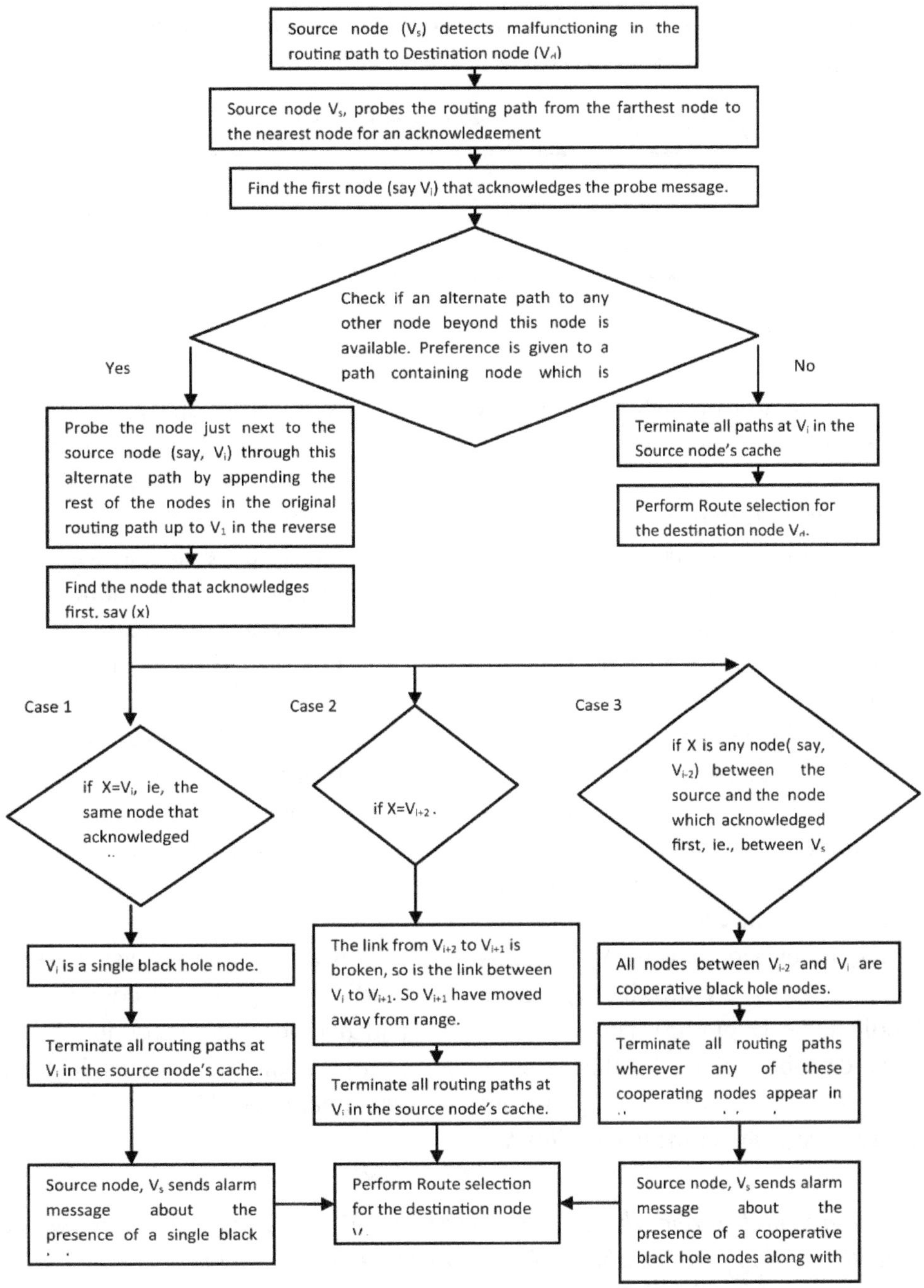

Figure 3. The source node is V_s and the destination is V_D. The routing path shows a single black hole node V_i, which drops all the packets given to it for forwarding

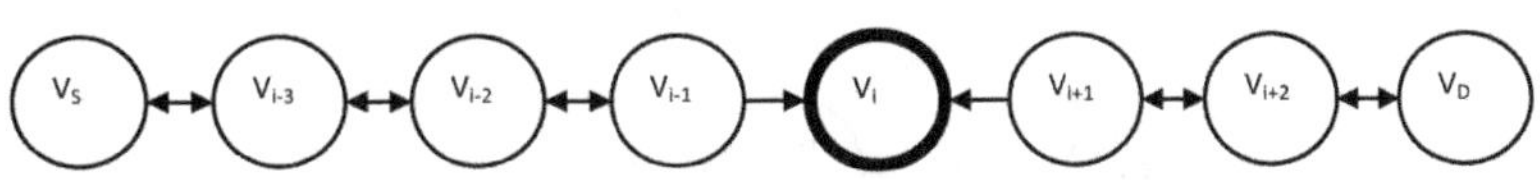

Figure 4. The source node V_s probes the node just beyond the suspected single black hole node (ie, V_{i+1}) in the original routing path to check for its reach ability through an alternate path from its cache. V_s finds that both V_{i-1} and V_{i+1} are reachable when probed directly but not through V_i, thus suggesting the presence of a single black hole node ie, V_i.

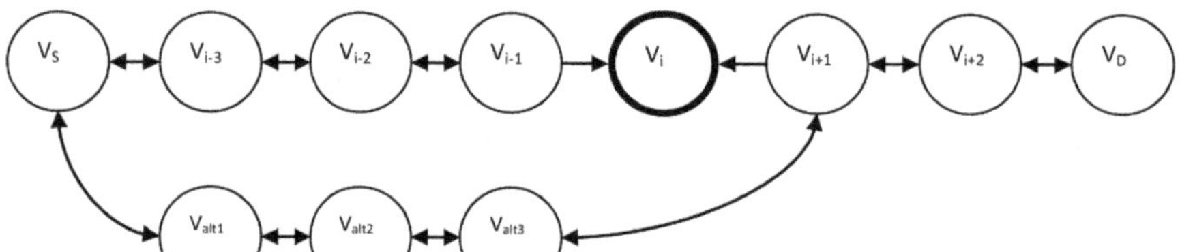

nearest node. Depending on the node that responds first, the following observations are made:

1. If V_i is the first node to respond in this path, then V_i is diagnosed to be a single black hole. The reason is, the forward probe shows that the link V_{i-1} to V_i is good, but now the reverse probe shows that the link V_i to V_{i-1} is not good thus violating the basic assumption of bidirectional link thus leading to the diagnosis that V_i is a single black hole. Now a broadcast message is sent which announces that V_i is a black hole, and all other nodes receiving this message would terminate all its routes in its route cache at V_i.
2. Else if any other node beyond V_i responds say V_{i-2}, then V_{i-2}, V_{i-1}, V_i are all cooperating black hole nodes (Figure 5). The reason is, the forward probe showed that the links V_{i-3} to V_{i-2}, V_{i-2} to V_{i-1}, V_{i-1} to V_i are all good, but now the reverse probe shows that the link V_{i-2} to V_{i-3} is not up while the links V_i to V_{i-1}, V_{i-1} to V_{i-2} are all good, thus the only possibility would be that the nodes V_{i-3} and V_{i+1} are the normal nodes while V_{i-2}, V_{i-1} and V_i cooperate and drop the packets within their links at the exit point of this cooperating chain. These nodes do not reveal to the external world that one of them member is dropping the packet. This is accomplished by sending an acknowledgment to the probing node for the packet it has not really received. Thus this situation reveals the presence of cooperative black hole in the path, the cooperating nodes being V_{i-2}, V_{i-1}, V_i.

Now send a broadcast message announcing that V_i, V_{i-1}, V_{i-2} are all cooperating black holes, and all other nodes receiving this message would terminate all their routes in their route cache wherever nodes V_i, V_{i-1} or V_{i-2} is available.

Figure 5. The source node probes the node V_{i-1} for reach ability using an alternate path available in its cache and identifies that V_i is reachable in the forward path and V_{i-2} in the alternate path suggesting the presence of cooperative black hole nodes in the routing path. ie, V_{i-2}, V_{i-1} and V_i.

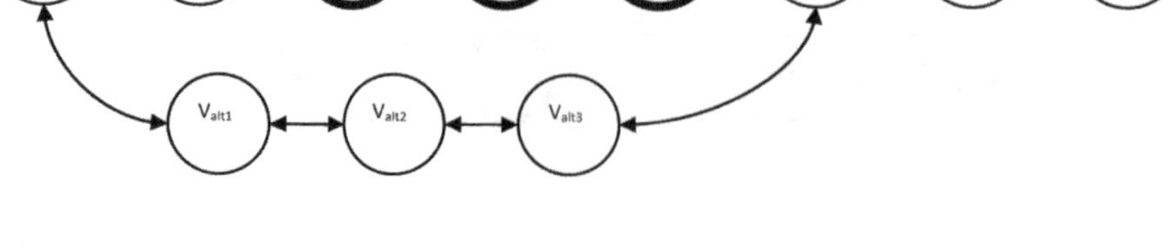

3. Else if V_{alt3} is the first node to respond, then it implies that V_{i+1} is neither reachable from the earlier probe nor by this new probe through the alternate path, thus indicating that V_{i+1} would have moved away from the accessibility of both V_i and V_{alt3}. Now traverse through the routes available in the cache of V_s and remove all paths beyond V_{i+1} inclusive of V_{i+1}.

Case 2: An alternate path to V_{i+1} does not exist, but an alternate path to some other node beyond V_i exists, say V_{i+2} as in Figure 6. Now append the sub path in the order V_{i+2}, V_{i+1}, V_i till the node just next to the Source node, V_s, i.e., probing node. So the new path would be V_s,V_{alt1},V_{alt2}, V_{alt3},V_{i+2}, V_{i+1}, V_i, V_{i-1},V_{i-2},V_{i-3}. Now V_s sends a new probing message through this new path, probing from the farthest to the nearest node. Depending on the node that responds first, the following conclusions are made:

1&2. The first two situations are the same as mentioned above for case 1.
3. Else if V_{i+2} (here) or any other node just before V_{i+1} responds then it is the same situation as condition 3 of case 1.
4. Else if any other node in the path before V_{i+2} is the first to respond then there are possibilities for this newly responding node to be a single black hole or the nodes in between would have moved away from accessibility. A route discovery process would sort it out, but since this would be a time consuming process, it is concluded that V_{i+1} is not reachable and so search through the cache and terminate all the paths beyond V_{i+1} inclusive of V_{i+1}.

SIMULATION AND RESULTS

In this section, the description of the simulation environment and the report of the simulation results are presented. Computer simulation is done using Network Simulator (NS-2) to build the simulation environment. The distributed coordination function (DCF) of IEEE 802.11 for wireless LANs has been used to simulate the MAC layer and set the traffic source to Continuous Bit Rate (CBR). The CBR traffic is generated with the rate of four packets per second. A flat network is constructed for simulation purpose and monitored for a number of parameters. The simulation models 50 mobile nodes, moving over a rectangular flat space. Simulation time is 1300 seconds. The random waypoint model is selected as a mobility model in a rectangular field (1000 x 1000 m^2) with a node's speed uniformly between 0 and a maximum value of 20 m/s. The values used for the various parameters in simulation are shown

Figure 6. The source node does not have an alternate path directly to V_{i+1}, ie, the node just next to the suspected black hole node and hence uses the alternate path to any other node (here, V_{i+2}) which is between V_i and the destination node V_D to probe for the reach ability of V_{i-1}. Now it identifies that V_{i-1} is not reachable through the alternate path and hence suggests the presence of a single black hole node i.e., V_i.

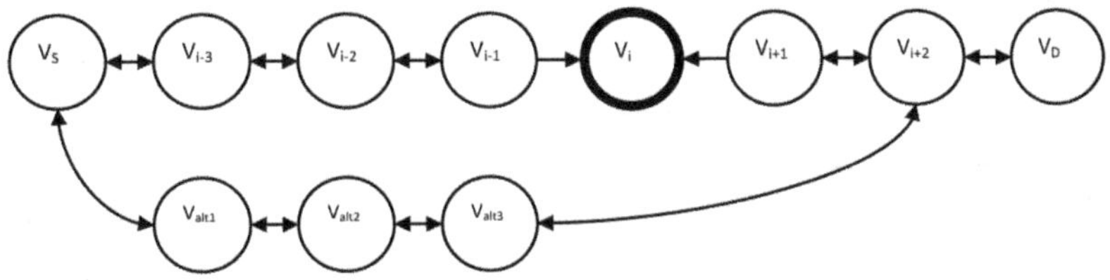

in Table 2. The function of this method is added to the DSR routing protocol to detect the black hole attacks.

To validate this algorithm, the Fidelity Table based approach (Tamilselvan & Sankaranarayanan, 2008) of the trust based technique and the simple probing technique (Just & Eranakis, 2003) have been taken for comparison. The following metrics have been used to evaluate this method, which eliminates the single and cooperative black holes attacks in MANETs:

1. Data Packet Delivery Ratio
2. Control Overhead
3. Detection Time
4. False Negative Probability

- *Data packet delivery ratio*: This is the percentage of the data packet sent, that are actually received by the destination Figure 7 shows data packet delivery ratio as a function of the number of cooperative black hole nodes. The proposed method maintains relatively high data packet delivery ratio by avoiding black hole nodes.
- *Control overhead*: The overhead of routing control packets to detect the black hole attacks. Transmissions are counted instead of packets since the goal is to compare the data transmission to routing related transmissions. The result of the routing control overhead is shown in Figure 8. The results show that the control overhead of the proposed scheme is better than the simple probing technique.
- *Detection time*: This is the time to detect the network which has black hole attack. It is measured by the attack detection time –

Table 2. Simulation parameters

Parameters	Values
Simulator	NS-2
MAC layer Protocol	IEEE802.11
Mobility model	Random waypoint
Node placement	Random, uniform
Terrain range	1000x1000 m^2
Transmission range	250m
Examined routing protocol	DSR
Channel bandwidth	2Mbps
Maximum speed	20 m/s
Application traffic	CBR
Simulation time	1300 s
Propagation mode	Free space
Packet size	512bytes
Packet rate	2 packets/s
Number of mobile nodes	50
Maximum connections	30
Number of single black hole nodes	Varied from 1 to 10
Number of cooperative black holes	Varied from 2 to 10

Figure 7. Packet delivery ratio vs number of cooperative black hole nodes

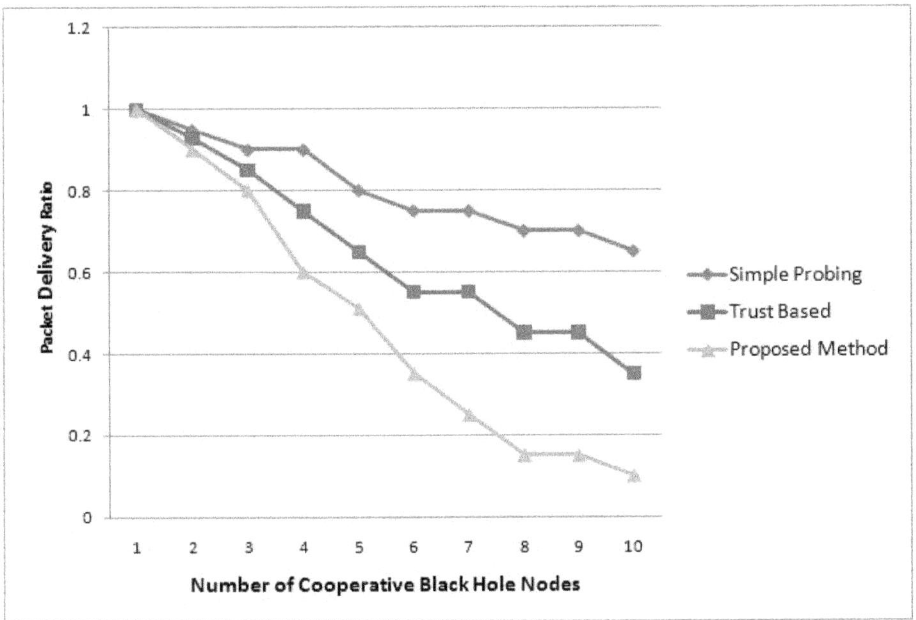

Figure 8. Control overhead vs simulation time

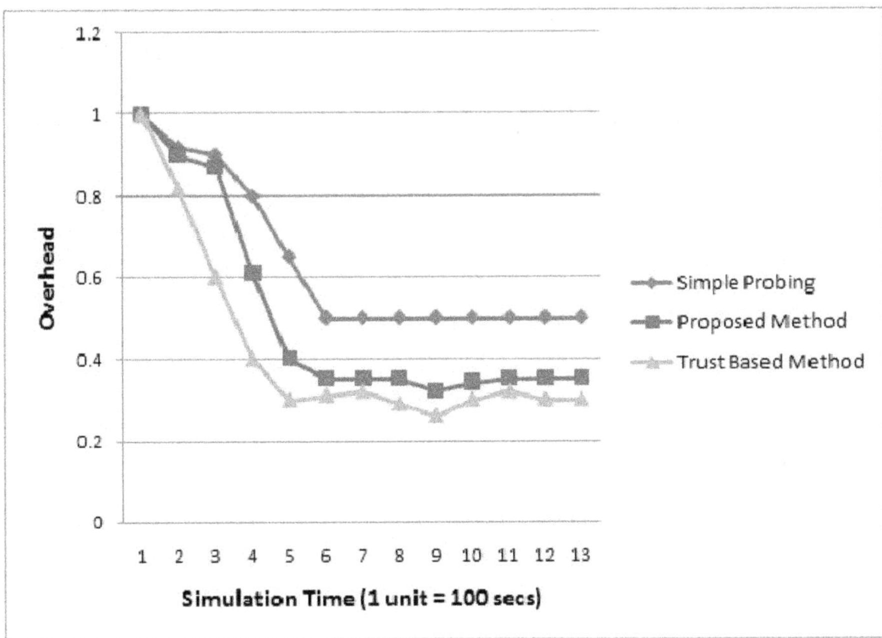

the traffic start time. To validate this algorithm, the algorithm is run in the presence of single black hole nodes alone and it was found that the results of all three techniques are comparable. Figure 9 shows the results. But the detection time in the pres-

Figure 9. Detection time vs number of single black hole nodes

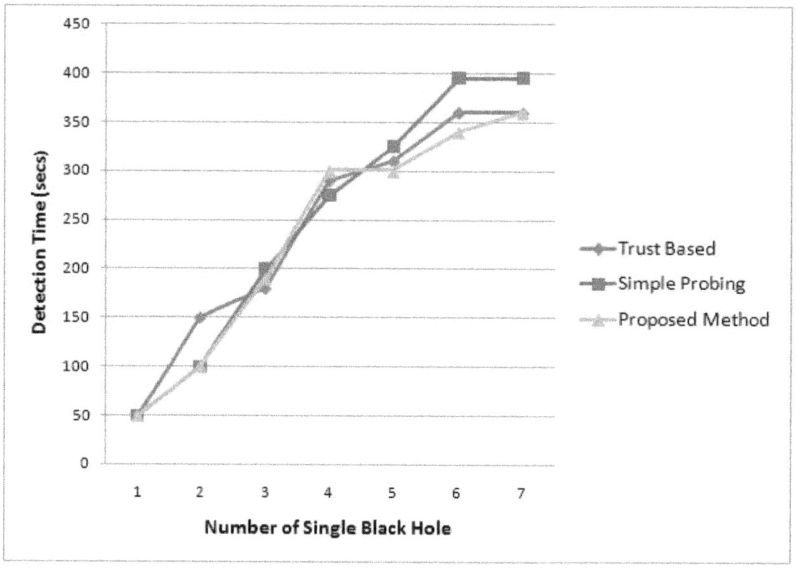

ence of cooperative black hole nodes is much better in the proposed scheme when compared to the rest of the techniques. Figure 10 depicts the results under this situation.

- *False negative probability*: The probability with which the source node does not detect the black hole attacks is studied and measured. Simulation was done in the presence of five random single black holes and 5 random cooperative black hole nodes. The result of false negative probability is shown in Figure 11. The detection probability increases as the node simulation time in-

Figure 10. Detection time vs number of cooperative black hole nodes

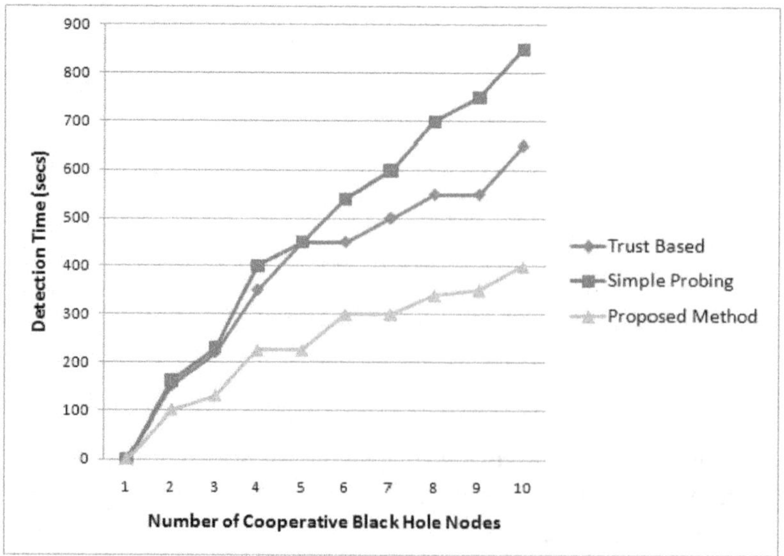

Figure 11. False negative probability vs simulation time

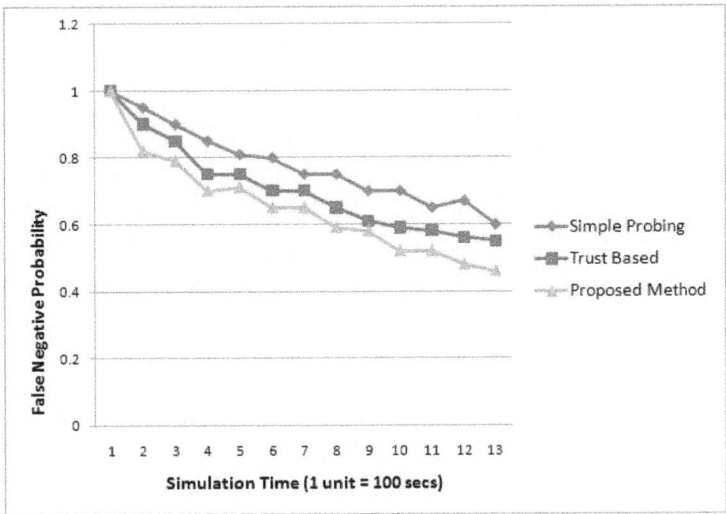

creases. The proposed method is better than the other two methods in this aspect also.

CONCLUSION

MANETs present different threats due to their different properties. These properties open up very different security risks from conventional wired networks, and each of them affects how security is provided and maintained. Due to the node mobility and dynamic nature of MANETs, link breaks are likely to be common rather than rare. This means that the network layer, in which routing takes place, is vitally an important focus for MANET security measures. Moreover, MANETs exhibit new vulnerabilities to malicious attacks of denial of cooperation. When designing protocols for these networks, special care has to be taken to include capabilities to handle each type of attack and also include fairness mechanisms for the increased requirements in this environment.

This article studies routing security issues of MANETs, describes single and cooperative black hole attacks which are specific types of internal attacks against it, and consolidates various works related to black hole attack detection methods and also proposes a feasible solution for it on top of the DSR protocol. Rather than generously applying cryptography to an existing protocol to achieve security, a new probing scheme which uses an alternate path to detect the presence of such types of attacks is introduced. The proposed solution is a light weight methodology that can be applied to:

1. Identify single black hole nodes and multiple black hole nodes cooperating with each other in a MANET ; and
2. Discover secure path from source to destination by avoiding single and cooperative black hole nodes.

The performance of the proposed approach has been examined through numerous simulations. The effects of various performance factors such as number of attacking nodes, detection time, packet delivery ratio and control overhead and compared it with trust based and probe based techniques are assessed. In this way, a quantitative study of the performance impact of the single and cooperative black hole attack on MANETs. From the results it has been observed that the proposed algorithm provides a reasonable control overhead and increased packet delivery ratio while minimizing the effects of misbehaving nodes. Moreover, all these can be done without a prior trust. Thus the results have demonstrated the effectiveness and accuracy of the method devised. Study on the impact of GRAY hole nodes (nodes which selectively drop packets), additional mechanism to support QoS and mechanisms to increase the fairness in the network can be considered in future.

REFERENCES

ANSI/IEEE Standards. (1999). *Wireless LAN medium access control (MAC) and physical layer (PHY) specification*. Retrieved from http:// standards.ieee.org/ getieee802/802.11.html

Awerbuch, B., Folmes, D., Nita-Rotaru, C., & Rubens, H. (2002). An on demand secure routing protocol resilient to Byzantine failures. *Proceedings of the 3rd ACM workshop on Wireless Security* (pp.21-30).

Bhalaji, N., Banerjee, S., & Shanmugam, A. (2008). A novel routing technique against packet dropping attack in ad hoc networks. *Journal of Computer Science*, *4*(7), 538–544. doi:10.3844/jcssp.2008.538.544

Bhalaji, N., & Shanmugam, A. (2009). Association between nodes to combat black hole attack in DSR based MANET. *International Conference on WOC '09, Wireless and Optical Communication Network 2009* (pp. 1-5).

Buchegger, S., & Le Boudec, J. (2002). Performance analysis of the confident protocol. *Proceedings of the 3rd ACM Symposium on Mobile Ad Hoc Networking & Computing* (pp. 226-236).

Deng, H., Li, W., & Agarwal, D. P. (2002). Routing security in wireless ad hoc networks. . *IEEE Communications Magazine*, *40*(10), 70–75. doi:10.1109/MCOM.2002.1039859

Hu, Y. C., Perrig, A., & Johnson, D. B. (2002). Ariadne: A secure on- demand routing protocol for ad hoc networks. *Proceedings of the 8th ACM International Conference of Mobile Computing and Networking* (pp. 12-23).

Hubaux, J. P., Buttyan, L., & Capkun, S. (2001). The quest for security in mobile ad hoc networks. *Proceedings of the 2nd ACM International Symposium on Mobile Ad hoc Networking and Computing* (pp.146-155).

Johnson, D., Maltz, D. A., Fu, Y. C., & Jetcheva, J. G. (2002). *The dynamic source routing protocol for mobile ad hoc networks* (DSR) (pp 1-25). Retrieved from http://www.monarch.cs.rice.edu/monarch-papers/dsr-chapter00.pdf

Just, M., Eranakis, E., & Wan, T. (2004). Resisting malicious packet dropping wireless ad hoc networks. Ad-Hoc. *Mobile and Wireless Networks, 2865*(2003), 151-163.

Liu, K., Deng, J., Varshney, P. K., & Balakrishnan, K. (2007). An acknowledgment based approach for the detection of routing misbehavior in MANETs. *IEEE Transactions on Mobile Computing*, *6*(5), 536–550. doi:10.1109/TMC.2007.1036

Luo, J., Fan, M., & Ye, D. (2008). Black hole attack prevention based on authentication mechanism. *Proceedings of ICCS 2008, IEEE 2008* (pp.173-177).

Milanovic, N., Malek, M., Davidson, A., & Milutinovic, V. M. (2004). Routing and security in mobile ad hoc networks. *IEEE Computer*, *37*(2), 61–65. doi:10.1109/MC.2004.1266297

Miranda, H., & Rodrigues, L. (2002). Preventing selfishness in open mobile ad hoc networks. *Proceedings of Seventh CaberNet Radicals Workshop*.

Papadimitratos, P., & Haas, Z. J. (2002). Securing the Internet routing infrastructure. *IEEE Transactions on Communications*, *10*(40), 60–68.

Patcha, A., & Mishra, A. (2003). Collaborative security architecture for black hole attack prevention in mobile ad hoc networks. *Proceedings of RAWCON, 03*, 75–78.

Rahman, A. A., & Hailes, S. (1997). A distributed trust model. *Proceedings of the Workshop on New Security Paradigms*, (pp. 48-60).

Ramaswamy, S., Fu, H., Sreekantaradhya, M., Dixon, J., & Nygard, K. (2003). Prevention of *cooperative black hole attack in wireless ad hoc networks* (pp. 1-7). Retrieved from www.cs.ndsu.nodak.edu/~nygard/ research/BlackHoleMANET.pdf

Rerney, L. M., & Nilsson, M. (2001). Investigating the energy consumption of a wireless network interface in an ad hoc networking environment. *Proceedings of IEEE INFOCOM'01*.

Sergio Marti, T. J. Giuli, Lai, K., & Baker, M. (2000). Mitigating routing misbehavior in mobile ad hoc networks. *Proceedings of The 6th Annual International Conference on Mobile computing and Networking*, (pp. 255–265).

Tamilselvan, L., & Sankaranarayanan, V. (2008). Preventive black hole attack in MANET. *Journal of Networks, 3*(5), 13–20. doi:10.4304/jnw.3.5.13-20

Weerasinghe, H., & Fu, H. (2008). Preventing cooperative black hole attacks in mobile ad hoc networks: Simulation implementation and evaluation. *International Journal of Software Engineering and Its Application, 2*(3), 39–54.

Yang, H., Shu, J., Meng, X., & Lu, S. (2006). SCAN: Self organized network layer security in mobile ad hoc networks. *IEEE Journal on Selected Areas in Communications, 24*(2), 261–273. doi:10.1109/JSAC.2005.861384

Zhao, M., & Zhou, J. (2009). Cooperative black hole attack prevention for mobile ad hoc networks. 2009 *International Symposium on Information Engineering and Electronic Commerce*, (pp.26-30).

Zhou, L., & Hass, Z. J. (1999). Securing ad hoc networks. *IEEE Network Magazine, 13*(6), 41–50.

This work was previously published in International Journal of Business Data Communications and Networking, Volume 6, Issue 1, edited by Debashis Saha and Varadharajan Sridhar, pp. 38-57 , copyright 2010 by IGI Publishing (an imprint of IGI Global).

Chapter 7
VANET Authentication Using Trust Distribution in a Clustered Environment

S. Sivagurunathan
Gandhigram Rural Institute – Deemed University, India

P. Subathra
Thiagarajar College of Engineering, India

ABSTRACT

Security is a prime concern in providing communication between vehicles in a Vehicular Ad-Hoc Network (VANET), where they share information among themselves and nearby fixed equipments. Unlike wired networks, the unique characteristics of Vehicular Ad hoc Networks (VANETs) pose a number of non-trivial challenges to security design. This chapter presents a threshold security mechanism with a mobility based clustering for Inter Vehicle Communication (IVC). Nodes that have similar moving pattern are grouped into a cluster. Unlike other clustering algorithms, it takes the moving pattern of the vehicles into consideration along with the driver's intention. The stability of clusters is estimated based on relative mobility of cluster members. A threshold cryptographic scheme is employed on top of the clusters to protect routing information and data traffic. To ensure distributed trust in the clustered environment, the private key (k) is divided into n pieces in such a way that k is easily reconstruct able from any p number of pieces. Even complete knowledge of (p-1) pieces reveals absolutely no information about k.

INTRODUCTION

MANET finds an important application in the Vehicular Ad-Hoc Network (VANET) (Munoz & Syracuse, 2002).Vehicular Networks are envisioned of the Intelligent Transportation Systems (ITS). Vehicles communicate with each other via Inter Vehicle Communication (IVC) as well as with roadside base stations via Roadside-to-Vehicle Communication (RVC). The optimal goal is that vehicular networks will contribute to safer and more efficient roads in the future by providing timely information to drivers and concerned authorities.

DOI: 10.4018/978-1-61350-477-2.ch007

Some of the key applications of VANETs are: SOS services, Stolen vehicle tracking, map download/ update, intersection collision warning, vehicle based road condition warning, emergency vehicle signal preemption and work zone warning. The main focus of this paper is on VANET to provide safety and comfort for passengers. To this end a special electronic device will be placed inside each vehicle which will provide Ad-Hoc Network connectivity for the passengers. This network tends to operate without any infra-structure or legacy client and server communication. Each vehicle equipped with VANET device will be a node in the Ad-Hoc network and can receive and relay others messages through the wireless network (Soyoung Park, et al., 2009). Collision warning, road sign alarms and in-place traffic view will give the driver essential tools to decide the best path along the way (Klaus et al., 2006). With these information, the drives can predetermine the nature of the traffic in the read ahead and act accordingly.

Security of the messages exchanged for the collaborative driving to speed up the traffic is a prime concern. This paper provides an easy way for vehicle assisted secure information exchange without any additional roadside infrastructure and special technologies. The fundamental vulnerability of VANET comes from open peer to peer architecture. Unlike wired networks that have dedicated routers, each vehicle in VANET may function as a router and forwards packets to other nodes. The wireless channel is accessible to both legitimate network users and attackers. The attack may range from passive eavesdropping to active impersonation. Since compromising a vehicle is possible, trust relationship among them is very important in case of co-operative driving. As a result there is no clear line of defense in VANETs from the security design perspective. The salient features of VANETs pose both challenges and opportunities in achieving the above security goals.

The VANET not only routes the exchange messages but also influence the behavior of the devices. Depending on the information they get, they will e.g., drive very carefully, and slowly in case of a glaze warning or choose an alternate route in the case they are informed about the traffic jam on their desired route. Adversaries could exploit this by injecting wrong messages and slowing down traffic or getting a vehicle free road. To prevent this kind of misuse security is very important in VANETs (Klaus Plobi, et al., 2006).

This article focuses on the state of the art for VANET security, discusses the techniques adopted for clustering and security, throws light on the mobility based clustering algorithm for VANET, deals with routing security, key management services and system model. This also explains the concept of threshold cryptography and its usage in key management.

LITERATURE REVIEW

The ultimate goal of designing the VANET is to provide safety and efficiency in transportation systems. They comprise network nodes, that is, vehicles and Road side Infrastructure Units RSU), equipped with on-board sensory, processing, and wireless communication modules. Vehicle-to-Vehicle (V2V) and Vehicle-to-Infrastructure (V2I) communication can enable a range of applications. Among these, primarily safety will be enabled, as numerous research and development initiatives indicate, by vehicles frequently beaconing their position, along with warnings on their condition or environment. Nonetheless, VANETs can be vulnerable to attacks and jeopardize user's privacy. For example, an attacker could inject beacons with false information, or collect vehicle's messages, track their locations, and infer sensitive user data. To thwart such attacks, security and privacy enhancing mechanisms are necessary or, in fact, a prerequisite for deployment.

The most prominent industrial effort in this domain is carried out by Car 2 Car Communication Consortium, the IEEE 1609.2 working group, the

NoW project and the SeVeCom project with all of them developing VANET Security architecture. Their common basic elements include the use of Certification authorities (CAs) and public key cryptography to protect vehicle to vehicle (V2V) and Vehicle to infrastructure (V2I) messages. It has now become an established consensus that Public Key cryptography is the way to go about for VANET. This is mainly due to the fact the messages are broadcast and one-to one communication is not the norm. Due to this fact symmetric key cryptography will incur huge costs in frequent key establishment procedures and they are also difficult to implement as the nodes are constantly on the move. Therefore here on we will concentrate on public key methods only. The primary requirements in providing a secure protocol are identified to be Message Authentication, integrity, non-repudiation and protection of private user information.

There are very few publications describing the security architecture of VANETS, (Sampigethaya, et al, 2005) but not many of them propose specific protocols that consider all the practical requirements needed to secure VANET safety applications. Gerla (Gerla & Vanese, 2005) describes the security concepts for vehicular networks. Hubaux (Hubaux et al, 2004) take a different perspective of VANET security and focus on privacy and secure positioning issues. Parno (Parno & Perrig, 2005) discuss the challenges, adversary types and some attacks encountered in vehicular networks; they also describe several security mechanisms that can be useful in securing these networks. The use of digital signatures in the vehicular environment is discussed by Gollan (Gollan & Meinel, 2002). Meanwhile, Boneh (Boneh & Shacham, 2004) mention VANET as an application for group signature, that is, cryptographic primitives for anonymous authentication. This is a stronger property than pseudonymous authentication, as any two group signature generated by a node cannot be linked.

Bellare (Bellare, et al, 2005) propose a dynamic group signature scheme which does not depend on the number of group members. In a Group signature scheme a member of a group to anonymously sign a message on behalf of the group. Zhou (Zhou, et al, 2007) propose another dynamic Group Signature approach based on elliptic curve cryptography. However, Calandriello (Calandriello, et al, 2007) propose a scheme for VANET security that relies on the concept of pseudonym authentication. It also assumes the presence of certification authority which is vested with legal power to disclose node identities and is required to certify the keys of vehicular nodes. Baber (Baber & Zou, 2009) propose a scheme based on spatial and temporal restricted certificates, which are issued upon user's request and can be used for VANET applications.

BACKGROUND AND MODEL

Since the VANET exhibits the adhoc nature of MANET, every vehicle (node) VANET has to rely upon the other vehicles for routing their packets as in a MANET. The need to store complete routing details for an entire network topology raises scalability issue. The flat hierarchy adopted by most of the existing VANET routing protocols may not be able to support the routing function efficiently, since their routing tables can grow to an immense size, if each vehicle has a complete view of the network topology. To address the scalability issues in VANET, the nodes can be clustered using the clustering algorithms.

Clustering algorithms can be dynamically performed to adapt to change in topology due to node mobility (McDonald & Znati, 1999; Sung-Ju Lee, et al., 2001). VANET is dynamically organized into groups called clusters to maintain a relatively stable effective topology (Lin & Gerla, 1997). By organizing vehicles into clusters, topology information can be aggregated.

The following observations have been made in VANET which motivated the application of the group concept in the proposed solution:

1. Vehicles in geographical proximity often share redundant information such as road and traffic conditions.
2. The mobility of the vehicles is spatially restricted and spatially dependent. Hence, vehicles in geographical proximity can navigate as a group, with the same average velocity, due to the spatial dependency, and with similar direction due to the spatial restrictions over a period of time.

Making use of the above observations, a proposal has been made to enable vehicles to form clusters.

Clustering algorithm in VANETs should be able to maintain its cluster structure as stable as possible while the topology changes (Charalampos Konstantopoulos, et al., 2008). Since vehicles in a cluster will move relative to each other, and on average have the same velocity, a group can be represented by a single vehicle which is referred as the *clusterhead.* This is a clustering algorithm that forms clusters based on mobility metric suggested Jeremy Blum (Jeremy Blum, et al., 2003). The formation of the clusters is determined by the mobility pattern of vehicles to ensure maximum cluster stability (Hao Yang, et al., 2004). It has been observed that vehicles in VANET may move in groups. This is known as group mobility. Hence, mobility based clustering would be a better option than the other available clustering algorithms (Jane & Peter, 2005).

The VANET is portioned into a number of clusters. In each cluster, exactly one distinguished node – the Cluster Head (CH) - is responsible for establishing and organizing the cluster. Gateways (GWs, which need not necessarily be CHs) manage communication with adjacent clusters. The CHs are responsible for sending CH beacons in their clusters, containing administrative information for the cluster members, e.g., list of vehicles and GWs in the cluster. Also, GWs periodically transmit GW beacons to inform their respective clusters about adjacent clusters. Figure 1 shows a scenario

Figure 1. Cluster formation

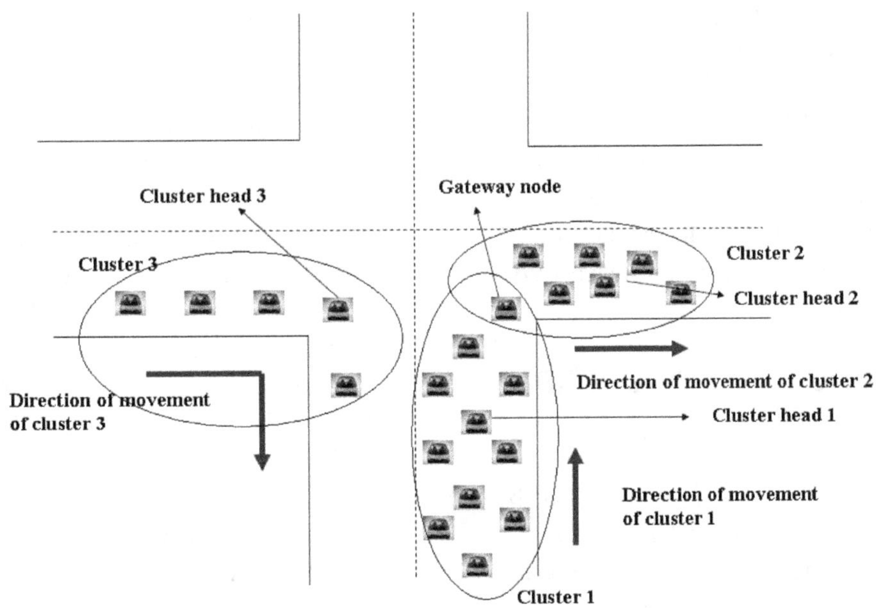

where the nodes of the VANET are grouped into three different clusters based on their mobility pattern, Cluster Heads are selected in each of the clusters and the node which is in the proximity of two clusters is elected as the Gateway node.

Security mechanism can be incorporated inside clusters in cluster based routing protocols. In case no clusters are given from outside the security part, they are formed as needed: Nodes finding no existing clusters create some themselves and the existing clusters are merged or split on demand. The following section describes the mechanism.

Providing authentication and maintaining the secrecy of the information in VANETs is a Herculean task and to do this, a key management service using threshold security technique can be used (Papadimitratos, & Haas, 2002) on top of the clustering algorithm. Due to the distributed nature of the key management scheme the difficulties in coordinating the vehicles and vulnerabilities such as single point of failure are avoided.

The threshold security technique uses the *(n, p)* threshold scheme where some secret data *D* is divided into *n* pieces $D_1, D_2 .. D_n$ in such a way that:

1. Knowledge of any *p* or more D_i pieces makes *D* easily computable.
2. Knowledge of any *(p-1)* or fewer D_i pieces makes *D* completely undetermined

(Adi Shamir, 1979).

Thereby it can maintain the secrecy of the information exchanged. The novelty here is the implementation of threshold cryptography in a mobility based clustered environment for VANET.

CLUSTERING FOR VEHICULAR COMMUNICATION

Vehicular mobility is highly constrained by the layout of road network, by traffic control devices, and by surrounding vehicles. So the generic MANET algorithms which have been tested under random nodal mobility, does not reflect vehicular movement and is not suitable for VANETs. Vehicle movement is characterized by high rates of speed, producing very high relative velocities. Driver behavior has a significant impact of mobility patterns both in the near term and in the long term. In the near term, vehicle movements vary very dramatically based on individual lane changing, braking, and passing behaviors. In the long term, mobility is affected by the variations in the intended destination of a driver.

Motivation for Clustering Algorithm

Though the existing MANET clustering approaches address vehicular mobility, they have limited applicability for inter-vehicle networks. The clustering algorithms that consider nodal mobility produce unstable clusters only. Ideally, the relative mobility between a cluster head and a member node should be low, so that they remain in radio contact as long as possible. Low mobility nodes will often have a very high relative mobility with nearby vehicles and therefore fleeting radio contact. Consider a car that is parked, but idling. It will have lower mobility than the vehicles passing it, but it is an unsuitable cluster head because the cluster membership will not be stable. Similarly, if there is a stop and go traffic on one side of a highway, free moving nodes on the other side should not choose these low mobility nodes as cluster heads. However, the use of previous observations of low relative mobility will not necessarily yield a stable clustering. Consider two vehicles that have traveled together on the same road for some time. One of the vehicles merges into the exit lane. Here the driver's intention to exit conveys much more information about their future relative mobility, than the previous observations of this value.

Finally, Inter Vehicle Communication (IVC) clustering stability will degrade with the requirements that no two cluster heads communicate with one another and that all nodes be in constant

Figure 2. Cluster head selection and gateway node identification

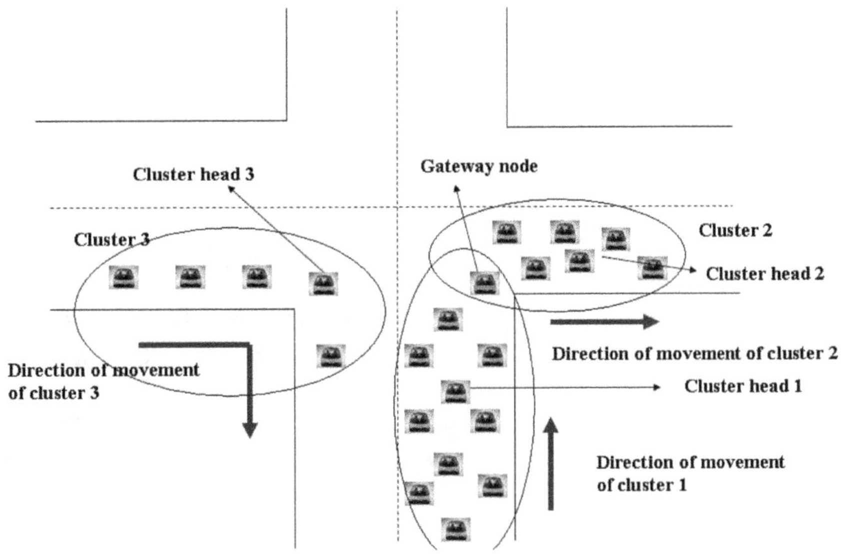

radio communication with their- cluster heads. The distances between vehicles with low relative mobility will often exhibit oscillatory behavior. Examples of this phenomenon can be found in stop and go traffic; as vehicles pass through four-way stop signs, or as vehicles slow to navigate curves in the roadway.

The Clustering algorithm attempts to build stable clusters by extending the previous work in MANET clustering and seeks to address these limitations. Like the previous algorithms, it attempts to preserve cluster head election and uses mobility information for clustering. One of the contributions of the algorithm is its use of driver intentions as input to the clustering algorithm. Furthermore, to accommodate oscillatory inter-vehicle distance, the algorithm allows for limited situations in which radio links can exist between cluster heads. The size of the cluster is decided dynamically by the cluster head during the formation of the cluster. Nodes which are able to receive the information from both of the cluster heads can act as a gateway between both of the clusters.

Figure 2 shows the initial cluster formation and the selection of cluster heads and gateway nodes using Clustering and in Figure 3 the change in cluster structure due to the mobility of nodes is depicted at a later time.

Network Model

The following section describes the model of the Inter Vehicle Communications network. In addition to describing the Inter Vehicle Communications capability, it serves as the mathematical foundation for this protocol.

The network is represented by a graph G = (V, E') which is comprised of a set of vertices V and time varying edges E'. For each pair of vertices (u, v) € V, (u,v) € E' if the distance between these two nodes is less than or equal to the radio range. In addition, at time t, now V is partitioned into two sets - our set of cluster heads C^t, and the set of ordinary nodes V - C^t. The concept of the neighborhood of v at time t, $\Gamma^t(v)$ is also introduced here. Two nodes are in the same neighborhood at time t if they are within radio range and are traveling on the same road and in the same direction. Each ordinary node is associated with a cluster head that is in its neighborhood.

In order to minimize the effect of oscillatory distances between nodes, the inter-cluster head

Figure 3. Change in cluster structure

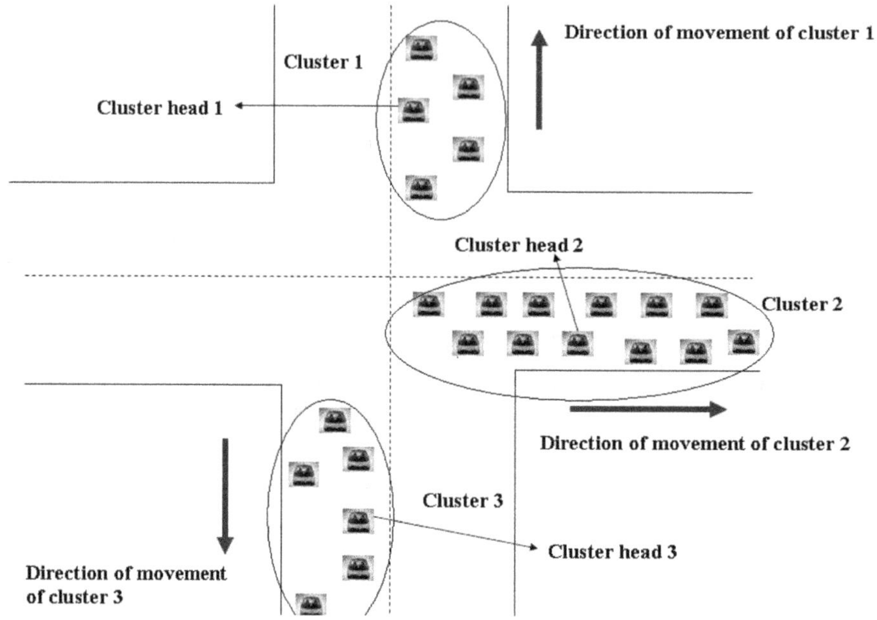

minimum, IC_{min}, is defined to be the minimum distance between two Cluster heads. Thus, IC_{min} should be less than or equal to the radio range. Each node v has an associated set of nodal properties, defined as follows.

- ID_v unique ID for node v.
- Loc^t_v location of node v at time t.
- $exit_v$ the point where node v will exit the current road.
- w^t_v weight of node v - its suitability for becoming cluster head at time t.
- *W'u~ v,* association weight of node u to node v -the suitability of node v to be the cluster head for node u at time t.
- C_v^t the set of nodes that are a cluster head for node v at time t
- Note that this set has either one node or is empty.

The protocol makes the following assumptions within this model:

1. $\forall u \in \Gamma^t(v)$ node v knows $loc^{t'}_u$ in the neighborhood of t.

A vehicle's location, obtained from GPS equipment, is likely to be transmitted for collision avoidance purposes. Thus, this assumption imposes no additional communication requirements.

2. $\forall_u \in \Gamma^t(v)$, node v knows $exit_u$ and $exit^{t'}_u$ for t' in the dt , neighborhood of t.

Each node must periodically broadcast its exit point from the current road and its current cluster head. Its exit point could be predicted based on data from an on-board route planning system or from a statistical analysis of previous trips on the current roadway.

Description of Protocol

The proposed clustering protocol is designed to satisfy the following clustering properties, where t' is a time occurring shortly after t:

1. $\forall_u \in \Gamma^T(v)$, $u \in C \Lambda (w_v > w_u V C^t_u \neq \theta) \Rightarrow v \in C^{t'}$
   v declares itself cluster head if it there are no cluster heads in its neighborhood, and it has the highest weight of all unaffiliated, nearby nodes.
2. $\forall u,v \in C^t, (dist^t(u,v) < IC_{min}) V (w_u < w_v) \Rightarrow u \notin C^{t'}$
   The distance between two cluster heads is greater than the IC_{min}. Otherwise, then the cluster head with the smaller weight resigns.
3. $C^t_v \neq G^t(v) \Rightarrow C^t = \theta$
   A node's cluster head must be in its neighborhood. As soon as a node loses radio contact with its cluster head, it becomes an unaffiliated node.
4. $C^t_v = \theta \Rightarrow C^{t'}_v = u \in \Gamma^{t'}(v) \cap C^t: \max(W^{t'}_{u\sim v})$
   If a node does not, have a cluster head, it will associate with the cluster head with the highest association weight.

In addition to the messages that transmit a node's cluster head and exit, there are three clustering messages to support this protocol:

1. Elect(v): Node U elects itself cluster head
2. Resign(v): Node U resigns as cluster head.
3. Associate(v,u): Node v joins the cluster headed by node u.

Each node v runs the following procedure at time t:

```
proc CLUSTER {
.// Checks to see if cluster head is
still in range
If (Cᵗ (v) ≠ Γᵗ(v)) Cᵗ (v) ≠ θ
// If the node is a cluster head
if (v ∈  Cᵗ(u)) {
// Closer than ICmin, to a higher
weight cluster
if (∃u,u ∈ Cᵗ Λ distᵗ (u,v) < ICmin  Λ
wᵗu > wᵗv)
Resign(v)
// If the node does not have a clus-
ter head
) else if (Cᵗ(v) = θ) {
// Join the cluster head with largest
weight
if (∃u, u ∈' Γᵗ(v)UCᵗ)
Associate(u.v) for u: max(wᵗu,v)
// No cluster heads nearby, I have
highest weight
else if (∀ u ∈ Γᵗ(v), wᵗv > wᵗuV Cᵗu ≠
θ)
Elect(v)
}}
```

THRESHOLD SECURITY

Secure Routing

Nodes of an adhoc network exhibit dynamically changing topology and moreover prone to malicious attacks. Routing protocols should be robust against these factors in order to provide availability. Routing protocols proposed for MANET cope well with the dynamically changing topology (Jane & Peter, 2005). However, none of them seem to have accommodated mechanisms to defend against malicious attacks. There is no single standard routing protocol that addresses this issue. Hence, the aim is to capture the common security threats and to provide guidelines to secure routing protocol.

The routing protocols are prone to two different sources of threats. The first comes from external attackers. By injecting erroneous routing information, replaying old routing information or distorting routing information, an attacker could successfully partition the network or introduce excessive traffic load into the network. The second and also the most sever kind of threats come from compromised nodes, which might advertise incorrect routing information to other nodes. Detection of such incorrect information is difficult. Merely requiring routing information to be signed by each node would not work because compromised nodes are able to generate valid signatures using their private keys.

Detection of compromised nodes through routing information is also difficult in MANET because of its dynamically changing topology: when a piece of routing information is found invalid, the information could be generated by a compromised node, or it could have become invalid as a result of topology changes. It is difficult to distinguish between the two cases. These features demand a security mechanism which is more robust under such situation.

Key Management Service

A cryptographic scheme such as digital signature is employed, to protect both routing information and data traffic. Use of such schemes usually requires a key management service. A public key infrastructure is adopted because of its superiority in distributing keys and in achieving integrity and non-repudiation. Efficient secret key schemes are used to secure further communication after nodes authenticate each other and establish a shared secret session key. In a public key infrastructure, each node has a public/priva te key pair. Public keys can be distributed to other nodes, while private keys should be kept confidential to individual nodes. There is a trusted entity called Certification authority *(CA)* (Gannaro, Karawczyk, & Rabin, 1996) for key management. The *CA* has a public/private key pair, with its public key known to every node, and signs certificates binding public keys to nodes. The trusted *CA* has to stay on-line to reflect the current bindings, because the bindings could change over time.

A public key should be revoked if the owner node is no longer trusted or is out of the network; a node may refresh its key pair periodically to reduce the chance of a successful brute-force attack on its private key. Usage of a single CA in a VANET has its own limitations. The *CA*, responsible for the security of the entire network, is a vulnerable point of the network: If the CA is unavailable, nodes cannot get the current public keys of other nodes to establish secure communication with others. If the *CA* is compromised and leaks its private key to an adversary, the adversary can then sign any erroneous certificate using this private key to impersonate any node or to revoke any certificate. A standard approach to improve availability of a service is replication. But a naive replication of the *CA* makes the service more vulnerable: compromise of any single replica, which possesses the service' private key, could lead to collapse of the entire system. The trust is distributed to a set of nodes by letting these nodes share the key management responsibility, which may solve the above mentioned difficulty.

System Model

The service, as a whole, has a public/private key pair. All nodes in the system know the public key of the service and trust any certificate signed using the corresponding private key. Nodes, as clients can submit query requests to get other client's public keys or submit update requests to change their own public keys. Internally, this key management service, with a (n, p) configuration $(n>=2p)$, consists of n special nodes, called servers which are nothing but the cluster heads, present within the clusters of a VANET. Each server has its own key pair and will store the public keys of all the nodes in the network. In particular, each server knows the public keys of other servers. The service is correct if the following two conditions hold:

- (Robustness) The service is always able to process query and update requests from clients.
- (Confidentiality) The private key of the service is never disclosed to an adversary.

In Figure 4, the initial selection of value (n, p) by cluster head 3 in cluster 3 is shown. Figure 5 shows the reselection of (n, p) value by cluster

Figure 4. Initial (n, p) value selection by cluster head 3 in cluster 3.

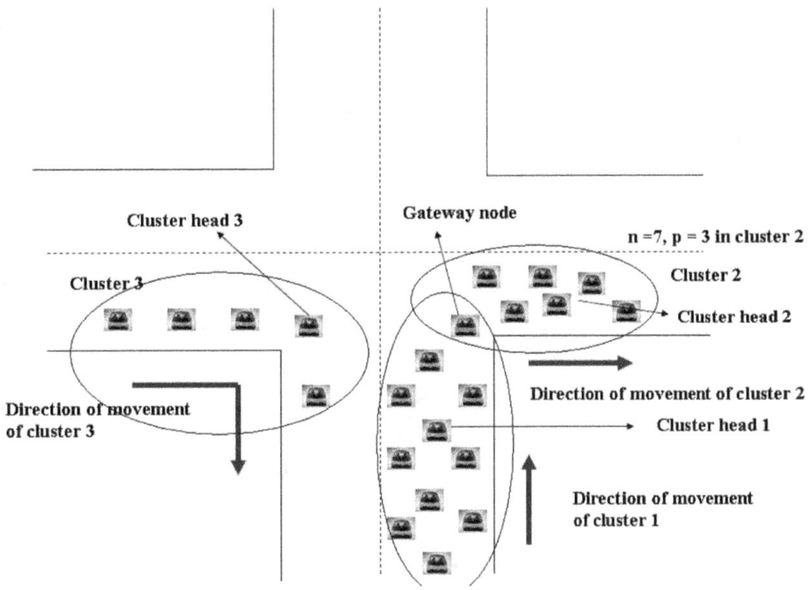

head 3 due to change in the number of cluster members in that cluster.

Threshold Cryptography

Distribution of trust in our key management service is accomplished using threshold cryptography (Desmedt, 1994; Desmedt & Frankel, 1990; Shigefusa Suzuki & Kazuhiko Nakada, 1991). A *(n, p)* threshold cryptography scheme allows *n* parties to share the ability to perform a cryptographic operation (e.g., creating a digital signature), so that any *p* parties can perform this operation jointly, whereas it is infeasible for at most *(p-1)* parties to do so, even by collusion.

Our scheme is based on polynomial interpolation: given *p* points in the 2-dimensional plane $(x_1,y_1), \ldots, (x_p, y_p)$ with distinct x_i's, there is one and only one polynomial $q(x)$ of degree $(p-1)$ such that $q(x_i)=y_i$, $\forall$ *i*. Without loss of generality, it can be assumed that the data *D* is (or can be made) a number. To divide it into pieces D_i, pick a random number *(p-1)* degree polynomial $q(x)= a_0+a_1x+\ldots+a_{p-1}x_{p-1}$ *in which* $a_0=D$, and evaluate: $D_1=q(1), \ldots D_i=q(i) \ldots D_n=q(n)$.

Given any subset *p* of these values (together with their identifying indices), Find the coefficients of *q (x)* by interpolation and then evaluate *D=q (0)*. Knowledge of just *(p-1)* of these values on the other hand does not suffice in order to calculate *D (*Ad*i* shamir, 1979*)*.

To make this claim more precise, modular arithmetic has been used instead of real arithmetic. The set of integers modulo a prime number p forms a field in which interpolation is possible. Given an integer valued data *D*, a prime p which is bigger than both D and *n* is chosen. The coefficients $a_0, a_1, \ldots, a_{p-1}$ in *q(x)* are randomly chosen from a uniform distribution over the integers in *[0, p)*, and the values *D1. .. .. Dn* are computed modulo *p*.

Let us now assume that *p-1* of these n pieces are revealed to an opponent. For each candidate value *D'* in *[0, p)* he can construct one and only one polynomial *q '(x)* of degree *p- 1* such that *q '(0) =D'* and *q '(0 =D~)* for the *p- 1* given arguments. By construction, these p possible polynomials are equally likely, and thus there is abolutely nothing the opponent can deduce about the real value of *D*.

Figure 5. Initial (n, p) value reselection by cluster head 3 in cluster 3.

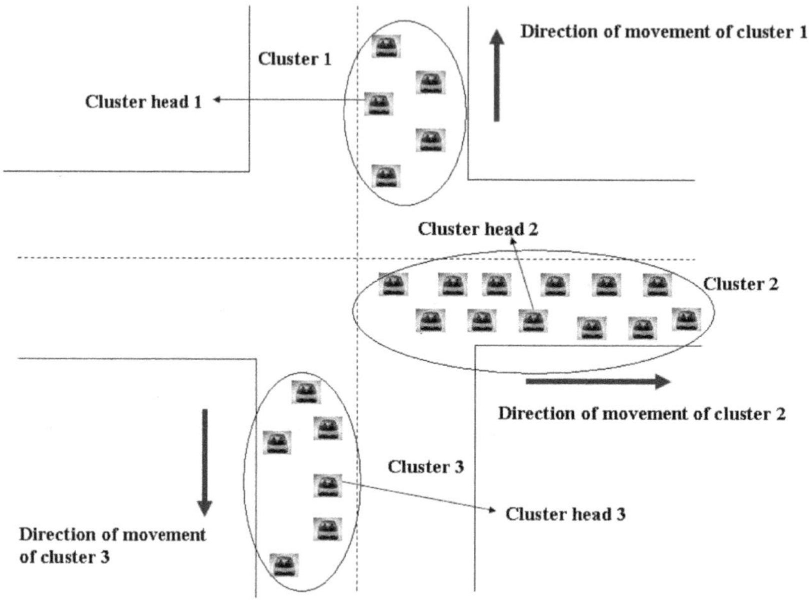

Efficient $O(nlog^2 n)$ algorithms for polynomial evaluation and interpolation are discussed by (Aho. A, et al, 1974) and (Knuth,D., 1969), but even the straightforward quadratic algorithms are fast enough for practical key management schemes. If the number *D* is long, it is advisable to break it into shorter blocks of bits (which are handled separately) in order to avoid multi precision arithmetic operations. The blocks cannot be arbitrarily short, since the smallest usable value of *p* is *n + 1* (there must be at least *n + 1* distinct arguments in *[0, p)* to evaluate *q(x)* at *x*). However, this is not a severe limitation since sixteen bit modulus (which can be handled by a cheap sixteen bit arithmetic unit) suffices for applications with up to 64,000 D~ pieces.

Some of the useful properties of this (k, n) threshold scheme (when compared to the mechanical locks and keys solutions) are:

1. The size of each piece does not exceed the size of the original data.
2. When *p* is kept fixed, D_i pieces can be dynamically added or deleted (e.g., when executives join or leave the company) without affecting the other D_i pieces. (A piece is deleted only when a leaving executive makes it completely inaccessible, even to himself.)
3. It is easy to change the D i pieces without changing the original data D, all that is needed is a new polynomial *q(x)* with the same free term. A frequent change of this type can greatly enhance security since the pieces exposed by security breaches cannot be accumulated unless all of them are values of the same edition of the *q(x)* polynomial.
4. By using tuples of polynomial values as Di pieces, a hierarchical scheme can be achieved in which the number of pieces to be determined D depends on their importance. For example, if the company's president is given three values of *q(x),* each vice-president two values of *q(x),* and each executive one value of *q(x),* then a (3, n) threshold scheme enables checks to be signed either by any three executives, or by any two executives one of whom is a vice-president, or by the president alone.

Key Management Service K/k

In our case, the *n* servers of the key management service share the ability to sign certificates. For the service to tolerate *(p-1)* compromised servers (Cluster Heads), an *(n, p)* threshold cryptography scheme is employed and the private key *k* of the service is divided into *n* shares $(s_1, s_2, .., s_n)$, assigning one share to each server. The $(s_1, s_2, .., s_n)$ is called an *(n, p)* sharing of *k*. figure 6 illustrates how the service is configured.

For the service to sign a certificate, each server generates a partial signature for the certificate using its private key share and submits the partial signature to a combiner (Here any server, holding the share can act as a combiner). With *p* correct partial signatures, the combiner is able to compute the signature for the certificate. However, compromised servers cannot generate correctly signed certificates by themselves because; they can generate at most *(p-1)* partial signatures.

When applying threshold cryptography, there must be a defense mechanism against compromised servers. For example, a compromised server could generate an incorrect partial signature. Use of this partial signature would yield an invalid signature. Fortunately, a combiner can verify the validity of a computed signature using the service public key. In case verification fails, the combiner tries another set of *p* partial signatures. This process continues until the combiner constructs the correct signature from *p* correct partial signatures. The loss of one or two partial keys will not affect the system as long as there are *p* number of partial signatures is present in the cluster.

The parameter *p* is selected by the cluster head based on the number of cluster members. To ensure proper security, the cluster head has to ensure that *n* is always greater than or equal to *2p ie., (n>= 2p)* where, n is the total number of nodes in the cluster. To do this, the cluster head keeps track of the number of nodes present within its range. And if there is any violation in the above condition, the cluster head goes for a key revocation scheme. In this scheme it reselects a new polynomial of this degree *(p'-1)*, where *n'* is the current number of cluster members and *(n'>=2p')*. The cluster head then constructs new shares of the secret key *k* and distributes them to its members. The members of the cluster update their shares. This key revocation scheme ensures the robustness of the Threshold cryptography when the number of vehicles go below the original *p* value in the cluster.

Figure 6. The configuration of a key management service: the key management service consists of n servers. The service, as a whole, has a public/private key pair K/k. The public key K is known to all nodes in the network, whereas the private key k is divided into n shares $s_1, s_2, .., s_n$, one share for each server. Each server i also has a public/private key pair K_i/k_i and knows the public keys of all nodes.

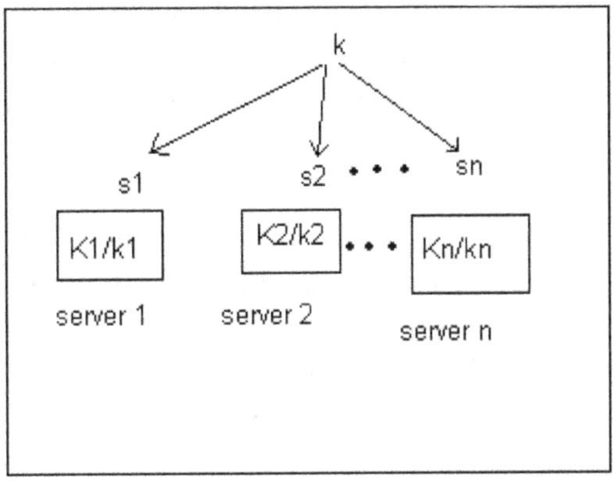

SIMULATIONS AND RESULTS

In order to gauge the improvement induced by the clustering algorithm, a simulation environment was created that represents both the specified communications model and realistic vehicle mobility on a highway.

Simulation Environment

The simulation of an IVC network was written based on the communications model described in system model. The nodal mobility data for this model was generated using CORSIM, the most widely used microscopic vehicle traffic simulation program in the US. As a microscopic traffic simulation it tracks each individual vehicle. The vehicle's mobility is determined by driver behavior, vehicle performance characteristics, and constraints imposed by the roadway geometry and surrounding vehicles.

The authors modeled the roadway geometry in CORSIM to roughly correspond to the 9.2 miles section of road, with 10 exits and 10 on ramps. Based on traffic data that was collected, a scenario was created for this roadway that models average traffic. The following parameterization was used for the clustering algorithm. The market penetration of IVC equipment was 100%. Though it will take time for this to occur, it is at this level that the scalability needs of the IVC network will be the greatest. The radio range was set to 500 feet. .The inter-cluster head minimum was set to 250 feet. The cluster head weight is proportional to the amount of time it will remain on the current road, and thus it captures a driver's intentions. It is defined to be the absolute value of the difference between a nodes current .location and its exit point. The association weight, W^{t}_{u-v} is inversely proportional to the distance between nodes *u* and *v*. The algorithm is run at each node once a second.

Results

In order to gauge the performance of Clustering algorithm, it has been compared against the following three approaches:

- *Lowest ID Clustering:* The cluster head is the node with the lowest ID, and the algorithm is rerun with change in network topology.
- *CGSRP Clustering:* This uses Lowest ID Clustering: however, cluster head election is preserved unless two cluster heads come into contact with one another or a node no longer is in radio contact with any cluster head.
- *Relative Mobility Clustering:* Like CGSRP clustering, cluster head election is preserved. Furthermore, only nodes with relative mobility below a certain threshold are clustered together. In these tests, only vehicles traveling in the same direction on the same road have a relative mobility below the threshold.

The performance of the clustering protocol was measured for l0 minute span on the highway scenario described earlier. The clustering algorithm produced significant improvements over all three of these approaches using two different metrics the average cluster head lifetime and the rate at which a node joins a different cluster.

1. Lowest ID,
2. CGSRP,
3. Relative Mobility,
4. Proposed clustering.

The average cluster head lifetime is the most direct measure of the stability of cluster head elections. As shown in Figure 7, This clustering algorithm improved the stability as measured by this metric 25.9 times better than Lowest ID Clustering, 9.3 times better than CGSRP Clustering, and 1.9 times better than Relative Mobility Clustering. The lifetime shown in this figure is adjusted for the rate at which the cluster heads leave the road network.

The re affiliation frequency, the seconds between cluster associations for a given node, represents the stability of cluster membership. As shown in Figure 7, using this clustering algorithm, nodes are associated with their cluster heads between 46% and 580% longer.

Figure 7. Stability of different types of clustering algorithms

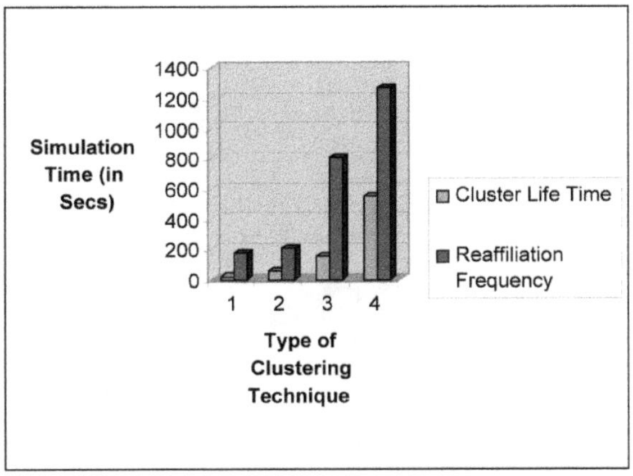

Figure 8 shows the strength of the proposed clustering for different number of hops. Figure 9 shows that the proposed scheme gives better results in terms of average number of cluster head changes.

The stability induced by Relative Mobility clustering and proposed algorithm comes at a cost of more clusters at any given moment. Relative Mobility Clustering has roughly twice the number of active cluster heads as Lowest ID and CGSRP clustering, since each direction of travel has its own set of cluster heads. This algorithm has even more; because more cluster heads may be within radio range as controlled by the inter-cluster head minimum distance parameter. Nonetheless, the total number of cluster elections was significantly lower using this algorithm. Due to the slower rate of cluster creation, the total number of cluster head elections in the other protocols quickly exceeds that of the new algorithm.

CONCLUSION

Inter-vehicle networks, when fully deployed, will be the largest open, mobile ad hoc network. The

Figure 8. Average number of cluster heads among 1-hop, 2-hop and 3-hop schemes.

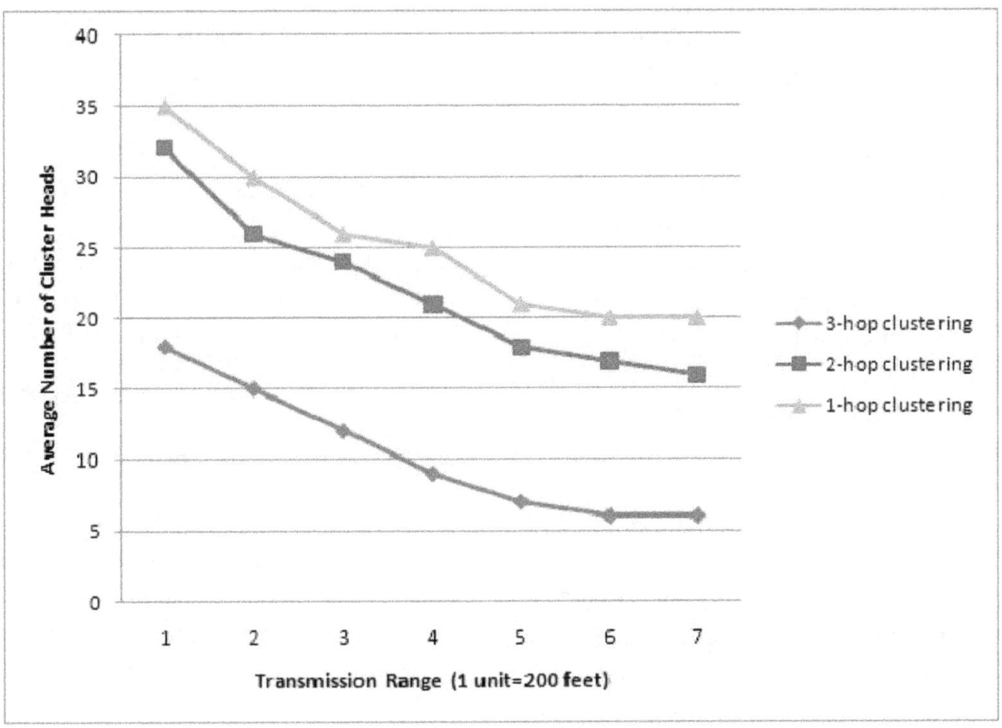

Figure 9. Comparison of cluster head changes

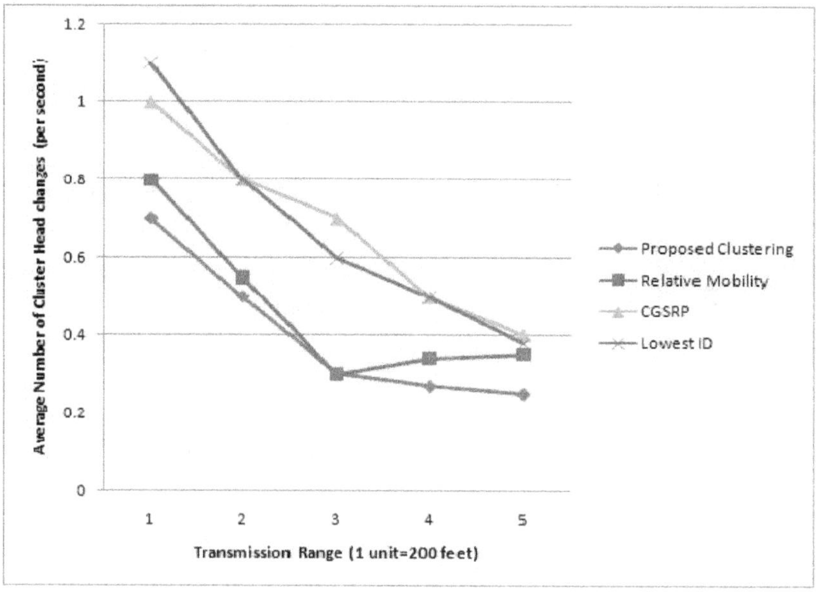

creation of a virtual network infrastructure is crucial for the scalability of media access protocols, routing protocols, and the security infrastructure. The stable clustering of nodes is the key to creating this infrastructure.

Existing MANET clustering protocols do not perform well in the IVC network due to the characteristics of vehicular mobility. This paper used the clustering algorithm, an extension of these protocols which accommodates oscillatory inter-vehicle distance and the inclusion of driver intentions. This clustering algorithm produced significant improvements in creating stable vehicle clusters.

The simulations reveal that the clustering algorithm works well with VANET and the threshold cryptography technique proves to be effective as well. The idea of using the threshold cryptography combined with mobility based clustering schemes is worth examining by further simulation, to establish its full potential.

The authors intend to perform future work to extend and analyze the protocol. Other instantiations of the protocol include different cluster head weighting and association weighting schemes and the relaxation of the definition of the nodal neighborhood. In order to further mitigate oscillatory inter-vehicle distances, the nodal neighborhood could include all of the nodes that had radio contact in the recent past. Moreover, the proposed algorithm should be tested on arterial roads, in rush-hour traffic, in off-peak traffic, and when an incident has occurred.

REFERENCES

Aho, A., Hopcroft, J., & Ullman, J. (1974). *The design and analysis of computer Algorithms*. Reading, MA: Addison-Wesley.

Aslam, B., & Zou, C. (2009). *Distributed certificate and application architecture for VANETs*. SIGCOMM- Project CPSC 534.

Basu, P., Khan, N., & Little, T. D. C. (2001). A mobility based metric for clustering in mobile ad hoc networks. *Proceedings of IEEE ICDCS 2001 Workshop on Wireless Networks and Mobile Computing*, (pp. 13-418). Phoenix, AZ.

Bellare, M., Shi, H., & Zhang, C. (2005). Foundations of group signatures: The case of dynamic groups, topics in cryptography. In Menezes, A. (Ed.), *CT-RSA Proceedings* (*Vol. 3376*). Lecture Notes in Computer Science Springer-Verlag.

Blum, J., Eskandarian, A., & Hoffman, L. (2003). Mobility management in IVC networks. *Proceedings of IEEE Intelligent Vehicles Symposium, 9*(11), 150–155.

Boneh, D., & Shacham, H. (2004). Group signatures with verifier- local revocation. In *CCS'04* (pp. 168–177). New York, NY: ACM Press.

Calandriello, G., Papadimitratos, P., Hubaux, J.-P., & Lioy, A. (2007). Efficient and robust pseudonymous authentication in VANET. In *Proceedings of the Fourth ACM International Workshop on Vehicular Ad Hoc Networks*, Canada.

Desmedt, Y. (1994). Threshold cryptography. *European Transactions on Telecommunications, 5*(4), 449–457. doi:10.1002/ett.4460050407

Desmedt, Y., & Frankel, Y. (1990). Threshold crypto system. *Advances in Cryptology-Crypto'89, 9th Annual International Cryptology Conference, 435.* (pp. 307-315). Springer.

Fonseca, E., & Festag, A. (2006). *A survey of existing approaches for secure ad hoc routing and their applicability to VANETS.* NEC Technical Report NLE-PR-2006-19, NEC Network Laboratories.

Gannaro, M., Jarecki, S., Karawczyk, H., & Rabin, T. (1996). Robust threshold DSS signatures. *Advances in Cryptology-Eurococrypt, 96*, 354–371.

Gerlach, M., Festag, A., Leinmuller, T., Goldacker, G., & Harsch, C. (2005). *Security architecture for vehicular communications.* In WIT2005, Hamburg, Germany.

Gollan, L., & Meinel, C. (2002). Digital signatures for automobiles. In *Proceedings of Systemics, Cybernetics and Informatics* (p. 02). SCI.

Hubaux, J. P., Capkun, S., & Luo, J. (2004). The security and privacy of Smart vehicles. *IEEE Security and Privacy Magazine, 3*, 49–55.

Jane, Y., Peter, H. J., & Chong. (2005). A survey of clustering schemes for mobile ad hoc networks. *IEEE Communications Survey & Tutorial, 7*(1).

Knuth, D. (1969). The *art of computer programming*: *Vol. 2. Semi numerical algorithms*. Reading, MA: Addison-Wesley.

Konstantopoulos, C., Gavalas, D., & Pantziou, E. G. (2008). Clustering in mobile ad hoc networks through neighborhood stability-based mobility prediction. *Computer Networks, 52*(9), 1797–1824. doi:10.1016/j.comnet.2008.01.018

Lee, S. J., Su, W., & Gerla, M. (2001). Wireless ad hoc multicast routing with mobility prediction. *Mobile Networks and Applications, 6*(4), 351–360. doi:10.1023/A:1011478628358

Lin, C. R., & Gerla, M. (1997). Adaptive clustering for mobile wireless networks. *IEEE Journal on Selected Areas in Communications, 15*(7), 1265–1275. doi:10.1109/49.622910

McDonald, A. B., & Znati, T. F. (1999). A mobility-based framework for adaptive clustering in wireless ad hoc networks. *IEEE Journal on Selected Areas in Communications, 17*(18), 1466–1486. doi:10.1109/49.780353

Munoz, J., & Syracuse, N. (2002). Proceedings of The 53 Internet Engineering Task Force Networks. *Wireless Networks, 9*, 443–460.

Papadimitratos, P., & Haas, Z. J. (2002). *Secure routing for mobile ad hoc networks.* SCS Communication Networks and Distributed Systems Modeling and Simulation Conference (CNDS 2002), San Antonio, TX.

Park, S., Zou, C. C., & Turgut, D. (2009). Reliable traffic information propagation in vehicular ad hoc networks. In Beyah, R., McNair, J., & Corbett, C. (Eds.), *Security in ad-hoc and sensor networks* (pp. 261–291). World Scientific Press. doi:10.1142/9789814271097_0009

Parno, B., & Perrig, A. (2005). Challenges in securing vehicular networks. In *Proceedings of ACM Workshop on Hot Topics in Networks* (HotNets-IV), College Park, MD, USA.

Petty, K. (1994). *Freeway service patrol (FSP) 1.1: The analysis software for the FSP Project. California PATH Research Report, UCB-ITS-PRR-95-20*. Berkeley: Institute of Transportation Studies, University of California.

Plobi, K., Nowey, T., & Mletzko, C. (2006). Towards a security architecture for vehicular ad hoc networks. availability, reliability and security. *Proceedings of the First International Conference on Availability, Reliability and Security*, April 20-22, ARES 2006 (pp. 374-381).

Sampigethaya, K., Huang, L., Li, M., Poovendran, R., Matsuura, K., & Sezaki, K. (2005). CARAVAN: Providing location privacy for VANET. In *Proceedings of the Workshop on Embedded Security in Cars* (ESCAR)'05.

Shamir, A. (1979). How to share secret. *ACM Transactions on Communications*, *22*(11), 612–613. doi:10.1145/359168.359176

Suzuki, S. C., & Nakada, K. (1991). *An authentication technique based on distributed security management for the global mobility network*. Symposium on Security and Privacy. IEEE Computer Society Press.

Yang, H., Luo, H., Ye, F., Lu, S., & Zhang, L. (2004). Security in mobile ad-hoc networks-Challenges and solutions. *IEEE Transactions on Wireless Communications*, *11*(1).

Zhou, X., Yang, X., Wei, P., & Hu, Y. (2007). Dynamic group signature with forward security and its application. In *Proceedings of the Sixth International Conference on Grid and Cooperative Computing* (pp. 473-480).

This work was previously published in International Journal of Business Data Communications and Networking, Volume 6, Issue 2, edited by Debashis Saha and Varadharajan Sridhar, pp. 1-18 , copyright 2010 by IGI Publishing (an imprint of IGI Global).

Chapter 8
A Distributed Secure Architecture for Vehicular Ad Hoc Networks

Tahani Gazdar
University of Manouba, Tunisia

Abdelfettah Belghith
University of Manouba, Tunisia

Abderrahim Benslimane
University of Avignon, France

ABSTRACT

In this paper, the authors propose a dynamic Public Key Infrastructure (PKI) for vehicular ad hoc networks to distribute the role of the central certification authority (CA) among a set of dynamically elected CAs. The election process is based on a clustering algorithm relying on trust levels and relative mobility. Furthermore, the authors have adapted the Dynamic Demilitarized Zones to protect the elected CAs from malicious nodes and enable them to act as registration authorities (RA). Extensive simulations are conducted to evaluate the performance of the clustering algorithm and investigate the impact of the vehicle speed, the vehicle average arrival rate, and the percentage of confident vehicles on the stability and efficiency of the security infrastructure. The authors demonstrate the percentage of confident nodes has a little impact on these performance metrics and that the minimum number of CAs to cover the entire platoon.

INTRODUCTION

A Vehicular ad hoc network (VANET) is an important and a special instantiation of Mobile ad hoc Networks (MANET) aiming essentially to increase road safety, yet it enables vehicles to share pertinent information on the road. To this end, vehicles must be able to route information and inform each other about potential ongoing events on the road. The inherent open architecture of VANETs, however, raises several network vulnerability and security issues. Unlike wired networks which can be protected by several layers of defense such as firewalls, attacks on

DOI: 10.4018/978-1-61350-477-2.ch008

vehicular networks emanate from different and various sources due to the constraints and requirements imposed on such networks. In particular, nodes or vehicles in VANETs are subject to a high mobility, the trajectories can be predictable and the environment is known as it can either be urban, rural or prescribed highways. Furthermore, obstacles and radio interference are predominant and affect the pattern of mobility and the quality of radio transmissions.

Communication security in VANETs has to take into account several stringent requirements (GWoch et al., 2006; Hubaux et al., 2006; Parno & Perrig, 2005). Firstly, vehicles must authenticate each others, so each driver must have a unique identity. The privacy and anonymity of drivers should, however, be preserved. For example, trying to prevent spoofing in a way that reveals the permanent identity of each vehicle violates privacy requirements and could not be tolerated. Secondly, security solutions in VANETs must insure non repudiation; namely a sender of a message cannot deny having sent such a message. The drivers responsible for an event occurring on the network, an accident for instance, should certainly be identified. Non repudiation is a stringent requirement that enables retracing events in case of an investigation for example. Thirdly, the security in VANETs requires data consistency. Data sent in similar messages should be consistent. The legitimacy of messages requires consistency with other similar messages.

Given that the characteristics of VANETs are different from those of MANETs, security solutions proposed for MANETs cannot be simply and directly transposed to VANETs. They need to be adapted to the nature of this type of networks. In this paper, we propose a secure architecture for VANETs based on previous works proposed for MANETs. Our solution consists on establishing a secure architecture based on an efficient trust model and a distributed clustering algorithm for designing a specific key management system properly adapted to VANETs. The distributed clustering algorithm uses, in particular, two metrics to select a cluster head: security and relative mobility. Furthermore, in order to enhance the security of a cluster, we propose a novel entity called the VANET Dynamic Demilitarized Zone (VDDZ) which is formed by a set of trusted nodes located at 1-hop from the CA in each cluster. The role of this set of nodes is to filter and pre-treat certification requests from unknown nodes thus shielding the CA.

The remainder of the paper is organized as follows. We discuss the related work on current security solutions proposed for VANETs. We then describe our network model, the trust model and the proposed distributed clustering algorithm and present simulation results of the proposed clustering algorithm.

RELATED WORK

PKI in VANET

The significant number of vehicles registered in different countries and their long distance routes, even beyond their regions of recording and licensing, requires a strong and an evolutionary structure for securing communications in VANETs. Therefore, the participation of the authorities in the recording phase of vehicles implies the need of certain level of centralization. Indeed, the communication via base stations is not sufficient for vehicular communications, because vehicles need to communicate with each other (without calling any server). As such, vehicles have to authenticate not only to base stations but also, to one another, which creates a problem of scalability. Hence, given the properties of a large scale and initially low penetration of vehicular communications infrastructure, Public Key infrastructure (PKI), is a good promising choice for enabling V2V security. In fact, PKI ensures secure communications by providing data integrity, confidentiality, strong authentication and non repudiation. It is based on asymmetric cryptography algorithms and Trust Third Parties (TTPs) called Certification Authorities (CA) which is responsible for certifying nodes' public keys. However, PKI cannot be applied directly in VANETs because it was intended for centralized networks. CAs must be continually available within the network and must be reachable by all nodes on the road. Another challenge in deploying a PKI in MANETs and particularly in VANETs amounts to the open nature of these networks which increases the risk of a single point of failure: if the CA is compromised the entire network will be too.

A large amount of research proposed solutions to these problems by decentralizing the CA functionalities. Ramaraj et al. (2009) proposed a self organized key management based on clusters. They proposed a model for key management where the VANET is divided into a number of clusters and the clustering algorithm is based on mobility. Since vehicles in a cluster move relatively to each other and on average they have the same velocity, a group can be represented by a single vehicle defined as the cluster head. Also they proposed a self organized PKI where any user can sign another user's public key. The set of signatures forms the network of trust relationships. Besides, each node generates its public key and the corresponding private key locally before joining the network. If a node u believes that a public key Pk_v belongs to a certain user v, it has to issue a certificate to node v signed by its private key Pk_u. This self-organized approach may have some drawbacks as it assumes that trust is transitive and, therefore, the system becomes more vulnerable to intrusions by malicious nodes. Another approach consists in distributing the CA functionalities among others parties in VANETs. Raya et al. (2006) proposed a distributed scheme for PKI in a VANET managed by many Certification Authorities (CA), each corresponding to a region (country, state, metropolitan, etc). Other potential candidates for assuming the role of CAs are car manufacturers. In either case, the different CAs will have to be cross-certified so that vehicles from different regions and different manufacturers can authenticate each others. This requires that each vehicle stores the public keys of all CAs whose certificates are needed to be verified. Blum and Eskandarian (2004) used a PKI with virtual infrastructure where the cluster heads are responsible for reliably disseminating messages after digitally signing them from the CA server and vehicles. This approach may have an advantage: the availability of the CA on the road at any time. Also, it can avoid the high cost of establishing infrastructures along the road.

Clustering in VANET

Deploying distributed PKI in VANETs will be more advantageous with an efficient clustering algorithm designed specifically for this type of networks. In fact, given the inherent specific characteristics of VANETs, it is rather not efficient nor adequate to transpose a clustering algorithms designed for MANETs into VANETs.

A host of researches were conducted to design new clustering algorithms which take into account the specific characteristics of a vehicular network. Papadimitratos et al. (2005) proposed a clustering method where vehicles are arranged into clusters (groups). Each cluster has a header and a trailer located in the front and rear of it and remaining nodes in the cluster are described as intermediate nodes. Aziz et al. (2006) proposed a location based approach to build clusters in VANETs. Roads are viewed as dissected into small area cells that define the clusters. A vehicle will automatically know to which group it belongs to by comparing its GPS position to a preloaded dissection of the area map into cells. The cluster head which is the vehicle closest to the center of the cell is determined dynamically. Cells, and hence clusters, overlap in such a way that any vehicle moving from one cell to the next remains in transmission range of both cluster heads. This approach may be efficient. In fact a vehicle will automatically know to which group it belongs to. Hence, group formation will not require any additional communication overhead or delay, however, the challenge of high speed of nodes on roads reduces the stability of the system and there will be an enormous number of cluster head changes.

Furthermore, Huang et al. (2005) assumed that vehicles are restricted to be in a group if each group member can hear broadcasts of every other group members. Authors assumed that, since vehicles in a group move relative to each others and on average they have the same velocity; a group can be represented by a single vehicle referred to as the group leader.

However, we notice that all these propositions are solely based on location metrics. Properly considering the behavior of vehicles and their cooperation with other vehicles in the network should be the crux of the cluster head election. An elected cluster head must have an assured trust level, since it will manage all security services in the cluster, yet it should have the lowest relative mobility within its vicinity. Furthermore, a cluster head must have at least one neighbor having a high trust level that can assume the role of the VDDZ.

THE PROPOSED CLUSTERING ALGORITHM

Network Model

We assume the existence of multiple central certification authorities (*Central CA*), each of which is responsible for a geographical region. The central CA manages all identities and credentials of the nodes registered with it. The road side units (RSU) are connected to the local servers and to the central CAs by wire line communications. In our network, we assume that these RSUs are certified by the central CAs of their regions as they will be, later, the intermediates between the central CA and other dynamic CAs elected dynamically on the road. Furthermore, each node upon registering in one *central CA,* will get a long term identity and many short term private/public keys pairs. If a node moves into a new cluster, it must change its pair of keys for anonymity purposes. Initially, we assume some vehicles registered in their regions as trusted vehicles. Figure 1 portrays the model layout of our network view.

General Framework

The basic idea of the architecture consists of establishing a dynamic distributed public key infrastructure where the CAs are the elected cluster heads. The election of a CA is based, in particular

Figure 1. Network model

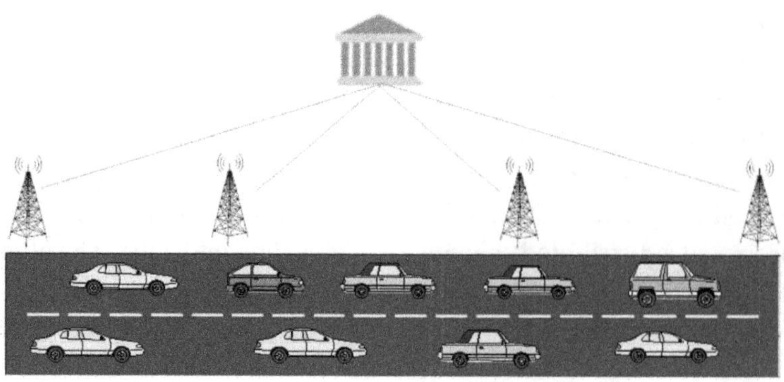

on two main metrics: relative mobility and security. Besides, we use the notion of the VANET Dynamic Demilitarized Zone (VDDZ) formed by all trust nodes within one hop of their CA. As such, a trust node can only be elected to be a CA if it has at least one trust neighbor node so that its VDDZ can be set. Recall, the role of the VDDZ is to prevent unknown vehicles to communicate directly with the CA node thus shielding the CA from any direct malicious node interaction. All guest vehicles are passed through the VDDZ to request certificates from the CA. The CA should also have the least relative mobility in its vicinity. For this we require that each vehicle in the platoon to be equipped with a Global Positioning System (GPS). Each node periodically transmits a Hello message containing among others, its instantaneous position. An elected CA manages all vehicles within *d* hops but the total number of served vehicles (member nodes) cannot exceed a predefined maximum. This is needed to balance the certification load among several CAs and to avoid the concentration of members in clouded areas, a stalling portion of a highway for instance. In the conducted simulations, we restrict our attention to vehicles travelling along the same road towards the same direction, although fast running vehicles surpass low running vehicles.

Trust Model

A trust metric, coined *Tm*, is defined as a continuous value in the interval [0-1]. A node with *Tm*=1 acquires the state of a confident or trust node. Nodes can obtain a high trust level when proving a good faith and adopting a good behavior. Each new unknown node starts with *Tm*=0.1. A node may assume one of the four following roles depending on its trust level:

CA: Certification authority of its cluster to certify public keys for nodes belonging to the same cluster. A CA has the highest trust level; *Tm* value must be equal to 1.

RA: Registration Authority of a cluster which protects the CA against attackers via the VDDZ in order to treat and filter the requests of certification. RA nodes must be confident with *Tm* =1, immediate neighbors of its CA and as such they form the VDDZ of this CA.

GW: Gateway which ensure connections between at least two different clusters. These nodes must be certified by all clusters in which it is member. GW nodes must have good trust level with *Tm* in [0.8, 1.0]. Furthermore, they can be members of the VDDZ (i.e., RAs) of their clusters if their *Tm*=1.

MN: Mobile Node in a cluster. It has an average trust level *Tm* in [0.1, 1]. A MN with a trust level *Tm* in [0.8, 1.0] can also be a GW if it ensures the connection between at least two different clusters.

All nodes having a *Tm* < 1 must prove a correct behavior and a good cooperation to increase their *Tm*. This is accomplished through the monitoring and the supervision of their behaviors at the MAC and Network layers. Rachedi and Benslimane (2010) proposed a hierarchical monitoring process. Each node with a high trust value monitors its neighbor nodes with lower trust values. A CA node monitors all the other CAs, RA, GW and MN. A RA monitors GW, and MN. A GW can supervise MN, but MN cannot supervise any other node. To ensure this task, they proposed a monitor process acting in two different network protocol layers: MAC layer and network layer. First, in the MAC layer, the monitoring nodes supervise the channel occupation by their neighbors. This consists of measuring the duration of the channel occupation by nodes with their own flows. However, in the network Layer, the monitoring nodes supervise the packet forwarding activities of their packets with respect to the total number of packets to be forwarded during a fixed time interval (monitoring period). The monitoring period is the observation period which consists

Figure 2. Finite states machine governing the transitions between roles

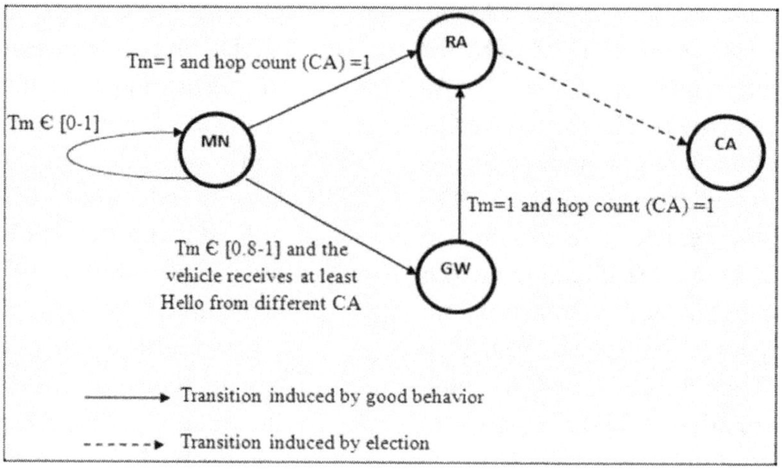

in collecting the information from nodes in order to calculate a reputation rating.

Figure 2 portrays the state transition diagram that governs the transitions among the four advocated roles, or states, as a function of the acquired level of trust. Each state nominates a role in a cluster, in addition, the links are labeled with the set of conditions which induces the transitions. As noted earlier, a GW can either be a MN or RA. This is not presently showed on the state diagram of Figure 2 as we concentrated here solely on transitions invoked according to the evolution of trust levels. Later when we explicit our distributed clustering proposal, we shall give a comprehensive finite state machine governing the transitions among all possible states in which a vehicle can be whether CA, RA, RA-GW, MN-GW, MN or some others to be further defined.

Secure and Distributed Clustering Algorithm

In our algorithm, we elect a cluster head (CH) to become a CA vehicle according to two different metrics: the security metric represented by both the trust level T_m of a candidate CA and the number of its confident neighbors, and the mobility metric which indicates the relative mobility of a vehicle in its vicinity.

Only confident nodes (T_m=1) can be candidate for CA. All clusters have a predefined maximum range d equals to the distance in hops of the farthest node from its cluster head, and a predefined maximum number of members. The algorithm is based on the exchange of periodic and specific messages. Moreover, to reinforce the security of the cluster, the algorithm selects the candidate having the largest number of trust neighbors (of RAs).

To this end, our distributed secure clustering algorithm defines 3 data structures, 5 different messages, 4 different timers and 12 states. Table 1 provides the data structures used by each node in the network independently of its state, to maintain an up-to-date and accurate view of the network region.

Table 2 defines the five types of messages. Independently of its current state each node in the network generates and transmits periodically a Hello message. This Hello message is forwarded by other nodes, also independently of their current states, up to (d+1) hops. This is needed to maintain and update the NT[] and CT[] tables at each

Table 1. Data structures

Structure Name	Use	Type and content
NT []	List of nodes located within (d+1) hops; d is the range of the cluster.	List; IPAddress, position, state, speed, T_m, hop-count
CT []	List of *CA_NODE* vehicles and *CA_NODE_CANDIDATE* located within (d+ 1) hops.	List; IP address, position, state, hop count, RM, NTN. RM: Relative Mobility NTN: Number of Trust nodes Neighbors
JT []	A List of nodes from which a JOIN message has been received.	List; IP address, position, hop-count

Table 2. Messages description

Message Name	Content
HELLO	IP address, current state, current position, current speed, NT[], sequence number, hop-count.
ELECTION_BEACON	CA (IP address), hop-count, NTN, RM, ID_num, MAC.
JOIN	Source IP Address, CA IP address, hop-count, sequence number.
ACCEPT	Source IP Address, CA IP address, hop-count, sequence number.
REJECT	Source IP Address, CA IP address, hop-count, sequence number.

node. At each forward operation, the hop-count is incremented. The other four messages are only sent at special events: ELECTION_BEACON, JOIN, ACCEPT and REJECT.

The ELECTION_BEACON is sent by a vehicle to propose itself as a CA vehicle. The Message Authenticated Code (MAC) field authenticates the beacon information encrypted with the private key *K* of the sender:

MAC=E_K (Hash (CA; Hop-count; NTN; RM; ID-num)

At the reception of a HELLO message, the receiving node calculates its relative speed to the vehicle originating of the HELLO. Just before sending the ELECTION_BEACON, the vehicle calculates the average of all previously calculated relative speeds with respect to all its neighbors; namely its relative mobility RM.

The ELECTION_BEACON message has to be forwarded up to (d+1) hops by receiving nodes. At each forward operation, the hop-count is incremented. At the reception of an ELECTION_BEACON, a vehicle can request membership to the candidate CA originating such a message by sending a JOIN message and waits for a response. The CA candidate accepts this membership by replying with an ACCEPT message, otherwise, it declines the membership by sending a REJECT message.

During the clustering process, each and every vehicle assumes one of a set of states. Recall that we have already defined four states; namely CA, RA, GW and MN states. In addition, eight more states are defined and are given in Table 3.

Finally, a set of four timers are needed to control the clustering process evolution. These are:

- **InitTimer**: The maximum lasting of the initialization step. In the INT_NODE state, a node generates periodically its HELLO messages, and does the forwarding of received messages.

Table 3. Vehicles states

State	Description
INT_NODE	A starting vehicle: a vehicle just entering the road.
OR_NODE	An orphan vehicle having no neighbor.
SEARCH_NODE	A vehicle looking for a CA.
CA _NODE_CANDIDATE	A vehicle candidate to be a CA.
WAIT_CA	A vehicle waiting a 'Hello' message from a CA node.
WAIT_ACCEPT	A vehicle waiting an 'ACCEPT' message from a CA_CANDIDATE_NODE.
MN_GW	A gateway, located at a distance more than one hop from the CA.
RA_GW	A gateway having the state of RA_NODE and connecting two clusters at the same time

- **ElectTimer**: The maximum lasting in the CA_NODE_CANDIDATE state where the node is eventually accumulating JOIN messages.
- **WaitAcceptTimer**: The maximum waiting time in the WAIT_ACCEPT state to receive an ACCEPT message.
- **HelloFromCATimer**: The maximum time period to wait for the next HELLO message. All nodes in states MN, RA, and GW should receive periodically a Hello from their owner CAs to which they are registered and from each served CA for which it is also acting as a gateway. Otherwise the relationship is relinquished.

Clustering Algorithm

In our secure clustering algorithm, each and every node assumes a given state at any given time. Upon entering the network, a vehicle gets into the INIT_NODE state and from there on it will participate in the clustering process and maintenance using its initialized data structures and relying on the exchange of appropriate control messages and the defined timers. Depending on occurred events, the node transits from state to state. Figure 3 provides the detailed finite state machine governing the execution and evolution of the proposed clustering scheme. Each one of our 11 defined states is put in a round corner rectangle. Transitions between these states are labeled by the events which must all happen to perform the transition. The definitions of the different required events are defined in Table 4.

Detailed Algorithms

To each and every state of the finite state machine (of Figure 3) corresponds an executable process. This process is automatically launched by a node upon entering the state. In the following, we detail the algorithm of each of these processes. Each process holds the name of the state it represents.

The function *SELECTCA()*, respectively the function *SELECTCACAND()*, used in the process *SEARCH_NODE*, selects the best CA_NODE, respectively the best CA_NODE_CANDIDATE, from the CT[] table according to the following criteria in this prescribed order: the relative mobility (RM), the number of trust neighbors (NTN), the hop count and finally the number of current members. The function *Update()* updates a data structure with the information found in the received message. Furthermore, the function *Delete()* simply deletes an entry from a given data structure and *send()* transmits the corresponding message to a given destination.

Figure 4 portrays an example of a network where clusters have a range that equals to one.

Figure 3. Clustering algorithm finite state machine

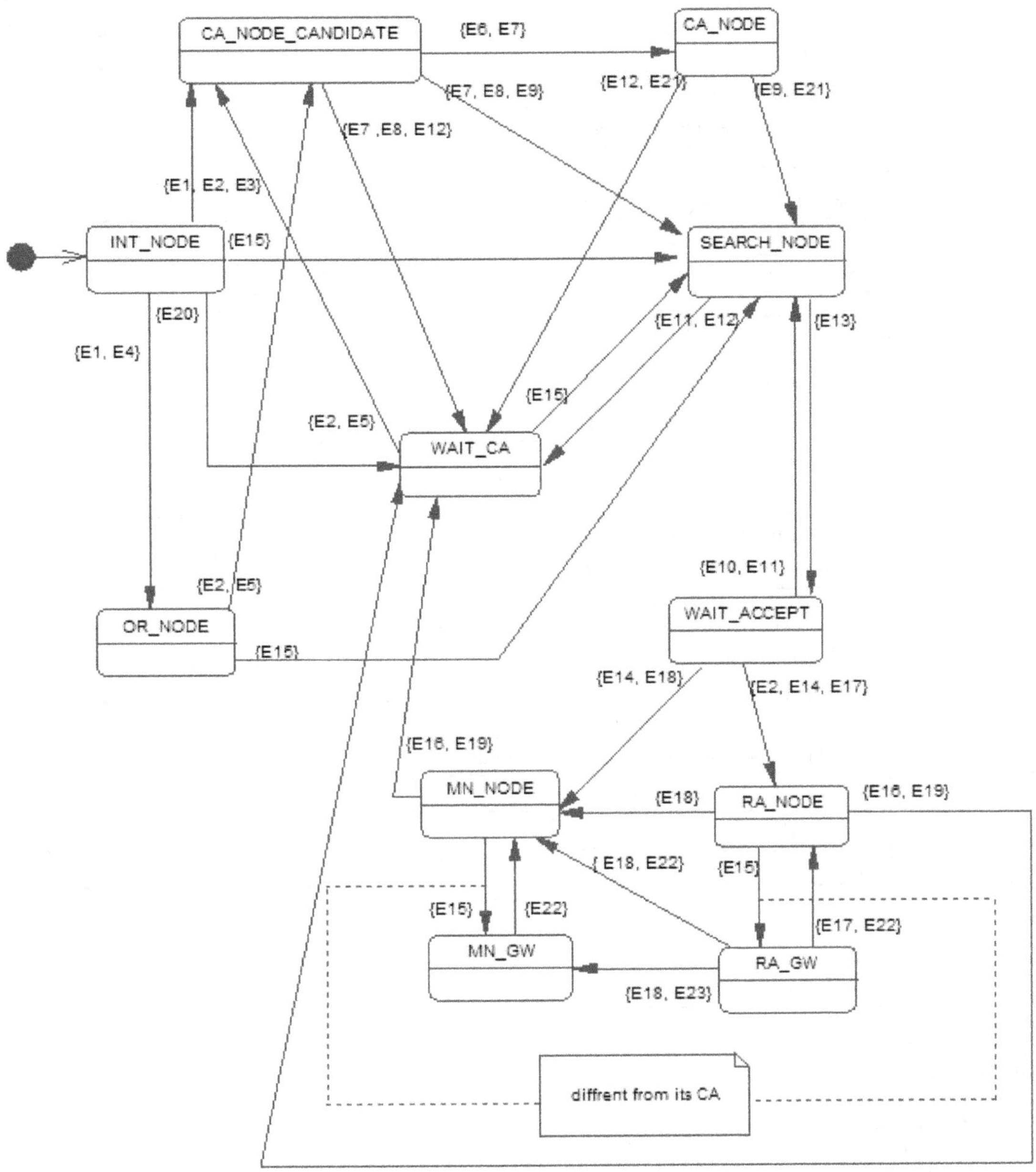

PERFORMANCE EVALUATION

Mobility Model

We consider a segment of road with a length L and 60m as width. It is composed by two lanes. All cars move along a linear trajectory toward the same direction (Figure 5).

The vehicles enter the road at a rate of λ vehicles per one second and are assigned to lanes in a round robin fashion. In addition, the vehicles come in an alternately way: one time on the first lane and the next time, on the second lane. Each vehicle assumes a speed of v meters per second (m/s) taken independently and randomly on the interval $[V_{min}, V_{max}]$ where $V_{min} = V_{aver} - V_{aver}/2$,

Table 4. Occurring required events

Notation	Event description
E1	InitTimer expires
E2	The node is confident
E3	The node has at least one confident neighbor
E4	No Hello is received
E5	A Hello from a confident node is received
E6	A Join message is received
E7	ElectTimer expires
E8	No Join message is received
E9	CT[] table is not empty
E10	WaitAcceptTimer expires
E11	No Accept message is received **OR** a Reject message is received
E12	CT[] table is empty
E13	A Join message is sent
E14	An Accept message is received
E15	A Hello from a CA node is received
E16	A HelloFromCATimer expires
E17	hop (CA) = 1
E18	hop (CA) > 1
E19	No hello from the CA is received
E20	No hello from a confident node is received **OR** the node is not confident
E21	No RA node still in the cluster
E22	size of CT[] equal to 1
E23	size of CT[] > 1

Algorithm 1.

```
PROCESS INIT_NODE()
        Initialize(CT[]);
        Initialize(JT[]);
        Initialize (NT[]);
        Initialize (InitTimer);
        During InitTimer do
For each Hello received do
                Update (Hello, NT[]);
If (Hello from CA) Then {Update (Hello, CT[]);
state = SEARCH_NODE;
Call SEARCH_NODE(); };
If (hop count < d+1) Then increment hop count and forward (Hello);
EndFor
At the expiration of InitTimer do
If (NT[] is empty) Then { state = OR_NODE; Call OR_NODE(); }
Else If (confident And at least one neighbor is confident)
Then { state = CA_NODE_CANDIDATE;
Call CA_NODE_CANDIDATE(); }
else { state=WAIT_CA; Call WAIT_CA(); }
End INIT_NODE
```

Algorithm 2.

```
PROCESS OR_NODE()
        AT the reception of a HELLO do
              Update (Hello, NT[]);
              If (hop count < d+1) Then increment hop count and forward (Hello);
              If (Hello from a confident neighbor node)
                          Then If (confident) Then { state = CA_NODE_CANDIDATE;
Call CA_NODE_CANDIDATE(); }
else { state=WAIT_CA; Call WAIT_CA(); };
If (Hello from CA node) Then { Update(Hello, CT[]);
state= SEARCH_NODE;
Call SEARCH_NODE(); };
End OR_NODE.
PROCESS WAIT_CA()
        AT the reception of a HELLO do
              Update (Hello, NT[]);
              If (hop count < d+1) Then increment hop count and forward (Hello);
If (Hello from CA node) Then { Update (Hello, CT[]);
state= SEARCH_NODE;
Call SEARCH_NODE(); };
If (Hello from a confident neighbor node)
                          Then If (confident) Then { state = CA_NODE_CANDIDATE;
Call CA_NODE_CANDIDATE(); }
End WAIT_CA.
```

Algorithm 3.

```
PROCESS CA_NODE_CANDIDATE ()
 int i;
send (ELECTION_BEACON);
 Initialize (ElectTimer);
       During ElectTimer do
For each message received do
if (Hello) Then Update(Hello, NT []) ;
                    if (Join) Then update(Join, JT []) ;
                    if (Hello from CA)Then Update(Hello, CT []) ;
if (ELECTION_BEACON) Then Update(ELECTION_BEACON, CT []) ;
                    If (hop count < d+1) Then increment hop count and forward (Hello);
At the expiration of ElectTimer do
if (JT [] is not empty) Then { state = CA_NODE;
 i=0;
while (i<Max numbers) do
 send (Accept, JT[i]);
i++;
 Delete(JT[i]);
End while
while ( i < size of (JT[]) ) do
 send (Reject, JT[i]) ;
 i++;
End while
 Call CA_NODE();
 }
Else If (CT[] is not empty)
Then { state=SEARCH_NODE ;
Call SEARCH_NODE();};
Else { state= Wait_CA;
Call Wait_CA();}
End CA_NODE_CANDIDATE.
```

Algorithm 4.

```
PROCESS SEARCH_NODE()
        If (CT[] is empty) Then { state= WAIT_CA; Call WAIT_CA(); }
                              Else { if (CA in CT[]) Then send (Join, SELECTCA());
Else send (Join, SELECTCACAND());
state=Wait_Accept;
Call Wait_Accept(); }
End SEARCH_NODE.
```

Algorithm 5.

```
PROCESS WAIT_ACCEPT()
Initialize (WaitAcceptTimer);
        Upon the reception of an Accept message from a CA
                if (node is confident and hop count to CA =1)
                        Then {Reset (HelloFromCATimer CA);
                                state=RA_NODE;
Call RA_NODE(); };
Else { Reset (HelloFromCATimer CA);
                                state=MN_NODE;
Call MN_NODE(); };
        Upon the expiration of WaitAcceptTimer OR a Reject is received do
state = SEARCH_NODE;
Call SEARCH_NODE();
End WAIT_ACCEPT.
```

Algorithm 6.

```
PROCESS CA_NODE()
        Upon loosing the last RA node do
If (CT[] is not empty) Then state=SEARCH_NODE;
                                        Call SEARCH_NODE();
                                Else state=WAIT_CA;
                                        Call WAIT_CA();
End CA_NODE.
```

Algorithm 7.

```
PROCESS RA_NODE()
        For each received Hello do
                If (Hello from owner CA) Then {Update information about CA;
                Reset (HelloFromCATimer of owner CA);
                If (hop count to CA > 1) Then state=MN_NODE; Call MN_NODE();}; If (Hello from another CA) Then { Update (Hello,
NT[] and CT[]);
                                        Reset (HelloFromCATimer of this CA);
        state=RA_GW;
        Call RA_GW(); };
                Update(Hello, NT[]);
                If (hop count < d+1) Then increment hop count and forward (Hello);
EndFor
        Upon the expiration of HelloFromCATimer do
Delete (CA, CT[]); state=WAIT_CA; Call WAIT_CA();
End RA_NODE.
```

Algorithm 8.

```
PROCESS MN_NODE()
        For each received Hello do
                If (Hello from owner CA) Then {Update information about CA;
                                        Reset (HelloFromCATimer of owner CA); };
                If (Hello from another CA) Then {Update (Hello, NT[] and CT[]);
                                        Reset (HelloFromCATimer of this CA);
state=MN_GW;
                                        Call MN_GW(); };
                Update(Hello, NT[]);
                If (hop count < d+1) Then increment hop count and forward (Hello);
        EndFor
        At the expiration of HelloFromCATimer of owner CA do
Delete (CA, CT[]); state=WAIT_CA; Call WAIT_CA();
End MN_NODE.
```

Algorithm 9.

```
PROCESS MN_GW()
        For each Hello received do
                If (Hello from served CA(i)) Then { Update information about CA(i);
                                        Reset(HelloFromCATimer of served CA(i));};
If (Hello from another CA) Then { Update(Hello, CT[]);
                                Reset (HelloFromCATimer of this CA); };
                Update(Hello, NT[]);
                If (hop count < d+1) Then increment hop count and forward (Hello);
        EndFor
        At the expiration of HelloFromCATimer of CA(i) do
Delete (CA(i), CT[]);
If (size-of CT[] == 1) Then {state=MN_NODE; Call MN_NODE();
End MN_GW.
```

Algorithm 10.

```
PROCESS RA_GW()
        For each Hello received do
        If (Hello from owner CA) Then {Update information about CA;
                                Reset (HelloFromCATimer of owner CA);
                                 If (hop count to CA > 1) Then state=MN_GW; Call MN_GW();};
                If (Hello from served CA(i)) Then {Update information about CA(i);
                                        Reset (HelloFromCATimer of CA(i)); };
                If (Hello from another CA) Then {Update(Hello, CT[]);
                Reset (HelloFromCATimer of this CA); };
                Update(Hello, NT[]);
                If (hop count < d+1) Then increment hop count and forward (Hello);
EndFor
At the expiration of HelloFromCATimer of CA(i) do
Delete (CA(i), CT[]);
If (HelloFromCATimer of the owner CA and size-of CT[] > 1) Then {
state=MN_GW;
                                                Call MN_GW();};
                If (HelloFromCATimer of a served CA(i) and size-of CT[] =1) Then { state=RA_NODE;
                 Call RA_NODE();};
End RA_GW.
```

Figure 4. Example of a network with d=1

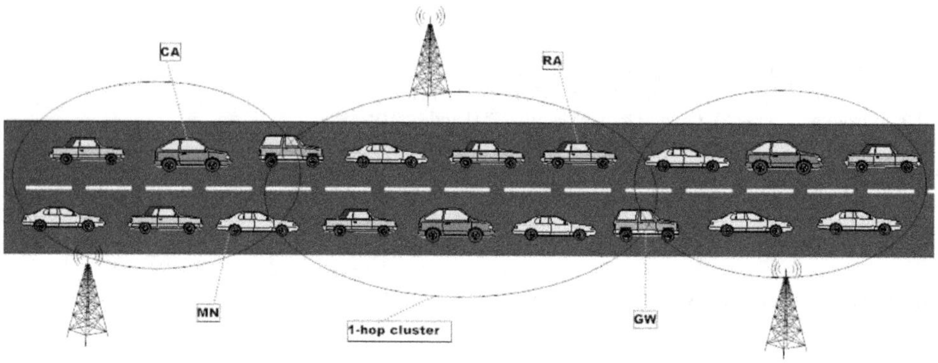

Figure 5. The considered road segment

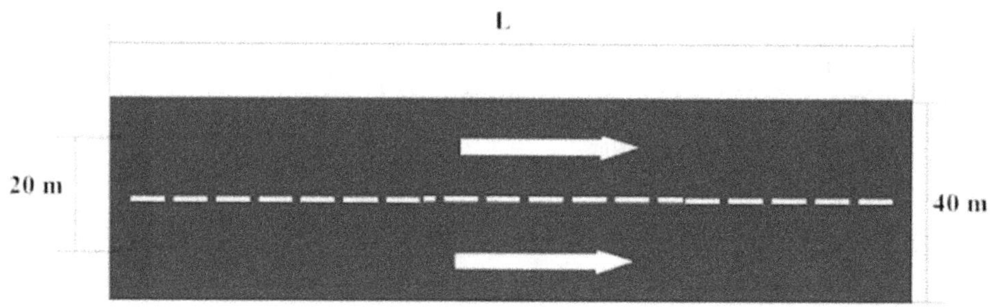

$V_{max}=V_{aver}+V_{aver}/2$ and V_{aver} is the average speed. All vehicles maintain their speeds constant along the road.

Simulation Setup

In order to study the performance of our clustering algorithm, we focus on the stability of the network in terms of both the life time of the elected CAs and on the efficiency of the clustering.

Let T_{sim} be the total simulation time which is composed of two successive sub periods $T_{transient}$ and T. $T_{transient}$ represents the time needed by the first entering vehicle to reach the end of the road (to travel the assumed road segment). That is for a speed of 10 m/s and L=10 km, T_c equals 1000 seconds. T represents the effective simulation time for which we gather our simulation statistics. $T_{transient}$ is considered as a transient phase needed to fill the road and set up the platoon for our simulations scenarios. We use the discrete event simulator OMNET++ and the IEEE802.11 Mac layer. Table 5 summarizes the different parameters use.

Furthermore, we consider different scenarios according to the percentage of assumed trust nodes (100%, 50% and 10% of the vehicles are confident), the average speeds and the arrival rates of vehicles as detailed in Table 6.

Let us consider an elected vehicle CA_i and let $\mathbf{t}_i$ be its election instant, v_i its speed, $\mathbf{x}_i$ its position at this instant, and $\mathbf{t}_i$**'** the instant at which CA_i loses its CA state which could be at the exit from the road segment, before the end of the road segment or at the end of the simulation time. We define the average relative time to live of CA vehicles, called TTLCA, as follows:

$$TTLCA = 100 \times (\sum\nolimits_{(1 \le number\ of\ CA\ during\ T)} (t_1^{'} - t_1) / ((L - x_1) / v_1)) / number\ of\ CA\ during\ \mathrm{T} \qquad (1)$$

Table 5. Simulation parameters

Parameters	Value
L	10000m
Transmission range	450m
Hello period	2s
InitTimer	5s
ElectTimer	5s
WaitAcceptTimer	5s
HelloFromCATimer	5s
The average speed	[10-40] m/s
T	$T = T_{transient}$
d = Maximum range of a cluster	3
Maximum number of vehicles in a cluster	300
n= number of observation instants in T	20

Table 6. Details of the different scenario

The average speed (m/s)		10	20	30	40
$T_{sim} = T_{transient} + T$ $T_{transient}$ T		2000 1000 1000	1000 500 500	666 333 333	500 250 250
The average total number of vehicles in the platoon at any instant in T	*λ1=0.5vehicles/s*	500	250	167	125
	λ2=1vehicles/s	1000	500	333	250
	λ3=1.5vehicles/s	1750	750	499	375
	λ4=2vehicles/s	2000	1000	666	500

Within the effective simulation period T, we define a set of n equally spaced observation instants called t_{oi} ($0 < i \leq n$). At each observation instant, we calculate the percentage of vehicles $NV(S_i)$ being in states S_i which can be either CA_NODE, RA_NODE, MN_NODE, GW_MN or GW_RA. That is:

$$NV_{t_{01}} = 100 \times (\textstyle\sum_1 \text{number of vehicles in state i}/ \text{Total number of vehicles at } t_{01}) \tag{2}$$

NV_{toi} gives then the percentage of vehicles within the platoon at the observation instant t_{oi} that have already acquired a defined status (CA, RA, GW or MN). We define the efficiency E of our clustering algorithm as the average of NV_{toi} over all observation instants.

$$E = (\textstyle\sum_{1 \leq n} NV_{t_{01}}) / n \tag{3}$$

Simulation Results

Effects of the Speed and Rate on the Number of CA and RA Vehicles

An efficient clustering algorithm should elect a minimal set of cluster heads. However, for security purposes the number of RAs per cluster head should be maximized. As such, and in the quest to study the appropriateness of our proposed secure clustering algorithm, we portray in Figure 6, Figure 7, Figure 8, Figure 9, Figure 10, and Figure 11, the average number of CAs in the platoon and the number of RAs per cluster as a function of the average speed and the arrival rate of vehicles for the cases of 10%, 50% and then 100% of confident vehicles.

First of all, let us comment on the minimal number of clusters that should be formed in the platoon without the restrictions of the fixed maximum of 300 vehicles per cluster and at least one RA per cluster. Recall that the transmission range is 450 meters, the length of the road segment is 10 Km and the maximum range of a cluster is d=3. Consequently, the minimum number of clusters that can accommodate the entire platoon is 4.

Now consider Figure 6 that is portraying the average number of CAs as a function of the average speed and for different arrival rates and 10% of the nodes are confident. At a speed of 10 m/s and *λ4=2vehicles/s*, we have 2000 vehicles in the platoon as specified in Table 6 Since the maximum size of a cluster is fixed to 300 vehicles, we need a minimum of 7 cluster heads to cover the entire platoon. At this very same speed but for *λ3=1.5vehicles/s*, we have instead 1750 vehicles and therefore 6 clusters are enough to cover the entire platoon. For *λ2=1vehicles/s* and *λ1=0.5vehicles/s*, we have respectively 1000 and 500 vehicles and therefore we require exactly the minimum number of clusters which is 4. Now, from the speed of 20 m/s and above until 40 m/s, the total number of vehicles is less than 1200, (see Table 6), for all assumed arrival rates and consequently just the minimum of 4 clusters is needed to cover the entire platoon. Here we should note that the requirement of having at least one RA per cluster is met for all the considered scenarios. However, when the average speed gets higher than 40 m/s, this requirement will not be met and hence we may not be able to form enough clusters to cover the entire platoon.

Now, let us consider Figure 7 and Figure 8 portraying the same as in the previous figure but for a percentage of confident vehicles equals to 50% and 100%. We found the exact same result as in Figure 6. Indeed, in these scenarios we just have more confident nodes which provide more RAs per cluster, but the case of 10% confident

Figure 6. The average number of CA vehicles, 10% confident vehicles

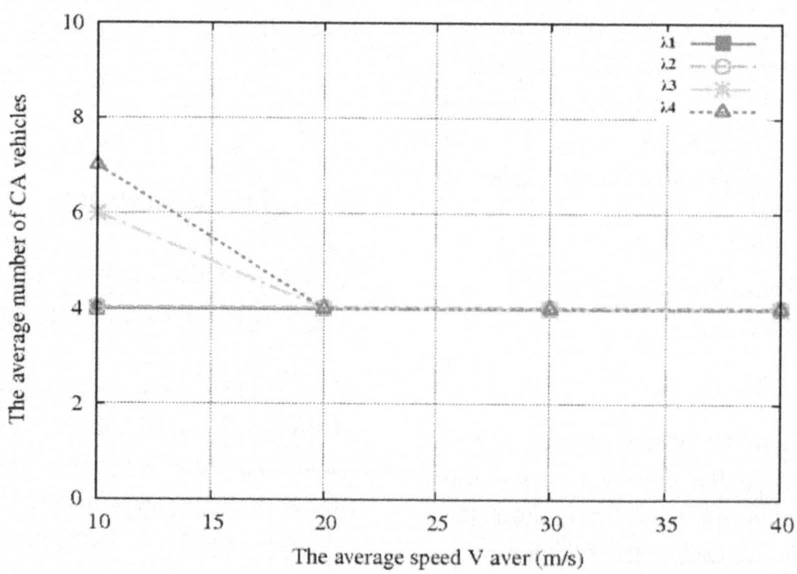

Figure 7. The average number of CA vehicles, 50% confident vehicles

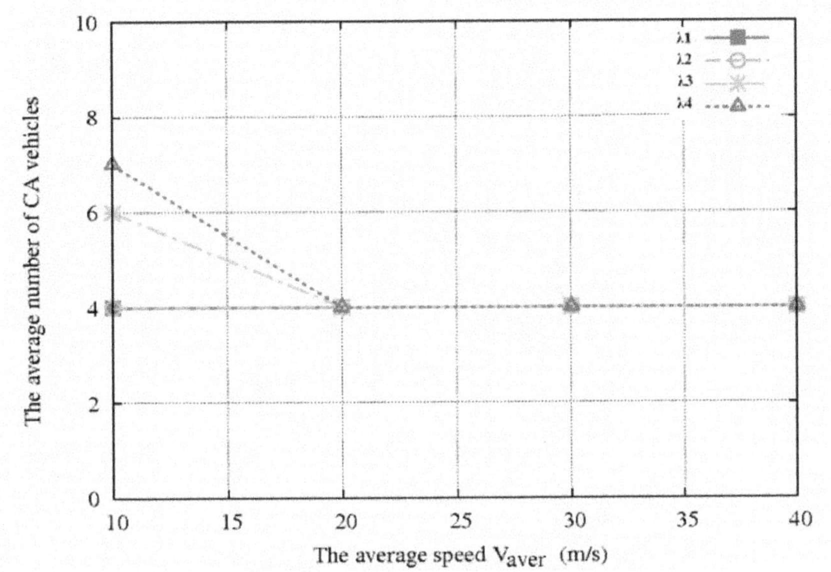

Figure 8. The average number of CA vehicles, 100% confident vehicles

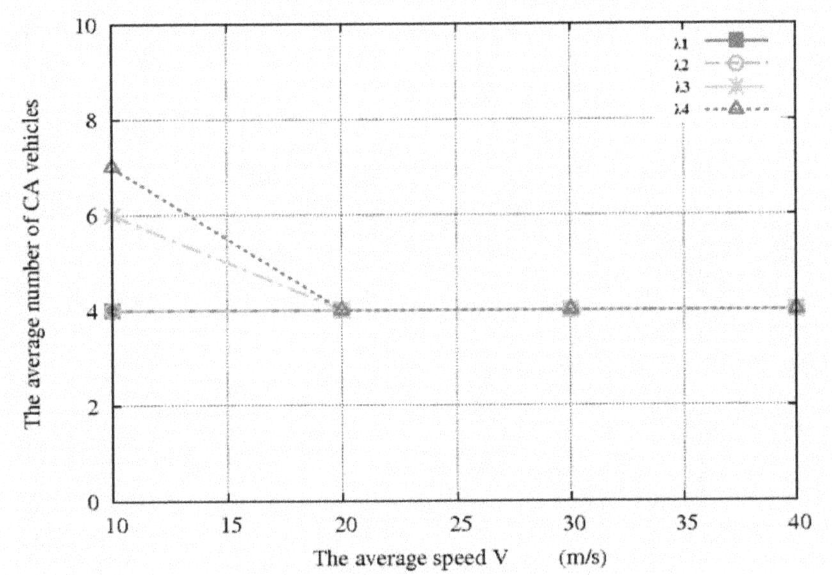

nodes satisfies already the requirement of having at least one RA per cluster as portrayed through Figure 9.

Let us investigate now the average number of RAs per cluster. Figure 9, Figure 10 and Figure 11 portray this average as a function of the average speed, the percentage of confident vehicles and the arrival rate. We clearly observe that the average number of RAs per cluster increases with both the percentage of confident nodes and the vehicle arrival rate but decreases when the average speed increases. Indeed, a higher percentage of

Figure 9. The average number of RA vehicles per cluster, 10% confident vehicles

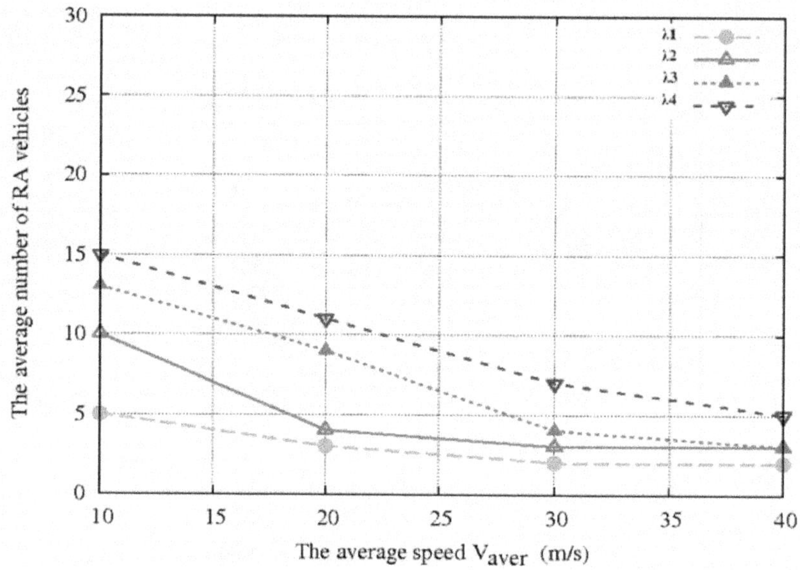

Figure 10. The average number of RA vehicles per cluster, 50% confident vehicles

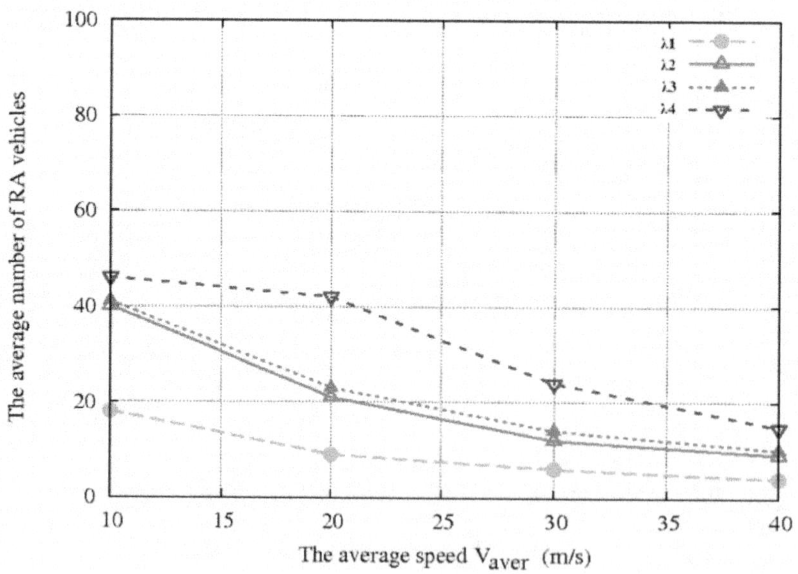

confident nodes provide more confident neighbors to any cluster head and consequently more RAs, a higher arrival rate provides more vehicles in the platoon but when the average speed gets higher, the total number of vehicles within the platoon gets smaller, see Table 6. More interestingly, we note that even for the case of 10% confident nodes, an average speed of 40 m/s and an arrival rate of $\lambda 1=0.5 vehicles/s$, each cluster head possesses an average of 2 RAs. This explains, once again, why our proposal uses just the minimum required number of clusters. For a lesser percentage of confident

Figure 11. The average number of RA vehicles per cluster, 100% confident vehicles

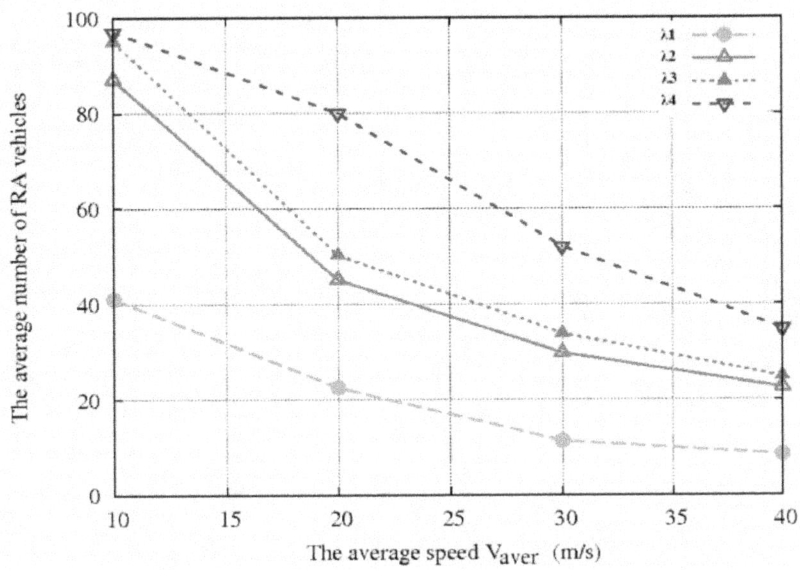

vehicles and or a higher speed and or a reduced arrival rate, we may not be able to form enough clusters to cover the entire platoon. A very small percentage of confident vehicles can be assumed as far as the average speed and the arrival rate are sufficient to provide at least one RA per cluster.

Effects of the Speed and Rate on the Efficiency of the Clustering Algorithm

The efficiency of our secure clustering algorithm in terms of the proportion of the vehicles in the platoon that acquired a definitive status, namely the proportion of vehicles in states CA, RA, GW or MN, is of utmost importance. This is given by Eq (3). In Figure 12, Figure 13 and Figure 14, we plot the efficiency E as a function of the average speed, the arrival rate of vehicles and for the three different assumed percentages of confident vehicles. We first observe an excellent efficiency way above the 90% for all scenarios and cases.

The efficiency E stays above the 97% for speeds below 20 m/s even for the case of 10% confident vehicles. It declines a little bit for higher speeds. It increases as the arrival rate increases. Indeed, for an arrival rate of $\lambda 4=2 vehicles/s$ or even $\lambda 3=1.5 vehicles/s$ the efficiency stays around the 100%. For smaller arrival rates and high speeds, the efficiency declines a little but since the average number of RAs per cluster becomes very small as shown earlier in Figure 9 and Figure 10. Remarkably yet, the efficiency is rather resilient to the decrease in the percentage of confident vehicles. Indeed Figure 13 and Figure 14 present roughly the same results despite passing from 50% to 100% confident vehicles.

Effects of the Speed and Rate on the Stability of the Network

The stability of the network under our secure clustering algorithm can be measured by the average Time To Live of the elected CAs, namely the TTLCA defined by Eq (1). The longer the elected CAs can maintain their status, the stronger is the stability of the different memberships and the better is the efficiency of our proposed secure clustering scheme.

Figure 15, Figure 16 and Figure 17 portray the TTLCA as a function of the average speed for the

Figure 12. The efficiency, 10% confident vehicles

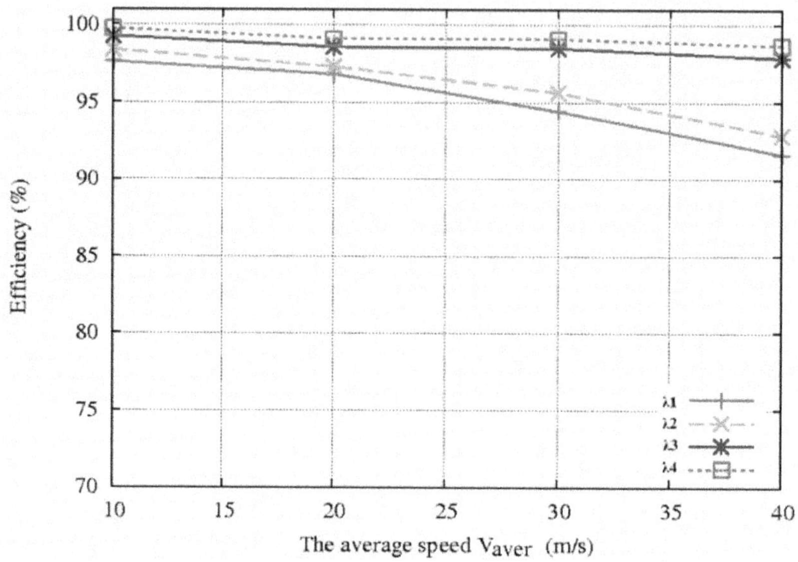

Figure 13. The efficiency, 50% confident vehicles

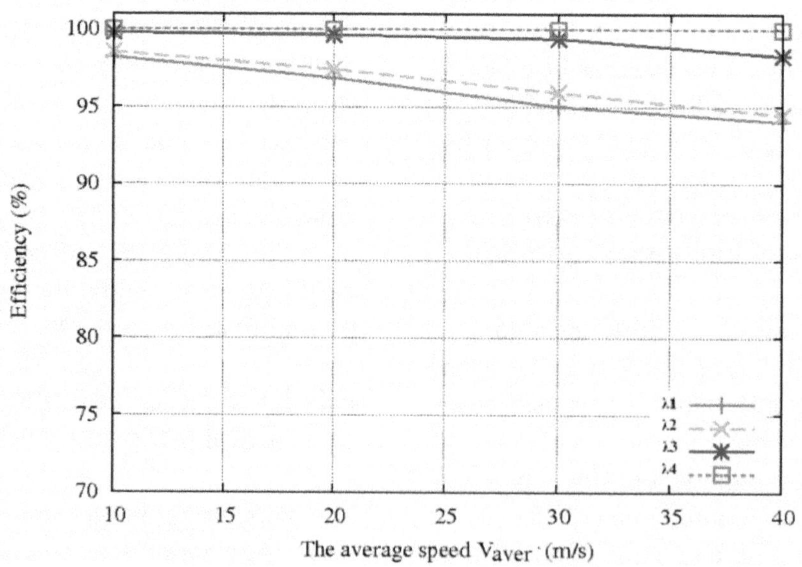

four different arrival rates and for the three different percentages of confident vehicles.

Figure 17 relative to 100% confident vehicles shows a TTLCA of 100% independently of the speed and the arrival rates of vehicles to the platoon. This means that any elected CA stays so until it exits from the assumed road segment, hence providing a maximum stability to its members and to its covered region within the platoon. Figure 15 and Figure 16 show the same behavior of maximum stability only when the vehicle arrival rate is high, namely for

Figure 14. The efficiency, 100% confident vehicles

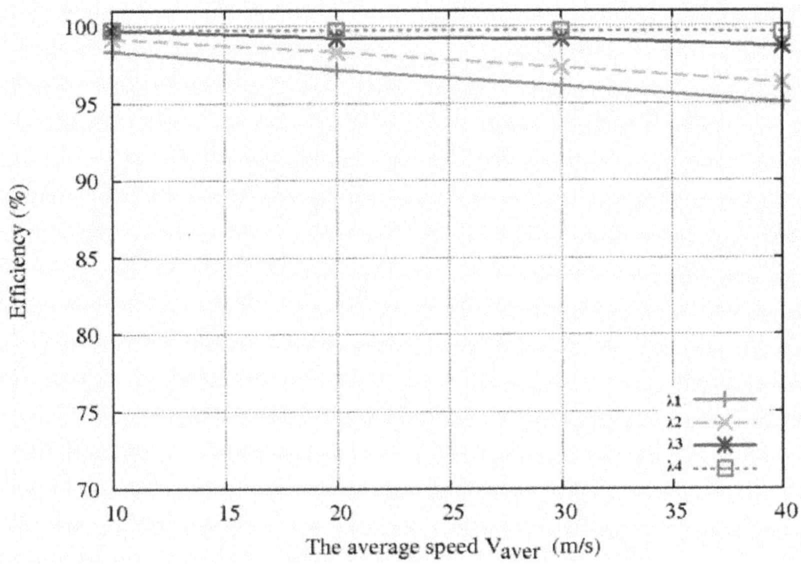

Figure 15. The average life time of CA, 10% confident vehicles

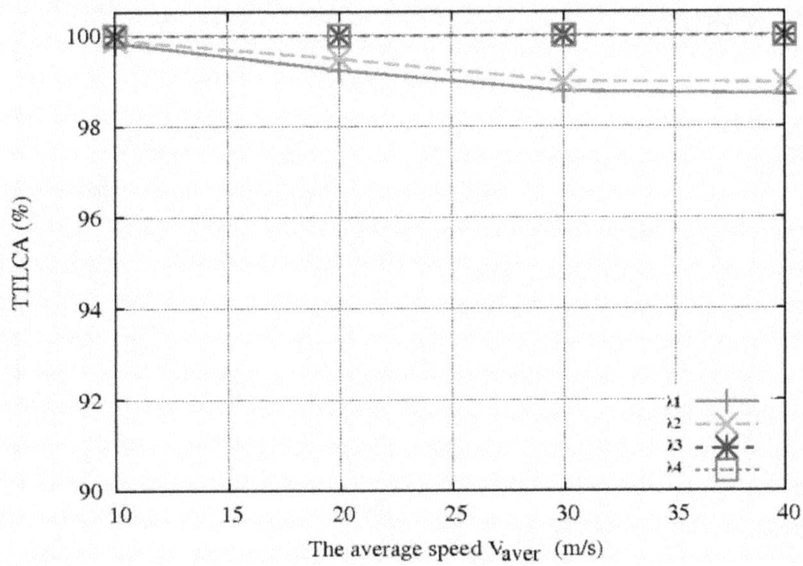

λ4=2vehicles/s and for *λ3=1.5vehicles/s*. However, for smaller arrival rates, namely for *λ1=0.5vehicles/sλ2=1vehicles/s*, the TTLCA declines a little bit as the average speed increases. This small decrease in the TTLCA is mainly due to the small average number of RAs per cluster at these points. However, the TTLCA stays above the 98% even for a speed of 40 m/s. On the other hand and more interestingly we notice once again that the percentage of confident vehicles has a very small impact on the stability of the clustering scheme. Indeed Figure 15 and

Figure 16. The average life time of CA, 50% confident vehicles

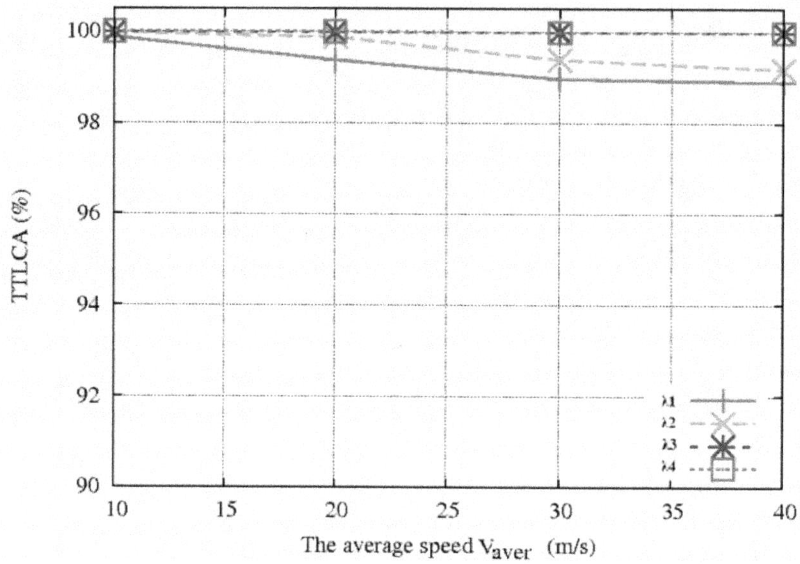

Figure 17. The average life time of CA, 100% confident vehicle

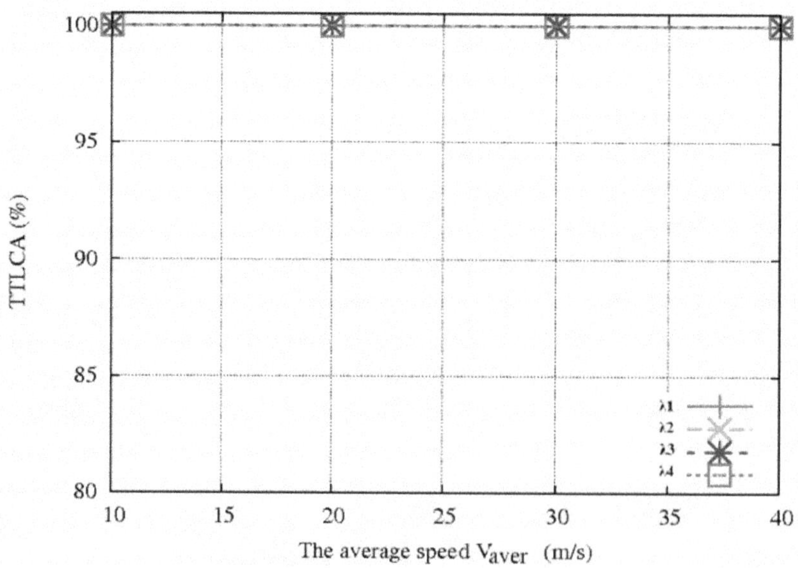

Figure 16 present roughly the same values of TTLCA despite the decrease in the percentage of confident nodes from 50% for Figure 16 to 10% for Figure 15. Moreover, even at the highest considered speed and the lowest considered arrival rate, only a gain of 2% in the TTLCA can be achieved when going from 10% to 100% confident vehicles within the platoon, testifying the great stability of our proposal.

Effects of the Speed and Rate on the Average Latency to Join a Cluster

We finally investigate the average delay required by a vehicle to acquire a defined status being a CA, RA, GW or MN. First, we recall that the Hello period is fixed to 2 seconds. Hence the average delay cannot be less than 1 second. Recall also that a vehicle upon entering the platoon sends its Hello and then waits either for a Hello from a CA that comes the latest in 2 seconds or the expiration of its InitTimer that happens after 5 seconds. In the latter it will eventually be candidate for a CA. In the former it enters the Search state where it sends a Join message and waits for an Accept message upon which it becomes a MN or a RA. In our studied scenarios, it is this former case that is the most frequent. Consequently, the average latency should be closer to 2 seconds rather than 5 seconds.

Figure 18, Figure 19 and Figure 20 portray the average latency as a function of the speed and for the four different vehicle arrival rates and the three percentages of confident nodes. These figures show that the average latency evolves from around 1.6 seconds (for the case of 100% confident vehicles, a speed of 10 m/s and an arrival rate of $\lambda 4=2$*vehicles/s*) to 3.3 seconds (for the case of 10% confident vehicles, a speed of 40 m/s and an arrival rate of $\lambda 1=0.5$*vehicles/s*).

The latency increases with the speed since when the speed increases, the number of vehicles in the platoon decreases. But the latency decreases when the arrival rate increases and this is more pronounced at the lowest considered speed since more CAs are present on the platoon as portrayed in Figure 6, Figure 7 and Figure 8, and hence a lower number of hops separate the newly incoming vehicle from its prospective CA.

Once again, the percentage of confident vehicles has a little impact on the latency as long as the cluster formation is possible and the elected clusters cover the entire platoon.

CONCLUSION

In this paper, we proposed a secure architecture for VANETs based on a trust model and an efficient clustering algorithm. The proposed architecture

Figure 18. The average latency, 10% confident vehicles

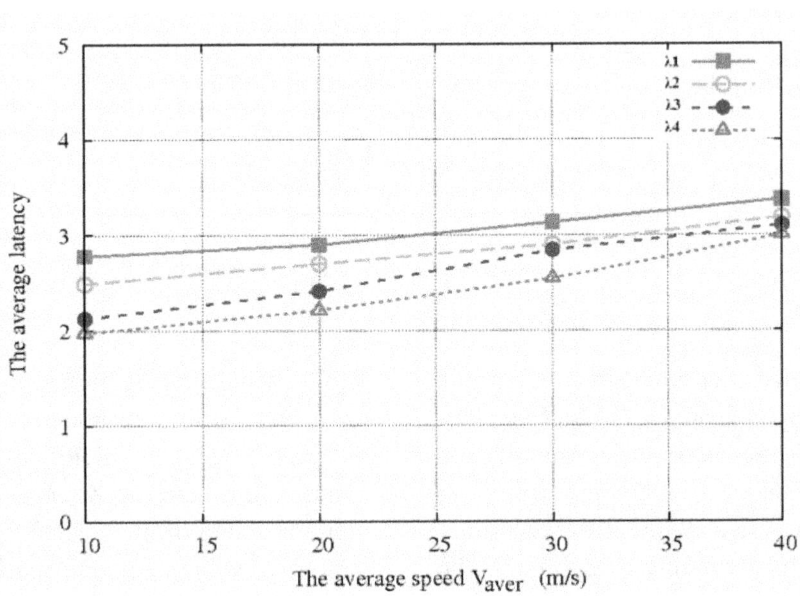

Figure 19. The average latency, 50% confident vehicles

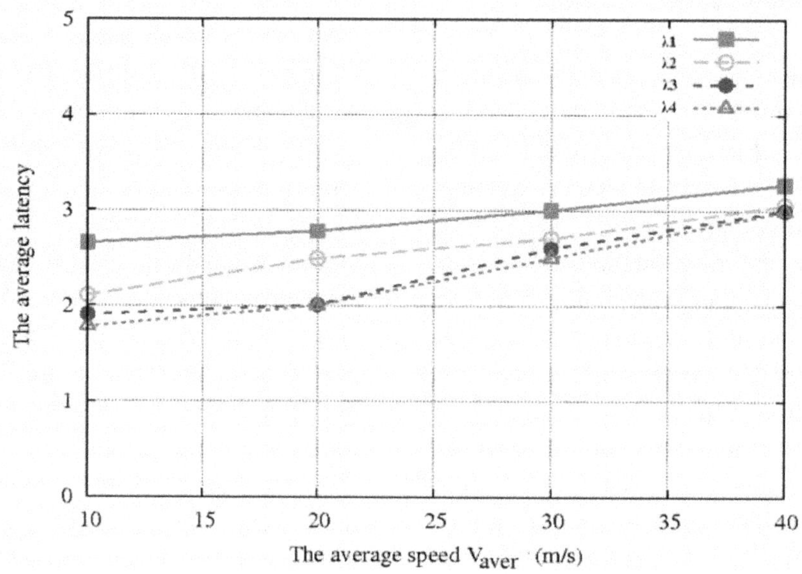

Figure 20. The average latency, 100% confident vehicles

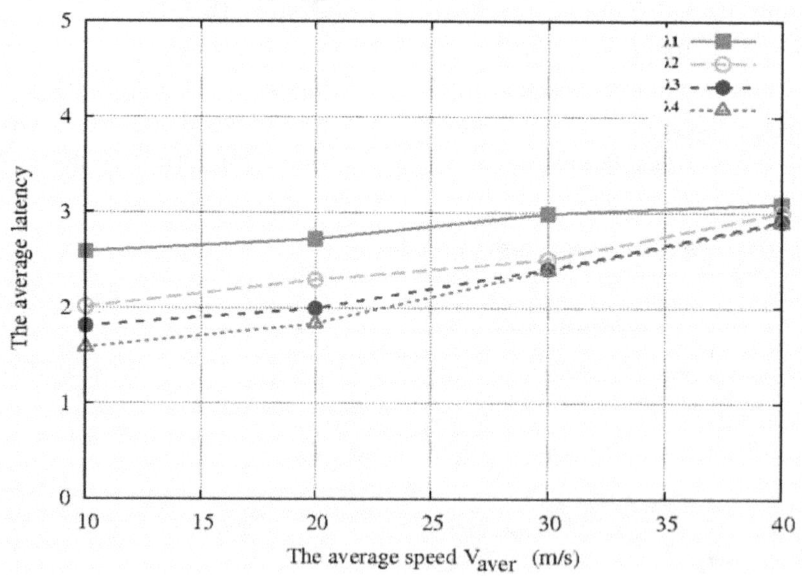

distributes the certification authority (CA) among mobile vehicles on the road. Our clustering algorithm is mainly based on two metrics: security and relative mobility. The security parameter relies on a trust metric which evolves dynamically according to a predefined behavior monitoring process. The mobility metric is related to the relative velocity of a node within its vicinity. Furthermore, we adopted a security shield, called the Vanet Dynamic Demilitarized Zone (VDDZ), within the neighborhood of the elected CAs to prevent any unknown vehicle to compromise their security.

Extensive simulations using various scenarios were conducted to investigate the performance of our proposed architecture in terms of efficiency, stability and membership latency. In particular, we showed that the percentage of confident vehicles has a little impact on these performance metrics as long as the assumed average vehicle speed and the vehicle arrival rate allow the formation of clusters that cover the entire platoon. Furthermore, we showed that our proposal ensures the election of a minimal number of CAs covering the entire road segment.

We are currently investigating the robustness of our proposal under more realistic mobility models and scenarios.

REFERENCES

Blum, J., & Eskandarian, A. (2004). The Threat of Intelligent Collisions. *IT Professional*, *6*(1), 24–29. doi:10.1109/MITP.2004.1265539

Di Crescenzo, G., Zhang, T., & Pietrowicz, S. (2007). Anonymity Notions for Public Key Infrastructure in Vehicular Networks. In *Proceedings of the IEEE International Conference on Mobile Ad hoc and Sensor Systems*.

Leinmueller, T., Buttyan, L., Hubaux, J. P., Kargl, F., Kroh, R., Papadimitratos, P., et al. (2006). *SEVECOM "Secure Vehicle Communication"*. Paper presented at the IST Mobile and Wireless Communication Summit (MobileSummit2006), Mykonos, Greece.

Li, C., Huang, M., & Chu, Y. (2008). A Secure and Efficient Communication Scheme with Authenticated Key Establishment and Privacy Preserving for Vehicular Ad hoc Networks. *Science Direct Computer Communications*, *31*, 2803–2814.

Little, T. D. C., & Agrawal, A. (2005). An Information Propagation Scheme for VANETs. In *Proceeding in 8th International IEEE Conference on Intelligent Transportation Systems (ITSC2005)*, Vienna, Austria.

Ma, X., Chen, X., & Refai, H. (2009). Performance and Reliability of DSRC Vehicular Safety Communication: A formal Analysis, EURASIP. *Journal on Wireless Communications and Networking*, *3*, 1687–1472.

Papadimitratos, A., Buttyan, L., & Holczer, T., GWoch, E., Freudiger, J., Raya, M., Ma, Z., Kargl, F., Kung, A., & Hubaux, J. P. (2008). Secure Vehicular Communication Systems: Design and Architecture. *IEEE Communications Magazine*, 100–109. doi:10.1109/MCOM.2008.4689252

Papadimitratos, P., Gligor, V., & Hubaux, J. P. (2006). Securing Vehicular Communications Assumptions, Requirements, and Principles. In *Proceedings of the Workshop on Embedded Security in Cars (ESCAR).*

Parno, B., & Perrig, A. (2005). Challenges in Securing Vehicular Networks. In *Proceedings of the Fourth Workshop on Hot Topics in Networks (HotNets-IV),* College Park, MD.

Rachedi, A., & Benslimane, A. (2010). A Secure and Resistant Architecture against Attacks for Mobile Ad Hoc Networks. *Journal of Security and Communication Network*, *3*, 150–166. doi:10.1002/sec.116

Raya, M., Aziz, A., & Hubaux, J. P. (2006). *Efficient Secure Aggregation in VANETs*. In *Proceedings of VANET06*, Los Angeles, CA.

Raya, M., Paradimitratos, P., & Hubaux, J. P. (2006). *Securing Vehicular Communications*. IEEE Wireless Communications.

Sampigethaya, K., Huang, L., Li, M., Poovendran, R., Mastsuura, K., & Sezaki, K. (2005). Providing Location Privacy for VANET. In *Proceedings of Embedded Security in Cars (ESCAR)*. CARAVAN.

Sivagurunathan, S., Subathra, P., Mohan, V., & Ramaraj, N. (2009). Authentic Vehicular Environment Using a Cluster Based Key Management. *European Journal of Scientific Research*, *36*(2), 299–307.

Taleb, T., Fadlullah, Z., Benslimane, A., & Letaief, K. B. (2009). *Towards an Effective Risk-conscious and Collaborative Vehicular Collision Avoidance System*. IEEE Transactions on Vehicular Technology.

Wang, N., Huang, Y., & Chen, W. (2008). A Novel Secure Communication scheme in Vehicular Ad hoc Networks. *Science Direct Computer Communications*, *31*, 2827–2837.

Zhou, T., Choudhury, R., Ning, P., & Chakrabarty, K. (2007). Privacy Preserving Detection of Sybil Attacks in Vehicular Ad hoc Networks. In *Proceedings of the Fourth Annual International Conference on Mobile and Ubiquitous Systems: Networking and Services (MobiQuitous)*.

This work was previously published in International Journal of Business Data Communications and Networking, Volume 6, Issue 4, edited by Debashis Saha and Varadharajan Sridhar, pp. 38-63 , copyright 2010 by IGI Publishing (an imprint of IGI Global).

Chapter 9
Security Aspects in Radio Frequency Identification Systems

Győző Gódor
Budapest University of Technology and Economics, Hungary

Sándor Imre
Budapest University of Technology and Economics, Hungary

ABSTRACT

Radio frequency identification technology is becoming ubiquitous, and as an unfortunate side effect, more and more authentication solutions come with more security issues. In former contributions, the authors introduced a solely hash-based secure authentication algorithm that is capable of providing protection against most of the well-known attacks and performs exceptionally well even in very large systems. The authors gave a theoretical analysis of Simple Lightweight Authentication Protocol (SLAP) protocol from security and performance point of view. This chapter gives a detailed examination of small computational capacity systems from the point of view of security. The authors define the model of attacker and the well-known attacks which can be achieved in these kinds of environments. Furthermore, the chapter gives a summary of the significant RFID authentication protocols which are found in literature. The authors present several lightweight authentication protocols and some novel elliptic curve cryptography based methods. Besides, the chapter illustrates the SLAP protocol's performance characteristics with measurements carried out in a simulation environment and compares with the theoretical results. The authors show the effects of numerous attacks and the system's different parameters on the authentication time. Finally, the chapter examines the performance and security characteristics of two other protocols chosen from the literature in order to compare to SLAP algorithm and give proper explanation for the differences between them.

DOI: 10.4018/978-1-61350-477-2.ch009

INTRODUCTION

In the last decade the RFID (Radio Frequency Identification) technology has developed apace. Accordingly, the usage of these systems is very widespread. RFID applications can be found in various fields of our common life, e.g. personal identification in passports, access control, thief prevention, payment systems, tickets for transportation, supply-chain management, asset tracking, retailing etc. Common RFID systems consist of three main functional units: Tags, Reader and Back-End. From these three parts the Tags possess the most vulnerable computational capacity. RFID systems operate via air-interface, which has numerous security risks. Since this medium can be easily accessed by everyone, the messages between the communicating parties can be observed, eavesdropped or forged and later the fake messages can be replayed. Furthermore, customers (possessing a Tag) can be tracked or their transactions can be recorded.

In order to prevent the unauthorized access to any kind of protected database or system, or any sensitive data, authentication and encryption have to be applied. As Back-End and Reader have higher performance and power properties, therefore the secure communication between them can be solved. However, Tags have very limited memory and very low computational capacity, therefore new solutions are needed, which require less operation and fewer messages. These methods are the so-called lightweight authentication protocols. These algorithms apply only hash functions and simple logical operations. There are a lot of lightweight protocols presented in the literature, however, each one has its drawbacks. The major weaknesses are the decrease of anonymity level and the necessity of very high computational capacity.

In (Gódor, Antal, & Imre, 2008) we have already introduced our SLAP protocol (Simple Lightweight Authentication Protocol) and we have given a detailed security analysis by which we verified the correctness of SLAP. Our solution provides an efficient mutual authentication method by which a given Tag and the Back-End can authenticate each other easily, quickly and securely even if large amount of Tags are stored in the database. The protocol can defy the well-known attacks and does not demand high computational capacity.

In (Gódor & Imre, 2009) we have already examined our proposed authentication method's performance characteristics with measurements carried out in a simulation environment and these results were compared with the theoretical results.

In this contribution, which is an enhanced version of our formerly published article (Gódor & Imre, 2010), we give a detailed examination of small computational capacity systems from the point of view of security. We define the model of attacker and the well-known attacks which can be achieved in these kinds of environments. Furthermore, we give a detailed overview about the well-known lightweight and elliptic curve cryptography based authentication protocols which have already been proposed in the literature. Additionally, we propose our SLAP protocol again briefly and we introduce its security and performance characteristics. We show the effects of numerous attacks and the system's different parameters on the authentication time. Finally, we examine the performance of two other protocols chosen from the literature in order to compare with SLAP algorithm and give proper explanation for the differences between them. For this reason we have implemented the SLAP protocol and the others in OMNeT++ so that we can make a comparison between them from the aspects of performance and security. We will prove that our protocol possesses better security and performance parameters than the other ones.

SECURITY ISSUES IN SMALL COMPUTATIONAL CAPACITY SYSTEMS

In this section, we introduce the security aspects of a small computational capacity environment; the attacker model will also be defined.

Attacker Model

Each person who influences the authentication process or by monitoring the messages later could influence the identification in the interest of gains or any malicious purposes is regarded as an attacker. Two different types of an attacker could be differentiated: active and passive attacker.

Usually an active attacker can be realized easier since in an identification process this kind of attackers take part actively, e.g., monitoring messages and later replay them, modifying messages or impersonating the parties and filling a transponder part. In case of a correctly developed authentication protocol, these attacks can be realized easily since when an attacker tries to modify a given message the used cryptographic functions permit the detection of them. For instance, if the identification number is not sent in plain but the hash value is sent, the receiver could identify any modifications of the message, thus the attack can be detected.

In case of a passive attacker the detection of an attack is not a giveaway. These attackers only monitor the messages and draw the inference by using the eavesdropped messages. Since a passive attacker does not modify the messages none of the parties could detect the presence of the attacker.

Security Aspects

Recently, RFID applications have been implemented for various areas and the technology is also improved. Security concerns and user privacy issues are potential risks for RFID proliferation. In the mean time, security problems are not considered as a big barrier in the real world. The reason is that the applications based on RFID are mainly developed for supply chain management, and user friendly Tags just start to emerge, which means the cost of RFID Tag makes it difficult to apply as a substitute for the existing barcode. However, RFID technology is growing very fast, and the cost of Tag for general purposes will be reasonable in the near future. Security issues should be protected for the admirable usage of RFID (Juels, 2006), (Dimitriou, 2006).

Several threats for security in RFID systems are obstacles to make RFID more popular, familiar, and widespread than before (Thompson, Chaudhry, & Thompson, 2006). In this section, we look into design goals (security and other) of RFID protocol. In the followings, the security goals in RFID systems are introduced:

- *Protection against Eavesdropping:* Any Reader can read the response from Tags without any knowledge of the Tag's owner. So, the response from Tags should be encrypted to be recognizable only to authorized ones and not to give any information gain to adversaries. The eavesdroppers can not take part in the RFID system protocol but can eavesdrop the communication between Readers and Tags. By eavesdropping, the eavesdropper may catch secret information and perform replay attack. So the RFID system should be designed that the eavesdroppers do not get any secret information from the eavesdropped information and it is secure against the replay attack.
- *Protection against tracing:* This is the most important goal, the adversary can collect the responses from all the Tags, the trackers know that when, from where and how much information is transmitted since trackers can perform traffic analysis. Then the trackers can track the location of the user. Therefore, the RFID system should

be designed that the trackers can not know that different responses of a Tag are from the same Tags.

- *Protection against spoofing:* An adversary can spoof Tags (or Readers) to Readers (or Tags). Spoofing Readers is the same meaning with cloning the Tag, which gives adversaries a chance to substitute genuine with counterfeit. An adversary who spoofs a Tag can get information about the Tag, since the adversary can be regarded as authorized parties. Thus, spoofing Tags are related to privacy which is defined that no unwanted information about the Tag has to be leaked from the system. By spoofing valid Tags, a thief could fool automated checkout or security systems into thinking a product was still on a shelf. Alternatively, a thief could rewrite or replace Tags on expensive items with spoofed data from cheaper items. For example, thief could change the identity of the product, making a very expensive stereo read as a cheap Walkman at the automated cash register at the exit of some retail store. Even worse scenario could be if you have a container full of goods intended for your warehouse and on its route a skilled port worker, where your container is offloaded, changes your data on the Tag with data from some bogus company that his criminal organization established. Then instead of delivering that container to you it will be delivered straight to the thieves.
- *Protection against Denial of Service attack:* An adversary can disturb the interactions between Readers and Tags by intercepting or blocking messages transmitted. Such an attack could cause a server and a Tag to lose synchronization (Schuba, Krsul, Kuhn, Spafford, & Sunda, 1997). For example, the server might update the shared data, while the Tag does not, in such a case they would no longer be able to authenticate each other.
- *Backward and Forward untraceability:* Given a set of readings between Tags and Readers, given the fact that all information stored in the involved Tags has been revealed at time t, the adversary must not be able to find any relation between any readings of a same Tag or set of Tags that occurred at a time $t_0 < t$. This is called "Backward untraceability" (Le, Burmester, & Medeiros, 2007), it requires that knowledge of a Tag's current internal state should not help identify the Tag's past interactions. Backward untraceability has been considered as the most important security requirement in strong RFID authentication, since otherwise the past transcripts of a Tag (e.g., from Readers' logs) may allow tracking of the Tag owner's past behaviors. Forward untraceability can be similarly defined as requiring that knowledge of a Tag's internal state at time t should not help identify the Tag's interactions that occurred at time $t_0 > t$.
- *Anti-cloning:* This is an additional security requirement, which means that adversary cannot clone Tag without tampering with Tag. When adversary tampers with Tag, adversary has same information as Tag itself, and then we cannot prevent Tag cloning. However, there are many ways to cloning without tampering with Tag, for example: the replay attack. Therefore, anti-cloning can be one of security requirements.
- *Availability:* This is also one of security requirements, Tag should be available all the time when a user wants. Even though physical attacks such that jamming responses of Tag or blocking responses of Tag by using metal mesh or foil cannot be prevented by logical operations, an authentication protocol should provides the data

recovery against the data loss such as DoS, message hijacking, power interruption, etc. during the authentication processes. Especially, the desynchronization attack by utilizing a man-in-the-middle attack must be prevented.

- *Ownership transfer:* Molnar dealt with ownership transfer at the first time (Molnar & Wagner, 2005), (Lim & Kwon, 2006). Ownership of RFID Tags may be changed frequently during their lifetime. For example, Tags are initially created and attached to objects by manufacturers and Tagged objects are then handed over to retailers, and finally consumers buy Tagged objects in shopping malls. The owner of a Tagged object may also transfer its ownership to another party (e.g., by buying an object and giving it to his friend as a birthday present or by selling or swapping used objects via a garage sale or in a swap meet). Ownership transfer of a Tag means transfer of authorization to read the Tag. Thus, such an ownership transfer must guarantee that once ownership is transferred to another party, the old owner should not be able to read the Tag any more.
- *Scalability:* Scalability means that how many RFID Tags can be accommodated in the Back-End server. To accommodate a large number of RFID Tags, computational time of identification should be reasonable, which means that the Back-End server have to identified a Tag fastest and most effective.

Privacy

Privacy is increasingly under threat from all kind of sources today (Thompson, Chaudhry, & Thompson, 2006). We have increasing numbers of cameras monitoring every our move, mobile phones eavesdropping, e-mail control and all kind of computer monitoring and surveillance. These and many other ways of intruding on our privacy today are leading to the feeling that we are not alone and safe even in our own houses. RFID and especially EPC's uniquely identifiable products concern and worry many people who believe that this technology will just add more pressure on individual privacy. That's why one of the main concerns for RFID systems is privacy. RFID systems use a shared radio medium which allows eavesdropping. Unprotected communications between Tags and Readers over a wireless channel can disclose information about the Tags and their positions. We identify the following two major privacy issues.

- *Tag Information Privacy:* In a typical RFID system, when an RFID Reader queries an RFID Tag, it responds by sending its identifier to the Reader, the Reader can then request further details by sending this identifier to a server. If unauthorized Readers can also get a Tag identifier, then they may be able to determine the additional information related to the Tag. Example with the gum, the purchaser could then be tracked if he/she ever entered that same store again, or perhaps more frightening, if they entered any other store with RFID reading capability. Because, unlike a barcode, RFID Tags can be read from much greater distances without knowing of the holder and the reading of such devices is non-directional. This means that if you enter a store with a pack of gum in your pocket, the Reader can identify that pack of gum, the time and date you bought it, where you bought it, and how frequently you come into the store. If you used a credit card or a frequent shopper card to purchase it, the manufacturer and store could also tie that information to your name, address, and e-mail. You could then receive targeted advertisements by gum companies as you walk down the aisle, or receive mailings through your e-mail or regular mail about other products. To protect against such information leakage, RFID systems need to be controlled so that only authorized Readers are able to access the information associated with a Tag.

- *Tag Location Privacy:* If the responses of a Tag are linkable to each other or distinguishable from those of other Tags, then the location of a Tag could be tracked by multiple collaborating Tag Readers. For example, if the response of a Tag to a Reader query is a static ID code, then the movements of the Tag can be monitored, and the social interactions of an individual carrying a Tag may be available to third parties without him or her knowing. If messages from Tags are anonymous, then the Tag tracking problem can be avoided.

Privacy protection of customers is one of the main concerns when using RFID technology. More and more privacy protection methods are developed and introduced in the literature. Each method has advantages and disadvantages, too. In the following we would like to present two of them, which are effective and feasible.

"Killing" and "Sleeping"

The most useful approach for the protection of consumer privacy is "Killing the Tag" before the goods are carried out from retailer (Stamatiou, 2007). For implement "Killing Tag", Reader sends a specified "Killing" commands to Tags, "Killing" command is specific PIN (32 bits long in the EPC Class-1 Gen-2 standard. When the Tag received a "Killing" command, makes itself completely unusable for good. A killed Tag is truly dead, and can never be re-activated.

This method is applied in many different environments, but it also has some disadvantages, such as when customers want to change back the goods or in the library and rental shop a good can borrow and pay multiple times. Instead of "killing Tag", why do not we just make the Tag "sleeping" means that the Tag only temporarily inactive? This concept is simple, but would be difficult to manage in practice, "sleeping" Tags will not really aim to protect customer privacy if any Reader can also "wake" the Tag, the activation of Tags should of course be allowed only to authorized Readers. A Tag could for example wake up only if it receives the correct passcode or key from a Reader.

This solution offers as effective privacy protection than the killing approach, but is more flexible. The biggest problem with this solution is that it requires the owner of an item to manage activation passwords for RFID devices. The user may have dozens of RFID tagged items in different states, in different places and possibly requiring different passwords for reading them.

Blocker Tags

Juels, Rivest, and Szydlo proposed a privacy-protecting scheme that they call blocker-Tag scheme (Juels, 2003). In this scheme, a blocker Tag is a RFID device that can simulate many ordinary RFID Tags simultaneously. When carried by a consumer, a blocker Tag thus "blocks" RFID Readers. It can do so universally by simulating all possible RFID Tags. Or a blocker Tag can block selectively by simulating only selected subsets of ID codes, such as those by a particular manufacturer, or those in a designated "privacy zone".

It is mentioned that in this scheme RFID Reader is only able to communicate with a single RFID Tag at a time. If more than one Tag responds to a query by the Reader (as will often happen naturally, for example, in a supermarket automated checkout), the Reader detects a "collision". In this case, it does not read accurately any of the information transmitted by the Tags. The Reader and RFID Tags then need to engage in some sort of protocol so that the Reader can communicate with the conflicting Tags one at a time. Such a protocol is called a "singulation protocol" - it enables RFID Readers to scan multiple Tags simultaneously. To

Figure 1. Blocking with tree-walking

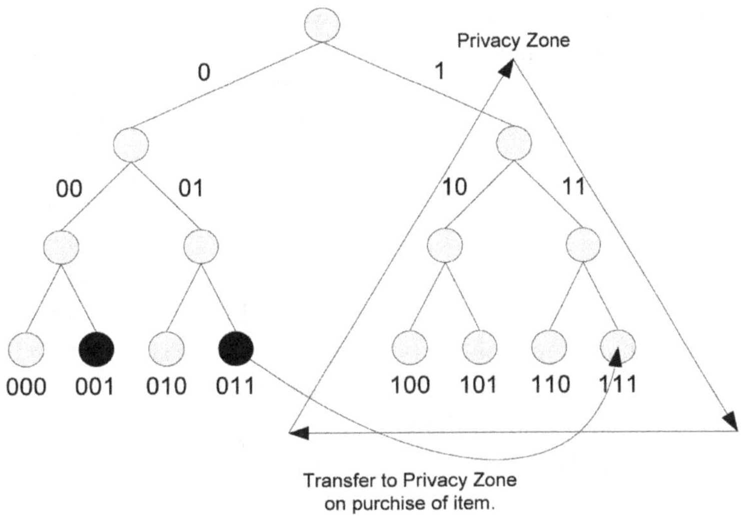

ensure that Tag signals do not interfere with one another during the scanning process, the Reader first ascertains what Tags are present and then addresses Tags individually.

One type of RFID singulation protocol is known as tree-walking. Figure 1 shows how blocking might work in conjunction with tree-walking.

The tree-walking singulation algorithm enables an RFID Reader to identify the serial numbers of nearby Tags individually by means of a bit-by-bit query process resembling a depth-first search of a binary tree (Juels, 2003). In the tree-walking, s-bit Tag identifiers can be viewed with the leaves of binary tree, which have the depth s. The root of the tree has depth 0. The left child of each x node has label $x \| 0$ while the right child of them has label $x \| 1$.

The search on the tree-walking is a recursive depth-first search performed by a Reader. The Reader will interrogate all Tags from the root of the tree. Each Tag responds with the first bit of its identifier, if the Reader only received a "0" bit, it means that the Tag is located on the left side of the tree, but if the Reader received a "1" bit, it means that the Tag is located on the right of the tree. Consequently, if there are Tags in both the left and right subtrees of tree, then the Tags together simultaneously broadcast both a "0" and a "1", creating a collision in the broadcast bit. In this case, when a collision is detected, the Reader recourses (sequentially in turn) beginning at its child nodes. This procedure eventually yields the leaves - and thus the s-bit identifiers - of responding Tags.

A blocker can prevent Reader scanning by simulating collisions in the singulation tree. For example if any attacker scanning of all Tags simply, the blocker can block the scanning by sending both "0" and "1" bit to the Reader and then Reader have to go all across the tree. If the Tag's identifier is long enough, then the trees will have many, many leaves. So, such a blocker would always cause a Reader to stall.

As we have explained, the aim of the blocker is not to disrupt the scanning process. Rather, the scheme is selective. Blocking here relies on designation of the leading bit of a Tag identifier as the privacy bit. The blocker only disrupts the scanning process when a Reader attempts to scan Tags in the privacy zone, i.e., in the right half of the singulation tree. The blocker does not interfere with the normal scanning of Tags with "0" privacy bits, i.e., those outside the privacy zone.

This method is currently considered as a practical solution for the existing RFID privacy and security protection. Problem is that additional blocker Tag is needed for every Tag and it is susceptible whether Tag bearers strictly follow to attach the additional Tag.

RELATED WORK

In the last decade there were numerous contributions and conferences which dealt with the security of RFID systems. Most of the suggested algorithms make allowance for small computational capacity and limited energy source. Due to the very low CPU performance and the limited memory capacity of RFID Tags in the very beginning of the development of RFID authentication protocols only mathematical and logical operations could be used. These protocols are the so-called lightweight authentication protocols.

After the computational capacity of Tags increased simple cryptographic functions, e.g. one-way hash functions could be used and many lightweight authentication protocols were developed. The well-known lightweight authentication protocols are the followings: Hash Lock Scheme (Weis, 2003), (Weis, Sarma, Rivest, & Engels, 2003), Ohkubo's Hash Chain (Ohkubo, Suzuki, & Kinoshita, 2003), Lightweight Mutual Authentication Protocol (Luo, Chan, & Li, 2005), Key-Tree Based Hash Protocols (Molnar & Wagner, 2004), (Nohara, Inoue, Baba, & Yasuura, 2006), (Nohl & Evans, 2006), (Buttyán, Holczer, & Vajda, 2006), a Lightweight RFID Protocol to protect against Traceability and Cloning attacks (Dimitriou, 2006), RFID Mutual Authentication Scheme based on Synchronized Secret Information (Lee, Asano, & Kim, 2006) and M2AP (Peris-Lopez, Hernandez-Castro, Estevez-Tapiador, & Ribagorda, 2006). We also proposed the Improved Lightweight Authentication Algorithm (Gódor & Antal, 2008) which is an enhancement of the LMAP protocol. Many contributions examined these protocols and have given detailed overviews about them (Stamatiou, 2007), (Thompson, Chaudhry, & Thompson, 2006), (Bakhtiari, Safavi-Naini, & Pieprzyk, 1995), and (Lim & Kwon, 2006).

Due to the evolution of the technology more efficiency RFID Tags were developed hereby allowing development of simple symmetric key cryptography based authentication protocols, like YA-TRAP: Yet Another Trivial RFID Authentication Protocol (Tsudik, 2006), A Lightweight RFID Protocol to protect against Traceability and Cloning attacks (Dimitriou, 2005), Simple Lightweight Authentication Protocol (Gódor & Imre, 2010), and An Advanced Mutual- Authentication Algorithm Using AES for RFID Systems (Toiruul & Lee, 2006).

In these days the RFID technology achieved that CPU speed on which PKI systems can be run. However, the well-known RSA algorithm uses very large keys and the method is very complex, thus researchers have turned to the elliptic curve cryptography (Miller, 1986). A big benefit of these cryptosystems that more shorter key guarantees the same security level than in RSA a big one. For instance a 160 bit long key in elliptic curve equals a 1024 bit long key in RSA. Such ECC based authentication protocols are the followings: EC Based Mutual Authentication Protocol (ECMA) (Braun, Hess, & Meyer, 2008), ERAP: ECC based RFID Authentication Protocol (Ahamed, Rahman, & Hoque, 2008), and CryptoGPS: Anti-counterfeiting Solution for RFID Tags (Robshaw, 2008), (Batina, Guajardo, Kerins, Mentens, Tuyls, & Verbauwhede, 2007).

Hash Lock Scheme and Randomized Hash Lock Scheme

The basic idea of the Hash Lock algorithm (Figure 2) (Weis, 2003), (Weis, Sarma, Rivest, & Engels, 2003) is that, for the challenge of the Back-End, at first not the identification number of the Tag is not sent, but the Back-End is verified whether it is legitimate or not. A pre-shared key is stored in the memory of the Tag, which is known only by the Tag itself and the Back-End. A hashed value

Figure 2. Hash lock protocol

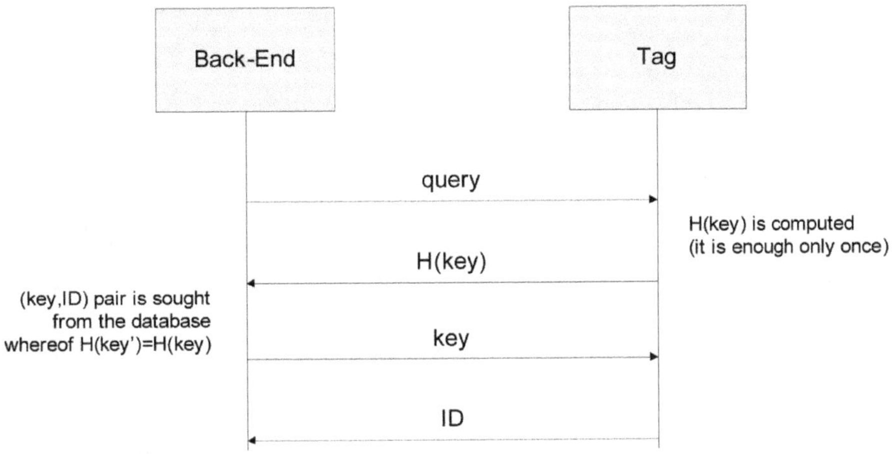

of this key will be sent, which needs to be checked by the Back-End whether the pair of this hash value is contained in its database or not. If the hash value exists in the database, the Back-End sends the key of the Tag in plain text, for which the Tag responds his identification number.

This scheme has many disadvantages. Since in the last message the ID is sent in plain, therefore it can be easily tracked. Another major problem is that both the hash value of the key and the identification number are static, thus by eavesdropping and replaying these messages the Tag can be impersonated.

Eliminating the weaknesses of this algorithm, the response message of the Tag is supplemented with a random number, see Figure 3. The hashed value of the *R* random number is concatenated with the ID, and they are sent to the Back-End, which performs a detailed search. In other words, each ID stored in the database is concatenated with this *R* value, and the Back-End has to compute their hash and compares them with the received one. In case of equality, the Tag is authenticated.

Indeed, the tracking problem is solved, but the replay attack is not eliminated, because the Back-End cannot be certain about which transaction

Figure 3. Randomized hash lock protocol

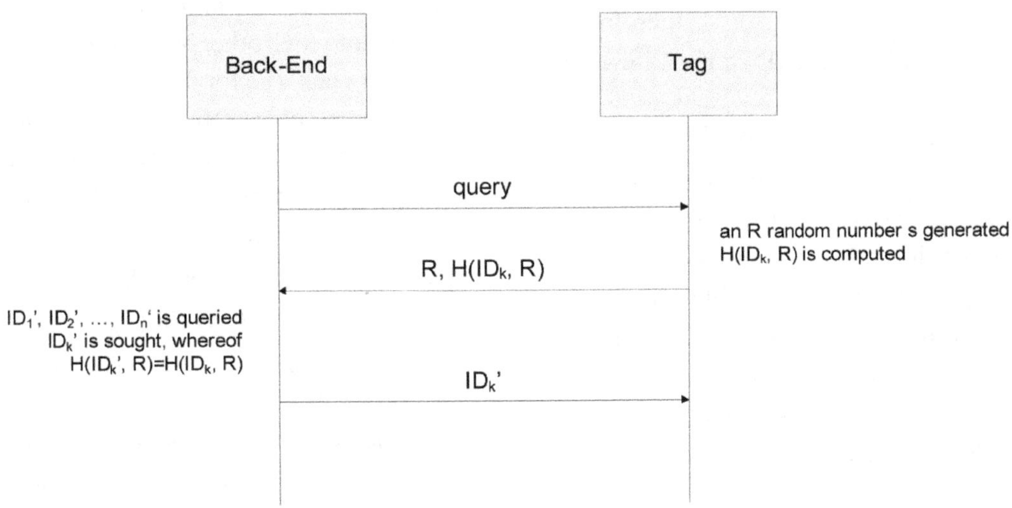

Figure 4. Ohkubo's hash chain

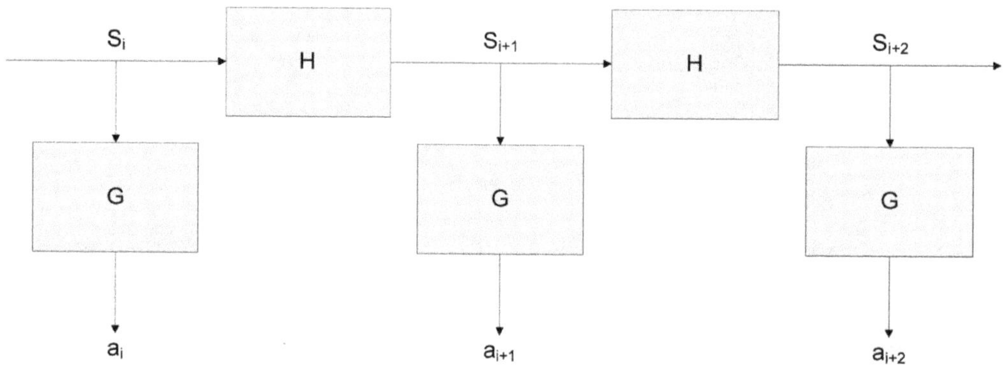

the given message belongs to. Thus by eavesdropping a message and replaying them the Back-End will not be able to decide that the message is a real or a fake one.

Ohkubo's Hash Chain

Figure 4 shows the protocol proposed by Ohkubo (Ohkubo, Suzuki, & Kinoshita, 2003). In this protocol each Tag generates a new s_i security identity mapped by an H hash function in every transaction, then derives the fingerprint of this secret by another G hash operation and this digest has to be forwarded to the Back-End. The Back-End knows the first element of the hash chain (the initial s_1 secret) and its hash values have to be derived, until the current map will be exactly the same as the fingerprint sent by the Tag. In the database (ID, s_1) pairs are stored, therefore in case of correspondence the ID of the Tag will be known.

The advantage of this protocol is that an attacker cannot become acquainted with the s_i secret, not even in case a lot of consecutive messages are eavesdropped, since in each transaction the varying S_i secret is mapped by the G hash function. In addition, when a Tag is compromised, an attacker is not capable of linking it to earlier transactions, since he would have to derive the s_{i-1}, s_{i-2} etc. pre-images. However, by using strong, one-way and collision-resistant H and G hash functions, it is very difficult to solve this problem (Bakhtiari, Safavi-Naini, & Pieprzyk, 1995).

However, the suggested algorithm has disadvantage, too. To authenticate the Tags, many hash operation has to be computed by the Back-End. The exact number of these mappings is i per each Tag. In case of a sufficiently huge database, this process has a very high computational capacity.

Lightweight Mutual Authentication Protocol

The Lightweight Mutual Authentication Protocol (LMAP) suggested by Luo, Chan and Li (Luo, Chan, & Li, 2005) is based on Ohkubo's Hash Chain and a simple XOR operation. In this protocol both the Tag and the Back-End have to authenticate each other before the Tag's ID is sent. In order to achieve forward privacy, the protocol uses the hash chain proposed by Ohkubo, by which the secret identity is altered in each transaction. So in the ith transaction the Tag forwards the $a_i = G(s_i)$ message to the Back-End and replaces the former secret with the new one using the $s_{i+1} = H(s_i)$ operation, where s_i can be computed recursively and G and H are one-way hash functions.

In the beginning the s_0 secret and the ID are stored in the memory of the Tag. Due to the pre-

Figure 5. Lightweight mutual authentication protocol

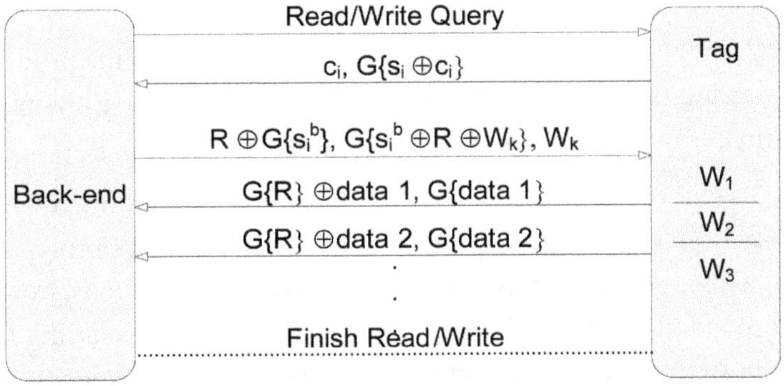

vention of the tracking, the Tag changes its secret by computing a one-way H hash and the mapped value of this fingerprint, derived by G, is sent to the Back-End.

Figure 5 shows the process of the algorithm. Every Tag has three read/write memory blocks (W_1,W_2 and W_3). Besides there is a fourth block, in which the c_i transaction identification number is held. The Back-End stores the s_0 secret, the contents of the W_1,W_2 and W_3 memory blocks and the Tag's ID.

The strength of the protocol is that encrypted data traffic is supported between the parties without the usage of any kind of cryptographic ciphering algorithms. Nevertheless, it has a major weakness, namely, an attacker can easily replay the second message of the protocol, thus that party can be impersonated. The replay attack would be prevented if the Back-End sent a challenge in the first message to verify the freshness of the response message. Another problem is that the Back-End can be overloaded, if an attacker forwards big amount of fake messages, which contains a very high value c_i transaction ID. Every s_0 secret stored in the Back-End's database has to be hashed c_i times and the fingerprint has to be compared with the received $G(s_{i'} \oplus c_i)$ value, which is randomly generated by the attacker.

Improved Lightweight Mutual Authentication Protocol

In our former contribution (Gódor & Antal, 2008) we have proposed an improvement for the LMAP protocol which can be seen on Figure 6.

Figure 6. Improved lightweight mutual authentication protocol

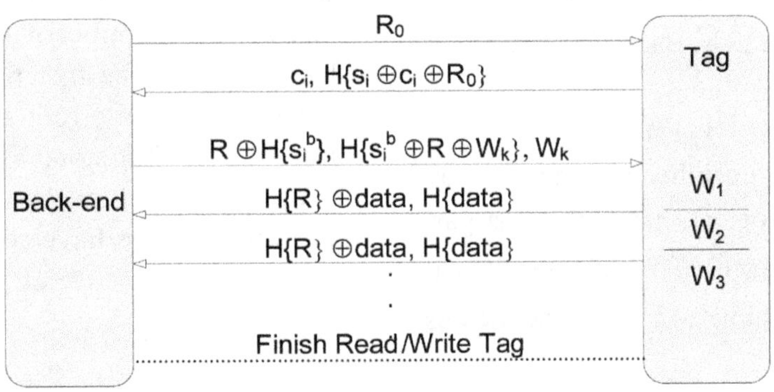

Instead of Ohkubo's hash chain, we use a simple iterative hash in every transaction. In the beginning let the secret value be s_0, which can be loaded into the Tag during production. After a successful authentication, s_{i+1} is calculated by hashing s_i.

During the ith transaction a challenge (R_0) is sent by the Back-End, which enables it to verify whether the Tag has responded recently. The Tag calculates the XOR value of the transaction counter (c_i), the current secret value (s_i) and the Back-End's challenge. The Tag sends the transaction counter and the hash value of the XOR's output. The challenge is needed to prevent adversaries from replaying the Tag's message at a later time.

Given c_i, the Back-End does an exhaustive search to find the Tag's current secret value in its database, for which:

$$H(s_i^b \oplus c_i \oplus R_0) = H(s_i \oplus c_i \oplus R_0)$$

where s_i^b is derived from s_{i-1}^b by hashing. Because of the cryptographic hash function's preimage collision resistant property described in Rogaways's and Shrimpton's work (Rogaway, & Shrimpton, 2004), this holds if and only if $s_i^b = s_i$. If it finds a match in the database, it means the Tag is authentic.

For mutual authentication the Back-End still has to prove it knows the Tag's current secret value. Also it is desirable to share a common random session number that can later be used by the parties to send encrypted data. Let this random number be denoted by *R*. The Back-End's response consists of three parts:

$R \oplus H(s_i^b)$, $H(s_i^b \oplus R \oplus W_k)$ and W_k.

The Tag starts to decipher the message by calculating the XOR of $H(s_i)$ and the first part of the received message ($R \oplus H(s_i^b)$), thus it learns the random number R^t. It authenticates the Back-End if and only if both $R^t = R$ and $s_i = s_i^b$ hold.

By the end of the protocol run the parties also agreed on which W_k memory block to read/write in the current transaction. The integrity of W_k is ensured by bounding it with s_i in the second part of the Back-End's reply.

Also the Tag may transfer encrypted data to the Back-End by sending $G(R) \oplus data_i$. In order to ensure data integrity it also sends the hash value of the data. Knowing the R random session number, the Back-End can easily decipher the message and check its integrity.

Key-Tree Based Hash Protocols

In order to guarantee the anonymity of the Tags, the identification of the Tag must not be forwarded in plain. This raises the question if the Back-End does not receive the Tag's ID in plain text, how does it know which ID to look for in its database? It would have to compare the received fingerprint with each hashed database value. In case of equality the authentication of the Tag would succeed. Nevertheless, the number of required hash operations is directly proportional to the number of Tags in the database. In case of a few million Tags the computational capacity will be extremely high.

Molnar and Wagner (Molnar & Wagner, 2004) proposed a tree based hash authentication protocol (Figure 7). The idea of this protocol is that in a tree with depth of *d* and a maximal branching factor of *k*, for finding a given Tag the number of the operations is proportional to $d \cdot k = d \cdot \sqrt[d]{N}$, where *N* is the number of the Tags in the database. The leaves of the tree represent the Tags, the edges between the root and the leaves are the secrets. Every Tag stores as much different secrets as the number of levels of the tree. The Tags in the same subtree have secrets in common, but there is at least one secret, which is different. This

Figure 7. Key-tree based hash authentication protocol

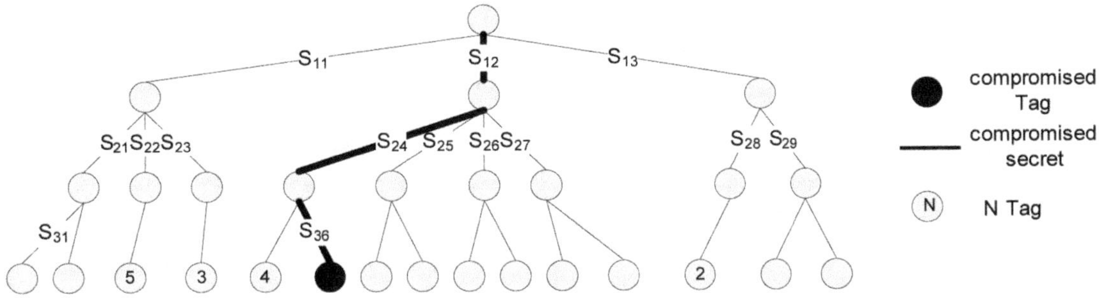

is because two different leaves of the tree cannot be located on the exact same branch.

The advantage of this protocol is the low computational complexity, the disadvantage is the decreased level of anonymity. If the sibling of a Tag under a given branch is compromised, then the anonymity of that Tag will be in danger. Since the Tag shares $d-1$ secrets with its compromised mate, the attacker knows all but one of its secrets. Therefore from *N* Tag contained in the tree an attacker can reduce the number of possible Tags to *k*, which are the direct neighbor of the compromised Tag. This means the attacker can guess the identity of the Tag with a probability of $\frac{1}{k}$ instead of $\frac{1}{N}$. Of course the level by which the Tag's anonymity decreases depends on which subtree the compromised Tag is in.

Numerous papers (Nohara, Inoue, Baba, & Yasuura, 2006), (Nohl & Evans, 2006) deal with the determination of information infiltration caused by the compromised Tags. Buttyán et al. (Buttyán, Holczer, & Vajda, 2006) have defined the optimal depth and branching factor for the tree in order to optimize computational capacity for the Back-End and minimize the decrease of anonymity in Tags.

Crypto-GPS

The Crypto-GPS authentication protocol is based on the GPS identification scheme. GPS is a public-key identification scheme which is proposed by Girault (Girault, 1992) and analyzed by Poupard, and Stern in (Poupard & Stern, 1998). Nowadays, the GPS scheme is standardized in ISO/IEC 9798-5, moreover it is listed in the final NESSIE recommendations.

There are many variants and optimizations of the GPS scheme which provide a flexible usage. The standard GPS protocol uses a number of public parameters and the identification consists of three steps. First, the Prover forwards a commitment *x* for which the Verifier sends a challenge *c*. Then the Prover sends a response *y* which is derived from both the challenge and the private key. Finally, the Verifier checks the validity of the answer.

In case of an RFID environment the Reader (the Verifier) and the Tag (the Prover) have different performance characteristics. The Verifier is more powerful than the Prover. However, if we examine the GPS scheme closely, it can be realized that there are two expensive computations, an exponentiation and a modular reduction which should be made in each identification round. These operations are not suitable for a low-cost RFID Tag in case of on-tag computations. That is why, these operations should be removed from the Tags' side, which can be achieved by using pre-generated coupons. These pre-computed coupons are feasible for the usage in a typical RFID system (e.g. in case of access management) because the storage capacity becomes more important rather that the computational capacity of Tags.

Figure 8. The Crypto-GPS authentication protocol

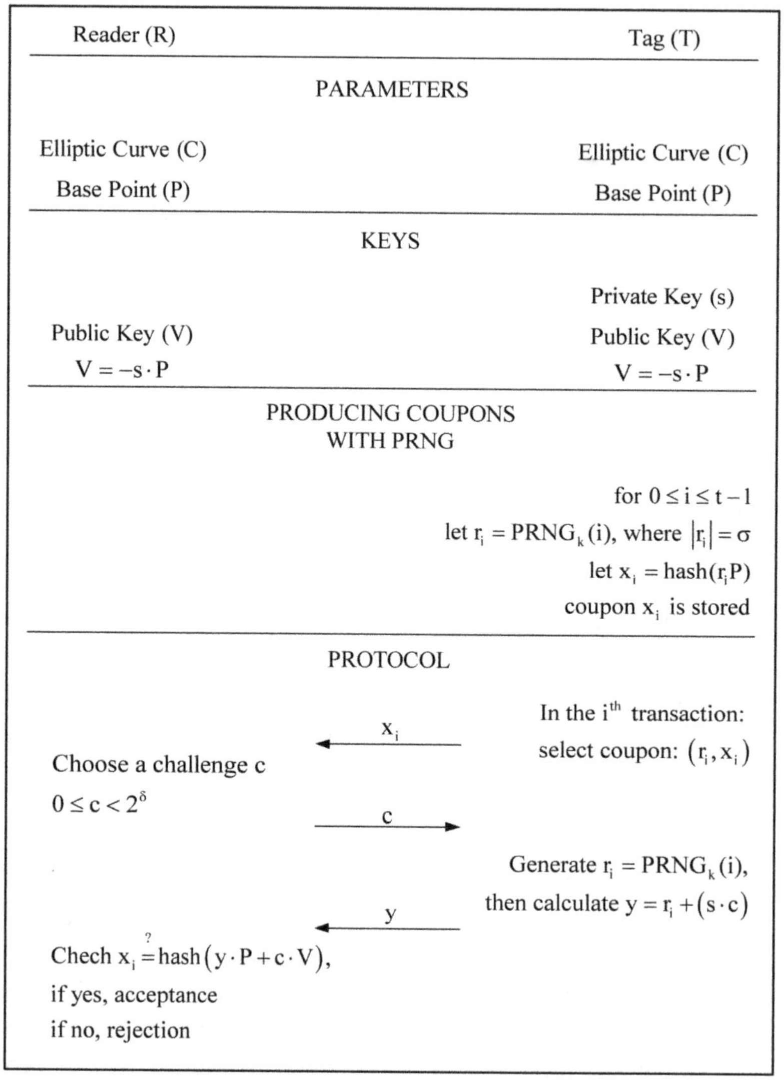

Crypto-GPS has two different variants, the first uses RSA-like moduli and the other bases on the usage of elliptic curve operations. The elliptic curve based variant fits for low computational capacity systems. Figure 8 illustrates the elliptic curve based Crypto-GPS authentication protocol with coupons (Girault, Juniot, & Robshaw, 2007), (McLoone & Robshaw, 2007), (Poschmann, Robshaw, Vater, & Paar, 2009).

As it can be seen the shared parameters of the Elliptic Curve (*C*) and the Base Point (*P*) on that curve. The Crypto-GPS private key (*s*), which is σ bits long, is stored on a given Tag. The public key $V = -s{\cdot}P$ is an elliptic curve point and this is stored by the Reader. At initialization, *t* coupons have to be generated and these have to be stored on a given Tag. In case of the ith coupon, the generation are the following: using a pseudo-random number generator (PRNG) r_i is generated randomly, then P has to be multiplied by r_i, and finally, a hash computation has to be operated on the derived value. On other words the ith coupon consists of the pair (r_i, x_i).

In order to achieve an expected security level several parameter sizes have to be set accordingly. The proposed parameter sizes are the followings: for a security level of 80 bits σ is set to 160 $(|s| = \sigma = 160)$, the length of the challenge *c* is denoted by δ where the particular value depends on the security levels and the various optimizations (usually $|c| = \delta = 20$). Finally, the third parameter, ρ, which denotes the length of r_i, is a requirement of Crypto-GPS:

$$(|r| = \rho = \sigma + \delta + 80 = 260).$$

The Messages of the Protocol

1. In the first step the Tag chooses a coupon (r_i, x_i), and sends the x_i to the Reader.
2. In the second step the Reader chooses randomly a challenge *c* and forwards it to the Tag.
3. In the third step the Tag computes $y = r_i + (c \cdot s)$, and sends *y* to the Reader.
4. Finally, the Reader checks whether the hash value of the received *y* corresponds with the x_i coupon which is sent by the Tag in the first message.

A major benefit of the usage of Crypto-GPS with coupons in an RFID environment is that on the Tags' side expensive operations are not required. The on-tag performance is dominated by the simple integer calculation $y = r_i + (c \cdot s)$ which is identical in both the classical and the elliptic curve variants of GPS. Hence this kind of authentication algorithm can be executed also on a low-cost RFID Tag.

The EC Based Mutual Authentication Protocol

The authors suggest the use of curves above $\mathbb{F}_{2^m}$ while implementing the protocol (Braun, Hess, & Meyer, 2008) shown in Figure 9. It is important to note, to use projective coordinate representation

Figure 9. The EC based mutual authentication protocol

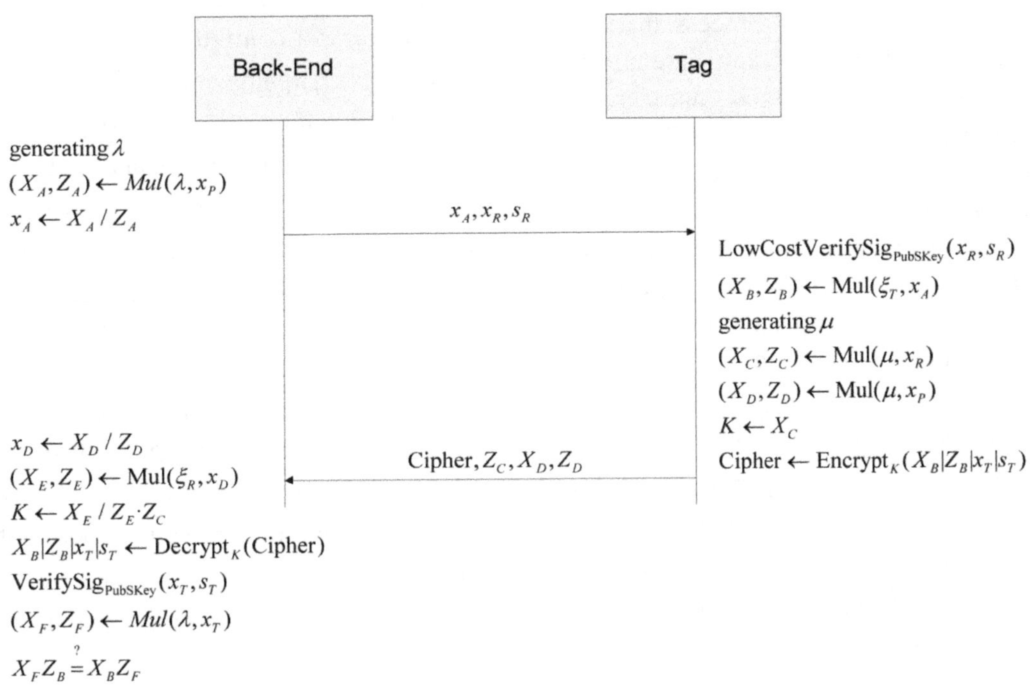

to describe the points of the curve, thus avoiding inverse calculations, which take significant processing time. The points of the curve are named by capital letters from the beginning of the English alphabet (*A, B, C, D, E, F*), the projective coordinates are named by capital letters from the end of the English alphabet (*X, Y, Z*), the affine coordinates are named by lowercase letters (*x, y*), and the lower index shows the curvepoint that the coordinate represents. This protocol is highly based on the assumption, that the Montgomery-ladder implementation calculates using only *x* coordinates. The input of the Montgomery-algorithm is not the complete point, but only its affine *x* coordinate, while its output is the emerged point's projective (*X, Z*) coordinate pair. This way, the *y* coordinates of the curvepoints are practically unnecessary during the runtime of the protocol, which lowers the size of the messages.

Knowledge of the Reader and the Tag

Let us assume *E* is an elliptic curve over $\mathbb{F}_{2^m}$. Let us assume $P(x_p, y_p)$ is a generator point over this curve, with a *q* order, where *q* is prime. Let us assume also that there is a process matching GenSig and VerifySig public key signing algorithms (where the necessary signing and signature verification keys are PrivSKey and PubSKey). The LowCostVerifySig is a more simple, less resource dependent implementation of the algorithm, which can also be run suitably by the more simple RFID Tag with less processing capacity (hereby the algorithm suggested by the authors of the protocol is not discussed).

Every Tag has a secret key (generated upon the first upload of the Tag), named ξ_T. ξ_T is selected from the $1 \leq \xi_T < q$ range. The Tags also store a so called *certificate*, which is a structure containing two elements (x_T, s_T), where x_T is the public key related to the secret key, and the affine *x* coordinate of the $T = \xi_T \cdot P$ point. s_T is the signature generated for x_T, so $s_T = \text{GenSig}_{\text{PrivSKey}}(x_T)$. Then the certificate verification is simple by checking whether s_T is a valid signature on x_T.

Every Reader has a randomly chosen secret key, named ξ_R, and a PubSKey public signing key. Every legitimate Reader has a pregenerated certificate, alike the Tags. The signature of the certificate can be generated using e.g. the ECDSA signing algorithm.

The Messages of the Protocol

The Reader chooses a value of $1 \leq \lambda < q$ and calculates the value of the x_A affine coordinate (let us call this x_A a random challenge) from the $A = \lambda \cdot P$ point. Then it forwards this to the Tag with its own certificate.

The Tag checks the certificate using the PubSKey key of the Reader. If the certificate is invalid, the Tag discards the message and breaks the running protocol. If the certificate is valid, the Tag uses the value of ξ_T as the scalar and the value of x_A challenge as the affine coordinate and performs a point multiplication. The output is the (X_B, Z_B) pair, which is the X_B / Z_B affine *x* coordinate of the $B = \xi_T \cdot (\lambda \cdot P)$ point. Then it generates a random μ value and scalar-multiplying it with the x_R point coordinate it gets a cipher key, which is the X_C projective coordinate of the scalar-multiplication. Using this key and a symmetric key cipher it encrypts its certificate and the response to the challenge by x_A (the "|" sign is the operation symbol for concatenation of the bit sequences). The message sent to the Reader contains this encrypted part, the Z_C projective coordinate, and the projective coordinates (X_D and Z_D) of the point generated by the scalar multiplication of the random μ and the generator point.

Figure 10. The ERAP protocol

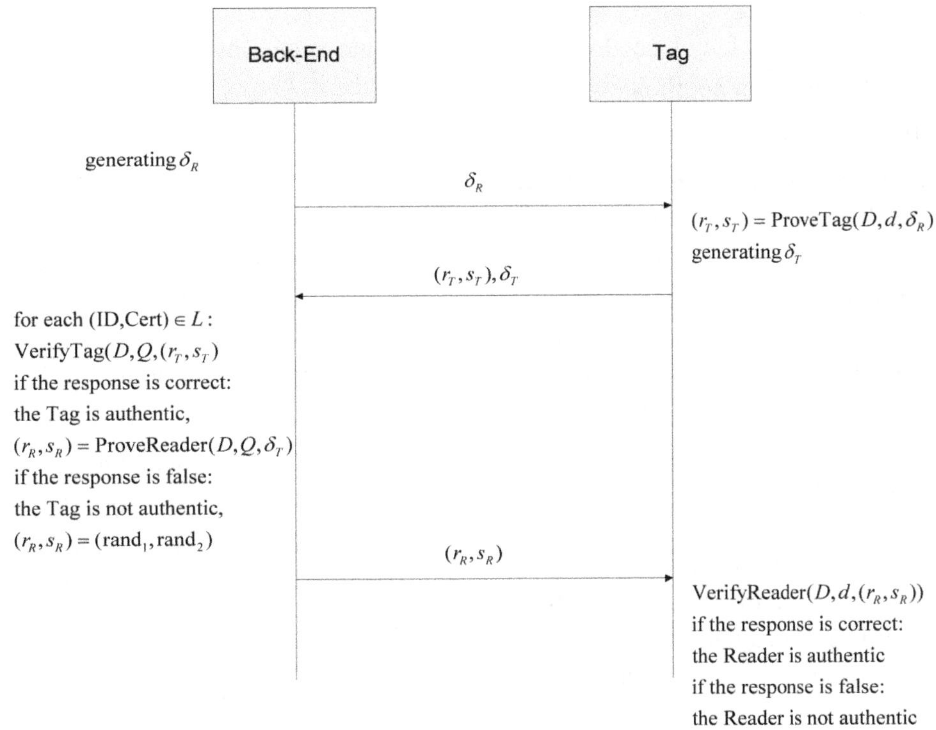

When receiving the response message the Reader calculates the X_D affine coordinate from X_D and Z_D projective coordinates. The reason that this is not done by the Tag in the previous step, is that inverse calculation is a resource consuming operation, so it is done by the Reader with more resources. The Reader calculates the coordinates of an E point using the x_D point coordinate and its own secret key. Transforming the *x* coordinate of the E point to an affine coordinate and multiplying by Z_C it gets the correct key to decrypt the encrypted message piece. After decryption it can check the given Tag's certificate. As the final step in the authentication process it calculates the X_F and Z_F projective coordinates using the $F = \lambda \cdot (\xi_T \cdot P)$ equation, then it checks, if the X_F / Z_B and X_B / Z_F affine coordinates are equal (by checking the last equation in the figure above). If they equal, authentication is successful, otherwise the Tag is unauthorized, invalid.

The ERAP Protocol

The ERAP (ECC Based RFID Authentication Protocol) (Ahamed, Rahman, & Hoque, 2008) can be seen in Figure 10. As a matter of fact the ERAP Protocol bases on a small compass modified ECDSA signing scheme. In this case the ProveTag, VerifyTag and the ProveReader, VerifyReader operate a signing (Prove*) and a signature verification (Verify*) function.

Knowledge of the Reader and the Tag

Knowledge of the Reader:

- domain parameters: $D = (q, FR, S, a, b, G, n, h)$
    - q: the size of the finite field ($\mathbb{F}_q$)

- *FR*: the representation of the elements of the finite field
- *S*: at least 160 bit length seed value
- $a, b \in \mathbb{F}_q$: the two coefficients which define the elliptic curve over $\mathbb{F}_q$
- *G*: a prime ordered base point which is an element of the curve but $G \neq \mathcal{O}$
- *n*: the order of the base point *G*
- *h*: cofactor, $h = \# E(\mathbb{F}_q) / n$

- In that case when a Reader is successfully authenticated itself towards the CA, the list of (ID, *D*, *Q*) triplets (Tag ID, Domain Parameters of a Tag and Private Key of a Tag) will be sent by a CA.

There are some requirements for the CA: it cannot be compromised, in the authentication protocol it is a trusted third party, moreover the following processes are executed by them:

- domain parameter generation algorithm which means that for each Tag a *D* is generated, in other words it generates random curves
- verification of the domain parameters which means that the correctness of *D* is checked
- key generation algorithm which means that from a *D* input value a (*Q*, *d*) pair is derived where *Q* is the public key and *d* is the private key
- public key verification algorithm which means that by using a domain parameter *D* and a public key *Q* the CA determines whether the *Q* is valid or not considering a given *D* parameters

Knowledge of the Tag:

- (ID, *D*, *d*) triplet where ID is the Tag's identification number, *D* is the Tag's domain parameter set and *d* is the private key of the Tag

The Messages of the Protocol

The message flow chart diagram of ERAP is depicted in Figure 10. In the first step of the protocol the Reader generates a δ_R random challenge, where $1 \leq \delta_R \leq n$, which message is sent to the Tag.

After the Tag received the δ_R challenge, by using its own domain parameters *D* and his private key *d* Tag executes the ProveTag algorithm. The derived (r_T, s_T) pair and the Tag's randomly generated nonce δ_T as a challenge, where $1 \leq \delta_T \leq n$, will be sent back to Reader.

The received (r_T, s_T) pair should be checked by using the VerifyTag algorithm for which the input values are the Tag's *D* domain parameters and its public key *Q*. Back-End executes an exhaustive search in order to find that Tag's record which sent the response message. In that case the result of the algorithm is "acceptance", the Tag is successfully authenticated and the Reader calculates the (r_R, s_R) pair using the ProveReader algorithm. For this algorithm the necessary input parameters are the given Tag's domain parameters *D*, his public key *Q* and the δ_T challenge. In that case the output of the VerifyTag is "rejection", the Tag is not authentic and the (r_R, s_R) pair will be chosen randomly. The derived pair will be sent to the Tag.

The Tag checks the authenticity of Reader by the VerifyReader algorithm, of which the input parameters are the received (r_R, s_R) pair, the Tag's domain parameters *D* and his private key *d*. In case of the result of the algorithm is "acceptance", the Reader is authentic else not.

The ProveTag algorithm can be seen in Figure 11 and the VerifyTag algorithm is shown in Figure 12. The ProveReader and the VerifyReader algorithms are the very same as the Tag's ones except that in this processes the secret key of the Reader should be used.

Figure 11. The ProveTag algorithm

Input : domain parameters $D = (q, FR, S, a, b, G, n, h)$
private key d; challenge δ_R
Output : ordered pair (r_T, s_T)

1. Select a random integer k, where $(1 \le k \le n-1)$
2. Compute $P = k \cdot G$ and convert the field element x_P to integer $\overline{x}_P$
3. Compute $r_T = \overline{x}_P \bmod n$
   If $r_T = 0$, then goto step 1.
4. Compute $s_T = k^{-1}(\delta_R + dr_T) \bmod n$
   If $s_T = 0$, then goto step 1.
5. The result: (r_T, s_T)

Figure 12. The VerifyTag algorithm

Input : domain parameters $D = (q, FR, S, a, b, G, n, h)$
private key Q; (r_T, s_T) pair for δ_R
Output : acceptance or rejection

1. Verifying: $1 \le r_T \le n-1$ and $1 \le s_T \le n-1$
   If any verification fails, then the response should be rejected.
2. Compute $w = s_T^{-1} \bmod n$
3. Compute $u_1 = \delta_R w \bmod n$ and $u_2 = r_T w \bmod n$
4. Compute $P' = u_1 \cdot G + u_2 \cdot Q$ value
   If $P' = \mathcal{O}$, then the result is rejection.
5. Convert x_P to integer $\overline{x}_P$, and compute $v = \overline{x}_P \bmod n$
   If $v = r_T$, then the result is acceptance, otherwise rejection.

As it can be seen in the ProveTag algorithm there is another randomly generated k value besides δ_R challenge, which provides that for the very same challenge the reply message will be different.

SIMPLE LIGHTWEIGHT AUTHENTICATION PROTOCOL

In this section the Simple Lightweight Authentication Protocol (SLAP) is introduced briefly, we confine ourselves to only the protocol steps and the initial assumptions. The detailed security analysis presented in (Gódor, Antal, & Imre, 2008). Figure 13 shows the messages of SLAP. In the followings we describe the basic assumptions of the protocol, namely the possessions of the Back-End and the Tags, and the initial values.

Back-End's capabilities:

- ID_{i-1} : the previous transaction identification number which is not disclosed in plain text

Figure 13. Messages of the simple lightweight authentication protocol

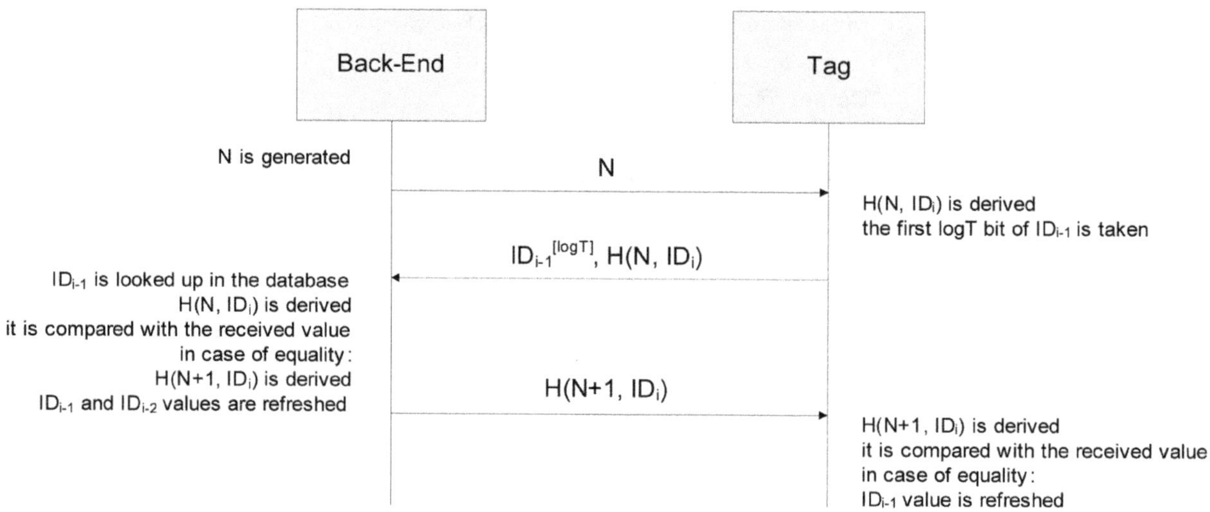

- ID_{i-2}: the one before the last transaction ID which has an important role in the protocol since by using this ID the de-synchronization attack can be eliminated
- *N*: random number generated by the Back-End right before the communication the used operations: random number generation, HASH calculation, database query

Tag's capabilities:

- ID_{i-1}
- the used operations: HASH calculation

The initial values:

- ID_0: the initial transaction ID which is a randomly generated value, stored in every Tag and also in the database of Back-End
- ID_{-1}: its value is set to 0
- *i*: the transaction identifier. Let the initial transaction identifier be $i = 1$, which indicates the state when the system is set up.

1. In the first message the Back-End sends a random number *N*. Let us suppose the Back-End is able to generate random numbers in a way that nobody else can predict it beforehand.
2. After receiving the first message, the RFID Tag derives the ith transaction identifier by using the following equation: $ID_i = H(ID_{i-1})$. It is very important to note that this transaction identifier is not written into the memory of the Tag yet. Thereafter ID_i is appended to *N* and the Tag computes its hash.
3. In the reply message, beside this hash value, the first $\log T$ bits of the ID_{i-1} are sent in plain text, where *T* denotes the number of Tags in the whole system. In case *T* is so big that the value of $\log T$ is higher than the quarter of the output length of the used *H* hash function, then the Tag forwards only the first quarter of the hash value in the reply message. If at least the order of magnitude of the value of *T* can be estimated at the phase of system planning, a suitable hash function can be chosen, where the output space is sufficiently large. The limitation of the length of ID_{i-1} is essential, because if too much of ID_{i-1} is revealed, an attacker

may succeed in finding the missing part by guessing.

4. Having received the Tag's reply and possessing the ID_{i-1}, which is stored in its database, the Back-End can check the identity of the given Tag. In order to do so, the first $\log T$ bits of each stored ID_{i-1} should be compared with the incoming $\log T$ length bitstream. For each database record where the equality exists the Back-End derives its own ID_i and computes $H(N, ID_i)$, which is compared with the hash value sent by the Tag. If both are the same, the Tag is authenticated and the Back-End stores the new identification number in the database, which means that ID_{i-1} is overwritten by ID_i, and ID_{i-2} is overwritten by ID_{-1}.
5. In the last step of the SLAP protocol, the Back-End has to prove that it possesses the ith transaction identifier. For this reason it sends $H(N+1, ID_i)$ to the Tag. Then the Tag computes this hash value similarly and compares it with the received one. In case of equality the Back-End is authenticated and the Tag stores the new identifier in its memory.

SECURITY ANALYSIS

In the following section we analyze our SLAP protocol from the point of view of principal security aspects.

Data Security and Integrity of Messages

Tags do not store any valuable information apart from the identification number of the $(i-1)^{th}$ transaction. In case an attacker is able to crack a Tag and is capable of reading out its content, the attacker is unable to get any information about other Tags. Since messages always contain the hash of the identity of a Tag, moreover they differ in every transaction, therefore confidential information is unobtainable by an eavesdropper. There is no integrity check for the messages, but it is unnecessary, because any kind of changes can be detected by the parties of the communication. If one bit differs the authentication is immediately failed. The chance that an attacker can alter a message in a way that the pre-image of the hashed value included in the message is the same as the identity of another Tag tends to zero.

Mutual Authentication

The transaction identifier used in the $(i-1)^{th}$ transaction is only known by the Tag and the Back-End. During the authentication process this ID is never sent in plain text, and by using a one-way hash operation an attacker is not able to derive this identity. Therefore the parties of the communication can be certain that the other partner is authentic, is not an attacker, as an attacker cannot gain knowledge of ID_i during the transaction.

Protection against Replay Attack

Replaying the second message is beyond all reason, since the N random number (nonce) is generated at the beginning of each protocol run by the Back-End. Therefore the $H(N', ID_i)$ hash value will not be equal to the $H(N, ID_i)$ value computed by the Back-End in the ith transaction, as N' was a nonce in the kth transaction, where $k < i$.

Anonymity of a Tag

The usage of the H hash and the varying transaction identifier in each message sent by the Tag makes it undistinguishable for a spectator from a random variable bit string. Since there is no confidential data sent in plain text in the air-interface the Tag can remain anonym before a third party.

Undistinguishable Messages from a Random Bit String

The content of each message – in case a strong hash function and random number generator are used – cannot be differentiated from a random bit string. The value of N is random, therefore by using a random input, the output of a hash operation and thus the first $\log T$ bits of ID_{i-1} will also be random, since this value is the first $\log T$ bits of the output of the hash function.

Protection against Impersonation Attack

An attacker has information neither about ID_{i-1} nor about ID_i. However, he can try to send such an X message in which the $H(X)$ hash value is precisely equal to $H(N, ID_i)$. But in practice it is impossible to do this, because if a collision-resistant hash function (Bakhtiari, Safavi-Naini, & Pieprzyk, 1995), (Weis, Security and Privacy in Radio-Frequency Identification Devices, 2003) is used, the probability of inventing such X is negligible.

Prevention of Tracking

In the ith transaction the $ID_{i-1}^{[logT]}$ and $H(N, ID_i)$ values are sent by the Tag, where both the $ID_{i-1}^{[logT]}$ and the ID_i are valid only in this transaction, thus an eavesdropper has no information about which message belongs to which Tag. Supposing that an attacker manages to crack a Tag, to associate the gained ID_i with a former ID_k identity (where $0 < k < i$), due to the usage of a hash function, the attacker has to be capable of deriving the pre-image of the ID_i mapped by the H hash function (ID_{i-1}), later computing the pre-image of this ID, and so on. However, this attack cannot be realized due to the collision-resistant property of the hash function.

Forward Privacy

Let us suppose that an attacker was able to crack the Tag after the ith transaction. For the attacker to be able to verify that the Tag had taken part in a kth transaction (where $k < i$), ID_k value would have to be computed. To do this, the pre-image of the ID_i mapped by H has to be derived one or more (in the worst case: $i-1$) cases. But as a consequence of the above mentioned reasons, it is impossible to carry out.

Defenses against the Man-in-the-Middle Attack

The essence of this kind of attack is that an attacker shows himself as Back-End towards the Tag and as a Tag towards the Back-End. While forwarding messages to the appropriate destination the attacker gains valuable additional information. By using our SLAP protocol parties cannot be impersonalized, because neither of them accept the messages from the other party if it does not have the knowledge of ID_{i-1}.

Resistance against DoS Attacks

It is pointless to send messages filled up with discretionary content to the Back-End, because this kind of attack cannot overload the system to such an extent that would render it impossible for Tags to reach it. In case of the naive methods, in order to determine that a given Tag is not authentic, the H hash operation has to be computed on each identity stored in the database. If an attacker deliberately sends such messages repeatedly in rapid succession, it can overload the server so much that it has no time to serve the other Tags.

Even in such cases the proposed protocol does not overload the Back-End. On average the Back-End has to compute no more than one or two hash operations in order to determine that the Tag is not authentic.

Defenses against the De-synchronization Attack

The purpose of this kind of attack is that the participants of the transaction come to different states by blocking some messages of the protocol. An attacker can make an attempt to block the first or the second message of the protocol by physical devices or jamming the radio signals, so the other partner cannot learn the messages. At this time, though the authentication fails, both participants hold possession of the same ID_{i-1} transaction identity. In this way each party can successfully be authenticated. Let us see what happens when the third message is blocked. Since the $H(N+1, ID_i)$ hashed value, which is sent by the Back-End, is not received by the Tag, it seems to the Tag that the authentication process failed. So the previous identification number (ID_{i-1}) is not overwritten with the new one (ID_i). Meanwhile, from the Back-End's point of view, the authentication process succeeds, therefore the stored IDs are refreshed. Next time, when a Tag would like to authenticate himself, the Back-End will not be able to find a database record corresponding with the incoming ID, since both the first $\log T$ bits of the ID_{i-1} and the ID_i contained in the hash are shifted one transaction ahead.

This is why the Back-End is implemented in such a way that when the first $\log T$ bits of the received ID_{i-1} do not correspond with the ID_{i-1} stored in the database (probably because of the

Figure 14. Synchronization of the parties

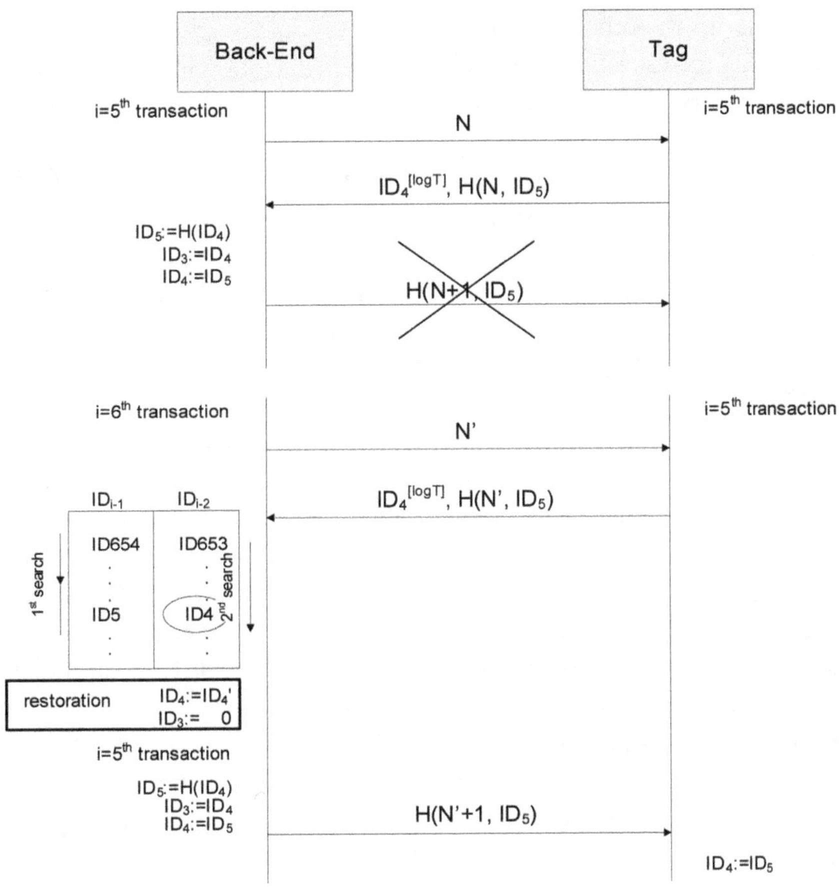

above mentioned attack), then ID_{i-2} is checked. If the received value is equal to one of the stored ID_{i-2}, then using the $ID_i \leftarrow ID_{i-1}$, $ID_{i-1} \leftarrow ID_{i-2}$ replacement each ID for that Tag has to be restored to the former value. After that the protocol can operate in the usual way. Figure 14 shows the method how to reestablish the synchronization between the Back-End and the Tag.

It is easy to see that the proposed solution can guarantee the synchronization between the parties in every case. It is impossible that an attacker can cause more than one ID shifting in this authentication process, for after desynchronizing the transaction IDs the Tag will send the hash value of the same ID_i transaction identifier and a new nonce (N'). The Back-End will realize that this fingerprint does not correspond to any data stored in the database, thus the operation of the authentication protocol will be interrupted.

Similarly it is impossible that the synchronization of the parties shifts to the other direction, which means that the Tag refreshes its ID, while the Back-End does not. The Back-End sends back the $H(N+1, ID_i)$ message only when the Tag has already been authenticated and its ID has been refreshed. Similarly the Tag refreshes its ID only when the response has arrived.

PERFORMANCE ANALYSIS BY SIMULATION

In order to measure the performance characteristics of the SLAP and to make a comparison with other proposed protocols we implemented the SLAP and two other algorithms in OMNeT++ (OMNeT++) which is an object-oriented modular discrete event network simulator. The simulator can be used for traffic modeling of telecommunication networks, protocol modeling, modeling multiprocessors and other distributed hardware systems or modeling any other system where the discrete event approach is suitable.

Figure 15. The model of the RFID system in OMNeT++

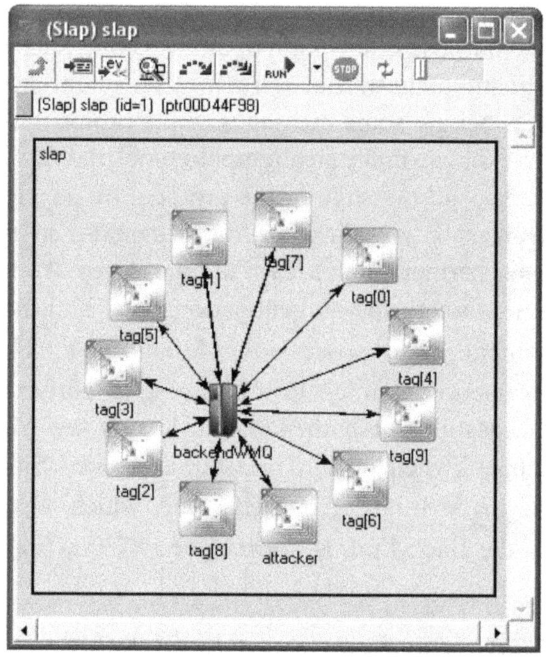

In the followings we describe the measurement scenarios and introduce simulation results.

The Model of the RFID System

Figure 15 shows the model of the RFID system. The model contains four various modules which are the followings:

- Back-End module
- Tag module
- Message Queue module
- Simulation Control module

The functioning of the Back-End can be described briefly: at stated intervals a random challenge is broadcasted to the neighboring Tags. The replay messages of the Tags have to be verified that a given response is sent by a legal Tag or not. In that case when the Tag is legal the ID_{i-1} and ID_{i-2} transaction identifiers are refreshed.

The function of the Tag module is more simple than the Back-End's. At the beginning of the simulation this module asks its own transaction identification from the Simulation Control module and waits for the messages from the Back-End. After receiving the challenge the actual transaction identification has to be derived and appended to the random value sent by the Back-End and the module computes its hash. In case that the Back-End being authenticated the Tag module refreshes the transaction identification.

The Message Queue module generates the real delay of the Tags' responses for the simulation.

The Simulation Control module initializes the simulation and generates various statistics about the performance of the authentication algorithms.

The messages of the simulation are almost the same as the messages of the real RFID systems. However, in order to measure the needed processing time of the authentication process, additional internal messages between the Back-End and the Message Queue is needed. Figure 16 shows the messages and the events between the modules.

Scenarios

In the first simulation run 10 Tags took part of the experiment and there were 10000 elements in the database. This scenario meets the requirements of common RFID system, because generally there is no more authenticated entity than 10000 in an average application and in a given time there are no more Tags nearby a Reader than 10. During the simulations the Tag sent the first three byte of ID_{i-1} by helping for the computation.

According to Luo et al. (Luo, Chan, & Li, 2005), that the necessary time of an SHA-1 hash operation's performance for a Tag is 0.2 second, in our simulation we also used this interval. Since in the SLAP protocol three hash operations (0.6 sec) are executed by a given Tag, therefore the expiration time and the period of an authentication process have to be chosen larger than this value, let these values are 1 sec. In each test case we examine 1000 authentication periods.

In the further scenarios when an examined parameter was changed, the simulation was executed with 10000, 100000 and 1000000 cardinality of database in order to examine how to influence the

Figure 16. The messages and events between the modules

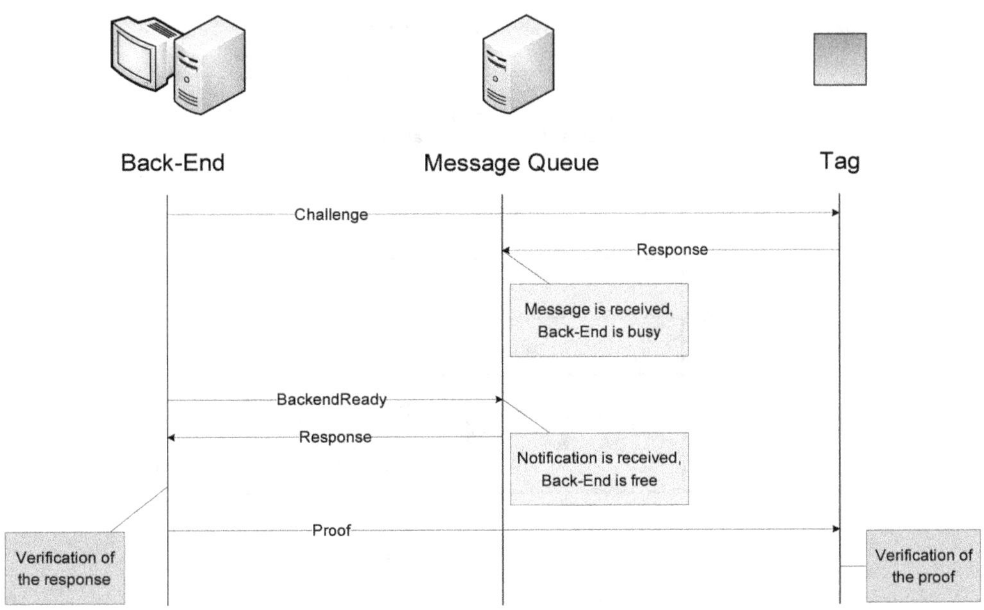

efficiency of the algorithm against the dimension of database by a given parameter. In the first case the number of Tags which took part of the simulation was increased 10 to 100. After that we have used MD5 hash function instead of SHA-1 for the authentication process. Then the duration of the Tags' hash operation is become doubled. As well we examined how the system performance is influenced by setting the expiration time of challenge to 2 seconds.

Finally, we examined the resistance of SLAP against various attacks. In the first instance the third message of our SLAP protocol was blocked in 5% of the cases. In another test case fake response messages were sent back to the Back-End by an attacker. Then the Back-End was flooded with false messages at stated intervals (e.g. 0.1 sec). These test cases were repeated in such case when the messages sent back by an attacker were previously eavesdropped $\log T$ bits instead of randomly generated values.

Simulation Results

In the first scenario there were 10000 successful authentications – 10 Tags are authenticated 1000 times–, while 30012 hash operations are computed by the Back-End and the processing time was 5.89 sec. The average authentication time of a Tag was 0.6032 sec and the dispersion of the measured durations is 0.0017. This means that the time used for authentication was mostly contributed by the Tags' hash calculation, the Back-End needed much less time to complete its task.

Then we examined how the above mentioned measurement results are altered by changing the size of the database. Live up to our expectations there is no significant relation between the number of hash operations computed by Back-End and the size of database, the authentication can be performed in $O(1)$ steps. As it can be seen in Figure 17 in case of the third instance when the database contains 1000000 Tags, the number of hash operation required for the successful authentication is increased a little. The reason for this is that several Tags of 1000000 can be found the first three byte of which are the same so extra hash operations have to be done. Figure 18 shows the average authentication times which are very similar in all cases.

In the next test case the number of Tags which have to be authenticated at the same time is increased to 100. This traffic is called Burst Traffic. In this instance, when the size of database was

Figure 17. The number of hash operations in the first scenario

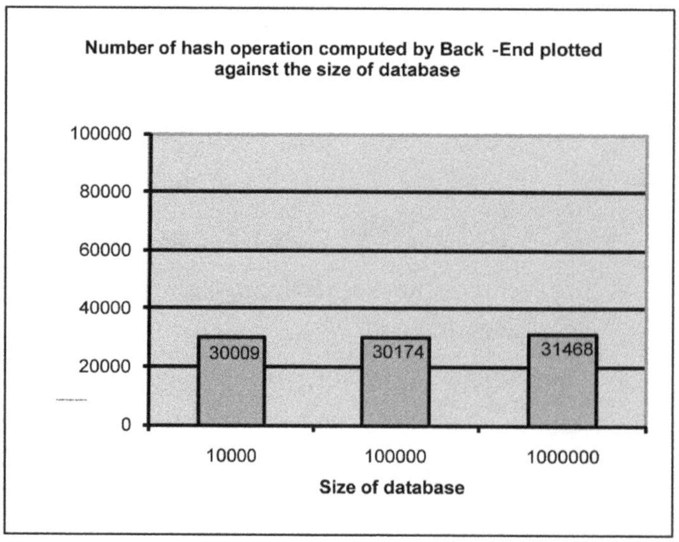

Figure 18. The average authentication time in the first scenario

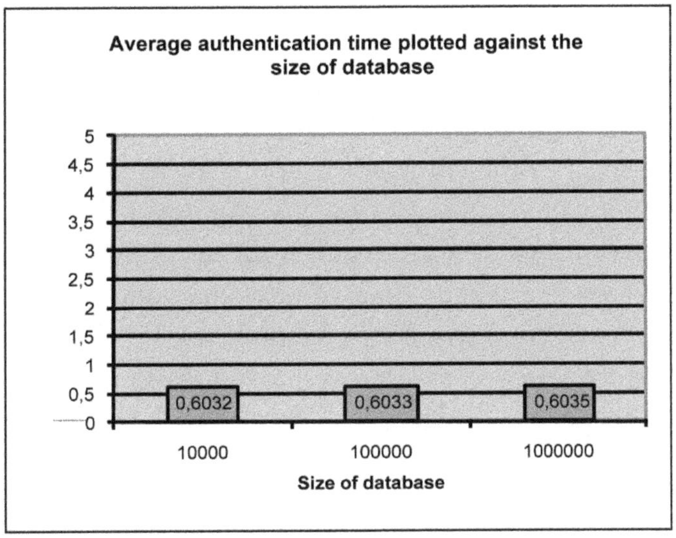

increased 10000 to 1000000 and at the same time the number of authenticated Tags was increased 10 to 100, the average serving time was increased approximately 35 msec. These two instances can be seen in Figure 19 and Figure 20.

In that case when we used MD5 hash functions instead of SHA-1 there was no significant difference. The difference of the average serving times is 132 μsec from which the time needed for the computation of the three hash operation is no more than 0.3 μsec. The remained time difference can be attributed to the comparison of the additional four bytes – the output of an SHA-1 hash is 20 bytes and an MD5 hash is 16 bytes – or the extra memory allocation etc.

The increment of the duration of the Tags' hash operation did not bring in radical changing, the serving time is increased by a constant value in that case when the Tag could respond within the time of the expiration time of challenge.

Finally, we have analyzed the protection of our algorithm against various attacks. It is obvious that if one of the messages is blocked, the authentication process is failed. The question is that when the third message is intercepted whether the Back-End was able to realize and re-synchronize the parties in the next authentication period. Our SLAP protocol resists this kind of attack, the desynchronization attack can be fixed by one database query and two hash computations (Figure 21). In this measurement the probability of blocking was 0.1, thus the authentication failed. In the next authentication period the ID of a given Tag had to be searched among the latest and the previous transaction IDs, apart from this the Back-End's response time changed hardly as it can be seen in Figure 22.

The next test case was more interesting. In this instance an attacker sent fake responses to the Back-End at 100 msec intervals. In the other test case we assumed that the first $\log T$ bits of Tag's ID were eavesdropped and by using these bits an attacker generated the false responses. By using the first $\log T$ bits the Back-End can be forced to calculate hash because at least one hit would be in the database, thus it has to be verified that the $H(N, ID_i)$ value is equal to that one sent by the Tag. This experiment brought some surprising results namely by using the first $\log T$ bits for the DoS attacks the time of the authentication process is decreased instead of increasing what is expected considering the additional hash com-

Figure 19. The average serving time when the number of authenticated tags is 10 and the database contains 10000 tags

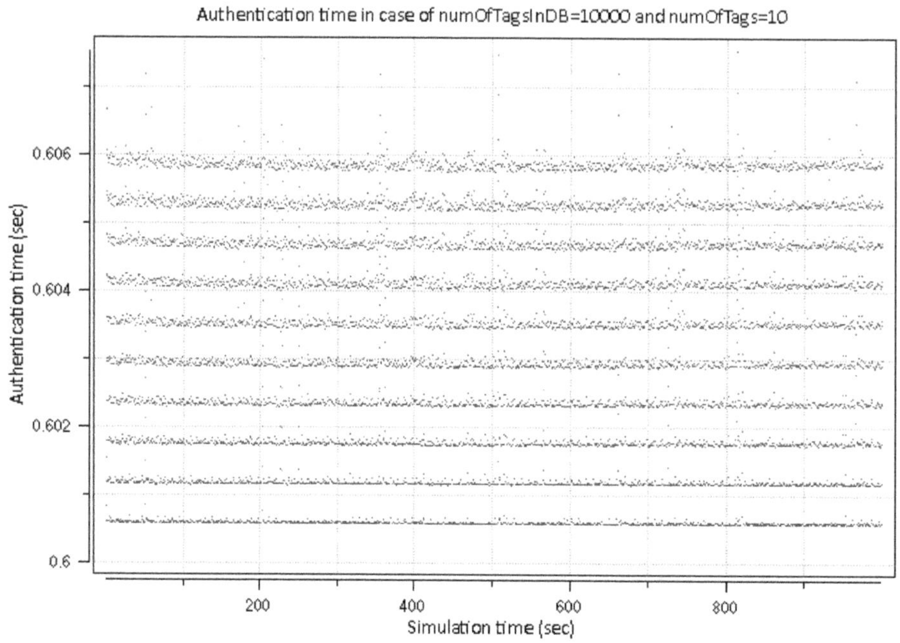

Figure 20. The average serving time when the number of authenticated tags is 100 and the database contains 1000000 tags

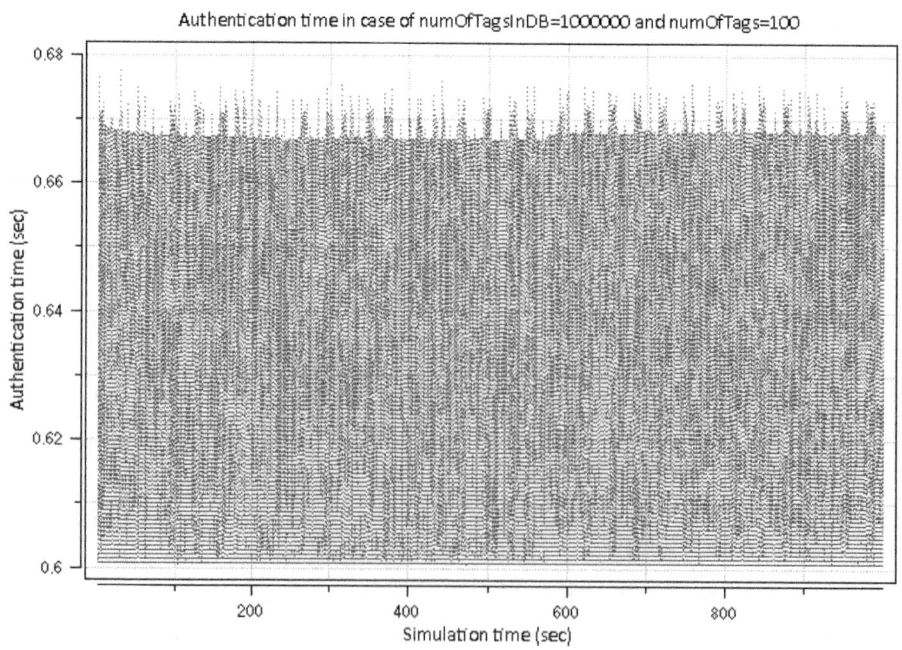

Figure 21. Re-synchronization

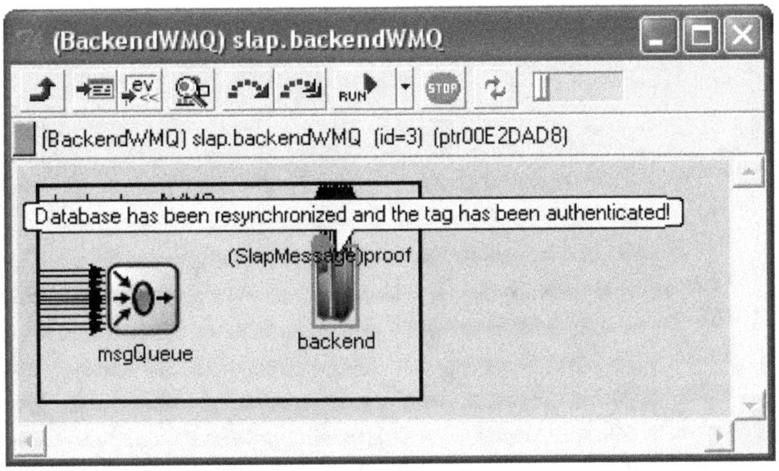

Figure 22. The back-end's response time in case of normal operation and de-synchronization attack

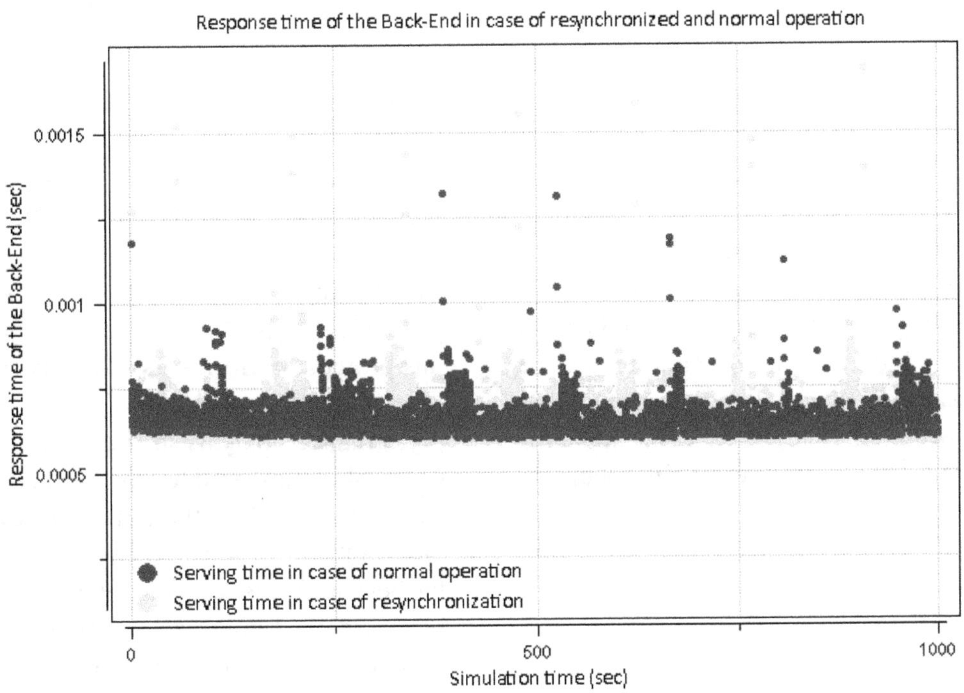

putations. It can be explained that when an attacker uses randomly generated $\log T$ bits in a fake message instead of intercepted ones, this bit string will very likely to correspond with none of the first $\log T$ bits of the Tags' ID stored in the database, for this reason the Back-End has to be searched among ID_{i-2} identifications which is an additional database operation. The difference of the authentication times can be seen in Figure 23 in case that an attacker uses the previously eavesdropped first $\log T$ bits or not. Table 1 shows the efficiency of the SLAP protocol in case

Figure 23. DoS attack with and without using previously eavesdropped first logT bits

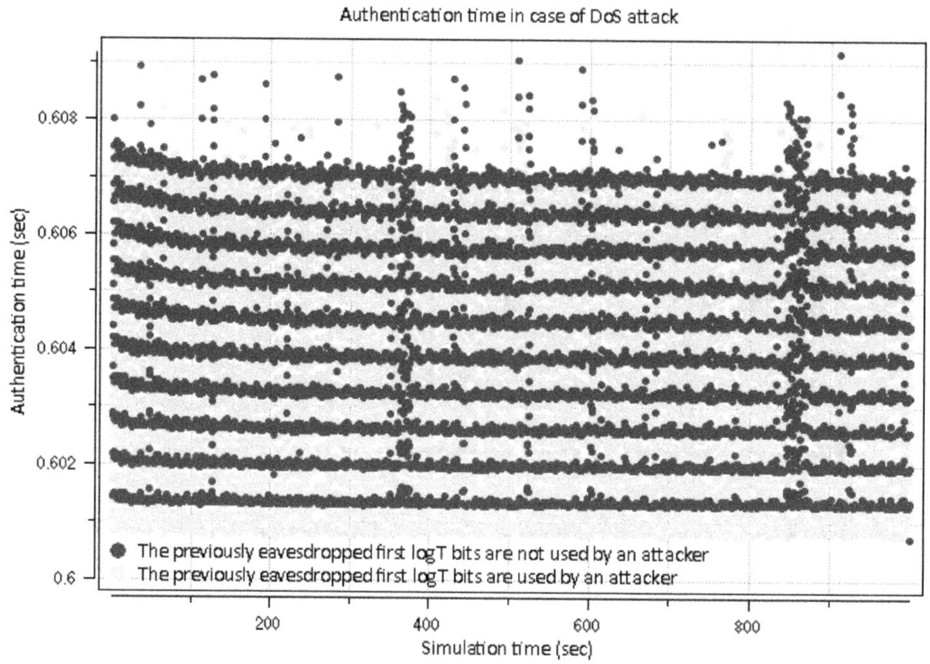

of DoS attacks and such database which contains 1000000 Tags.

Finally, we examined how the system performance is influenced by changing the expiration time of challenge. As it can be seen in Figure 24 the number of hash operations computed by the Back-End is hardly changed in case of increased challenge expiration time. The average authentication time remained 604 ms even if the database contained 1000000 Tags.

COMPARISON WITH OTHER AUTHENTICATION PROTOCOLS

Considering the earlier introduced performance parameters of SLAP protocol it can be said that this protocol performs its task very well in such cases like DoS attack. For the sake of complete performance analysis in this section the SLAP protocol is compared to two another lightweight authentication algorithms. Therefore a Key-Tree based authentication algorithm, which possessed the best performance parameters, and the LMAP, proposed by Luo et al., were implemented in OMNeT++. We considered these protocols from

Table 1. The performance characteristics in case of DoS attack

	No attacker	DoS attack	First logT bit
Average authentication time	604 ms	604 ms	604 ms
Maximum authentication time	608 ms	609 ms	609 ms
Standard deviation	1.8 ms	1.8 ms	1.9 ms
Number of hash operations	31468	34168	53832

Figure 24. The number of hash operations computed by back-end in case of various challenge expiration times

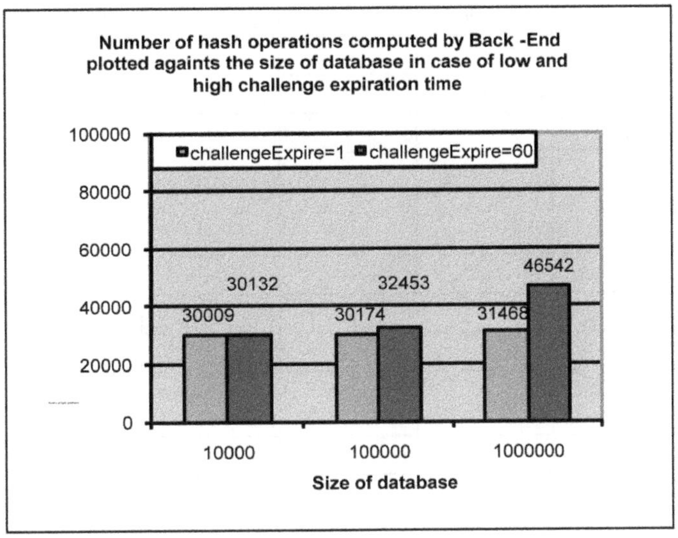

the point of view of efficiency. In a little while we will see that our SLAP protocol is a match for other protocols but exceeds them pretty much.

Security Aspects

In Section 3 we have already examined the well-known lightweight authentication protocols published in the literature and in Section 5 we have already given a security analysis about our SLAP protocol.

Considering the limitation of the length of this contribution only the security aspects will be summarized enumeration-wise in case of systems with limited memory and small cpu capacity. In the followings the main challenges and major attacks, which have to be achieved and protected against, are defined:

- Data Security and Integrity of Messages
- Mutual Authentication
- Protection against Replay Attack
- Anonymity of a Tag
- Undistinguishable Messages from a Random Bit String
- Protection against Impersonation Attack
- Prevention of Tracking
- Forward Privacy
- Defenses against the Man-in-the-Middle Attack
- Resistance against DoS Attacks
- Defenses against the De-synchronization Attack

The security parameters of the above mentioned protocols are summarized in Table 2. In the table + indicates that case when a given protocol resists against the given attack, – indicates the vulnerability, –/+ means the ambiguous cases.

Performance Analysis

At first we examined a Key-Tree based protocol whether the number of hash operations computed by Back-End is directly proportional to the logarithm of the database's size. It can be seen in Figure 25 while the size was multiplied tenfold and once more increased tenfold, the number of hash operations was incremented to a much lesser degree. In each

Table 2. Comparative table of the protocols

Security Aspects	Hash-lock	Ext. hash-lock	LMAP	Ohkubo	Key-Tree based	SLAP
Data Security and Integrity of Messages	+	+	+	+	+	+
Mutual Authentication	–/+	–/+	+	+	+	+
Protection against Replay Attack	–	–	+	–	+	+
Anonymity of a Tag	–	–	+	+	–/+	+
Undistinguishable Messages from a Random Bit String	–	–	–	+	+	+
Protection against Impersonation Attack	–	–	+	–	+	+
Prevention of Tracking	–	+	–/+	+	+	+
Forward Privacy	–	–	+	+	+	+
Defenses against the Man-in-the-Middle Attack	–	–	+	+	+	+
Resistance against DoS Attacks	–	–	– – –	– – –	+	+ + +

phase of the authentication process at least two hash operations have to be computed: one of them for deriving the hash of the secret in a given tree level and the other for generating the proof which will be sent for the Tag. In this instance there were 14 secrets which determined the size of database ($2^{14} = 16384 \approx 10000$). So $14 \cdot 2 = 28$ hash operations are needed for a successful authentication.

Figure 26 shows the diagram of the average authentication times. While the bottle-neck of the system is the capacity of Tags thus the whole authentication time depends on them. In case of 14 secrets 28 hash operations have to be calculated, thus the whole period will be $28 \cdot 0.2 = 5.6$ sec. In case of SLAP these parameters are better by a magnitude of 1.

Figure 25. The number of hash operations – SLAP vs. Key-Tree

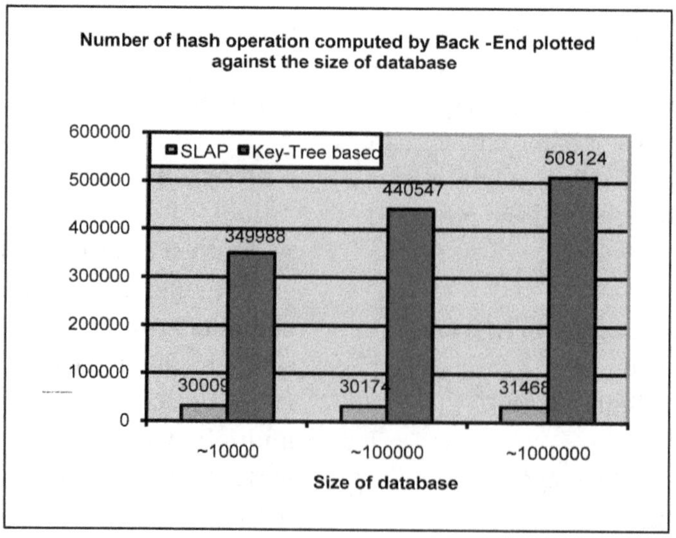

Figure 26. The average authentication times – SLAP vs. Key-Tree

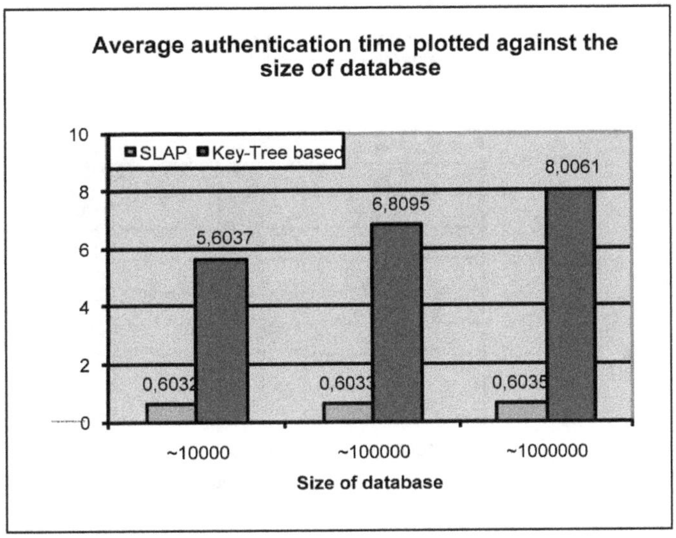

In case of Burst Traffic the Key-Tree based protocols perform quite good, the serving time of the Tags was very low and highly changed. Figure 27 shows the difference of the average authentication times in case of the SLAP protocol and the Key-Tree based protocol. It can be seen that in case of the Key-Tree based protocol the Tags' serving time is less than in case of the SLAP, however the difference is no more than 15-20 msec.

Now let us consider the LMAP protocol. Unfortunately the protocol has much worse performance than the others. This is probably because who designed the protocol did not think about the fact that the Tags can not identify whether the message from the Back-End is for them or for other nearby Tags. This problem is so severe that

Figure 27. The difference of the average serving time of the normal and burst traffic

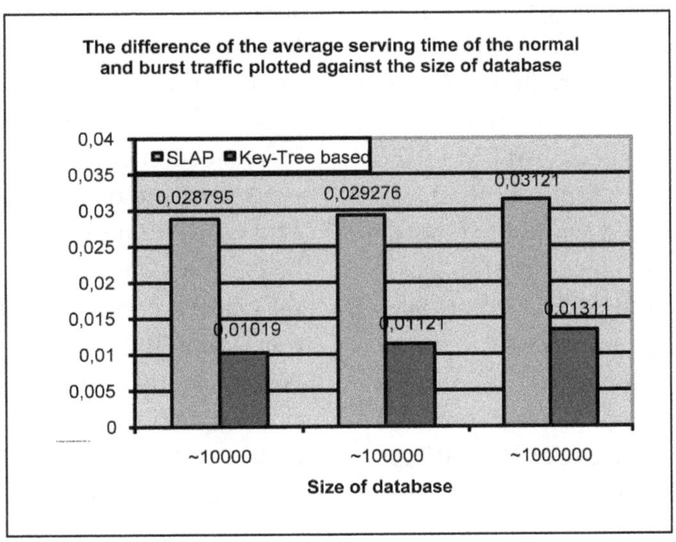

Figure 28. The number of hash operations – SLAP vs. LMAP

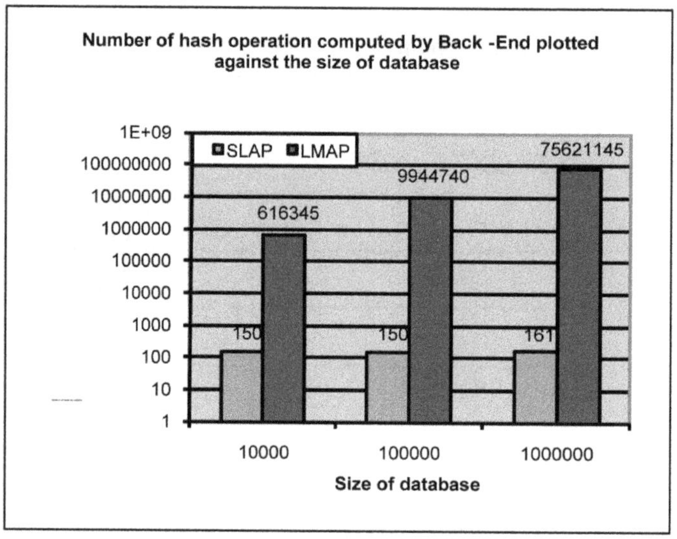

more Tags' parallel authentication can paralyze the entire system.

The protocol does not give any support for the Back-End to decide how to decode the last message's data. So it tries all possibilities, until the hash trace of the decoded data and the $G(data)$ sent by the Tag does not match. This consumes further resource which increases the Back-End's response time slightly as there are already many $O(N)$ hash calculations.

Because of the weaker efficiency of the LMAP protocol the simulation runs with former parameters would have been for a very long time so the simulation runs were carried out 5 times instead of 1000. Figure 28 shows the number of hash operations versus the size of database. As it can be seen in case of LMAP the number of hash operations required by Back-End was multiplied tenfold while the records of the database were increased tenfold, too. On the other hand in case of SLAP performing 5 authentications the number of required operations were 150, 150 and 161.

The average authentication times (Figure 29) indicate also a bad notion. In case of 100000 Tags the serving time was increased considerably, but in case of 1000000 Tags this duration would be unbearably high. So the practical applicability of LMAP can be queried.

In case of LMAP the degradation of the system performance is directly proportional to the number of authentications, because when the transaction number of a given Tag is 10, the whole database has to be hashed ten times by the Back-End in order to complete the authentication process. In case of the SLAP there is no such a problem because in each transaction the transaction ID has to be computed by the Back-End and the Tags. Figure 30 illustrates that when the transaction number is increased the authentication time increased in case of LMAP, however the average authentication time was not influenced in case of SLAP.

As we mentioned formerly, the major weakness of the LMAP protocol is that the Back-End can be carried away by DoS attack. For the sake of the cause it is enough that a message with a high transaction ID is sent to the Back-End by an attacker. In this simulation the Back-End was flooded fake messages which contained false authentication information at 1 sec intervals. At first the transaction number was 1, after that it

Figure 29. The average authentication times – SLAP vs. LMAP

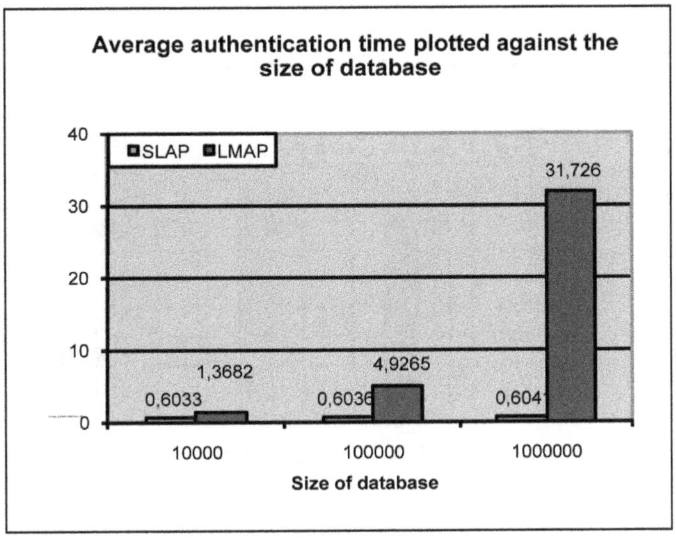

was 10. The number of hash operations computed by the Back-End can be seen in Figure 31 when there was no attacker, the transaction number was 1 and 10. As it can be seen the number of hash operations increased tenfold in case of DoS attacks. As it was mentioned earlier this problem is not exist in SLAP protocol because in our solution the transaction ID is computed in each transaction.

Figure 30. The authentication times in case of SLAP and LMAP protocols

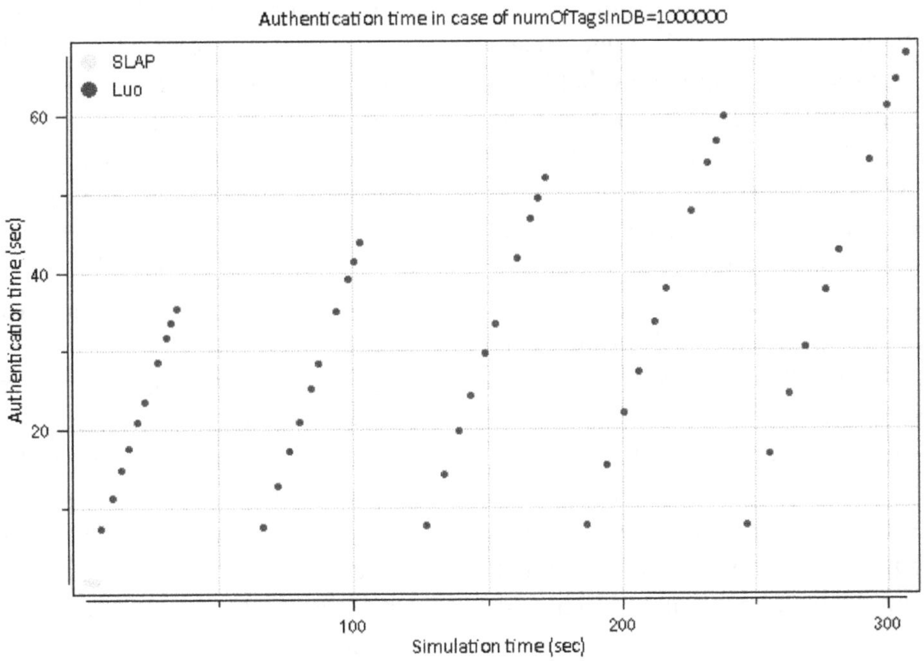

Figure 31. The number of hash operations in case of DoS attack

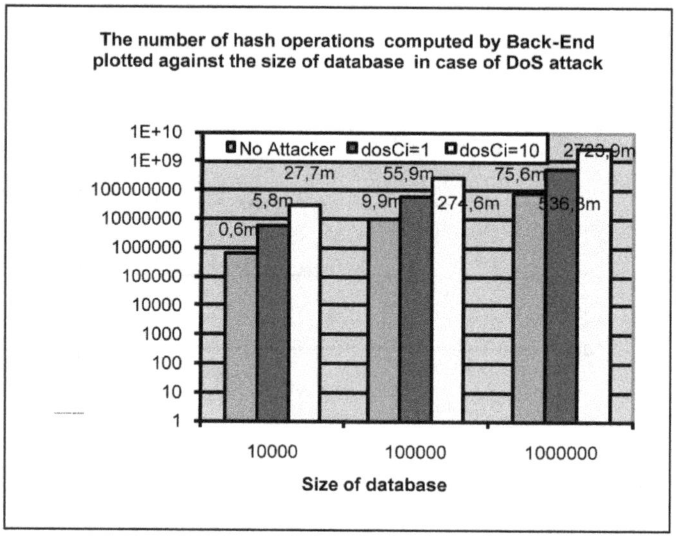

According to the above it can be said that the SLAP protocol possesses better security and performance parameters than the other protocols presented in the literature.

CONCLUSION

In this contribution we showed that although the usage of RFID technology is quite widespread, there are numerous security problems that have to be solved. The principal issue is the mutual authentication with the parties not having to send their ID in plain text. We have given a good overview about some well known authentication protocols that appeared in the literature and we drew attention to the weaknesses of these protocols. The major disadvantages are the decrease of anonymity level and the necessity of very high computational capacity.

We introduced the security features of the small computational capacity environments and we showed numerous attacks which can be achieved. Additionally we have given a detailed overview about the well-known lightweight and elliptic curve cryptography based authentication protocols already proposed in the literature.

We proposed our SLAP protocol which meets the security requirements of small computational capacity RFID environments. Our solution provides an efficient authentication method for the Back-End and Tags, by which a given Tag can be authenticated easily, quickly and securely even if big amount of Tag IDs are stored in the database. Using the SLAP protocol the decrease in the Tag's anonymity level is prevented when a Tag gets compromised.

The protocol was also analyzed from the performance's and security's point of view both for the Back-End and Tags. In both cases the results are very promising. The protocol can defy the well-known attacks and does not demand high computational capacity.

Furthermore we implemented the SLAP protocol and two other lightweight protocols in order to make a comparison between them from the point of view of security and efficiency. From the simulation results it appeared that our proposed SLAP protocol possesses better security and performance parameters than the other protocols published in the literature.

Our future consideration is to create a test implementation to analyze the SLAP protocol in real circumstances in terms of security and performance.

REFERENCES

Ahamed, S. I., Rahman, F., & Hoque, E. (2008). *ERAP: ECC based RFID authentication protocol.* 12th IEEE International Workshop on Future Trends of Distributed Computing Systems.

Bakhtiari, S., Safavi-Naini, R., & Pieprzyk, J. (1995). *Cryptographic hash functions: A survey.* Technical Report 95-09.

Batina, L., Guajardo, J., Kerins, T., Mentens, N., Tuyls, P., & Verbauwhede, I. (2007). Public-key cryptography for RFID-tags. *IEEE International Conference on Pervasive Computing and Communications Workshops*, (pp. 217-222).

Bosselaers, A., Govaerts, R., & Vandewalle, J. (1996). Fast hashing on the Pentium. *Lecture Notes in Computer Science*, 298–313. doi:10.1007/3-540-68697-5_23

Braun, M., Hess, E., & Meyer, B. (2008). Using elliptic curves on RFID tags. *IJCSNS International Journal of Computer Science and Network Security, 8*(2).

Buttyán, L., Holczer, T., & Vajda, I. (2006). Optimal key-trees for tree-based private authentication. *Workshop on Privacy Enhancing Technologies - PET 2006.* Cambridge, United Kingdom.

Dimitriou, T. (2005). A lightweight RFID protocol to protect against traceability and cloning attacks. *Proceedings of the First International Conference on Security and Privacy for Emerging Areas in Communications Networks - in SECURE-COMM'05,* (pp. 59-66). Washington, DC: IEEE Computer Society.

Dimitriou, T. (2006). *A secure and efficient RFID protocol that could make big brother (partially) obsolete*. International Conference on Pervasive Computing and Communications – PerCom 2006.

Girault, M. (1992). Self-certified public keys. *Proceedings of Eurocrypt'91. LNCS, 547*, 490–497.

Girault, M., Juniot, L., & Robshaw, M. J. B. (2007). *The feasibility of on-the-tag public key cryptography*. Workshop on RFID Security, RFIDSec'07.

Gódor, G., & Antal, M. (2008). *Improved lightweight mutual authentication protocol for RFID systems*. 13th IFIP International Conference on Personal Wireless Communications. Toulouse, France: Springer.

Gódor, G., Antal, M., & Imre, S. (2008). *Mutual authentication protocol for low computational capacity RFID systems. IEEE Globecom 2008 CCNS*. New Orleans, LA, USA: IEEE.

Gódor, G., & Imre, S. (2009). *Performance analysis of the simple lightweight authentication protocol*. The 7th International Conference on Advances in Mobile Computing and Multimedia – MoMM2009, Kuala Lumpur, Malaysia, 14-16 December 2009.

Gódor, G., & Imre, S. (2010). Simple lightweight authentication protocol: security and performance considerations. *International Journal of Business Data Communications and Networking, 6*(3), 66–94. doi:10.4018/jbdcn.2010070104

Juels, A. (2003). *The blocker tag: Selective blocking of RFID tags for consumer privacy.* 8th ACM Conference on Computer and Communications Security.

Juels, A. (2006). RFID security and privacy: A research survey. *IEEE Journal on Selected Areas in Communications, 24*(2), 381–394. doi:10.1109/JSAC.2005.861395

Le, T. V., Burmester, M., & Medeiros, B. D. (2007). *Forward-secure RFID authentication and key exchange*. Cryptology ePrint Archive, Report 2007/051.

Lee, S., Asano, T., & Kim, K. (2006). *RFID mutual authentication scheme based on synchronized secret information.* The 2006 Symposium on Cryptography and Information Security – SCIS.

Lim, C. H., & Kwon, T. (2006). Strong and robust RFID authentication enabling perfect ownership transfer. *Lecture Notes in Computer Science,* Springer-Verlag, North Carolina, USA, December 2006.

Luo, Z., Chan, T., & Li, J. S. (2005). A lightweight mutual authentication protocol for RFID networks. *International Conference on e-Business Engineering*, (pp. 620-625).

McLoone, M., & Robshaw, M. J. B. (2007). *Public key cryptography and RFID tags. LNCS 377/2006* (pp. 372–384). Berlin, Germany: Springer.

Miller, V. (1986). *Use of elliptic curves in cryptography*. International Cryptology Conference.

Molnar, D., & Wagner, D. (2004). Privacy and security in library RFID: Issues, practices, and architectures. *Conference on Computer and Communications Security - ACM CCS*, (pp. 210-219). Washington, DC, USA.

Molnar, D., & Wagner, D. (2005). *A scalable, delegatable pseudonym protocol enabling ownership transfer of RFID Tags*. Springer.

Nohara, Y., Inoue, S., Baba, K., & Yasuura, H. (2006). Quantitative evaluation of unlinkable ID matching schemes. *Workshop on Privacy in the Electronic Society - WPES*, (pp. 55-60). Alexandria, Virginia, USA.

Nohl, K., & Evans, D. (2006). Quantifying information leakage in tree-based hash protocols. *Conference on Information and Communications Security – ICICS'06*, (pp. 228-237). Raleigh, North Carolina, USA.

Ohkubo, M., Suzuki, K., & Kinoshita, S. (2003). *Cryptographic approach to "privacy-friendly" tags*. RFID Privacy Workshop. MA, USA: MIT.

OMNeT++. (n.d.). *Community site.* Retrieved from http://www.omnetpp.org

Peris-Lopez, P., Hernandez-Castro J. C., Estevez-Tapiador J. M., & Ribagorda, A. (2006). *A minimalist mutual-authentication protocol for low-cost RFID tags*.

Poschmann, A., Robshaw, M., Vater, F., & Paar, C. (2009). *Lightweight cryptography and RFID: Tackling the hidden overheads*. 7th International Conference on Information Security and Cryptology 2009 – ICISC 2009, Seoul, Korea.

Poupard, G., & Stern, J. (1998). Security analysis of a practical "on the fly" authentication and signature generation. *Proc. of Eurocrypt'98*. *LNCS, 1403*, 422–436.

Robshaw, M. (2008). *cryptoGPS: Anti-counterfeiting solution for RFID tags*. Orange Labs. Tech. Rep.

Rogaway, P., & Shrimpton, T. (2004). Cryptographic hash-function basics: Definitions, implications, and separations for preimage resistance, second-preimage resistance, and collision resistance. *Lecture Notes in Computer Science*, 371–388. doi:10.1007/978-3-540-25937-4_24

Sarma, S., Weis, S., & Engels, D. (2003). Radio-frequency identification: Security risks and challenges. *Cryptobytes. RSA Laboratories, 6*, 2–9.

Schuba, L., Krsul, V., Kuhn, G., Spafford, H., & Sunda, A. (1997). *Analysis of a denial of service attack on TCP.* IEEE Symposium on Security and Privacy 1997.

Stamatiou, P. (2007). *Privacy implications of RFID tags.* Retrieved April 9, 2010, from http://paulstamatiou.com/wp-content/uploads/2007/11/pstam_rfid_paper.pdf

Thompson, D., Chaudhry, N., & Thompson, W. (2006). *RFID security threat model. Proceedings Acxiom Laboratory for Applied Research (ALAR)*. Conf.

Toiruul, B., & Lee, K. (2006). An advanced mutual-authentication algorithm using AES for RFID systems. *IJCSNS International Journal of Computer Science and Network Security, 6*(9).

Tsudik, G. (2006). YA-TRAP: Yet another trivial RFID authentication protocol. *Fourth Annual IEEE International Conference Pervasive Computing and Communications Workshops, 2006. PerCom Workshops 2006, 6*(9B).

Weis, S. (2003). *Security and privacy in radio-frequency identification devices.* Massachusetts, USA: Massachusetts Institute of Technology (MIT).

Weis, S., Sarma, S., Rivest, R., & Engels, D. (2003). Security and privacy aspects of low-cost radio frequency identification systems. *International Conference on Security in Pervasive Computing - SPC 2003*, (pp. 454-469). Boppard, Germany.

This work was previously published in International Journal of Business Data Communications and Networking, Volume 6, Issue 3, edited by Debashis Saha and Varadharajan Sridhar, pp. 66-94, copyright 2010 by IGI Publishing (an imprint of IGI Global).

Section 3
Network Data Services and Applications:
Improving Interoperability and Performance

Chapter 10
Deployment of Live Audio Services on FTTx Networks

Sergio Cabrero
University of Oviedo, Spain

David Melendi
University of Oviedo, Spain

Roberto García
University of Oviedo, Spain

Xabiel G. Pañeda
University of Oviedo, Spain

Víctor García
University of Oviedo, Spain

ABSTRACT

Most live audio/video services deliver information which is produced just-in-time; thus, the deployment of these services is very challenging: there is never a second chance to broadcast the contents. Therefore, aspects such as the selection of the most suitable architecture or the impact of the underlying network technology need to be carefully studied. This chapter presents an evaluation of different architectures that may be used to deploy a live audio service on cable networks. This evaluation has been performed thanks to the development of a model of a real service that accurately reflects the behavior of the users, devices, and protocols involved in the service. The results of the evaluations can be applied to any deployment planned on a real cable network.

INTRODUCTION

Since the development of early web pages for scientific purposes, the Internet has suffered huge changes mainly due to the improvement of access technologies and the widespread access of users from the home environment. Many services have been designed, developed and deployed, making the web a vast repository of resources for entertainment, research, study, etc. The ideas behind most of these services have been taken from other activities, e.g., e-mail from snail mail, audio conference from telephone services, etc. Audio and video have also taken advantage of

DOI: 10.4018/978-1-61350-477-2.ch010

this adoption of ideas and, in the 90s, these services were deployed taking concepts from tape and video recorders and TV and radio services. There are two types of audio/video services on the Internet: live-audio/video and audio/video-on-demand. Live audio/video services deliver information with is produced just-in-time for the user who is connected. Information could have been previously recorded and stored, or can be captured and broadcast live; in either case, the information can only be seen once. Only one type of interaction is possible, the pause, and its behavior is similar to that of a TV broadcast. On the other hand, the principles of audio/video-on-demand are totally different. Users request the information at any time and the servers deliver it exclusively. Also, this system allows users to interact with information: pauses, backward and forward jumps are allowed. Its behavior is similar to that of a videotape or DVD player. Nowadays, most audio/video services are based on streaming technology. This technology permits the distribution of continuous streams of multimedia data that can be reproduced as they are received, and without the need of initial storage.

The deployment of live services is very challenging as they usually broadcast live information. Thus, testing is very limited and there is never a second chance to broadcast the contents. This implies that the quality of service must be under control in order to avoid problems in the delivery of contents and achieve maximum user satisfaction. Quality can be guaranteed by reserving network resources, but at the same time increasing the cost: cost and quality need to be balanced (Luna et al, 2003). Thus, aspects such as the selection of the most suitable architecture or the impact of the underlying network technology need to be carefully studied.

In addition, each of the companies involved in the deployment of one of these services may have its own requirements. Thus, the requirements of content providers may differ from those of the users or the network operators. For example, they would like to generate their contents in their own headquarters, or even use different locations for different shows. This situation conditions the architecture of the service and is a challenge for the network infrastructure, as an incorrect configuration may lead to a poor performance and affect the behavior of other services. Therefore, from a business perspective, it is essential to avoid the risk of losing clients due to poor quality.

To control quality of service, proper engineering techniques need to be used, including the usage of methodologies such as Melendi et al (2007) or Pañeda et al (2007). From the detailed analysis of the service to the execution of advanced tests, several steps must be followed to achieve the desired quality of service. In this context, models and workload tools are important elements for service managers. These tools are important to execute tests and draw conclusions about the future behavior of the service, the qualities which best fit the preferences of the users, the benefits of certain changes in the architecture of the service, the impact of the service on other services running in the same network, the impact of these services on the radio service, the behavior of the traffic as in Gopal et al (2006), etc. These conclusions can be used in the coordination between the different companies involved in the business model (Van der Raadt et al, 2005). Furthermore, service models can be used in the development of business models and feasibility analyses for the deployment of new services (Gordijn et al, 2006).

In the generation of this paper, the following steps have been carried out. First a real service has been analyzed and the behavior of its users and the traffic generated has been statistically characterized. These results were published in Melendi et al (2006). Then, a model of an Internet radio service has been developed and implemented in a simulation tool. To check its accuracy, the model has been configured with the results of the previous characterization and the results of its simulations compared with the empirical data. Finally, the model has been used to evaluate the possible architectures that may be requested by a content provider to deploy an Internet radio service on a cable network. These architectures essentially differ on the location of audio generation devices and streaming servers. The results of the evaluations are presented in this paper.

On the one hand, administrators of Internet radio services may use the results of this paper to achieve a proficient setup. They would be able to predict what will happen depending on the location of the devices that form part of the architecture of the service. Furthermore, if an inefficient architecture needs to be used due to client requirements, the paper sets out the most suitable parameters to configure the network infrastructure. On the other hand, the results may be meaningful for service consultants to assess content providers and select the most suitable configuration for their services.

The rest of the paper is organized as follows. Other related work is discussed in Section 2. The details of the modeled service are provided in Section 3. In Section 4 a summary of the analysis and statistical characterization of the service can be found. The developed model and its validation are provided in Section 5. The evaluations and their conclusions are documented in Section 6. Finally, conclusions and future work are presented in Section 7.

RELATED WORK

Some interesting papers on streaming service analysis have appeared during the last few years including, in some cases, model descriptions and results of test-beds.

Most of the papers where a real service is analyzed provide information of on-demand services and study different elements of content distribution, user behavior, quality of service and content popularity, such as Cherkasova and Gupta (2004) or Costa et al (2004). Cherkasova and Gupta (2004) analyze an enterprise media server workload using access logs from two different servers in HP Laboratories, to study video popularity distributions and the behavior of the users. Costa et al (2004) analyze four real workloads gathered from education and entertainment services, to obtain information about the behavior of the users.

On the other hand, only a few papers have been published with results of the analyses of live services. Veloso et al (2002) present a characterization of live streaming workloads from a server used by one of the top ten content providers in Brazil. More than 3.5M accesses to two live services are analyzed considering different aspects such as the origin of the users, on and off periods, transfers per session, average bitrates, etc. They also present a model for the generation of workloads that is later used in the platform GISMO (Jin and Bestavros, 2001). Sripanidkulchai et al (2004) analyze the live streaming workload from a large content delivery network. The authors study the workload, characterizing popularity, arrival process, session duration and transport protocol used. Van der Merwe et al (2002) analyze the traces of two services, one of them being live and having registered more than one million reproductions. One interesting aspect of this paper is that service data is integrated with routing information gathered from the underlying networks. The results reflect preferences of the users regarding the technology or the quality of the contents, access patterns, transport protocol usage, etc. Hei et al (2006) analyze the accesses to two popular live video channels in China in order to evaluate the behavior of a p2p distribution system as well as the behavior of the users, access profiles, origin of the users, etc. Finally, Melendi et al (2006) analyze the behavior of the users and the traffic interchanged between all the devices in the architecture of the service.

Apart from the behavior of the users, in order to develop an accurate model of the service, it is also very important to study the behavior of distribution protocols and streaming traffic. These have been studied by authors such as Mena and

Heidemann (2000), Lan and Heidemann (2001), Kuang and Williamson (2002), Van der Merwe et al (2002), Boyden et al (2005) or Melendi et al (2006). All of these authors study the distribution of multimedia contents with RealNetworks technologies (RealNetworks, 2002), used in the service used as a reference in this paper. Mena and Heidemann (2000) analyze traces of RealAudio traffic and compare them with typical telnet, FTP and HTTP traffic. One interesting conclusion is that audio streams exhibit a significant amount of regularity in packet lengths, bit rates and interpacket arrivals. Lan and Heidemann (2001) analyze real RealAudio traces and show the main characteristics of this traffic in terms of individual flow and aggregated traffic. Kuang and Williamson (2002) perform a measurement study of RealMedia traffic. One of their conclusions is that RealAudio traffic exhibits on-off patterns when distributed with RealVideo traffic. As mentioned previously, Van der Merwe et al (2002) analyze the traces of two services, showing the aggregated behavior of the traffic and comparing Microsoft and RealNetworks technologies. Boyden et al (2005) present a high-level characterization of RealVideo traffic, focusing on the impact of bottleneck links, media encoding bitrates and RealNetworks' TurboPlay option. Finally, apart from analyzing the behavior of the users, Melendi et al (2006) also analyzes the behavior of the traffic interchanged between all the devices in the architecture of a real Internet radio service. In the aforementioned papers the authors present their results with the necessary detail, but without describing the underlying protocols, as some of these protocols are proprietary and still unknown.

Similar to the approach of this paper, other studies and tools have appeared which are able to simulate service behaviors and carry out performance evaluations. Chawathe (2000) proposes a new architecture for live services called scattercast that is evaluated using simulations. An Internet radio is also used as a case study for this architecture. As mentioned previously, Lan and Heidemann (2001) present a characterization of RealAudio traffic and develop a model for simulation which is used to test changes in the behavior of the server in order to improve loss rates. Deshpande et al (2001) evaluate the performance of a peer-to-peer distribution of live contents using simulations. This is also the case of Xu et al (2002), who rely on simulation to design solutions for peer-to-peer streaming services, and the case of Xiaofei et al (2006), who use simulations to evaluate their solution AnySee and contrast its performance with other live streaming systems. Nguyen and Zakhor (2002) also relay in simulations to evaluate the performance of an architecture to broadcast live video from multiple senders to a single receiver. Padmanabhan et al (2002) carry out a performance evaluation of a new solution, CoopNet, based on simulations driven by traces of live content served on a real service. Schripanidkulchai et al (2004) follow the same approach as the current paper, as they characterize the behavior of users watching live audio and video streams using traces from a real service and evaluate the performance of several design alternatives by means of simulation.

Some of the aforementioned papers aim to analyze and characterize the behavior of users or the underlying protocols. Other papers present models of audio and video services, using information gathered from real traces. A third group of papers pursue to test innovative service architectures, propose changes to existing technologies or design totally new technologies. The focus of our paper is novel as it deals with the limitations of available technologies showing what will happen if a certain architecture is selected and how network devices can be configured to obtain the best performance possible. In contrast with previous papers, we do not try to propose totally new technologies or improvements to existing ones, but we aim to offer information that can be used in a real deployment. To our knowledge, no papers have addressed this issue in the past, comparing the alternatives available to deploy a radio service on a cable network. Thus, this paper is very use-

Table 1. Summarized details of the Asturies.com radio service

Life cycle	1,188 days	Volume of delivered information	156,998 Mbytes
Requests	42,168	Volume of stored contents	5,114 Mbytes
Different users	10,235	Length of contents available	864,273 seconds
Stored audio files	3,924	Length of delivered information	66,268,392 seconds

ful for system administrators who would like to deploy one of these services or who are planning a change of architecture.

SERVICE DESCRIPTION

As an initial reference for this paper, the Internet radio service provided by the digital newspaper Asturies.Com has been used. Table 1 shows some general information that reflects the status of the service.

This service is based on RealNetworks technologies, and its architecture is shown in Figure 1. It is mainly supported by a Helix Universal Server (RealNetworks, 2002) connected to the network of the cable operator Telecable de Asturias S.A.U. which is based on the DOCSIS standard (Cablelabs, 2005). The server broadcasts 24/7 contents formed by articles of news, programs, music and advertisements. Stored contents are used instead of live audio streams. The streams distributed by the server are generated just-in-time by a simulated live transfer agent, or SLTA, provided with Helix servers (RealNetworks, 2002).

Figure 1. Architecture of the service

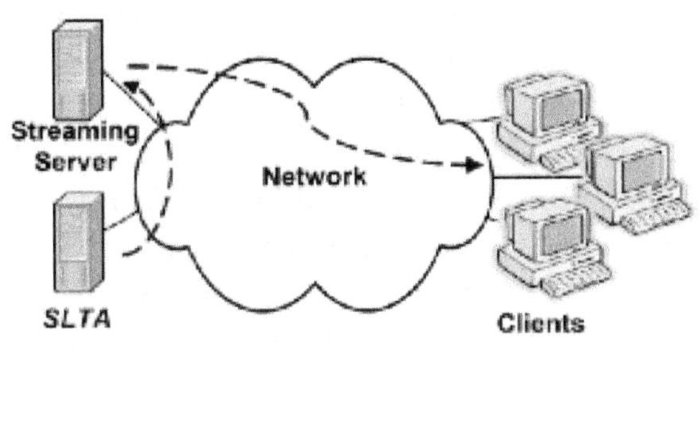

The SLTA is a process used to build a continuous stream of multimedia data packets using a set of pre-encoded files and a playlist with the order of these files and some metadata.

Both the streaming server and the SLTA have been deployed in the head-end of the cable operator. The service offers its contents using the surestream technology (RealNetworks, 2002) with encoding qualities of 11, 16 and 20 kbps. The distribution of these contents is only unicast as multicast is not allowed in the network.

ANALYSIS AND STATISTICAL CHARACTERIZATION OF THE SERVICE

In this section a summary of the characterization of the Asturies.Com radio service is provided. The complete characterization is available in detail in the previous research Melendi et al (2006).

Apart from other aspects of great interest, the traffic interchanged between every service device was analyzed in the previous research. Providing three applications intervene in the process of generation and delivery of contents, two network segments were analyzed in detail: the connection between the SLTA and the streaming server and the connection between this server and the clients. For these connections, RealNetworks uses two protocols: the standard protocol RTSP -Real Time Streaming Protocol- (Schulzrinne et al, 1998) for control purposes and the proprietary protocol RDT -Real Data Transport protocol- for the delivery of audio and video information. There

Figure 2. Traffic between SLTA and server

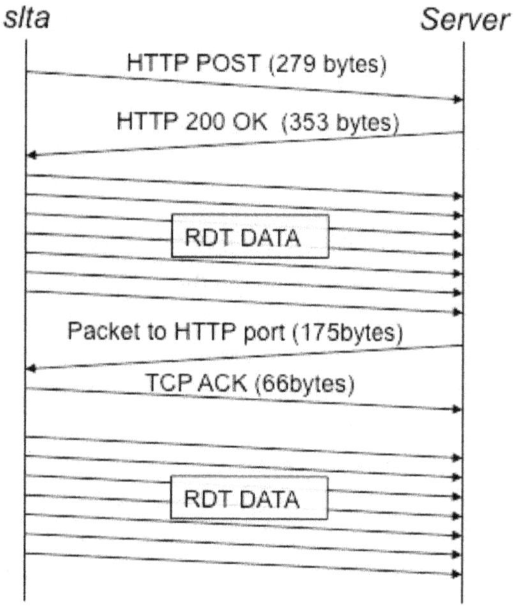

Figure 3. Traffic between client and server

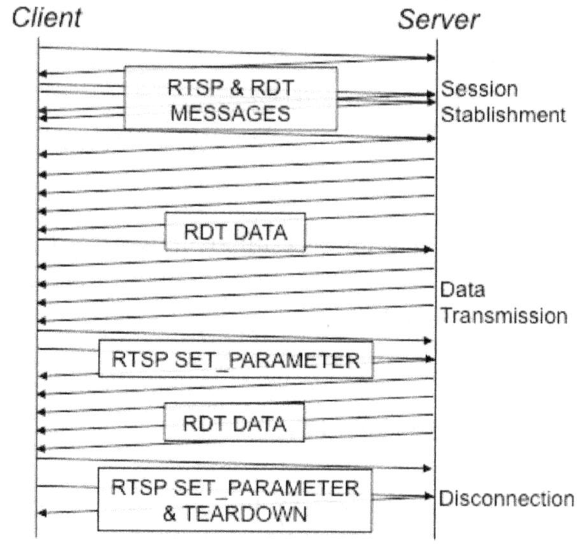

is no documentation available for RDT so this protocol has been reverse engineered.

First we will briefly describe the connection between the SLTA and the streaming server, graphically depicted in Figure 2.

There are two phases in the communication between these elements: session establishment and data transmission. While HTTP is used during the establishment of the session for authentication and configuration purposes, during the data transmission phase RDT is used to deliver audio information in packets of 392 bytes for the 11 kbps quality, 604 bytes for the 16 kbps quality and 640 bytes for the 20 kbps quality. Packet inter-arrival times are also different for each of the available qualities and the average values are 254.77 milliseconds for the 11 kbps quality, 278.27 milliseconds for the 16 kbps quality and 231.72 milliseconds for the 20 kbps quality. Also, control messages are sent by the server to the SLTA's HTTP port to report about the transmission every 10 seconds.

In the case of the connection between the streaming server and the clients, shown in Figure 3, three different phases can be identified in the communication: session establishment, data transmission and disconnection.

RTSP is used as control protocol, and during the data transmission phase the server uses RDT to deliver data to the client. In this case, the analyzed data shows that the size of the audio packets remains stable at 362 bytes for the 11kbps quality, 574 bytes for the 16kbps quality and 610 bytes for the 20kbps quality. This coincides with the results published by other authors such as Mena and Heidemann (2000) or Kuang and Williamson (2002). Again, packet interarrival times are different for each of the qualities with average values of 244.94 milliseconds for the 11 kbps quality, 273.70 milliseconds for the 16 kbps quality and 230.19 milliseconds for the 20 kbps quality.

The behavior of the users has also been analyzed. It is graphically shown in Figure 4 and described in Table 2.

From a total number of 30,105 requests, 4.32% are requests where no multimedia data was delivered and so considered erroneous, while 95.68% correspond to requests where multimedia data was successfully sent. Pauses have also been

Figure 4. Conceptual model of a user of the Internet radio service

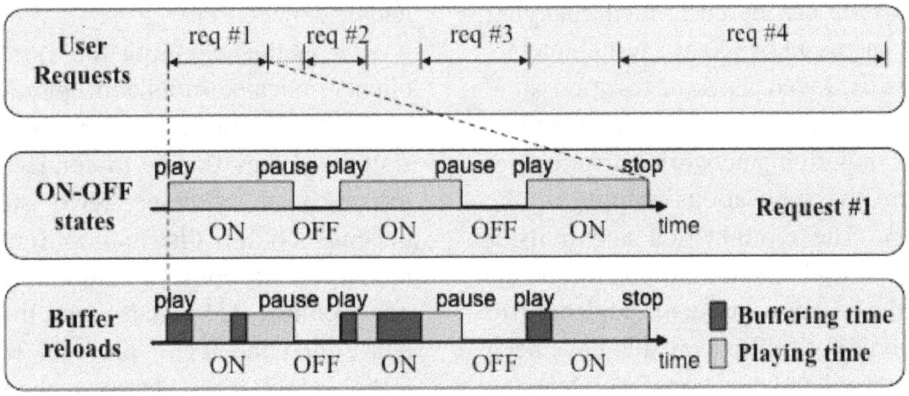

Table 2. Behaviour of the users of the service

	Average Value	Distribution Function	Parameters
Length of requests	1584.490 sec	Weibull	$\alpha = 0.06114$ $\beta = 0.43457$
Length of pauses	593.987 sec	Weibull	$\alpha = 0.19291$ $\beta = 0.35896$
Position of first pause	1238.401 sec	Weibull	$\alpha = 0.05268$ $\beta = 0.462$
Time between pauses	1161.341 sec	Gamma	$\alpha = 0.2328$ $\beta = 5004.06$

analyzed as they cause the so-called OFF periods (the server does not send any data to the client during a pause). Regarding the number of pauses per request, only 6.06% of the requests have this type of interaction. Moreover, the number of pauses in these requests is also relatively small. 81.48% of these requests present only 1 pause, 11.33% have two and 3.77% have three. The other 3.42% present a heterogeneous number of pauses, ranging from 4 to 47. In any case the number of pauses in a request has a discrete nature and the number of requests where 4 or more pauses are performed is insignificant.

MODEL OF THE SERVICE

Taking into account the elements of the architecture and the protocols of the analyzed service, a model of an Internet radio service has been developed. This section briefly describes the functionality of the model as well as the validation performed to prove its accuracy, by comparing the results obtained in different tests with those extracted from the service.

The model has been developed using the OPNET Modeler simulation tool (OPNET, 2009). The model is based on discrete-event simulation (DES) in order to obtain a more accurate model of the whole service: its architecture, the behavior of the users, transport and control protocols and streaming traffic performance. Moreover, this eases the integration of the DES model with the network devices available in the libraries of the simulator. These devices can be used for the execution of advanced tests with different network configurations. The model can easily be integrated into models of different network technologies.

Furthermore, once a technology has been chosen, different tests can be performed changing the topology of the network or the parameters of the devices deployed. This makes it possible to evaluate the performance of the model and its effects on the underlying network infrastructure and the rest of the applications running on the same network. The reproduction and transfer periods between each of the elements in the architecture of the service have been modeled. Also, the different types of traffic available have been considered (control and data transfer). Thus, the results obtained with the model accurately reflect the behavior of a real service.

Basically, the model has three elements that have been developed ad-hoc: client, streaming server and SLTA.

The Client Process

The behavior of the model of the client is configurable, but for validation and testing purposes we have configured it to behave in the same way as the clients described in the analysis and characterization section.

A summarized state diagram of this process is shown in Figure 5. The client passes through three main phases: initialization, transmission and disconnection.

During the beginning of the simulation all the clients indicated in its configuration are created at the same time, and they remain in an inactive state until they decide to connect to the server moving to the Connect RTSP state. This time depends on the distribution function used to model the temporary evolution of the requests. The Connect RTSP state starts the initialization phase, and the RTSP protocol is used for the initial negotiation between client and server. Afterwards, depending on the length of the request given by the corresponding function, several round trip time RDT messages are interchanged between the client and the server in order to estimate the available bandwidth. Once the available bandwidth has been estimated the client starts the transmission phase by moving to the Playing transitional state. From the Playing state the client moves to other states depending on the situation: sending synchronization messages (Sync), sending RDT acknowledgements (Ack), reproducing audio (Reprod.) or paused. It must be taken into account that the model of the client includes a modeled buffer and, thus, when network problems

Figure 5. Summarized state diagram of the client

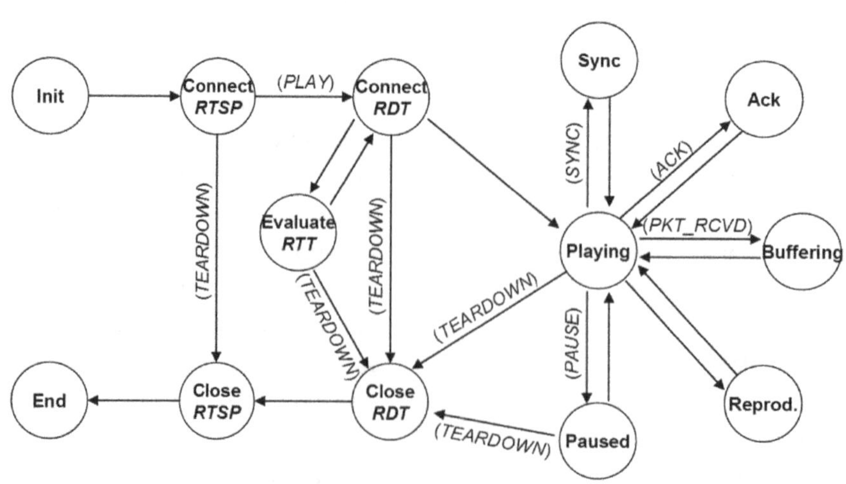

arise, buffer reloads are performed in the Buffering state.

Once the end of the request has been reached according to the corresponding distribution function, an RTSP TEARDOWN message is sent to the server and both the RDT and the RTSP connections are closed. After closing the connections the client ends and the memory is freed.

The Streaming Server

The model of the Internet streaming server (a summarized state diagram is shown in Figure 6) allows concurrent clients to connect and request the same audio contents. Also, the streaming server is connected to the SLTA process where the original information is generated.

Apart from the establishment of the RTSP over TCP connections, the server delivers audio data to all the clients connected by using the RDT over the UDP protocol, receiving from these clients the corresponding RDT acknowledgements and synchronization messages. As in the real service, three qualities have been considered for the available contents: 11 kbps, 16 kbps and 20 kbps. The server selects the most appropriate depending on the available bandwidth for each of the clients connected.

As shown in Figure 6, the model of the server is made up of a main process named manager, and a set of slave processes each of them corresponding to a client connected to the server.

The manager starts and ends with the simulation. During the beginning of the simulation, the manager creates a slave process that remains in the Listen state until a client connection request is received. Once the slave accepts this request, the manager creates a new slave process to wait for another connection coming from another client. The manager is also connected with the SLTA

Figure 6. Summarized state diagram of the server

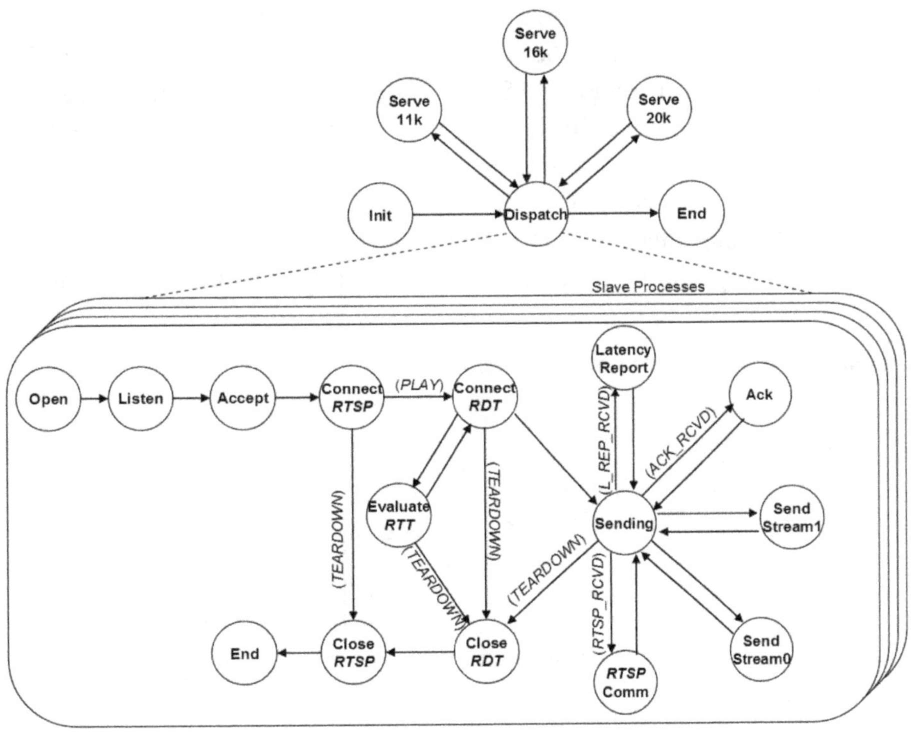

process, therefore, it has three states to model the reception of data packets in each of the three available qualities (Serve 11k, Serve 16k and Serve 20k states). Periodically, the manager enters into each of these states, gathers buffered data packets and delivers these packets to the proper slave processes (Dispatch state), depending on the quality they have chosen for their corresponding clients.

Every slave process that has accepted a connection request from a client in the Accept state follows the same sequence. There is an initialization phase that concludes with the selection of an encoding quality for the client in the Connect RDT state. This phase starts with a negotiation based on the interchange of several RTSP control messages (Connect RTSP state). If this negotiation is successful, several round trip time RDT messages are interchanged between the slave and the client in order to estimate the available bandwidth and select the most appropriate encoding quality. After the initialization phase, a data transfer phase starts when the slave moves to the transitional state Sending, moving to other states depending on the situation. The Send Stream 0 and Send Stream 1 states have been designed to deliver the streams of data detected in the characterization of the real traffic. The RTSP Comm state manages RTSP messages issued by the client such as PAUSE, PLAY, or SET_PARAMETER, interrupting or resuming the delivery of data when needed. The Ack and Latency Report states manage RDT acknowledgements and latency reports, performing the corresponding actions on RDT data packets. RDT and the RTSP connections are closed when an RTSP TEARDOWN message is received, causing the slave process to end.

The SLTA Process

Finally, the model of the SLTA delivers the streaming server data in the three available encoding qualities for its delivery to the clients. Thus, the model of the SLTA opens the appropriate connections with the streaming server and transfers the three traffic streams concurrently with the format specified in the characterization of the service.

Figure 7 shows the model used to simulate the behavior of this process. Although the SLTA is executed since the beginning of the simulations, it is only activated when the connection is established with the server in the Open state. Once this connection has been established, audio data fragments are generated by the SLTA in the aforementioned qualities and sent in RDT packets. In the Serve 11k state, a data packet of 392 bytes is delivered every 0.25477 seconds, incrementing timestamps by 0.255 seconds each time a new packet is issued. In the Serve 16k state, every 0.27827 seconds a packet of 604 bytes is generated increasing timestamps by 0.279 seconds. Finally, in the Serve 20k state, every 0.23172 seconds a packet of 650 bytes is generated increasing timestamps by 0.232 seconds. The process finishes when the end of the simulation is reached.

Figure 7. Summarized state diagram of the SLTA process

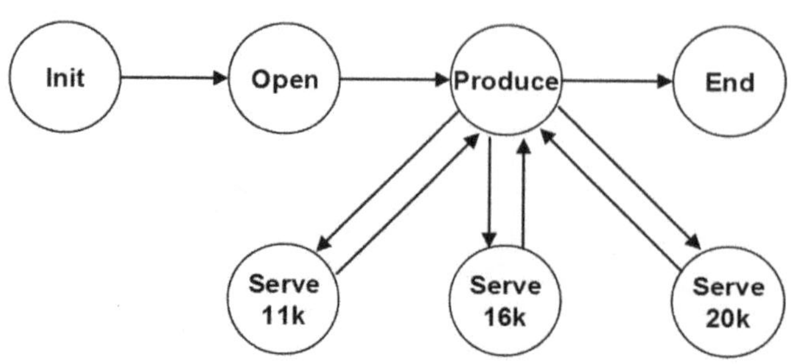

Validation of the Model

We have validated the model by comparing its results with those extracted from the real service. For this purpose, a simulation was executed with 100 audio clients connecting to the service through the cable network. The SLTA and the server were installed in the head-end of the network and no background traffic was considered, as it is not important to validate the model. After the simulations, the results were compared with the data extracted from the real service and, for that purpose, Kolmogorov-Smirnov or K-S tests were executed with a confidence interval of 95%. These tests compare the values of the distribution functions calculated during the characterization phase with those extracted from the execution of the simulations, evaluating the existing differences (Law and Kelton, 2000). Some of the results from K-S tests are shown in Table 3, and their low values reflect the accuracy of the model.

EVALUATION OF DIFFERENT ARCHITECTURES OF THE SERVICE

Once the model has been validated we have used it to perform simulations of different scenarios of the service, working over a DOCSIS cable network (Cablelabs, 2005). In a network of this kind, users are connected to a cable-modem, and all the cable-modems are controlled by a CMTS –Cable Modem Termination System- which behaves as a router, connecting the cable infrastructure with the backbone of the network operator.

In our case, the scenarios consider possible requests from a content provider regarding the deployment of the elements of the service. In this context, four scenarios are considered where the elements that produce (SLTA) and offer (server) audio are placed in different locations of the network.

There are two main goals behind the evaluation:

- Carry out a performance evaluation of the service with different architectures.
- Configure cable network elements to achieve an adequate performance of the service without affecting other services running in the same network.

The first simulation scenario has both the streaming server and the SLTA connected in the head-end of the operator and the clients connect from the cable network. This is the architecture used in the radio service of Asturies.Com. It is very common, as the providers usually contract a hosting service to run these programs in a server managed by another company, usually a network operator. Contents need to be previously produced, uploaded and published. For example, this may be the case of a top music radio station offering only the latest hits and advertisements.

The second scenario goes one step further because the server has been placed in the head-end of the network operator and the SLTA runs on the cable network. This configuration is also

Table 3. Results of K-S tests

	Real average value	Simulated average value	Result of K-S tests
Length of requests	1584.49 s	1529.76 s	0.16519
Length of pauses	593.98 s	612.14 s	0.15288
Pauses per request	0.0798	0.0769	0.01597
Position of first pause	1238.40 s	1298.48 s	0.08816
Time between pauses	1161.34 s	1588.21 s	0.12674

interesting, as it can be suitable for providers who would like to generate their audio contents on-site, while paying another company to broadcast these contents from a server installed in a suitable location, such as the head end of a network operator. With this configuration a content provider may offer live programmed from its own site, but still delegate the distribution of contents.

The third scenario has both the streaming server and the SLTA connected in the cable network. This configuration is possible for small companies with a reduced budget who would like to broadcast audio contents using their own connection to the Internet. Also, a content provider may choose it in order to have full control of the service and thus more freedom to fine tune service devices.

Finally, in the last scenario the SLTA runs on the head-end of the network operator while the streaming server has been connected to the cable-network. As will be further explained, this is the configuration with the worst performance. Nevertheless, it is interesting to have a complete view of the different combinations when placing the elements in the network.

As the CMTS controller is a key device in a DOCSIS cable network, the traffic delivered through this network device is studied in detail in all the scenarios.

This evaluation is only technical and does not consider any cost issues derived from the selection of a given architecture. The four alternatives have a similar complexity, but their deployment has implicit costs derived from housing or hosting contracts, hiring personnel, acquiring equipment, etc. that are outside the scope of this paper.

Simulation Scenario #1: Server and SLTA in the Head-End of the Network

First, both the SLTA and the server have been installed in the head-end of the network and 100 clients connected to the bus of the cable network try to access the service. In this scenario, the traffic interchanged between the different devices has been studied, with the results shown in Figures 6 to 9.

Figure 8 shows the traffic, in bits/sec, delivered from the SLTA to the CMTS controller. It remains nearly constant due to the characteristics of the audio traffic pattern and it is the result of the transmission of the three audio qualities (11, 16 and 20 kbps).

In Figure 9, the bandwidth consumption between the CMTS and the server is shown. The

Figure 8. Traffic generated by the SLTA

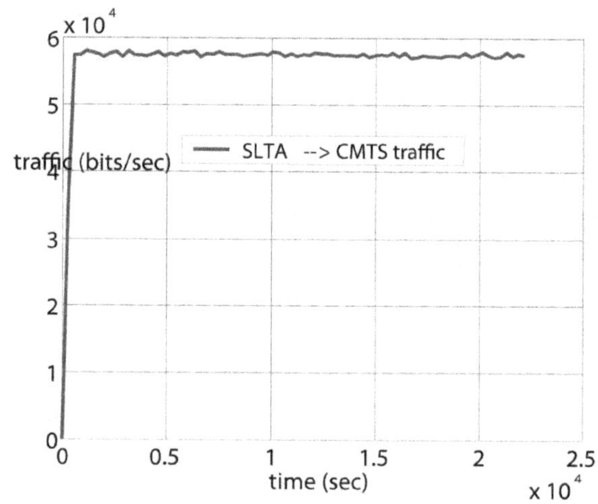

Figure 9. Traffic between the streaming server and the CMTS of the cable network

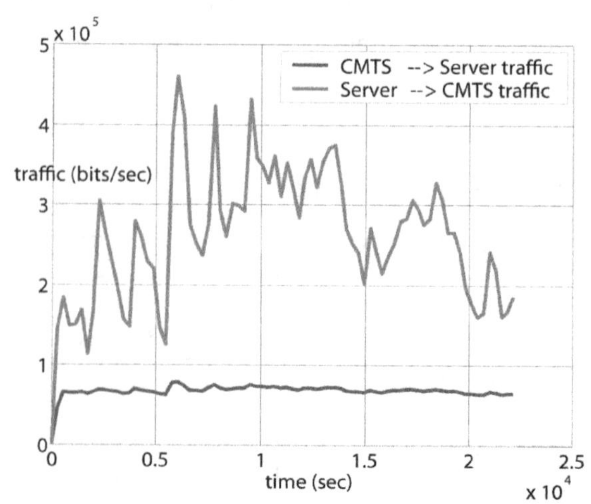

Figure 10. Clients connected to the service

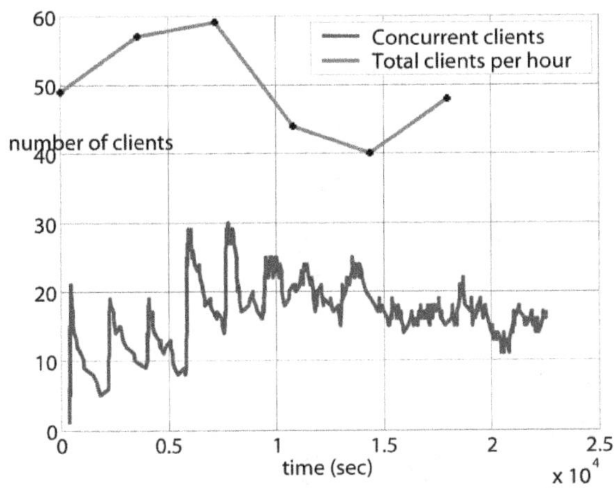

bidirectional behavior of this traffic is due to the fact that the CMTS resends the traffic delivered by the SLTA and the streaming server delivers the audio traffic to the users connected to the radio service.

The traffic generated in the streaming server responds both to the changes in the number of connected users, shown in Figure 10, and the selected audio quality.

Figure 11 displays the traffic in the bus of the cable network, in both upstream and downstream directions. Due to the distribution of the elements in this scenario, the throughput of the downstream bus is clearly higher than the throughput of the upstream bus. In this scenario, small user requests are sent in upstream direction and the audio data from the server is delivered in downstream direction. Moreover, downstream management traffic increases the bus throughput. The difference between the DOCSIS load and the downstream throughput traffic is due to the MAC -Medium Access Control- mechanism of the DOCSIS protocol. In order to control the access in upstream direction, the CMTS sends bandwidth allocation MAP messages in downstream direction. These MAP messages, have a typical size of 64 bytes and are sent every 2 milliseconds. Consequently, MAP messages in downstream direction produce an overload of, approximately, 256 Kbps.

Also, the delays of the audio packets received by the clients of the service have been represented. As shown in Figure 12, data packets present very reasonable delays below 3 milliseconds.

Furthermore, the effects of the transmission of audio data packets on the DOCSIS protocol have also been analyzed. As the majority of the traffic is transmitted using the downstream chan-

Figure 11. Traffic in the DOCSIS bus

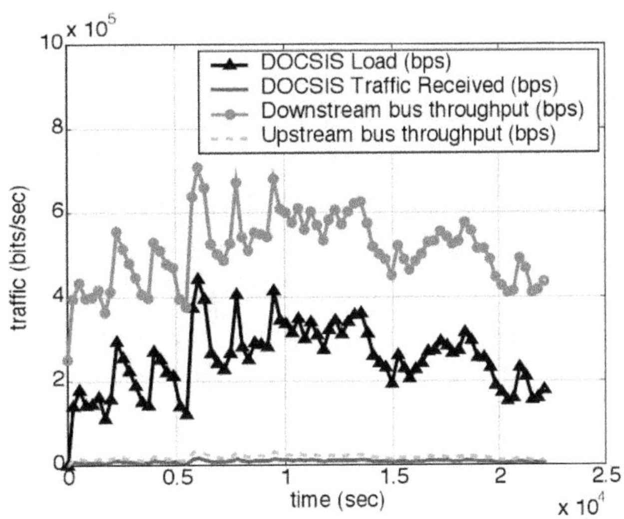

Figure 12. Delays of packets received by clients

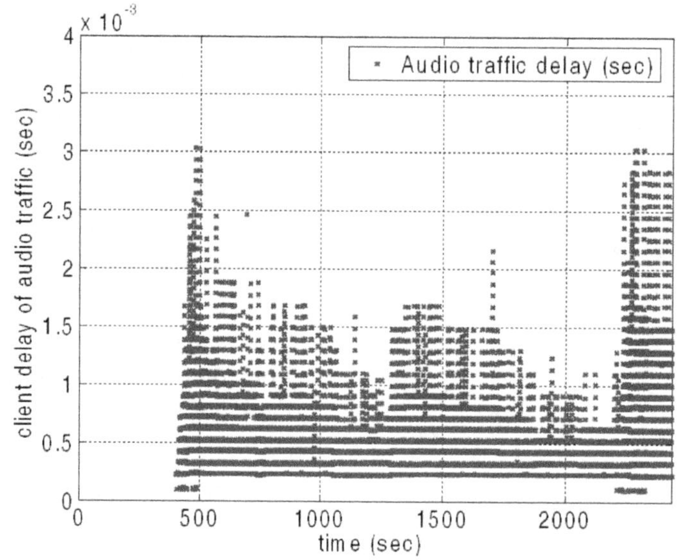

nels, as shown in Figure 9, no significant problems have been detected. This is also demonstrated by the performance of the upstream channel. Figure 13 shows the end-to-end delays in the upstream channel. Their low values (below 5 milliseconds) indicate that there are no transmission problems.

This is also confirmed by the length of the MAP messages request queue for bandwidth allocation, shown in Figure 14.

To finish the analysis of this scenario, the inter-packet times for each of the qualities of the data packets sent by the SLTA to the server are shown in Figure 15. The values are similar to those in Melendi et al (2006).

Figure 13. Delays of DOCSIS frames in channel 0 of the CMTS controller

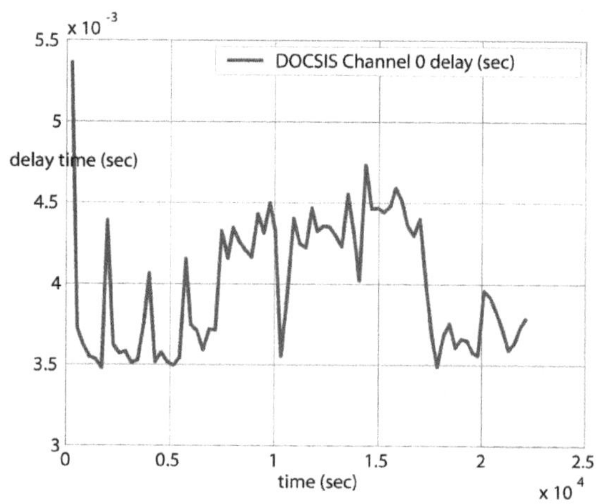

Figure 14. Length of MAP request queue in DOCSIS

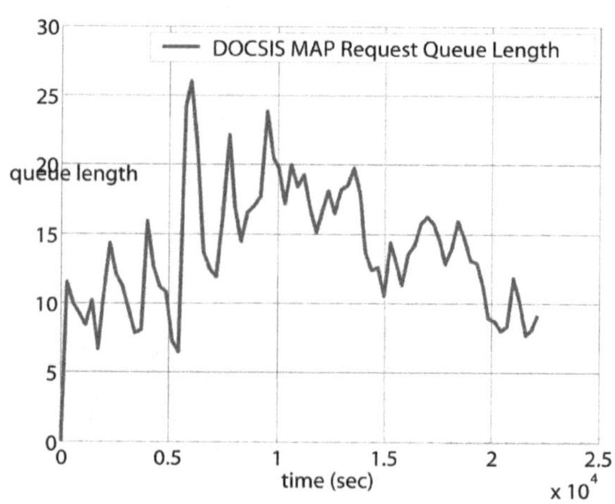

Figure 15. CDF of the time between packets sent by the SLTA to the server

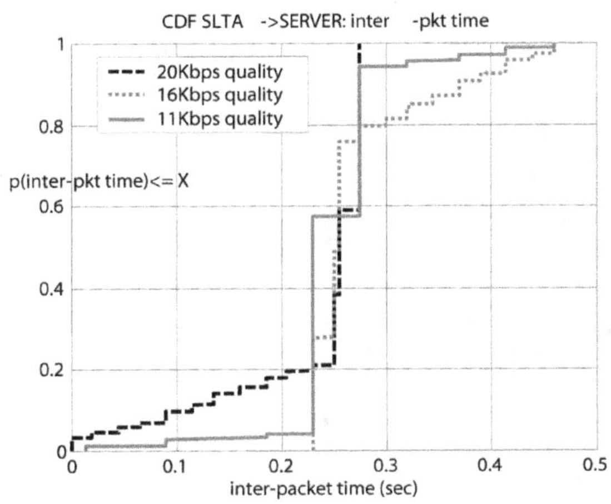

Simulation Scenario #2: Server in the Head-End of the Network and SLTA in the Cable Network

In this new scenario, the SLTA has been installed in the cable network while the server remains connected to the head-end. This is a suitable scenario for content providers who want to manage their own radio channels on-site but use a streaming server owned by a network operator or content distribution network manager.

In the cable network, apart from 100 cable-modems installed for each of the clients of the service, another cable-modem has been installed to connect the SLTA with the network. With this scenario, 2 hours and 22 minutes have been simulated to evaluate the traffic interchanged between every service device, obtaining the results shown in Figure 16 and Figure 17.

The explanation for Figure 16 is the same as that for Figure 9. The traffic in the output of the SLTA behaves as shown in Figure 8 except that in this new scenario it is delivered in upstream direction from the cable network. This new situation is reflected in Figure 17. It indicates that the SLTA connected to the bus of the cable network generates non-negligible traffic in the upstream

Figure 16. Traffic between server and CMTS

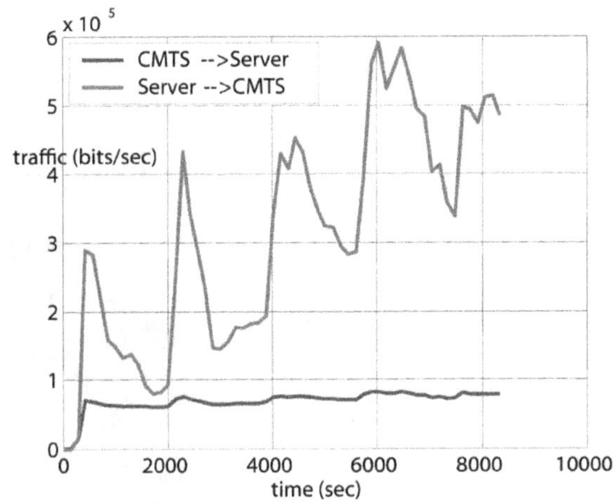

Figure 17. Traffic in the DOCSIS bus

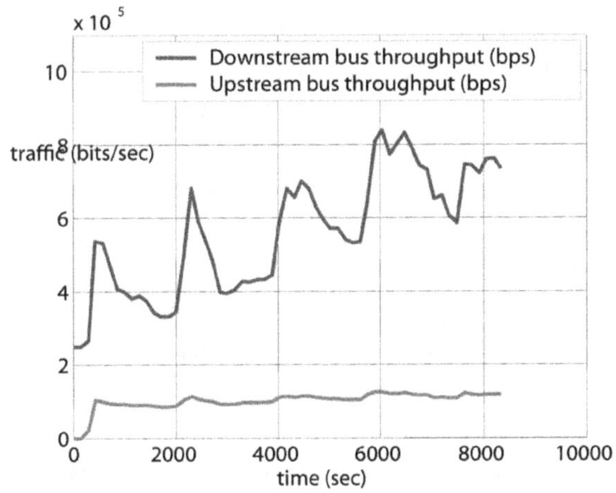

channel. Again, the downstream bus throughput is increased by the delivery of MAP messages.

The delays of the data packets received by the clients of the radio service have also been studied. As shown in Figure 18, the vast majority of these delays remain under 600 milliseconds.

The effects of the audio transmission on the DOCSIS protocol have also been studied. As most of the traffic is again transmitted using the downstream channel (from the streaming server to the clients), no significant problems have been detected. This is also demonstrated by the perfor-

Figure 18. Delays of data packets

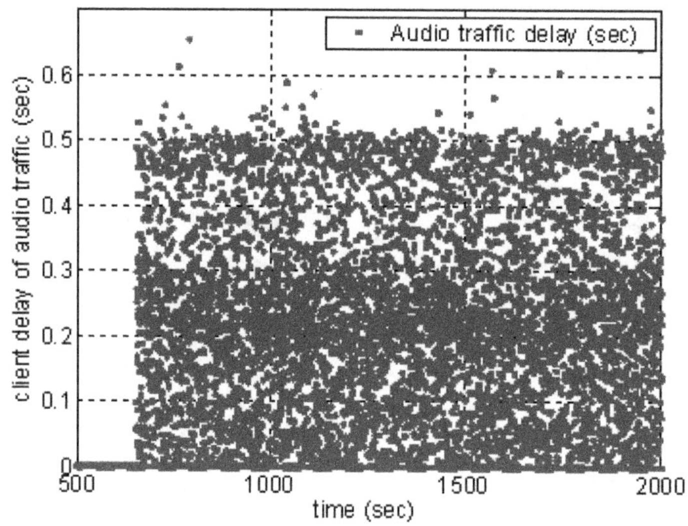

mance of the upstream channel. Figure 19 shows the end-to-end delays in the upstream channel. Their low values (below 9 milliseconds) indicate that there are no transmission problems. This is also confirmed by the length of the MAP messages request queue for bandwidth allocation, shown in Figure 20.

If we compare the results of this scenario with those obtained in the first scenario, we can see how the SLTA connected to the cable bus produces a considerable increase of the traffic in the upstream channel. This causes a greater number of MAP bandwidth allocation messages, higher end-to-end delays in the upstream channel and, thus, an increase in the delay of the audio packets received by the clients.

Figure 19. Delays of DOCSIS frames in channel 0 of the CMTS controller

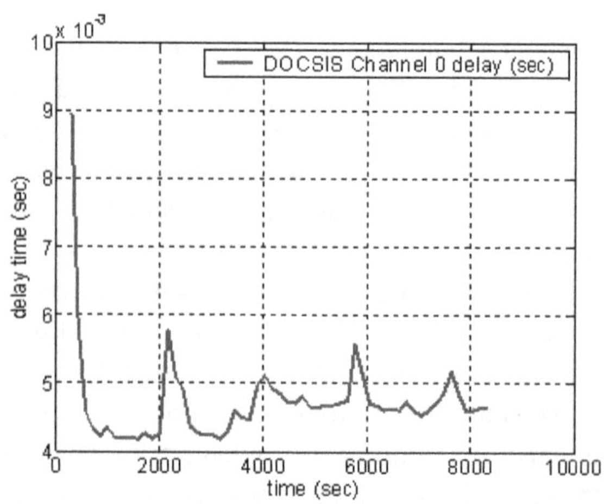

Figure 20. DOCSIS MAP request queue length

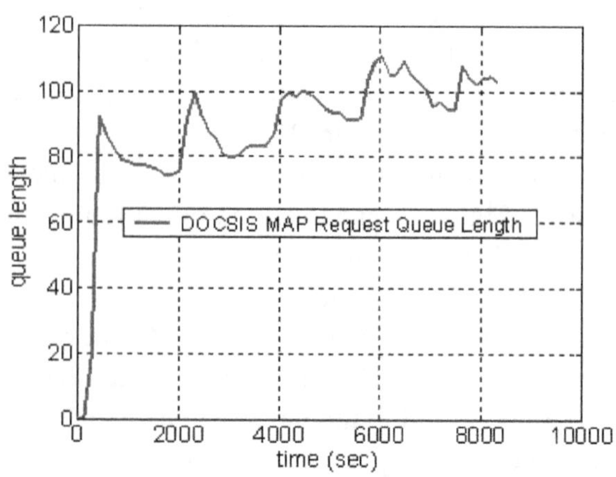

Table 4. Configuration of cable-modems for scenario # 3

Cable-modem SLTA		Cable-modem streaming server	
Quality of service	UGS	Quality of service	UGS
Grant size	1024 bytes	Grant size	1024 bytes
Nominal grant interval	0.1 sec	Nominal grant interval	0.01 sec

Simulation Scenario #3: SLTA and Server in the Cable Network

In this case, both the streaming server and the SLTA are connected to the bus of the cable network. In the cable network, apart from 100 cable-modems installed for each of the clients, two cable-modems have been installed to connect the server and the SLTA.

In this case, all the traffic of the service (server and SLTA) must be delivered using the upstream channel. If the cable-modems of the server and the SLTA are configured with a Best-Effort quality, the clients of the service suffer transmission problems because the CMTS is unable to assign enough transmission slots. To improve this situation and allow the clients to receive the requested multimedia data with a quality similar to that reflected in scenarios 1 and 2, the cable-modems of the streaming server and the SLTA have been configured with a UGS quality in order to obtain periodical transmission slot assignments. The configuration of DOCSIS for both cable-modems is shown in Table 4.

To obtain these configurations we have considered the size of the data packets delivered by the SLTA and the server and the volume of information generated. Thus, the data packets have a maximum size of 640 bytes, so the grant transmission assignments must be higher in the configuration of DOCSIS. To select the time between assignments the volume of traffic generated by the devices was considered, based on the simulations performed in the previous scenarios. With these configurations, all the aforementioned problems reported by the CMTS controller were avoided, i.e., the CMTS was able to assign all the needed transmission slots to all the cable-modems connected to the cable network.

With the mentioned configuration, 4 hours of simulation were executed and the traffic interchanged between all the elements of the service was analyzed. Some of the results are shown in Figure 21 and Figure 22.

The traffic in the output of the SLTA behaves as shown in Figure 8 and the explanation for Figure 21 is the same as that for Figure 9, except that in this new scenario both the SLTA and the server are in the cable network and their traffic is delivered in upstream direction. The throughput of the cable bus is shown in Figure 22, where the effect of the delivery of MAP messages can again be observed.

The delays of all the data packets received by the clients of the radio service have also been studied as in the previous scenarios. Apart from

Figure 21. Traffic between server and CMTS

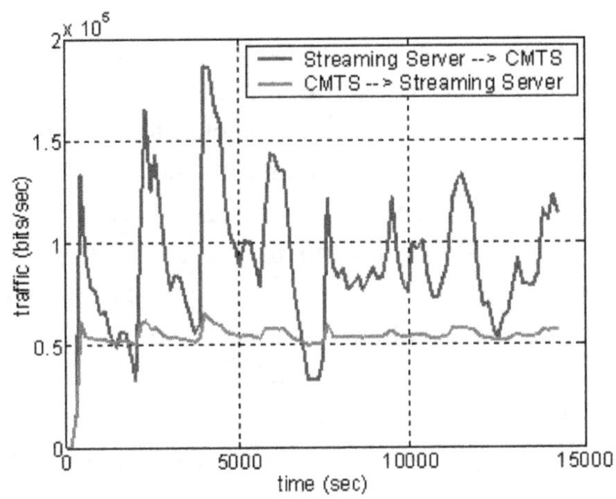

Figure 22. Traffic in the DOCSIS bus

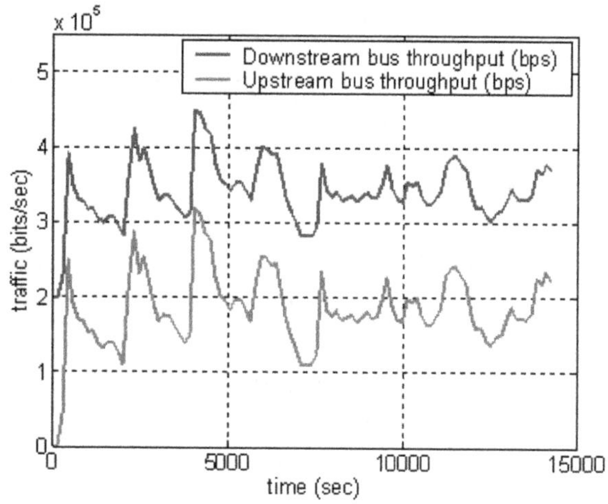

isolated packets with high values the vast majority of the delays are under 1.8 seconds.

If we compare these results with those obtained in the rest of the scenarios, it is interesting to note the increase in the volume of upstream traffic caused by the installation of the SLTA and the server in the cable network. This increase causes a higher number of MAP bandwidth allocation messages and an increase in the delays suffered in the delivery of data packets to the clients of the service.

Simulation Scenario #4: SLTA in the Head-End of the Network and Server in the Cable Network

As in the previous scenario, this is not the most adequate configuration for an Internet radio service as the server is connected to the bus of the cable network and all the traffic it generates needs to be delivered using the upstream channel of the network. This may lead to situations with a high delay as the controller does not have enough bandwidth to guarantee the quality of the transmission in that channel.

For the analysis of this scenario two situations have been analyzed with different configurations of DOCSIS to prevent problems in the allocation of bandwidth. In the first situation, the cable-modem used by the server has been configured with a Best-Effort quality, which has been complemented with the activation of fragmentation and concatenation. This has been done to avoid problems in the delivery of data to this cable-modem as the CMTS is unable to assign the number of slots needed for the transmission. Nevertheless, a lot of errors were produced in the simulations as the CMTS was still unable to assign the number of slots needed to deliver the data using the upstream channel. In the second simulation, the quality of service UGS option of DOCSIS 1.1 (Unsolicited Grant Service) was activated. With this option, the cable-modem of the server is assigned with periodical transmission slots without the need to compete for upstream transmission intervals with the rest of the cable-modems in the cable network. The results are similar to those obtained for the simulation scenario 3, so they will not be commented on here.

Remarks

Analyzing in detail the results obtained in the documented scenarios, we can make the following observations:

- The best results are obtained when both the SLTA and the streaming server are installed in the head-end of the network. All the multimedia traffic is transmitted using the downstream channel through the CMTS controller, which produces the lowest delays, and, hence, a better quality perceived by the users. Even if all the cable-modems are configured with a Best-Effort quality the clients will receive the audio data without any problem.
- It is especially noteworthy the difference in the delays registered for the analyzed scenarios which range from 3 milliseconds in the first scenario to 1.8 seconds in sce-

narios 3 and 4, showing the importance of choosing a proper architecture.

- The output bandwidth in the SLTA remains stable under 60 kbps.
- As expected, the output bandwidth in the server varies with the number of clients connected to the service.
- There are situations where the content provider would prefer to have the SLTA running locally and not in the head-end of the network, so this device may be connected to the cable network by means of a cable-modem. A UGS configuration in the cable-modem used to connect the SLTA would guarantee the transmission of the three available qualities using the upstream channel. The traffic of the server would then flow to the clients of the service through the downstream channel of the cable network without any significant problem.
- The worst scenario for the service is when both the SLTA and the server are connected to the cable network. This may be the case of a company offering the audio contents through a regular connection to the cable-network. In this case, all the traffic of the service flows via the upstream channel to the CMTS controller and from this controller to its destination via the downstream channel. It is then necessary to make the proper reservation of resources to guarantee the quality of service, reducing the transmission possibilities of the rest of the cable-modems in the network.
- After several tests to obtain the most suitable parameters for every service device, it has been proved that an adequate analysis of the interchanged traffic may help to find the best configuration possible for the service. If the devices need to be deployed in the cable network, the best configuration is the following:
    - Cable-modems of the clients of the service: Best-Effort quality
    - Cable-modem of the SLTA: UGS quality with a grant size of 1024 bytes and a nominal grant interval of 0.1 seconds
    - Cable-modem of the streaming server: UGS quality with a grant size of 1024 bytes and a nominal grant interval of 0.01 seconds

CONCLUSION

In this paper, the results of the evaluation of different architectures typical in the deployment of Internet radio services on cable networks have been presented.

In order to perform the evaluations, a model has been designed and implemented on a simulation tool. It accurately reflects the characteristics of a typical Internet radio service and, thus, the results of the evaluations can be applied to any deployment planned on a cable network.

To check the accuracy and guarantee the results that may be obtained in the simulations, the model has been validated by comparing its results with the data gathered in a real service. For this purpose, the log files from a streaming server and the traffic interchanged between every service device have been studied in detail. The collected information has permitted the execution of different analyses and diverse service aspects have been precisely characterized. The model was configured following this characterization and the results of several simulations were compared with the empirical data.

Several architectures have been simulated using the implemented model. They have been selected as the most probable architectures to be used, depending on the preferences of the provider of the contents. The evaluations show the performance of each of these architectures and the effect they have on the underlying network infrastructure. Moreover, for the least efficient architectures, the most suitable parameters to configure the network infrastructure have been selected in order to improve the behavior of the service as much as possible.

This paper may be very useful for service administrators to achieve the optimal configuration for their services in a cable network, making them aware of what will happen depending on the location of each device. Even in the case of a forced selection of an inefficient alternative, the parameters shown for network devices can be negotiated with the network operator in order to achieve the best performance possible. The results can also be useful for service consultants to assess their customers and avoid performance problems that may arise depending on the selected architecture.

As the model of the service has been designed to run independently of the underlying network technology, future work will be undertaken to carry out performance evaluations with different network access technologies, such as xDSL, WiFi, WiMAX or UMTS/HSPA. Also, we plan to carry out emulation experiments with modeled clients performing requests through a simulated network to a real server. Thus, real and simulated traffic will be integrated in order to obtain more accurate performance evaluations of the service.

ACKNOWLEDGMENT

This research has been financed by the network operator Telecable de Asturias S.A.U. and the Spanish National Research Program within the project FUTURMEDIA (TSI2007-60474). It has also received the collaboration of the company Asturia Activa, owner of the digital newspaper Asturies.Com.

REFERENCES

Boyden, S., Mahanti, A., & Williamson, C. (2005). Characterizing the behaviour of real video streams. In *SCS Symposium on Performance Evaluation of Computer and Telecommunication Systems* (pp. 783-791). Philadelphia, PA: SCS.

Cablelabs. (2005). *Data-over-cable service interface specification DOCSIS 2.0.* Radio Frequency Interface Specification CM-SP-RFIv2.0-I08-050408, 8th release, Cablelabs.

Chawathe, Y. (2000). *Scattercast: An architecture for Internet broadcast distribution as an infrastructure service.* Doctoral dissertation, University of California at Berkley.

Cherkasova, L., & Gupta, M. (2004). Analysis of enterprise media server workload: Access patterns, locality, content evolution and rates of change. [TON]. *IEEE/ACM Transactions on Networking, 12*(5), 781–794. doi:10.1109/TNET.2004.836125

Costa, C., Cunha, I., Borges, A., Ramos, C., Rocha, M., Almeida, J., & Ribeiro-Neto, B. (2004). *Analyzing client interactivity in streaming media.* In *13th International Conference on World Wide Web* (pp. 534-543). New York, NY: ACM.

Deshpande, H., Bawa, M., & García-Molina, H. (2001). *Streaming live media over a peer-to-peer network (Tech. Rep.).* Stanford University.

Gopal, D., & Abdulkadir, T. (2006). Self-similarity and Internet performance. *Journal of Research and Practice in Information Technology, 38*(2), 171–180.

Gordijn, J., Yu, E., & Van der Raadt, B. (2006). e-service design using i* and e3 value modeling. *IEEE Software, 23*(3), 26–33. doi:10.1109/MS.2006.71

Hei, H., Liang, C., Liang, J., Liu, Y., & Ross, K. W. (2007). A measurement study of large-scale P2P IPTV system. *IEEE Transactions on Multimedia*, *9*(8), 1672–1687. doi:10.1109/TMM.2007.907451

Jin, S., & Bestavros, A. (2001). GISMO: A generator of Internet streaming media objects and workloads. *ACM SIGMETRICTS Performance Evaluation Review*, *29*(3), 2–10. doi:10.1145/507553.507554

Kuang, T., & Williamson, C. (2002). A measurement study of RealMedia streaming traffic. In *SPIE ITCOM* (pp. 68–79). Boston, MA: SPIE. doi:10.1117/12.473407

Lan, K. C., & Heidemann, J. (2001). *Structural modeling of realaudio traffic* (Tech. Rep. ISI-TR-544). USC/Information Sciences Institute.

Law, A. M., & Kelton, W. D. (2000). *Simulation modeling and analysis* (3rd ed.). Boston, MA: McGraw-Hill Higher Education.

Luna, C. E., Kondi, L. P., & Katsaggelos, A. K. (2003). Maximizing user utility in video streaming applications. *IEEE Transactions on Circuits and Systems for Video Technology*, *13*, 141–148. doi:10.1109/TCSVT.2002.808439

Melendi, D., Pañeda, X. G., Vilas, M., García, R., & García, V. (2007). A methodology for the deployment of live audio and video services. In *2nd International Conference on Signal Processing and Multimedia Applications* (pp. 15-21). Barcelona, Spain: INSTICC Press.

Melendi, D., Vilas, M., García, R., Pañeda, X. G., & García, V. (2006). Characterization of a real Internet radio service. In *32nd EuroMicro Conference on Software Engineering and Advanced Applications* (pp. 356-363). Dubrovnik, Croatia: IEEE.

Mena, A., & Heidemann, J. (2000). An empirical study of real audio traffic. In *19th IEEE International Conference on Computer Communications* (pp. 1001-1010). Tel-Aviv, Israel: IEEE.

Nguyen, T., & Zakhor, A. (2002). Distributed video streaming over Internet. In *Multimedia Computing and Networking* (pp. 186–195). California: SPIE/ACM.

Padmanabhan, V. N., Wang, H. J., Chou, P. A., & Sripanidkulchai, K. (2002). Distributing streaming media content using cooperative networking. In *12th International Workshop on Network and Operating Systems Support for Digital Audio and Video* (pp. 177-186). Miami, FL: ACM.

Pañeda, X. G., Melendi, D., García, R., Vilas, M., & García, V. (2007). A methodology for performance, content analysis, and coniguration of audio/video-on-demand services. *International Journal of Business Data Communications and Networking*, *3*(4), 17–46. doi:10.4018/jbdcn.2007100102

RealNetworks. (2002). *Helix universal server administration guide*. USA: RealNetworks.

Schulzrinne, H., Rao, A., & Lanpier, R. (1998). *Real time streaming protocol (RTSP). RFC2326*. USA: The Internet Society.

Sripanidkulchai, K., Ganjam, A., Maggs, B., & Zhang, H. (2004). The feasibility of supporting large-scale live streaming applications with dynamic application end-points. *ACM SIGCOMM Computer Communication Review*, *34*(4), 107–120. doi:10.1145/1030194.1015480

Sripanidkulchai, K., Maggs, B., & Zhang, H. (2004). An analysis of live streaming workloads on the Internet. In *ACM SIGCOMM Internet Measurement Conference* (pp. 41-54). Sicily, Italy: ACM.

Technologies, O. P. N. E. T. (2009). *OPNET modeler: Making networks and applications perform*. Retrieved December, 2009, from http://www.opnet.com

Van der Merwe, J., Sen, S., & Kalmanek, C. (2002). *Streaming video traffic: Characterization and network impact*. In 7th International Workshop on Web Content Caching and Distribution, Boulder, Colorado.

Van der Raadt, B., Gardin, T., & Yu, E. (2005). Exploring Web services from a business value perspective. In *13th IEEE International Conference on Requirements Engineering* (pp. 53-62). Paris, France: IEEE.

Veloso, E., Almeida, V., Meira, W., Bestavros, A., & Jin, S. (2002). A hierarchical characterization of a live streaming media workload. In *2nd ACM SIGCOMM Internet Measurement Workshop 2002* (pp. 117-130). Marseille, France: ACM.

Xiaofei, L., Jin, H., Liu, T., Ni, L. M., & Deng, D. (2006). AnySee: Peer-to-peer live streaming. In *25th IEEE International Conference on Computer Communications* (pp. 1-10). Barcelona, Spain: IEEE.

Xu, D., Hefeeda, M., Hambrusch, S., & Bhargava, B. (2002). On peer-to-peer media streaming. In *22nd International Conference on Distributed Computing Systems* (pp. 363-371). Vienna, Austria: IEEE.

Yoshimura, Y., & Masugi, M. (2004). A QoS monitoring method for video streaming service based on presentation timeline detection at user clients. In *10th IEEE Asia-Pacific Conference on Communications, vol. 2* (pp. 681-685). Beijing, China: IEEE.

This work was previously published in International Journal of Business Data Communications and Networking, Volume 6, Issue 2, edited by Debashis Saha and Varadharajan Sridhar, pp. 52-74, copyright 2010 by IGI Publishing (an imprint of IGI Global).

Chapter 11
Scalable P2P Video Streaming

Majed Alhaisoni
University of Essex, UK

Mohammed Ghanbari
University of Essex, UK

Antonio Liotta
Technische Universiteit Eindhoven, The Netherlands

ABSTRACT

P2P networks are a technology able to deliver real time and video-on-demand services over IP networks. Layered video coding techniques are being introduced due to their ability to deliver streams at different scales (temporal, spatial and SNR) that solve the heterogeneity problem. This eases transmission in the case of limited bandwidth, as the devices can pick and decode the minimum bit rate base layer. Existing work examines layered video in client-server scenarios. In contrast, this paper analyzes scalable coding H.264/SVC over P2P networks based on an SNR-temporal Codec. Due to the interdependency between the different SVC layers, issues of reliability and quality of experience arise unless proper measures are taken to protect the base layer. The authors explore the effectiveness of a combination of P2P strategies, for example, hybrid P2P architecture, P2P locality, and P2P redundancy, to assess the viability and benefits of scalable video coding over P2P. The resulting performance is compared with a state-of-the-art P2P TV platform.

INTRODUCTION

Streaming of video over Peer-to-Peer (P2P) networks is a potential approach to the ever-growing problem of distributing video over the Internet. In this system, each terminal relays the stream as soon as it becomes available, for the benefit of the other users, acting in fact as a stream-distribution hub for any incoming data. This will significantly reduce the bottlenecks normally generated by the traditional server-based video distribution (Liu et al., 2008). The most striking strength of P2P based streaming over conventional IPTV, is its ability to better adjust to the number and distribution of users over the network. In turn, this leads to higher resource utilization and, more importantly, contributes to a better overall quality of experience. Thus, in contrast to the conventional shared

DOI: 10.4018/978-1-61350-477-2.ch011

media applications, where the network resources (e.g., bandwidth) are shared among the users and more users exhaust the network degrading the quality, here it is the content that is shared. This creates a fascinating dilemma for Internet video. With P2P streaming, as more users want to consume content, the very same users will act as distribution relays, contributing to increasing the overall throughput. When more users buffer the same streaming video content, the distribution process can happen in parallel and redundant video feeds lead to an immediate increase in video quality and, with it, to the overall user's quality of experience.

In this evolution, overlay networks play a vital role. The users' terminals (or peers) interconnect with each other, forming an overlay network where the intercommunications are virtual, transient and dynamic. As peers join or leave the network, the distribution overlay re-adjusts in order to maintain video feeds, despite the churn of users. Because peers (terminals) work as clients and servers simultaneously the stream is relayed efficiently, despite increasing the number of receivers.

On the other hand, different architectures have been proposed in order to create and manage overlay networks, such as Tree-based (Diot et al., 2000) and Mesh-based (Qiu & Srikant, 2004). These approaches have been studied and each approach has its own advantages and disadvantages (Marfia et al., 2007). In this paper, a mesh-based approach is considered. Many hosts can be supported by a P2P multimedia system, possibly in excess of hundreds or even millions, with miscellaneous heterogeneity in bandwidth, capability, storage, network and mobility. It is desirable that, once the stream is injected into the overlay network and starts circulating, different types of devices (i.e., Mobile Phones and PDAs) are able to access and relay the stream.

Another important factor of the media distribution process is the video encoding technique. In a normal media production system, the stream is encoded before the transmission and then decoded and played back at the destination client. Different video coding schemes are available, including those based on single layer and multi layer coding. However, it appears that a P2P strategy well correlates with the principles of scalable video coding (SVC). In this method, video is coded into a scalable bit stream, where users use the portion of the stream that suits their needs best. In a heterogeneous network, where link capacity and terminal capability may vary considerably, more users have the potential to connect. Having more users in the network makes P2P streaming more efficient; but it is also essential to properly manage the heterogeneity of the distribution overlay. Most of conventional P2P systems are based on single-layer video, which makes them limited when it comes to optimizing the distribution process over heterogeneous systems.

To overcome such limitations, in this paper we explore the combination of P2P streaming with scalable video coding. In our system, those peers that can only handle the lowest video quality (i.e., the base layer) will feed each other in the same way as in conventional single-layer P2P streaming. This of course occurs at a much reduced transmission cost, since base layer rates are at a fraction of the single layer stream, albeit the video quality is lower.

On the other hand, peers demanding higher video quality will get the extra data from the enhancement layers of other peers. In this case with a fraction of extra data, those users having sufficient network and computational resources to render video at higher quality will be able to achieve a higher quality through the necessary enhancement layers. As for the base-layer overlay, when more high-capacity users will be present, the enhancement-layer overlay will become more effective too.

For the sake of our experimental-based evaluation, in this article we focus on H.264/SVC (scalable video coding), which has the ability to be decoded at different levels of granularity, hence satisfying our heterogeneity requirement.

We have chosen H.264/SVC for this purpose but, contrary to existing studies, we are scrutinizing this video scheme in combination with hybrid organization of P2P distribution approach. Early results of P2P streaming of SVC coded video were presented in (Alhaisoni et al., 2009). Herein we considerably extend the experimental analysis, incorporating a number of new features. First, we add network locality into the P2P distribution mechanism to improve network utilization and, consequently, increase the quality of the stream delivered to the end user. We do so in order to alleviate a well-known problem of the existing P2P streaming systems, which have a high foot-print on the network arising from sub-optimal stream delivery paths.

Multi-layer (SVC) streaming over P2P is a hard problem since it is essential to make sure that packet loss at the base layer is minimal (which would have detrimental effects on the stream quality). It is therefore important to investigate techniques that minimize the probability of packet loss. With this target in mind, we adopt the techniques of semantic layering (hybrid overlay organization), multiple-streaming (base-layer redundancy) over P2P, and forced handovers (to increase computational load-balancing).

Our results show that adopting the abovementioned techniques with SNR-temporal scalability leads to a sensible reduction in packet loss and end-to-end delay (latency), with a remarkable improvement on network utilization and of video quality (measured via PSNR). Results are compared with Joost-Like, a commercial P2P TV system.

RELATED WORK

Since we are dealing with network quality of service (QoS), P2P locality awareness, and layered coding we first give an overview of different studies that have looked at these topics individually.

Ways to pursue efficiency between overlay and underlay networks have been investigated only recently. Authors in (Liu et al., 2004) propose a technique, where the peers on the overlay are chosen based on their mutual physical proximity, in order to keep traffic as localized as possible. A similar approach is described in (Xu et al., 2003), where they measure the latency distance between the nodes and appropriate Internet servers called landmarks. A rough estimate of awareness among nodes is obtained to cluster them altogether, as in (Banerjee et al., 2002; Zhao et al., 2002).

P2P streaming of object based layered coded video (MPEG4) has been already proposed by (Ahmed & Mushtaq, 2007) where layers are organized in the overlays and the best peer from each layer is selected. This has been based on pinging and tracking a set of peers to select the best peer. Moreover, they organize the peers on a tree-like structure, which creates difficulty in maintaining the peers in case of failures or high churn. In contrast, in our work, we have selected H.264/Scalable Video Codec (SVC) with SNR scalability and the option to output one stream consisting of base layer and enhanced layer and maintain the inter-connections among the peers according to the proposed algorithm. This is combined with hierarchical hybrid organization of peers based on locality and semantic layering (base and enhancement layers). Besides these techniques, multi-stream is introduced into the base layer which forms the lower bit rate against other layers.

Overlay locality is studied also in (Nguyen and Zakhor), where the authors make use of network-layer information (e.g., low latency, low number of hops and high bandwidth). We use though a different distance metric, based on RTT (round trip time) estimations, to prioritize overlay transmissions. Additionally, we use a cluster management algorithm based on locality and semantic layering whereby intercommunicating peers are forced to periodically handover, in order to pursue computational as well as network efficiency (as explained in Alhaisoni et al., 2010).

Hefeeda et al (Hefeeda et al., 2003) have proposed a mechanism for P2P media streaming using Collectcast. Their work was based on downloading from different peers. They compare topology-aware and end-to-end selection based approaches.

The latter approach is also the subject of (Mushtaq & Ahmed, 2006), which employs a simpler version of our RTT approach based on continuous pinging of peers. Similarly, we adopt clustering to limit the signalling overheads associated with this process and prevent bottlenecks. Other studies such as (Hefeeda & Bhargava, 2003), propose relevant methods to serve multiple clients based on utility functions or clustering. On the other hand, a number of works proposed layers selection schemes under unicast and multicast setting (Rejaie & Ortega, 2003). Furthermore, SVC is studied by (Dai et al., 2007) to maximize the throughput over P2P. On the other hand, authors in (Xiao et al., 2008) have studied the scheduling mechanisms in layered video by introducing four objectives that should be achieved by data scheduling for layered streaming: high throughput, high layer delivery ratio, low useless packets ratio and low subscription jitter. Additionally, authors in (Lee & Guan, 2003) proposed a centralized peer-to-peer model for video streaming in order to provide higher perceived quality of service. To evaluate their proposal, they designed a simple layered video coding using 3D-DCT with cascaded vector quantization tables, which are generated using self-organizing tree map. Furthermore, layered video is studied in (Shen et al., 2005) by investigating streaming layered encoded video over peers. Each video is encoded into hierarchical layers which are stored on different peers. Clients request are served by streaming multiple layers of the requested video from different peers. Unequal error protection is provided for different layers according to their importance by varying the number of copies stored for each.

Looking at previous studies, we can say that our main contribution is:

1. To study a new combination of existing techniques (cross-layer optimization, localization, semantic layering, hybrid organization, switching over, forced handovers, redundancy and layered video-streaming).
2. To take the perspective of the network operator, in trying to harmonize overlay and underlay networks.
3. To quantify how the network-aware approach affects on the SVC recital.
4. Introducing effective ways to protect the most important parts of SVC such as base layer.
5. Developing new ways of gauging the packet loss ratio, end-to-end delay, and the network utilization by introducing upper and lower bounds.

PROPOSED APPROACH PRINCIPLES

Video Coding Schemes

Video coding schemes play a vital role in the streaming application as the video is encoded and then transmitted to the end-users. Hence different coding schemes have been developed, including advanced video coding, multiple description coding and layered coding (SVC).

We are interested in a scheme that addresses the requirements of a heterogeneous system at the network as well as terminal and peer level. Thus we focus on scalable video coding (SVC) which allows the delivery of a range of qualities. SVC offers different layers and different qualities with different bit-rates. This gives more chance for different devices under different downlink/uplink throughput to send, receive and decode different layers. This will help in distributing the multimedia content over Mobile 3G and PDAs (Wiegand & Sullivan, 2007; Schwarz et al., 2006).

In SVC, bit-streams are decodable at different rates and spatial quality and temporal resolution can be controlled to achieve a range of quality levels. We are exploring ways in which this flexibility can be best combined with P2P mechanisms.

This paper utilizes Quality (SNR) scalability combined with temporal scalability, as briefly described below:

- **Quality (SNR) scalability:** in this scalability, both layers have the same resolution and frame rate, but the base layer has lower quality and the enhancement layer carries the residual data to improve the quality of the base layer.

In this paper, Quality (SNR) scalability is employed. However, due to the vital dependences between the base and enhancement layers, in some case where the base layer is not received due to the dynamic nature of P2P networks, video can not be decoded easily as enhancement layers are dependent on their on-time arrival.

Hence, combination of SNR and temporal scalabilities not only can give a low bit rate base layer, but due to temporal layers (in the form of B-pictures) reduces the overall bit rate, which is marginally (around 10%) over the single layer. However, in this case due to inter-layer predictions, where some of the base or enhancement layers use their predictions from each other, interdependency between layer data increases. This interdependency can create picture drift. H.264/SVC defines two types of quality scalability, coarse-grain and medium grain scalability. For simplicity we are interested in coarse-grain scalability as shown in Figure 1.

Figure 1. Structure of a group of pictures (GOP) in a SVC encoded stream with coarse-grain SNR and temporal scalability

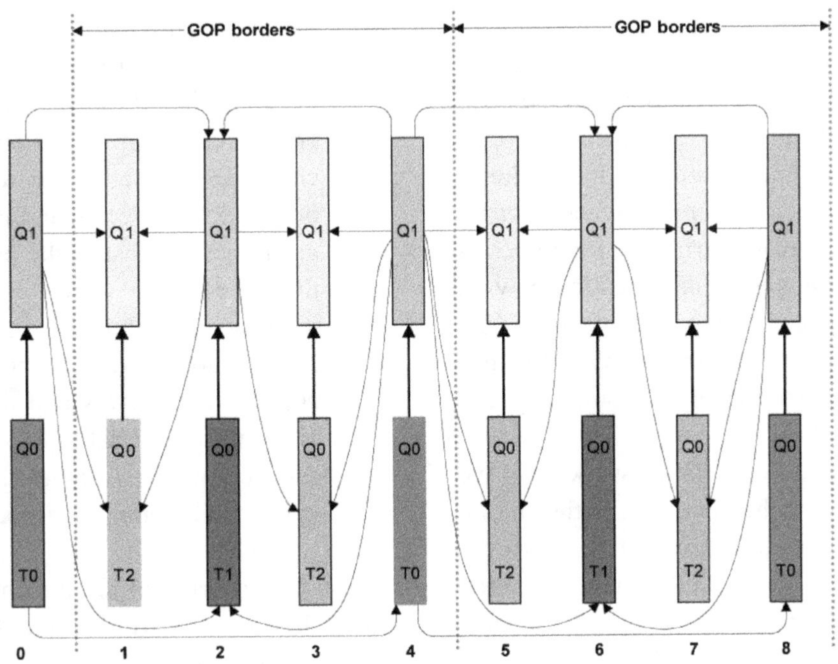

Overlay-Management

In this section, we show how the inter-connections among the peers on the overlay network is handled and managed. In order to initiate and maintain the inter-communication between receivers and senders, we have introduced various strategies where every method plays a key role in the video smooth delivery. These can be classified as follow:

Peers' Organization

In order to maintain the inter-connections, peers will be grouped into different clusters. Clustering may be based on locality or other semantic properties but we focus on the combination of both. Each cluster is accessed via a super-node, similarly to the approach adopted for instance in KaZaA (Kazaa). This eases the possibility of maintaining the intercommunication routes among the peers based on a hierarchical routing approach. In this paper, a hybrid hierarchical organization is proposed to better get use of SVC. Thus, peers on the overlay are grouped based on locality and semantic layering. These two properties can be described as follow:

- **Locality organization:** Network locality can be defined based on the geographical/physical or the mutual distance between the peers. Hence, in order to make clients locally-aware, different distance metrics such as RTT (round trip time), number of hops, and the geographic location may be adopted. In the proposed algorithm, RTT has been used as a metric to reflect the mutual distance (or locality) among the peers.

By monitoring RTT at the network layer we obtain a measure of the quality of paths among the nodes. Bringing this information up to the overlay network, allows a better mapping of the logical overlay (used to choose which peer streams to and from which other peer) to the physical network (or underlay). In this way transmission with cross-layer optimization strategies can be made more efficient.

The issue of building an efficient RTT monitoring tool is the subject of investigation in the network management community and is not the main target of this article. We adopt techniques published studied in (Liotta et al., 2002; Ragusa et al., 2005) and focus herein on the evaluation of P2P-SVC streaming.

- **Semantic layering organization:** In this mechanism, we have classified the peers based on the same quality tier (i.e., base and enhancement layers). Hence, we have identified that the most important layer which is the base layer to be clustered in a certain group and complemented with locality as well. This gives the opportunity to introduce better ways to secure the base layer over the network. Semantic-based peers' organization helps on the delivery of the important parts of the content to the end users. It can be considered as a good choice for scalable video coding and where media contents are distinguished according to priority.
- **Hybrid organization:** In this paper, we have combined locality and semantic layering mechanisms together. Hence, different groups in the overlay are formed based on semantic layering and locality. The first group will consist of the peers which have the base layers which are considered as the most important part in SVC. The second group will entail the peers with the enhancement layers and this also may include the discardable base layers as both can be considered as enhancement layers in this SNR-temporal scalability.

Such clustering offers the chance to allocate the video parts easily as some devices may not able to download and decode all the layers. In this

Figure 2. Hybrid organization of peers based on network locality and semantic layering

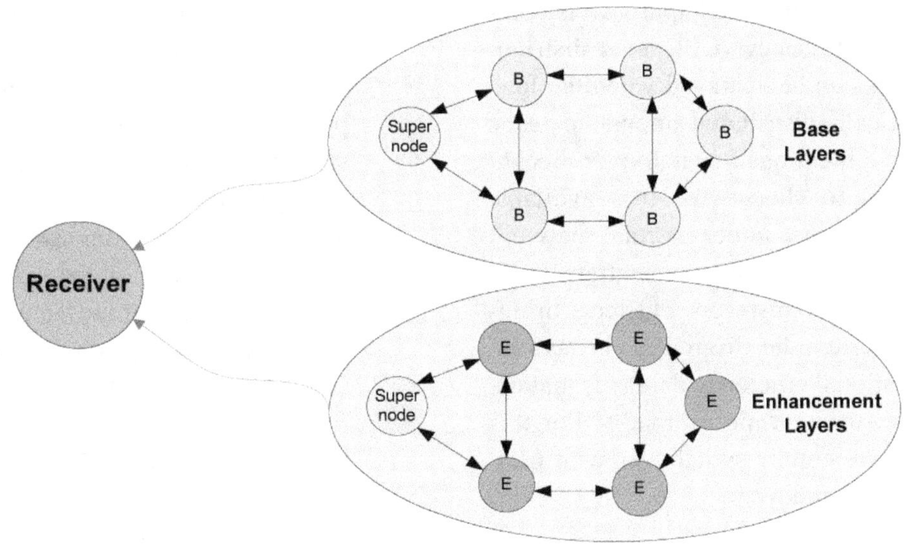

case, searching the content for the base layer which gives the basic quality will be much easier. Other devices with high bandwidth can also download the enhancement layers from other groups. Another viable favourable of this hybrid organization is to use efficient way to secure the most important layer (i.e., base layer). This approach is discussed in the experimental methodology. Figure 2 shows how peers are cluster in a hybrid form:

Load Distribution

In order to maintain the load balancing among the contributor peers, different handover techniques have been embedded into the proposed approach. Two conditions trigger handover among the interconnected peers:

- **Switching over:** Since the network may experience various constraints such as congestion, bottleneck and link failures, the RTT values will be severely affected and may not be reliable. Additionally, these stochastic conditions will drastically affect network locality and degrade quality of service (QoS) parameters such as throughput, packets loss, and end-to-end delay. There is also another important requirement arising directly from the adoption of P2P: peers are not reliable entities and cannot be assumed to be always connected. Nodes may leave and join at unpredictable times. So we must adopt a mechanism which allows receiving peers (in client mode) to maintain a smooth video, although the streaming peers (in server mode) are not constantly available.

One solution to this problem is that any intending client should regularly update the peers' list and re-order them based on the lower RTT values. In our implementation, we keep a ranking list based on RTT distances. Each peer streams to and from 3 others. Hence, when this list changes, we apply a switch-over to the new top 3 peers (those with minimal RTT to the peer under consideration). This approach has been chosen according to previous findings published in (Alhaisoni & Liotta, 2009) where we found that the average of the active peers that usually a node is downloading from is 3 to 4. Therefore, in this model, the maximum number of sender nodes has been set to be three.

- **Enforced Handover:** Another favourable property in the proposed approach is computational efficiency (CPU load distribution). This can be achieved when the load is periodically distributed among the peers. The above mentioned switch-over mechanism aims to choose the best available peers. However, under normal network conditions, this would concentrate communications in just few clusters, bringing the system far from P2P behaviour. Computational efficiency demands that the load is distributed more broadly. For this purpose, we couple switchover with periodic forced handovers.

EXPERIMENTAL METHODOLOGY

In this experiment SVC is combined with different techniques in order to gain sufficient quality over P2P networks. These mechanisms are as follow:

- **Hybrid Clustering of Peers**: This is based on the combination of locality-awareness and semantic layering (hybrid hierarchical organization). Hence, in this experiment peers who have the content of base layers will form a clique, and other peers who have the enhancement layers and close to each other will be grouped together.
- **Redundancy:** This technique can lead to positive advantages such as protecting the stream against packet loss. However, applying redundancy indiscriminately would lead to network congestion. In order to gain the benefits of redundancy (data integrity), limiting at the same time its negative effects (congestion), this technique is applied only for the base layer (low bit rate), which is instrumental for SVC delivery. To further reduce network load and increase load-balancing, redundancy is combined with network locality, semantic layering and periodical enforced handover. As mentioned, the number of redundant peers is set to three, but this may not be equally applied to both layers. Due to their importance, base layers are protected with a higher level of redundancy than the enhancement layers. Figure 3 shows a simple example, where 3 peers send redundant base-layer packets, whilst the enhancement packets are sent without redundancy.

Figure 3. Proposed architecture where the base layer (B) is transmitted from 3 peers {S1...S3} and the enhancement layer (E) is only from 1 peer {S4}

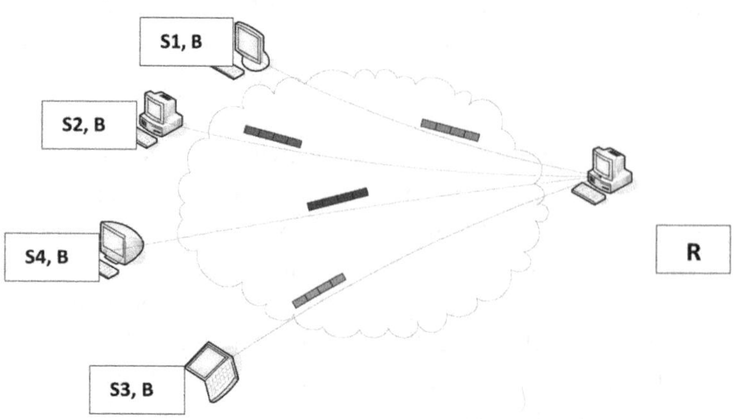

In this study the number of redundant base-layer packets streams varies from 1 to 3. For the base-layer stream, sources {S1…S3} are chosen based on peers' organization which is combined of locality and semantic layering. Additionally, the three sources are continuously forced to handover, choosing new sources from a pot of available sources in the base-layer cluster of the overlay. These are prioritized based on mutual inter-peer distances to ensure that traffic is kept as local as possible. Forced handovers ensure that the important feature of computational load-balancing is maintained. On the other hand, enhancement layers are transmitted only from 1 source (S4) to avoid excessive loading onto the network. It is to be noted that, contrary to the base-layer packets, enhancement-layer packets will marginally enhance the quality; but if those packets are lost quality degradation is also limited.

We have tested the quality of layered P2P streaming fewer than two scenarios of locality-awareness and randomized interconnections. The aim of this experiment is to examine the behaviour of layered coding (SNR combined with temporal) over P2P network and introduce efficient ways to maintain the quality the base layers. This technique is accompanied with the aim of introducing locality awareness, semantic layering and computational load balancing. Experiments and results are given in the following section.

PERFORMANCE EVALUATION

Simulation Environment

The proposed approach was implemented and tested on the ns-2 network simulator (http://isi.edu/nsnam/ns/). A sample of the used topology is shown in Figure 4. Although the used topology is fixed in the simulation, the two scenarios (*locality-aware and randomized*) are treated in a different way. So, for the randomized scenario, senders and receivers are chosen randomly for every run. On the locality-aware scenario, the senders are chosen based on locality and the receivers are selected randomly for every run. This gives the advantage of testing the localized scenario under different conditions over the used topology.

Moreover, various parameters were set on the used topology. First of all, each link has a band-

Figure 4. Simulation topology

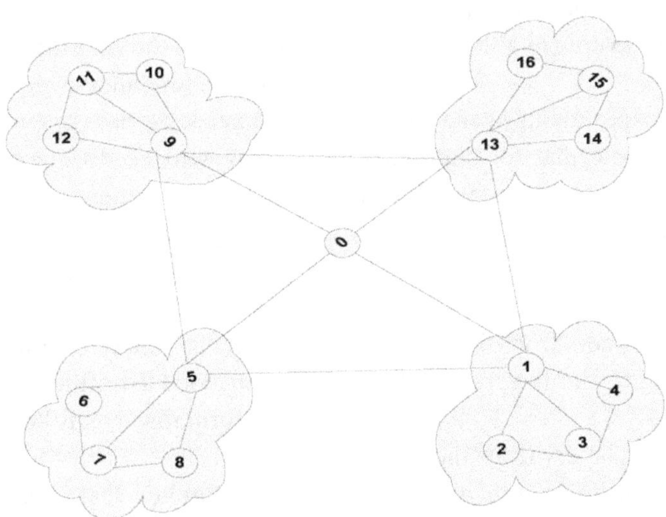

width of 2 Mbps with equal length (delay). However, the actual delay will be according to the nodes distance of each other; so, all the participants' peers have the same characteristics. IP as the network protocol and UDP as the transport protocol have been chosen. For simulation of video traffic, H.264/SVC has been adopted. Hence, the "Paris" sequence of CIF resolution with 4:2:0 format, coded by 30fps as Quality (SNR) with temporal scalability producing two layers, base and enhancement combined with temporal layers for each layer. The packets were sent from 3 peers for the base (redundancy) and 1 peer for the enhancement from various peers to the receiver to and are assembled on-the-fly by the decoder. This mimics the mesh-based approach on P2P TV streaming.

Secondly, in order to overload the network, it was vital to set the CBR background traffic to vary the network load and enable us to study the locality-aware approach under different loading conditions. The CBR traffic was setup from different sources to different destinations, with a 512 byte packet size. This background traffic operates during the whole duration of the simulations. This is added to the stream video on the running simulation.

For statistical significance, the locality-aware and randomized approaches were simulated independently and repeated 50 times. The presented results correspond to the average values of these simulations.

We randomized our experiments over a family of 40-node topologies (note that for simplicity Figure 1 represents a simpler topology). From the receiver side, different tasks will be carried out as follow:

1. Selection of sender peers from the overlay network according to the lowered offered RTT.
2. Selection of desired layers from the overlay network according to the available bandwidth.
3. Switching and hand over mechanism.
4. Checking the status of the peers periodically.

Simulation Scenarios

- **Locality-Awareness:** in this scenario, redundant stream packets are streamed from 1 up to 3 sources for the base layer, whereby every source is sending the same packets, to introduce redundancy. On the other hand, the enhancement layer is sent only by 1 source to avoid overloading the network. This technique is combined with hybrid organization and computational load balancing (periodical handover).
- **Randomized:** This scenario has the same characteristics of the locality-awareness. The difference between the two is that locality-awareness has a hybrid organization and load distribution mechanism, whereas the randomized is purely based on random connections and switchovers among peers (this approach mimics a popular Video-On-Demand application known as Joost, used here as benchmarking). Both scenarios are applied to the same network and loading conditions.

Evaluation Metrics

Owing to the introduced characteristics, specifically redundancy over the base layer, calculating packet loss, network utilization and end-to-end delay will be different from the common QoS definition for measuring the quality of communication system. In P2P networking generally and in this paper particularly, due to redundancy, where not every lost base packet can degrade the QoS but it has side effect on the network, new metrics are needed to define QoS for P2P systems. These are formalised as follow:

- **Packet loss ratio:** usually is defined as the ratio of the dropped over the transmitted data packets. It gives an account of efficiency and of the ability of the network to discover routes. However, in P2P communication, a new way of calculating the packet loss ratio needs to be defined to study the particular issues relating to redundancy. In our case a base packet is actually transmitted by several sources and is considered lost only if it is never received through any of the streams. This is formalized as follow:

Pi generic base packet (i) sent by all source nodes

d_j sending or source nodes

$$X_{ij} = \begin{cases} 1 & if\ P_i\ sent\,by\, d_j\ is\ lost \\ 0 & P_i\ \ is\ received \end{cases}$$

The decision as to whether a packet is lost or not will be according to the following Cartesian product:

$$PL_i = \prod_{j=1}^{d} x_{ij}$$

Therefore, if *P* is the total number of packets required to reconstruct a given stream, the packet loss ratio will be:

$$PL = \frac{\sum_{i=1}^{P} PL_i}{P} \tag{1}$$

- **Average end-to-end delay:** is average time span between transmission and arrival of data packets. In our case where the base layer packets are transmitted by 3 sources, delay of each received packet is the minimum delay among all the packets of the same type sent by all senders. This includes all possible delays introduced by the intermediate nodes for processing and querying of data. End-to-end delay has a detrimental effect on real-time IPTV.
- **Network utilization:** In point-to-point communication (i.e., Client-Server), it is common that network utilization is defined as the ratio of received packets over the sent packets, but in P2P paradigm, when some of the sent or received packets are not useful, this definition is not helpful. Also, since lost packets from any source may not have any ill effect on the received quality as long as these packets can be received from other peers. Thus, in the case of receiving the base layer packets from multi-source (redundancy), one can define an upper and lower bound to the network utilization, where the actual utilization stands somewhere in between.
- **Upper bound:** is defined as the ratio of received useful packets (base) over the total number of received packets (base) from all the sending peers. That is:

$$Ub = \frac{P - PL}{\sum_{i=o}^{n} R_i} \tag{2}$$

Where, Ub represents the upper bound utilization, P is the required number of base packets to reconstruct the video, n is the number of sending peers. Ri is the total number of received base packets from the ith peers and PL is the lost packets

- **Lower bound:** is defined as the ratio of received useful base packets over the total number of sent base packets from all the peers. That is:

$$Lb = \frac{P - PL}{\sum_{i=o}^{n} S_i} \tag{3}$$

Where, Lb represents the lower bound utilization, P is the number of required packets to reconstruct the video, n is the number of sending peers and Si is total sent packets by the ith peer and PL is the lost packets.

The difference between the upper and lower bounds is in fact due to some stray packets of the senders that are lost in the network. Although these stray packets may not be useful to the receiver, as the duplicate packets from the other peers will replace them, nevertheless their presence in the network can congest the network. Thus the difference between the lower and upper utilization is an indication of the side effect that multi-stream may have on the network. The lower is the difference; the better is the performance of the P2P stream. This difference, Nu is thus:

$$\text{Nu} = \frac{P - PL}{\sum_{i=o}^{n} R_i} - \frac{P - PL}{\sum_{i=o}^{n} S_i} \tag{4}$$

- **Peak Signal to Noise Ratio:** PSNR is an objective quality measure of the received video, taken as the user QoE. It is defined as the logarithm of the peak signal power over the mean squared difference (E^2) between the original and the received captured video. This is formulized as in equation (5) and Figure 5 shows how this is obtained through the simulation architecture.

$$PSNR = 10 \log \frac{255^2}{E^2} \tag{5}$$

SIMULATION RESULTS

Figure 6 shows the packet loss ratio for locality-aware scenario and the randomized scenario. Packet loss ratio in H.264/SVC can be considered as the most effective QoS factor as the two layers, base and enhancement, are interdependent. Particularly, losing the base layer in this SNR severely affects decoding ability of the bit stream and hence the decoded video quality. However, it is noticeable that the proposed scenario (locality-awareness) combined with sending the base layers by 3 senders are securing the base layer over the network. By disparity, the randomized connection of Joost-like approach cannot bring packet loss down to zero. In this case, even with sending the base layer by 3 senders (more redundancy), the

Figure 5. PSNR calculation

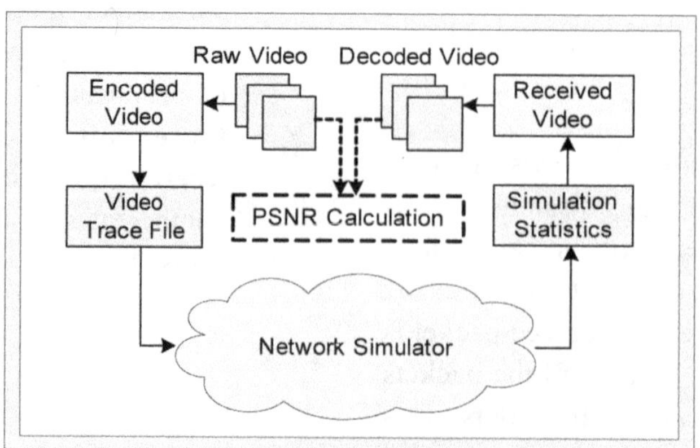

Figure 6. Packet loss ratio of all layers where base layer is sent by 3 peers and enhancement layer is sent by 1 sender

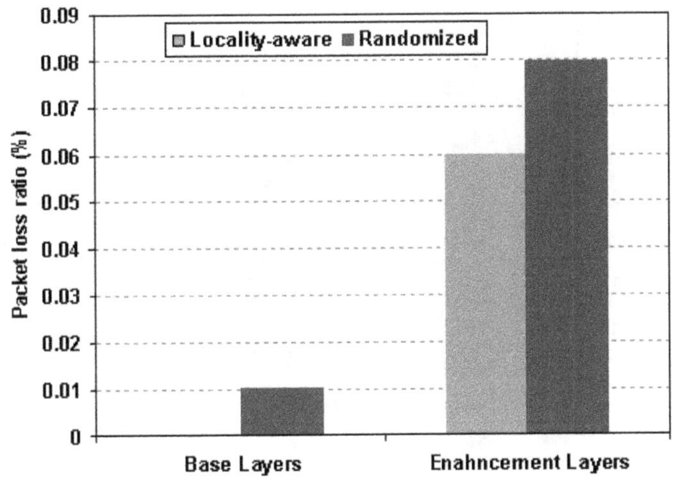

senders themselves create additional congestion such as bottleneck as they are selected randomly, so packet dropping increases over the base and enhancement as shown in Figure 6. This confirms that overlay and underlay networks harmonisation leads to improvement over the QoS and QoE factors. Furthermore, combining that with locality-awareness and other handover techniques has made the redundancy ill effects on the network to become only marginal. This is more clearly visible when we look at the network utilisation factor (following sections).

Looking now at the packet loss of the enhancement layers, it is still apparent that the classification of candidate peers based on network locality and semantic layering has given further advantages even without redundancy. Consequently, the packets loss in the locality-aware approach is

Figure 7. End-to-End Delay of all layers where base layer is sent by 3 peers and enhancement layer is sent by 1 sender

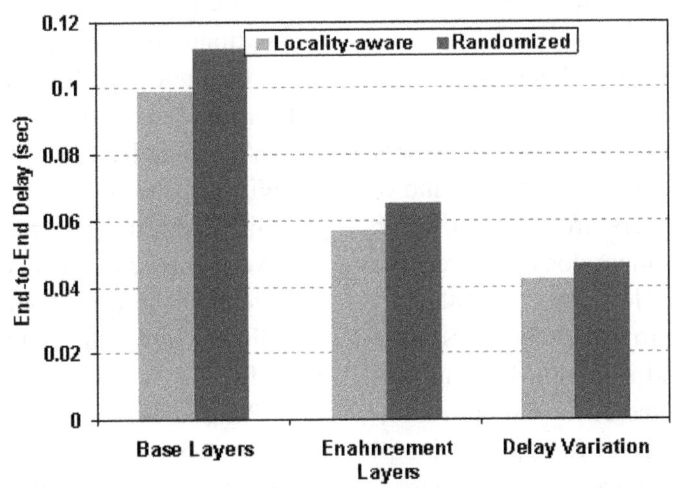

Figure 8. Network Utilization of the base layers where base layer is sent by 3 peers

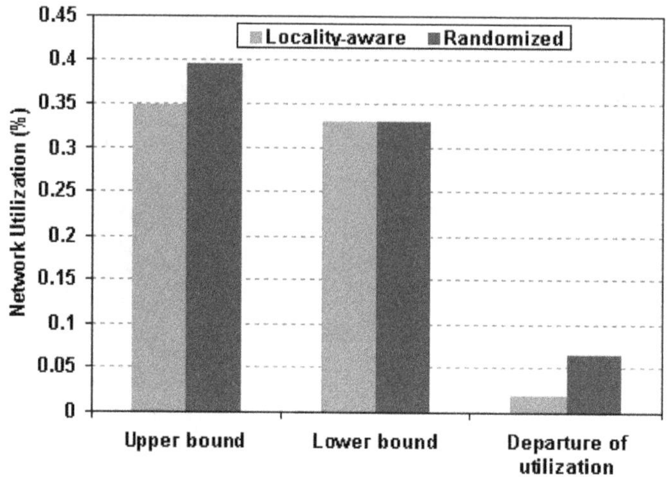

smaller than that of the randomised scenario and is not affected by the loss-free base layer. In contrast, the randomized approach has higher loss rate in both base and enhancement layers than the locality-aware scenario.

Delay is another important quality of experience parameter. A hybrid organization does have a significant, positive effect on communication latency. More specifically, when we looked at the variation of E2E delay, i.e., jitter (Figure 7); it can be noticed that locality-awareness combined with semantic layering (hybrid organization) is even more beneficial. It is important to note that, the end-to-end delay under localized connection has the least value at the redundancy level of 3 senders of base layer. Considering that packet loss rate is almost eradicated with just about 3 redundant senders (Figure 6).

Additionally, enhancement layers are sent only by 1 peer in both scenarios without any kind of redundancy as these layers are to enhance the quality and any drop of them does not affect the bit stream significantly. However, it is still obvious that the classification of candidate peers based on network locality and semantic layering has given positive advantages even without redundancy. Hence, the packets end-to-end delay is still negligible in the locality-aware approach. Moreover, delay variation between the layers in H.264/SVC is very important for decoding purposes as both layers need to be synchronized for proper decoding. However, it is visible that locality-awareness has the least value of variation.

In addition to the above discussed parameters, redundancy is known to have ill effects on network congestion; hence, the network utilization (upper and lower bounds) are defined and measured for both scenarios as base layer is transmitted by 3 peers and will introduce a redundancy to the network.

First of all, the upper network utilization is shown in Figure 8. The localized method shows lower utilization than the randomized method. As mentioned before this parameter on its own may not be a good quality indicator, but its deviation from the lower bound is a better indicator.

Additionally, the lower bound is shown in Figure 8; the localized method shows a better utilization than the randomized method. The lower bound can show actual network utilization better than the upper bound, since the lower bound indicates how much of the received packets are useful.

Thus, the best indication of network utilization is the departure of its higher (upper) and lower bounds, as shown in Figure 8. This is because the larger difference between them can indicate that some of the sent packets are lost in the network, congesting it at the cost of receiving packets from peers to improve video quality. Hence, the smaller the difference between upper and lower bounds of network utilization, the better is the performance, which is in favour of localized method, as shown in Figure 8.

Finally, the objective quality of the decoded video under locality-awareness and randomized are compared. The streams were coded and decoded with an H.264/SVC encoder/decoder of type JSVM15. Figure 9 shows the objective video quality as measured by PSNR. It is obvious that redundancy has played a vital role over the base layer as shown in Figure 9. This has insured that the most important parts of the video are delivered smoothly. The locality-aware approach does not show any packet loss over the base layer, so it maintains a constant level of QoE even with 3 injected redundant streams. Noticeably, the localized hybrid approach improves PSNR steadily up to a redundancy level of 3, where the quality reaches its maximum theoretically achievable value (base layer packet loss is zero at that point). Additionally, our hybrid approach is even improving the quality by the enhancement layers without any redundancy which adjudicate the benefit of the hybrid organization of peers. On the other hand, it was not feasible to decode all the frames of the randomized scenario due to the occurred lost over the base layers where other layers are getting their prediction from it. Hence, Figure 9 shows that the decoding has collapsed after decoding set of frames.

DISCUSSION AND CONCLUSION

In this paper, layered video was shown to be an efficient way to transport video clips over P2P networks. We have extended our P2P overlay management algorithm in order to make use of another P2P feature, namely redundancy, to protect the base layer. We have introduced a hybrid hierarchical organization of the peers based on locality and semantic layering. Accordingly, redundancy has been introduced to protect the base layer and ensure a smooth video delivery. This approach combines clustering with peer-channel prioritization, which

Figure 9. PSNR for locality-awareness compared to the randomized interconnections

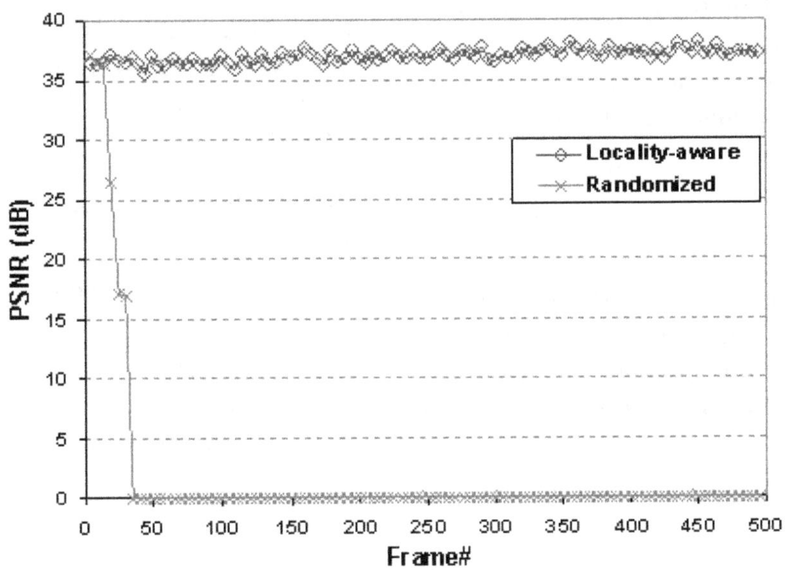

can be considered as a self-optimization approach based on cross-layer information. In fact the RTT values are made available to the overlay from the underlay, an approach that is in contrast with the classic concept of overlay isolation from underlay. These techniques were introduced with SNR-temporal. In this way a better trade-off solution is provided whereby the overheads associated with redundant streaming are outweighed by the benefits in terms of quality of experience. Results had shown improvements over various parameters including packet loss, end-to-end delay, network utilization and, ultimately, stream quality (PSNR).

The key strategy was to study ways to simultaneously target network and computational load-balancing. So it was essential to combine techniques aimed at computational efficiency (namely, periodic forced handover among inter-communicating peers) with strategies for network efficiency (specifically, we used RTT values to minimize communication costs). This study shows that there is a clear potential in P2P streaming but only if proper trade-off solutions are found and fine-tuned via in-depth studies. Our contribution was to illustrate a methodology to carry out a trade-off analysis of a non-trivial combination of factors and mechanisms. Just as any other complex systems, P2P streaming, particularly one based on layered coding, requires a formal calibration process which we have started investigating. Self-calibration and self-optimization of such complex P2P systems is in our immediate research plans.

REFERENCES

Ahmed, T., & Mushtaq, M. (2007). P2P Object-based adaptivE Multimedia Streaming (POEMS). *Journal of Network and Systems Management*, *15*(3), 289–310. doi:10.1007/s10922-007-9068-7

Alhaisoni, M., et al. (2009). Streaming Layered Video over P2P Networks. In *Proceedings of the 6th International Conference on Advances in Mobile Computing and Multimedia (MOMM2009)*, Kuala lumpurm, Malaysia. Washington, DC: IEEE.

Alhaisoni, M. (2010). Resource-awareness and trade-off optimisation in P2P video streaming. *International Journal of Advanced Media and Communication*, *4*(1). doi:10.1504/IJAMC.2010.030006

Alhaisoni, M., & Liotta, A. (2009). Characterization of signaling and traffic in Joost. *Peer-to-Peer Networking and Applications*, *2*(1), 75–83. doi:10.1007/s12083-008-0015-5

Banerjee, S. (2002). *Scalable application layer multicast*. New York: ACM.

Dai, L. (2007). *Maximizing Throughput in Layered Peer-to-Peer Streaming*. New York: ICC.

Diot, C. (2000). Deployment issues for the IP multicast service and architecture. *IEEE Network*, *14*(1), 78–88. doi:10.1109/65.819174

Hefeeda, M. (2003). *PROMISE: peer-to-peer media streaming using CollectCast*. New York: ACM.

Hefeeda, M., & Bhargava, B. (2003). On-Demand Media Streaming Over the Internet. In *Proceedings of the Ninth IEEE Workshop on Future Trends of Distributed Computing Systems*. Washington, DC: IEEE Computer Society.

Kazaa. (n.d.). *Home Page Kazaa*. Retrieved from www.kazaa.com

Lee, I., & Guan, L. (2003). Centralized peer-to-peer streaming with layered video. In *Proceedings of the 2003 International Conference on Multimedia and Expo* (Vol. 2, pp. 513-516). Washington, DC: IEEE Computer Society.

Liotta, A. (2002). Exploiting agent mobility for large-scale network monitoring. *IEEE Network, 16*(3), 7–15. doi:10.1109/MNET.2002.1002994

Liu, Y., et al. (2004). BLocation-aware topology matching in P2Psystems. In *Proceedings of INFOCOM.*

Liu, Y. (2008). A survey on peer-to-peer video streaming systems. *Peer-to-Peer Networking and Applications, 1*(1), 18–28. doi:10.1007/s12083-007-0006-y

Marfia, G., et al. (2007). P2P Streaming Systems: A Survey and Experiments. *ST Journal of Research,* 1-4.

Mushtaq, M., & Ahmed, T. (2006). *Adaptive Packet Video Streaming Over* (p. 2P). Networks Using Active Measurements.

Nguyen, T., & Zakhor, A. (n.d.). Distributed video streaming over Internet. *Citeseer.*

Qiu, D., & Srikant, R. (2004). Modeling and performance analysis of BitTorrent-like peer-to-peer networks. *ACM SIGCOMM Computer Communication Review, 34*(4), 367–378. doi:10.1145/1030194.1015508

Ragusa, C. (2005). An adaptive clustering approach for the management of dynamic systems. *IEEE Journal on Selected Areas in Communications, 23*(12), 2223–2235. doi:10.1109/JSAC.2005.857203

Rejaie, R., & Ortega, A. (2003). PALS: peer-to-peer adaptive layered streaming. In *Proceedings of the 13th international workshop on Network and operating systems support for digital audio and video,* Monterey, CA (pp. 153-161). New York: ACM.

Schwarz, H., et al. (2006, October). Overview of the scalable H. 264/MPEG4-AVC extension. In *Proceedings of the Conference on Image Processing,* Atlanta. Citeseer.

Shen, Y., et al. (2005). Streaming layered encoded video using peers. In *Proceedings of the ICME.*

Wiegand, T., & Sullivan, G. (2007). The H. 264/AVC video coding standard. *IEEE Signal Processing Magazine, 24*(2), 148–153. doi:10.1109/MSP.2007.323282

Xiao, X., et al. (2008). On optimal scheduling for layered video streaming in heterogeneous peer-to-peer networks. In *Proceedings of the 16th ACM international conference on Multimedia,* Vancouver, British Columbia, Canada (pp. 785-788). New York: ACM.

Xu, Z., et al. (2003). Building topology-aware overlays using global soft-state. In *Proceedings of ICDCS.* Citeseer.

Zhao, B., et al. (2002). Locality-aware mechanisms for large-scale networks. In *Proceedings of the Workshop Future Directions Distrib. Comput. (FuDiCo).* Citeseer.

This work was previously published in International Journal of Business Data Communications and Networking, Volume 6, Issue 3, edited by Debashis Saha and Varadharajan Sridhar, pp. 49-65, copyright 2010 by IGI Publishing (an imprint of IGI Global).

Chapter 12
Interoperable IPv6 Sensor Networking over PLC and RF Media

Cedric Chauvenet
INSA CITI Lab,Watteco Inc. and INRIA, France

Bernard Tourancheau
INSA CITI Lab, Universite Lyon1 and INRIA, France

Denis Genon-Catalot
LCI EA Grenoble INP-UPMF, France

Pierre-Emmanuel Goudet
Watteco Inc., France

Mathieu Pouillot
Watteco Inc., France

ABSTRACT

The evolution of technology has made the connection of all types of devices to IP networks possible. In this paper, the authors investigate the possible usage of IPv6 in sensor networks connected through the Power Line Communication (PLC) non-wireless medium and demonstrate possible interoperability. This work is based on the adaptation of the IEEE 802.15.4 standard protocol constrained by the low-power, lossy and low data-rate context of PLC transceiver that uses pulse modulation. The aim is to provide interoperability features with other media using a robust and reliable communication stack. The target application of such results ranges from smart metering and environment monitoring to home control and urban area energy efficiency applications. This paper proposes the first adaptation of the IEEE 802.15.4 standard commons for the PLC medium. Following this standard interface, the authors demonstrate data communication on PLC with low power energy requirement using the pulse PLC physical layer. This paper also presents an initial implementation of the Routing Protocol for Low power and lossy networks (RPL) setup proposed by the IETF working group. In this context, the authors demonstrate interoperability in a testbed between PLC and Wireless Sensor Networks (WSN).

DOI: 10.4018/978-1-61350-477-2.ch012

1. INTRODUCTION

The explosion of the connectivity that rose between 134% and 1,648% from 2000 to 2009 depending on the world region considered.

Since the emergence of the "Internet of things", the meaning of the word connectivity entered yet another dimension.

In this context there is definitely a need for open protocols, to enable the cohabitation between them and create a real interconnectivity through all these devices. We truly believe in the use of open protocols compared to proprietary protocols, because it provides easy interfacing facility between devices. The assumption of a proprietary approach sounds incompatible with the interoperability aim.

We assume that a single medium technology cannot fulfill every requirement, and that there will always be a need for different technologies together to cover the needs that will appear in the future.

With these assumptions, we present a reliable and robust interoperability in the low power and lossy networks (LLN) context. In particular we focus on the low rate wireless personal area networks (LR-WPAN), for which the IEEE 802.15.4 standard (IEEE, 2006) has been designed. The vast majority of these devices are wireless and we bring in the following an easy interoperability feature with another medium using Power Line Communication (PLC) medium.

PLC uses any electrical wires as a network medium and is therefore known as a "no-new wire technology". Indeed, all the electrical networks can be employed as data networks. Electrical power is transmitted over high voltage transmission lines, distributed over medium voltage, and used inside buildings at lower voltages. PLC can be applied at each stage.

Relying on PLC technology, the solution we present doesn't require any further infrastructures, see Figure 1. Furthermore, the electrical network is actually the most widely deployed network. We aim to use this existing medium in collaboration with classic RF solutions, thanks to the use of open standards such as IPv6 and get rid of the proprietary protocols on PLC that restrain its usage. Our goal is to make every node of the Internet of things connected in an interoperable way for it to be accessible from any Internet access point.

Figure 1. Applications targeted by PLC

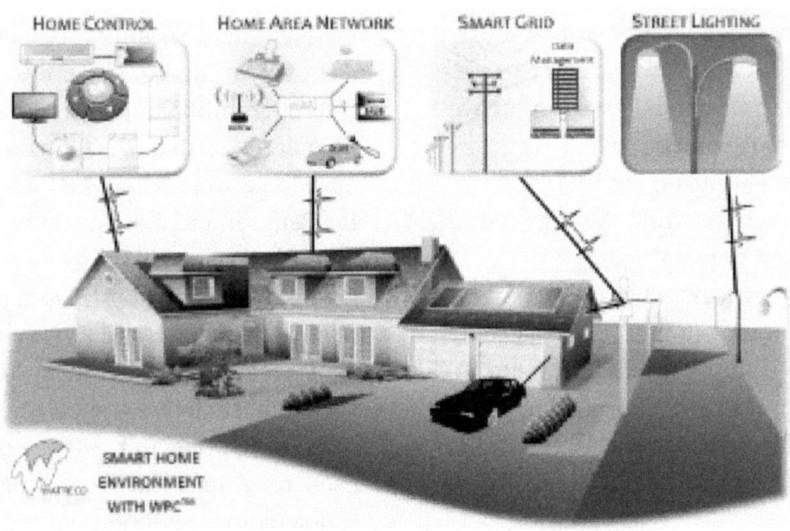

The paper is organized as follows: section 2 defines the context and brings a short presentation of the PLC transceiver of our testbed. It focuses on the main differences between RF and PLC from a MAC layer point of view, this justifies the need for an adaptation of the IEEE 802.15.4 standard over PLC. Section 3 explains the adaptation of this standard over PLC with the transceiver presented in the second section. Sections 4 and 5 describe respectively the network and the routing part of the communication stack. Section 6 shows an heterogeneous experiment of the 802.15.4 adaptation over PLC with RF nodes. Section 7 focuses on a interoperability networking testbed. Section 8 concludes on our work and presents the future work provided by such an implementation. Many applications can be addressed with this solution as shown in Figure 1.

2. RELATED WORK

In order to achieve IPv6 sensor networking, the IETF has defined the 6LoWPAN adaptation layer that includes a compression mechanism of the IPv6 header. 6LoWPAN provides header compression to reduce transmission overhead, fragmentation to support the IPv6 minimum MTU requirement, and support for layer-two forwarding to deliver IPv6 datagram over multiple hops. Although these developments were made for RF devices, most of the work done by 6LoWPAN for header compression for instance, would actually work quite well over other constrained media. Moreover, large development efforts based on the IEEE 802.15.4 standard in the sensor networks field have been made in the wireless domain.

However, like every medium, wireless communications have some limitations. In particular, the radio range depends on the environment, and some elements (such as grounded metal) make the RF transmission collapse (Bertoni, 1999). Reflections and absorptions may cause a reliable RF link become unreliable for a period of time and then being reusable again, creating lossy networks.

The grid is a good way to carry data through an entire building or town without being affected by obstacles such as walls, floors or even the weather, because there is always a physical link between each point of the network. From this observation, it seems meaningful to use PLC in sensor networks. Many PLC communication protocols exist, almost as many as manufacturers. Unlike PLC devices which aim to provide a high data rate (up to 200 Mb/s for Homeplug AV standard (Powerline Alliance HomePlug, n.d.)), the pulse transceiver that we studied works in the low data rate domain (less than 250kb/s). This transceiver aims to keep its power consumption, size and cost as low as possible. These features are rather similar to LR-WPAN.

Some PLC solutions provide rather high data rate such as Homeplug or G.hn, with the cost of high power consumption, and are targeting other applications than ours. Other solutions like X10 or Yitran have low data rates, but the size of the transceiver is large, and their power consumption is still high.

These solutions cannot keep power consumption, size and cost very low. Moreover, our approach is to use open and standardized protocols widely developed in WSN, enabling easy interoperability with RF. Some work has been made to enable interoperability between RF and PLC. All of them demonstrate the interest of such an interoperability, but they do not provide the IPv6 convergence that we focus on. An example of simple network mixing RF and CPL is presented by Pengfei and Jiakun (2009), enabling PLC to relay data sent with RF. The paper from Ha et al. (2007) insists on the strength of a heterogeneous network using RF and PLC with the design of the Cs-Layer convergence layer. Yamada et al. (2007) shows that a delivery success of 70% on PLC and 82% on RF rise up to 90% when both media are used together. The work from Yang and Geng (2008) compares three communication methods to demonstrate that the combination of PLC and RF increases significantly the communication quality. RF and CPL used together are an important issue but it needs a gateway that can be complicated because there is no narrow waist between these media. We want to tackle this problem and enable efficient collaboration with a minimal overload.

3. CONTEXT

The IEEE 802.15.4 standard has been originally written for RF devices, so it could be unusual to have a PLC device using this standard to communicate. To clarify this, a return to the standard rationale is needed: The goal of the IEEE 802.15.4 standard is to provide a common language through LR-WPAN. Unlike wireless local area networks (WLANs), WPANs involve little or no infrastructure. This feature allows small, power-efficient, inexpensive solutions to be implemented for a wide range of devices. A LR-WPAN is a simple, low-cost communication network allowing wireless connectivity applications with limited power and relaxed throughput requirements. The main objectives of an LR-WPAN are ease of installation, reliable datatransfer, short-range operation, extremely low cost, and a reasonable battery lifespan, while maintaining a simple and flexible protocol. IEEE 802.15.4 standard defines the protocol and compatible interconnections for data communication devices using low-data-rate, low-power, and low-complexity for short-range RF transmissions in a WPAN.

The power consumption of these devices has to be low, because they often run on battery or must not create an overload of power consumption either for the appliances where they are embedded on the grid. These assumptions have a big impact on the communication system design, because every frame sent has a cost that has to be considered.

Moreover, typical applications such as smart metering, home control or home area networks applications do not require a big throughput. A few 10kb/s is enough.

Our PLC transceiver is a low data-rate, low power and low cost communication technology, and many LR-WPAN features previously described correspond to our PLC environment. The main difference is the medium employed, which is used to communicate and either to power the node.

PLC nodes are therefore not constrained by the availability of energy, which is a major difference with wireless nodes. Though, it is essential to keep the consumption of every node as low as possible in smart home context. For instance, using a few watts consumption node to control a few watts device is unwise. Power lines used as a medium of communication provides a physical link between each node connected on the same phase of the grid, but the presence of a link does not means that communication is always possible considering power consumption, size, price and noise generated on the medium. Our approach is to keep these factors as low as possible, without trade-off. As well as WPAN, communication between any pairs of nodes on the grid is not always possible. Like RF nodes, which can reach only their neighbors in a restricted area, PLC nodes sometimes cannot reach some areas of the connected grid, depending on its topology and the loads connected.

There is local dynamic signal strength that provides a sort of mobility feature on the electrical grid, depending on the electrical activity on the network. We cannot avoid these disturbances due to real-life activity, and communication has to adapt to them. The nodes being statics, this mobility might better be called reachability, regarding the positioning on the grid and the loads that can be plugged/unplugged at any time. The plugged/unplugged loads continually change the grid response and impact the communication possibilities on the medium. Therefore, PLC is facing the same problem as WSN concerning the possible unreliability of the medium. IEEE 802.15.4 dedicated to PLC is a robust solution to low-rate communication over PLC. Conbinedwith IEEE 802.15.4, it becomes a good way to provide easy interoperability with LR-WPAN and merge the

Table 1. Wireless vs Powerline medium

	RF	CPL
Mobility	+++ (no cord)	Limited to the cord
Power	Limited to battery	+++ (grid powered)
Range/loss	Depends on nearby noise	Depends on power grid noise

better of these two worlds to carry data on a heterogeneous low-rate PAN with a betterreliability.

Wireless Medium in LR-WPAN

Wireless devices using the IEEE 802.15.4 standard provide good mobility features, roughly limited to the zone of reachability of at least one of their neighbors. The principal limitation of these devices is the management of their power resource that is limited to the battery capacity or energy scavenging. The management of the energy available is really a key point for these devices, and is directly linked to the lifetime of the network. If a node is managing a lot of paths, the depletion of its power resource can shut-off the node and result in a possible unreachability of the network routed through this particular node. Many works aim at optimizing the network's lifetime by optimizing the power management with the use of duty cycling (in various MAC protocols such as S-MAC (Ye, Heidemann, & Estrin, 2004), T-MAC (van Dam & Langendoen, 2003), or Z-MAC (Rhee, Warrier, Aia, Min, & Sichitiu, 2008)), efficient algorithm design (Shih et al., 2001) or sink positioning (Saad & Tourancheau, 2009).

PLC Medium in LR-PAN

The PLC medium cannot provide the same mobility feature as the wireless medium, because of the wire and the number of electrical outlets available. But the power resource management is not a problem, because the device is directly plugged in a power supply (except if the supply is not reliable). In PLC world, different coupling technologies exist that provide different networking characteristics such as data rates, transmission range, error rate with different needs in power load and different perturbation on the power line. In this paper, we study a low cost testbed, low power and low rate technology very suitable for sensor networking on LR-PAN. The range provided by the PLC medium can reach up to 1 km in an urban context, regardless of the environment, but depending on the electrical activity on the grid.

Table 1 compares these two media in a LR-PAN context.

Each medium can match different requirements. We see that they are both lossy but from different causes and they complete each other well on other criteria. With an efficient use of a combination of these two media, we should be able to provide a very reliable solution. This implies the need for interoperability and/or cooperation between the different media available, and this is our goal in the following.

PLC Technology

We first present the particular transceiver on which we rely. WPC™ (Watteco White Paper, 2008) stands for "Watt Pulse Communication". It is a technology developed by Watteco[2], which enables to carry out any type of low data transfer communication with a reliable propagation and which can be deployed on the same PLC network with other technologies without interfering. This transceiver is based on the transient behavior of electrical networks. By using network reaction respect to load change, it is possible to create high level, low energy pulses (compliant with EMC regulation). As a result, the pulse magnitude can be significantly higher than noise even after

Figure 2. WPC™communication principle

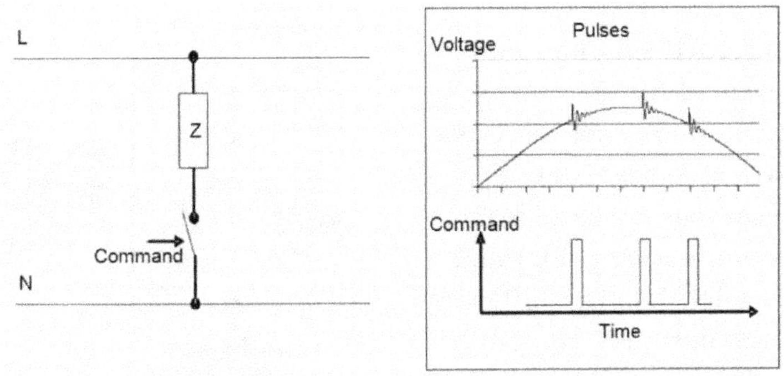

propagation and ensure a robust communication signal. The coupling device is very simple and the network reacts with its own resonance ensuring a very good propagation medium. This technology takes advantage of a physical natural phenomenon: the ignition pulse produced by appliances connected on an electric network. A pulse is a very short, a few nanoseconds, pulse produced by a load during its ignition or extinction and constitute an unambiguous signature. The transceiver includes a microcontroller driving an adapted load, producing the pulse when connected to the mains. This is represented in Figure 2. This pulse propagates over the power line at a long distance, up to 1 km in a public lighting environment. The emission of pulses can be triggered according to a controlled time schedule in order to communicate between two points in an electric network.

The chip has the following characteristics:

- A small size, approximately 500 mm2, induced by the network coupling method.
- A low power consumption, less than 10 mW, due to the coupling method and the simplicity of the reception floor.
- A reduced cost of electronics, i.e., in the dollar order range for large quantities, induced by the preceding items. This allows simplicity for broadcasting data and front-end analogic receiver, limiting electronic components and signal processing.

As a "PLC modem", an OEM module can re-use any existing protocol, providing a PLC adaptation of the communication stack. In this paper we focus on the adaptation of the IEEE 802.15.4 standard protocol and show the possibility it offers for uppers layers.

Table 2 outlines the mains features of this PLC transceiver.

Table 2. WPC Technical features

Data rate	**10kb/s bi-directional**
Digital Connection	Direct RX/TX or SPI serial mode
Frame Check and Correction	8 bits CRC and Hamming encoding
Input Voltage	110 to 230V, 50-60Hz
Power Supply	5V DC
Power Consumption	10mW average on communication

IEEE 802.15.4 Adaptation over PLC

The IEEE 802.15.4 standard (IEEE, 2006) defines the PHY and MAC layers of the OSI model. In this paragraph, we propose to adapt these two layers on PLC. The original standard defines two different types of devices that can participate in a network: a full-function device (FFD) and a reduced-function device (RFD). The FFD can operate in three modes serving as a PAN coordinator, a coordinator, or a device. A FFD can talk to RFDs or other FFDs, while a RFD can talk only to a FFD. A RFD is intended for applications that are extremely simple, such as a light switch or a passive infrared sensor. Our PLC nodes can operate as FFDs or RFDs, depending on the embedded software.

RF Transceiver Emulation Over PLC

Even if the scope of the standard is for wireless devices, we first investigated the idea of using it over raw PLC in Chauvenet (2010). Indeed we found that the definition of a LR-WPAN in the 802.15.4 standard was very close to our PLC testbed context[3].

For the adaptation of the IEEE 802.15.4 standard over PLC, an open IEEE 802.15.4 RF stack was used. As represented in Figure 3, the goal of this adaptation is to emulate a RF transceiver with our PLC transceiver. To achieve this, an enhanced SPI interface has been developed between the transceiver and the microcontroller to drive it as if it was a "regular" RF transceiver.

PHY Adaptation

The PHY layer of the IEEE 802.15.4 standard proposes the following services: activation/deactivation of the transceiver, Energy Detection (ED), Link Quality Indication (LQI), Channel Selection, Clear Channel Assessment (CCA), transmitting and receiving packets across the physical medium. The specific WPC™transceiver induces some timing adjustments because this transceiver transmits data only near the rising zero-crossing voltage while the reception is continuously possible.

Figure 3. Emulation of a classic RF transceiver by the PLC transceiver through an enhanced SPI interface

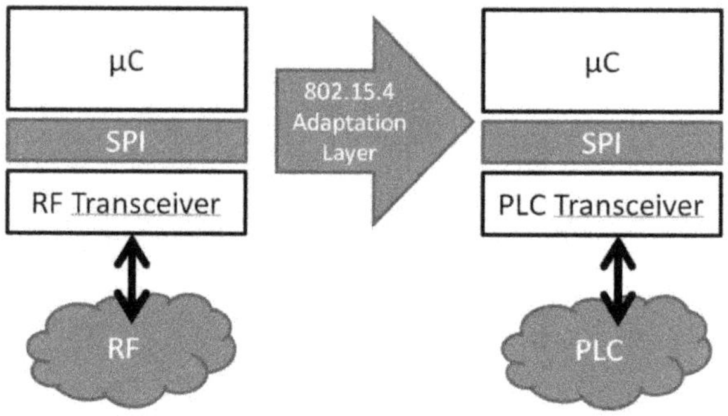

Our PLC transceiver fragments every IEEE 802.15.4 frame in smaller packets. The actual data rate is about 10Kbits/s over 50 Hz frequencies, meaning that 25 bytes are sent every voltage period. Therefore the transceiver fragments data in 25 bytes packets sent near the zero crossing of the voltage. This specific communication scheduling introduces timing adaptations compared to a classic RF transceiver which has a continuous transmission. The PHY specifications of the IEEE 802.15.4 standard fully describes the physical interface (i.e., frequency band, modulation and data rate). For our actual specific transceiver, these specifications are reduced to a minimum. The attributes of the physical interface are written in the PHY PIB (PAN Information Base). This information base has been adapted to fit with our PLC transceiver.

- phyCurrentChannel: Unlike RF, there is only one channel available for our physical interface. The normalization for the different PLC frequency band is still in progress but a time slot in the 2-4 Mhz band has already been reserved for our low rate pulse PLC by the IEEE P1901 working Group (IEEE, n.d.). This time slot is named "LRWBS" for Low Rate Wide Band Services.

- phyTransmitPower: This attribute is used to set the power of the transceiver transmission. The power transmission is constant for our PLC transceiver, so this parameter is set to a constant in accordance with PLC standards.
- phyCCAMode: This attribute specifies 3 different CCA modes, defined in the IEEE 802.15.4 standard. The CCA is performed with a specific method on our PLC transceiver.
- phyCheckCRC: This attribute is used to enable or disable the Cyclic Redundancy Check (CRC) checking. A CRC check is implemented in our PLC transceiver so you can enable or disable this attribute accordingly.
- phyGenerateCRC: This attribute is used to enable or disable the generation of the CRC. One generation is implemented in our PLC transceiver so you can enable or disable this attribute.

MAC Adaptation

The MAC layer used over our PLC transceiver is very close to the MAC layer defined by the standard However this transceiver provides a 10 kb/s data-rate using a specific modulation and is quite different from a classical RF transceiver. As a result, it induces some adaptation with the MAC part to deal with the latency of the transceiver (due to the lower data-rate than that of a classical 802.15.4 RF transceiver). A few timing adjustments are needed regarding the specific communication. The MAC layer of the IEEE 802.15.4 standard provides the following services: beacon management, channel access, GTS management, frame validation, acknowledged frame delivery, association/disassociation, security. Timing attributes defined in the MAC PIB were increased to fit with our PLC physical interface. The medium access we use is the carrier sense multiple access with a collision avoidance (CSMA-CA) as defined in the standard.

- macAckWaitDuration: This attribute determines the time limit for non-received acknowledge. This attribute has been incremented in order to respect the transceiver specificity. This incrementation is a result of the modification of the MAC constant "aUnitBackoffPeriod" which takes part of the "macAck-WaitDuration" attribute's calculation. This constant determines the number of symbols forming the basic time period used by the CSMA-CA algorithm, which is used in our adaptation layer.

Results of the Adaptation

LR-WPAN features defined by the IEEE 802.15.4 standard can be compared with our pulse PLC network implementing our 802.15.4 standard adaptation.

Table 3 shows that our PLC adaptation fits with the original definition of the IEEE 802.15.4 standard (IEEE, 2006) for most criteria.

Moreover, applications such as Home Area Networks, and Home monitoring targeted by PLC nodes are similar to those of WSN devices

This adaptation provides a communication on PLC with the 802.15.4 data format. Table 3 shows the comparison between the definition of a 802.15.4 network as defined in the standard and a PLC network using our adaptation.

4. NETWORK LAYER

In this section, we present the network layer of the communication stack and focus on the use of IPv6 (RFC2460) with the 6LoWPAN adaptation (RFC4944 (Montenegro, Kushalnagar, Hui, & Culler, 2007)).

Table 3. LR-WPAN vs WPC-PAN networks

	LR-WPAN (IEEE, 2006)	PLC-PAN
Data rate	250, 100, 40, 20kb/s	10kb/s
Topology	star, P2P, mesh	star, P2P
Adressing	16b or 64b	16b or 64b
GTS	optional	not managed
CSMA-CA	yes	yes
Full ack	yes	yes
Low power	yes	yes
RSSI	yes	yes (adapted to PLC)
Channels	16 in 2450 MHz band, 30 in 915 MHz, 3 in 868 MHz	1 channel in 2-4 MHz

IPV6

The IPv6 protocol is designed as the successor to IPv4 and enables the Internet to scale for decades to come. To overcome the declining unallocated address space and in anticipation that networked appliances and instruments will outnumber conventional computer hosts, IPv6 extends the IP address space from 32 to 128 bits and directly addresses very important issues such as auto-configuration, security, multicasting,….

This is especially welcome in the growing Internet of things we are speaking of. Recognizing the growth in link bandwidth in the wired Internet, IPv6 also increases the minimum MTU requirement from 576 to 1,280 bytes.

IP protocols are defined by the Internet Engineering Task Force (IETF) in the form of Request For comments open standard protocols (RFC). These specifications provide a complete service for IPv6 networks and they were recently proven adaptable and operational for networking on small devices (Hui, 2008; Dunkels, Voigt, & Alonso, 2004; Hui & Culler, 2008). The IETF working groups about sensor activities started from these significant developments, by defining a way to carry the IPv6 MTU into smaller frames with RFC4944, security in RFC4301, auto-configuration in RFC4862 and IETF drafts for routing and other aspects of the protocols.

As presented in Abeille, Durvy, Hui, and Dawson-Haggerty (2008), an IPv6 stack can be implemented in approximately 11.5 kilobytes of ROM and 1.8 kilobytes of RAM. With a complete run-time (timers, scheduler, etc.) as well as RFC-compliant UDP and TCP protocols above, an OS that provides a complete IPv6 network stack can be implemented within 35 kilobytes of ROM and 3 kilobytes of RAM. Hui and Culler (2008) pull down the memory requirement to 24 Kbytes of ROM and 3.6 Kbytes of RAM. So complete IPv6-based applications can fit in a micro-controller providing only 64 kilobytes of ROM and 4 to 8 kilobytes of RAM which is the order of magnitude of today's micro-controller such as the PLC nodes we consider.

More than providing an efficient networking solution, the use of IPv6 on top of a PLC technology tackles the recurrent problem of proprietary protocol on this medium. We aim to provide the IP standard as it stands, without using costly gateways to connect PLC networks to the IP world.

6LoWPAN

With RFC4944 (Montenegro, Kushalnagar, Hui, & Culler, 2007), the IETF has defined the 6LoWPAN adaptation layer that includes a compression mechanism of the IPv6 header. This mechanism is stateless which means that it creates no binding state between the compressor/decompressor pair. Most of link technologies designed for smart objects do not support the full 1280-byte MTU that IPv6 require. For instance, IEEE 802.15.4 only supports 127-byte MTUs so a full IPv6 packet does not fit in an IEEE 802.15.4 frame.

6LoWPAN provides header compression to reduce transmission overhead, fragmentation to support the IPv6 minimum MTU requirement, and support for layer-two forwarding to deliver IPv6 datagram over multiple hops.

802.15.4 protocol data units have different sizes depending on how much overhead is present (IEEE, 2006). Starting from a maximum physical layer packet size of 127 bytes (aMaxPHYPacketSize) and a maximum frame overhead of 25 (aMaxFrameOverhead), the resultant maximum frame size at the media access control layer is 102 bytes. Link-layer security imposes further overhead, which in the maximum case (21 bytes of overhead in the AES-CCM-128 case, versus 9 and 13 for AES-CCM-32 and AES-CCM-64, respectively) leaves only 81 bytes available. This is obviously far below the minimum IPv6 packet size of 1280 bytes, and in keeping with Section 5 of the IPv6 specification (RFC2460), a fragmentation and reassembly adaptation layer must be provided at the layer below IP. Furthermore, since the IPv6 header is 40 bytes long, this leaves only 41 bytes for upper-layer protocols, like UDP. The latter uses 8 bytes in the header which leaves only 33 bytes for application data. Additionally, there is a need for a fragmentation and reassembly layer, which will use even more bytes.

This computation shows that without compression mechanism, a 802.15.4 frames would provide less than a quarter of its total size to the application layer. The 6loWPAN mechanism is the results of a global reflexion about reducing the IPv6 header to comply efficiently with the 802.15.4 standard.

Drawing on IPv6 extension headers, it employs the header stacking principle to separate the orthogonal concepts and keep the header small and easy to parse.

To support the IPv6 1280-byte MTU, datagrams must thus be fragmented before they can be delivered to the link layer. Fragmented datagrams also provide additional opportunity for more effective buffering techniques when forwarding datagrams. For example, fragments can be delivered before a node reassembles the entire datagram, allowing nodes with severe memory constraints to forward complete datagrams.

5. ROUTING

The 6LoWPAN adaptation layer brings all the IPv6 mechanisms needed for large scale deployments and easy interoperability available for LR-WPAN with a minimum overhead. But these new IPv6-compliant nodes are very different from classic networking devices and routing over LLNs introduces requirements that existing routing protocols may not fully address.

Limited memory and communication capabilities constrain the routing state at each node as well as the routing information that might be communicated. These restrictions forbid the use of protocols that rely on complete link-state information. Traditional distance vector mobile ad-hoc networks (MANet) protocols are also unsuitable because most of them exchange route maintenance information at rates that exceed typical LR-WPAN communication and react to unreachability with expensive route-repair mechanisms. Instead, low power wireless PAN routing protocols must operate with incomplete information and tolerate some inconsistency.

The draft (Levis, Tavakoli, & Dawson-Haggerty, 2009) provides a brief survey of the strengths and weaknesses of existing protocols and examines whether existing and mature IETF protocols can be used without modification in these networks, or whether further work is necessary. It concludes that no existing IETF protocol meets the requirements of this domain, as existing protocols were not designed with all the constraints of our

Figure 4. The ROLL architecture

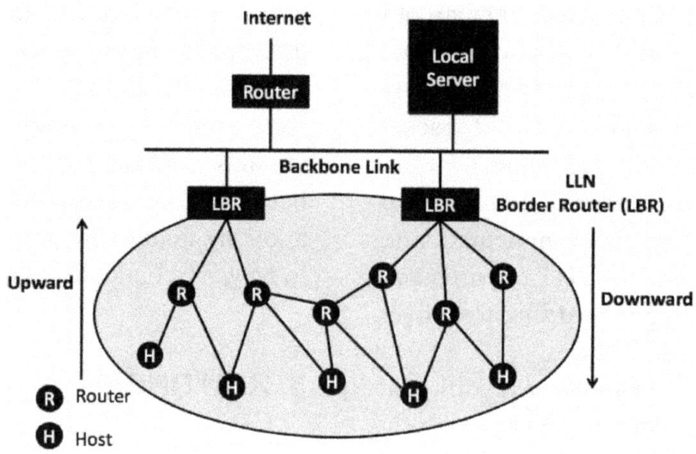

context. They have made trade-offs which may or may not be appropriate for LLNs.

To achieve the design of this new routing protocol, a new IETF working group called ROLL[4] was formed in 2008. This group has designed the RPL routing protocol[5] that should be able to operate over a variety of different link layers, including but not limited to low power wireless or PLC technologies. Note that there is no "wireless" word in the name of the protocol, because it does not rely on any particular features of a specific link layer technology. These features bring flexibility to the RPL protocol and fits with our goal of providing interoperability in LLNs.

The specification says that a network may run multiple instances of RPL concurrently and each instance may serve different constraints or performance criteria. A RPL instance may serve a specific objective function (OF), describing metrics, constraints, and an optimization objective to realize a desired objective in a given environment. This allows our network to define an OF for each specific application.

The ROLL working group has designed several drafts such as draft-ietf-roll-building-routing-reqs6 and draft-ietf-roll-home-routing-reqs (RFC 5826), tightening the use of the RPL protocol in the application field of our PLC devices.

Figure 4 shows an example of ROLL architecture. RPL enables Multipoint-to-Point (MP2P), Point-to-multipoint (P2MP) and a basic structure for point-to-point (P2P) traffic. The protocol model of RPL is based on the construction and the maintenance of one or several Directed Acyclic Graphs (DAGs). To achieve this, RPL defines DAG Information Object (DIO), DAG Information Solicitation (DIS) and Destination Advertisement Object (DAO), three types of a new ICMPv6 message called RPL Control Message.

In a nutshell, a DIO carries a number of metrics and other information that allows a node to discover a DAG Instance, select its DAG parents, and identify its siblings while employing loop avoidance strategies. DIO are periodically sent to maintain the DAG, and ensure reliability. A DIS message may be used to solicit a DIO from a RPL node (Its use is analogous to that of a Router Solicitation). DAO is an optional message that conveys destination information upwards along the DAG so that the root (or other intermediate nodes) can provision downward routes.

The RPL protocol is well suited for our PLC nodes for several reasons: This is a routing protocol specifically designed for LLNs, which is precisely our context. Indeed, it considers a set of constraints that were not in the motivation of other routing protocol (Levis, Tavakoli, & Dawson-Haggerty, 2009). This protocol is not related to a particular physical interface, so it is compliant with our PLC medium, and can also to be used over other media to create heterogeneous networks. The RPL protocol allows efficient MP2P traffic which is the main data flow in smart-grid application. For example, all nodes will report their electricity, gas or water consumption at periodic times. RPL also provides efficient P2MP traffic which can be useful in the same smart-grid context to inform all nodes about power pricing information and optimizing the energy consumption. It is also useful for street lighting applications here a master will command all the lights from the street. RPL empl oys the trickle timer technique (Levis et al., 2008) that adapts the management traffic to the stability of the networks, and avoids greedy traffic to manage the network. RPL also allows mesh topology which increases our reachability capacity, and ecreases the risks of unreachable zones on the power grid. The Objective Function (OF) concept offers a very wide range of applications, dealing with delay, power consumption or other metrics constraints independently from the physical layer used. Finally, RPL relies on the wide used IPv6 protocol which provide an easy connection with many web applications. The security aspect is currently in progress in ROLL, and it is a also an important feature needed for our applications.

6. EXPERIMENT

In this section, we present two different experiments relying on our adaptation of the IEEE 802.15.4 standard over PLC. First, a simple heterogeneous experiment between RF and PLC nodes. This is a first proof of concept of data transfers in a PLC/RF heterogeneous network. Such a MAC level experiment highlights a starting point of IEEE 802.15.4 interoperability. It doesn't implement the 6LoWPAN/IPv6/RPL part described above. The second experiment is an interoperability testbed, which is introduced in this section and will be presented in the next section. This experiment relies on the same 802.15.4 adaptation but implement the 6LoWPAN/IPv6/RPL part of the communication stack to show the extending possibilitiesof this adaptation.

Figure 5. Tmote SKY platform from Crossbow

WSN Node

Theses nodes are used in the first experiment. This node, represented in Figure 5, is a classic RF node Tmote SKY from Crossbow[7] composed of a TI MSP430 microcontroller (8Mhz microprocessor, 48k Flash, 10k RAM), a TI CC2420 RF transceiver (2.4GHz ISM-band carrier, 250kbps) IEEE 802.15.4 standard compliant, and an on-board antenna. It can be powered by USB, or with 2 AA batteries enabling the node to be autonomous and maybe mobile.

The USB interface was used as a serial link. This platform also includes sensors to measure temperature, humidity, and light. In this experiment, 2 Tmote SKY nodes were used. The first one acted as a sender. It read its own light sensor value and sent it in a IEEE 802.15.4 frame via its antenna. This frame was broadcasted every second. This node is battery powered and thus can be located anywhere within the RF range. The second one acted as a listener to relay the light sensor value received on its antenna to its USB interface. This node was connected to the gateway with a USB cable which powered it. This node acted as a RF/USB gateway.

Figure 6. The WPC DevKit from Watteco

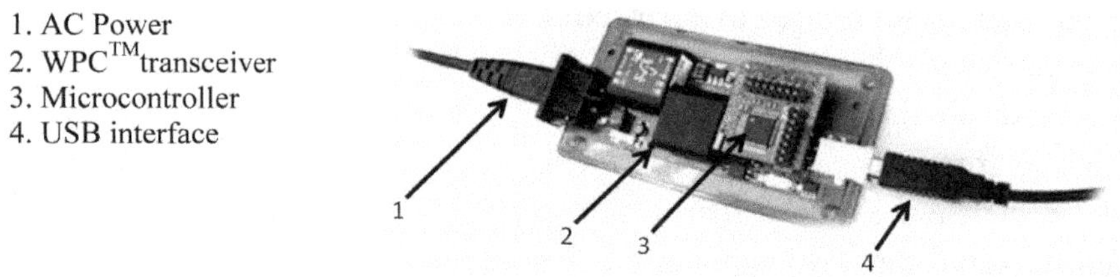

PLC Node

These PLC nodes are used in both experiments. The WPC™technology, is used through a "WPC Development Kit" from Watteco, shown in Figure 6. It enables users to easily integrate a WPC™transceiver and to use it as an original communication tool on PLC.

It is composed of 4 different parts:

The architecture is quite similar to a RF node which is basically composed of a micro-controller and a transceiver. Unlike RF nodes where the medium is the free space which does not need a connection, the WPCTMtransceiver needs to be connected to its medium through an electrical outlet. The USB interface provides a serial interface with the node. The USB part is optically-isolated from the high voltage part. In this first experiment, we used 2 PLC nodes with different configurations. The first one is a WPCTMDevelopment Kit connected to the PC gateway via its USB interface and connected to the grid, in order to send data over PLC. This device is used to relay the data sent from the gateway via the USB interface to the PLC medium.

In a way, this device acts as a USB/PLC translator. The second one is a special form factor of the first node, with additional features. This device is called a"smartplug". It embedded a WPCTMtransceiver to communicate, control and command an electric relay to switch ON/OFF an electric device up to 16 amps, and an ADE chip to measure the electrical consumption of the device plugged into it. A light bulb was plugged into this device to show the ON/OFF capability and create an electrical load to measure. This device acts as a light switch controlled by PLC, which is able to return electrical informations about consumption, the state of the switch, and an error message if the command fails.

Gateway

This gateway is used in the first experiment only. The gateway built an IEEE 802.15.4 bridge between Tmote SKY nodes and PLC nodes. Protocols used to build the gateway are represented in Figure 7. As this first experiment was a proof of the concept, the gateway ran on a traditional PC laptop.

Figure 7. Gateway

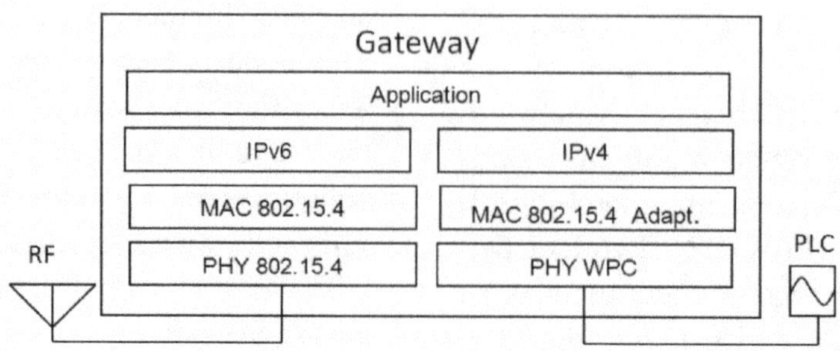

This gateway read the data sent by the Tmote SKY via the USB interface, compared it to a threshold and sent a command to the WPC™module via another USB interface. The threshold was set to define a darkness or brightness state from the light value collected. The state defined was then converted into a command to be carried over PLC, that finally switched ON/OFF the light bulb. To sum up, we built an elementary PLC light switch actuator, controlled by a RF sensor node, and running over the IEEE 802.15.4 standard protocol over heterogeneous media.

Low Power Development Platform

This platform is used in our second experiment. To test our implementation, we used a low power PLC platform from Watteco. This platform brings makes manipulating safer, avoiding the risk of high voltage shock and reduces the medium as a short and simple wire to put away the possible perturbations of a real PLC medium in order to focus specifically on the stack performance. As shown in Figure 8, our platform is made of a 220V/24V AC converter, 7 WPCDevKit and a 7 ports hub. Each PLC node of the platform has a strictly identical architecture as the WPCDevKit from Watteco. The WPC™transceiver of each PLC node is driven by an ATmega 1281 or AT90USB1287 micro-controller, depending on the function intended.

Figure 8. Low voltage PLC platform

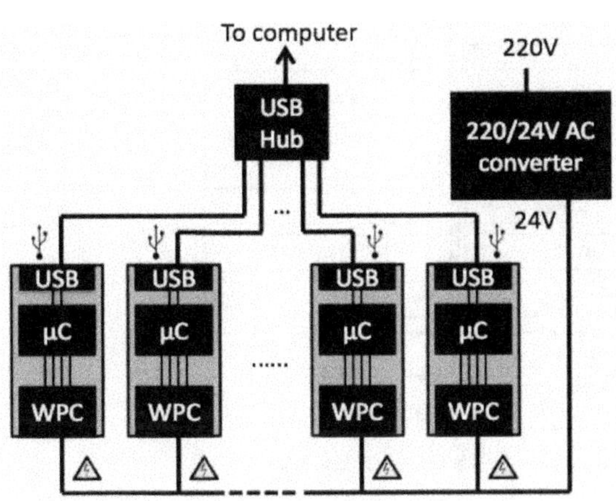

A connexion between all PLC nodes supplies the power and provides the PLC medium (on 24V here). All nodes have a USB port to provide a serial interface. We use this interface to act on each node (to control the topology of the network for example) and bring back some debug information.

This platform will be connected to a RF network through a router in the interoperability experimentation.

Ethernet to PLC Gateway

This gateway is used in the second experiment and is an evolution of the first gateway presented above.

To provide the connection with our local PLC network, we developed a PLC to Ethernet gateway based on the Atmel RZUSBSTICK architecture. This device creates an Ethernet over USB emulation. It is based on an AT90USB1287 micro-controller which drives the WPC™transceiver the same way as the ATmega1281.

This micro-controller is integrated on a separate PCB so that it can be plugged into a classic WPCDevKit.

It avoids having to re-create a whole PLC node. This card directly integrates the USB interface with high speed isolators. With this card on the WPCDevKit, it allows a PC or a router too see our PLC node as an Ethernet interface. Our PLC network is therefore seen as a subnet from the PC or the connected router.

This Ethernet emulation has 2 main contributions: It creates a link with other networking devices, and it can be used as a sniffer on Wireshark[8]. The second capability is a major achievement because it enables the use of a performant network analyzer on PLC medium.

Software for Heterogeneous Experiment

The 2 Tmote SKY nodes were running under the Contiki 2.3 OS (Dunkels, Gronvall, & Voigt,

Figure 9. Histogram of the switch state (upper part) and the corresponding power consumption (lower part)

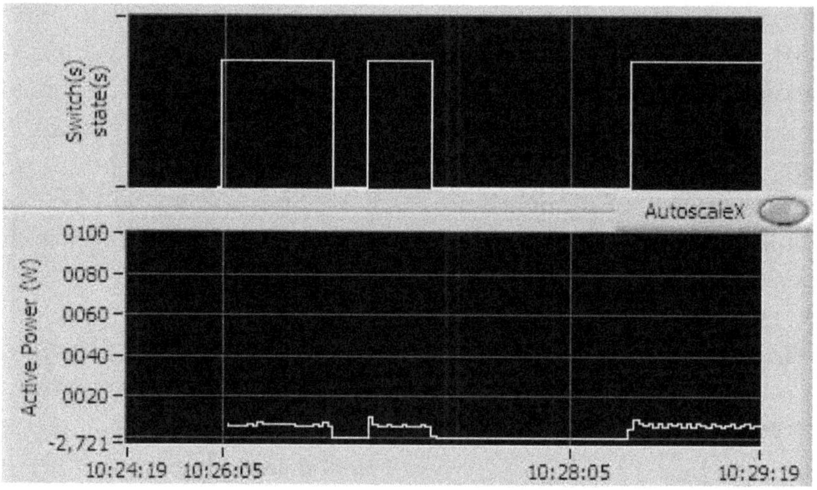

2004). This OS version includes an IPv6 stack with an implementation of 6LoWPAN. This brings the nodes the capability of exchanging IPv6 packets over IEEE 802.15.4 standard. Thus, the light sensor value transmitted between the 2 Tmote SKY nodes was carried in IPV6 packets. The 2 PLC nodes were running under Contiki 2.2 and the data transmitted between these nodes were carried in IPV4 packets. The first PLC node transmitted the command received from the gateway into UDP format. The Smartplug decoded the frame and performed an ON/OFF action on the bulb. This node then returned the electrical consumption of the load connected with a UDP frame. It was then decoded by the first PLC node which transmitted this information to the gateway. The gateway interface plotted the corresponding electrical consumption and the light value in a browser.

The gateway bridges PLC and Tmote SKY nodes at the application level as schematically shown in Figure 9. The software used in this experiment, see screenshots in Figure 10, showed

Figure 10. Communication layers

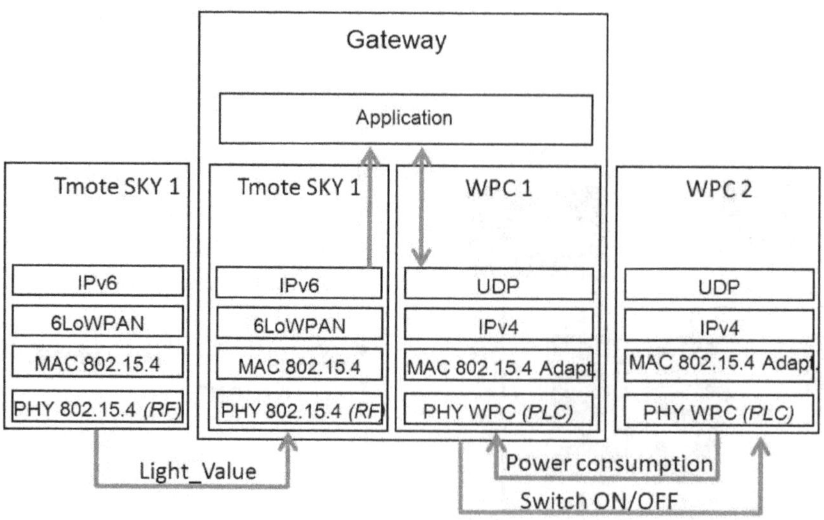

Figure 11. IEEE 802.15.4 frames exchange diagram

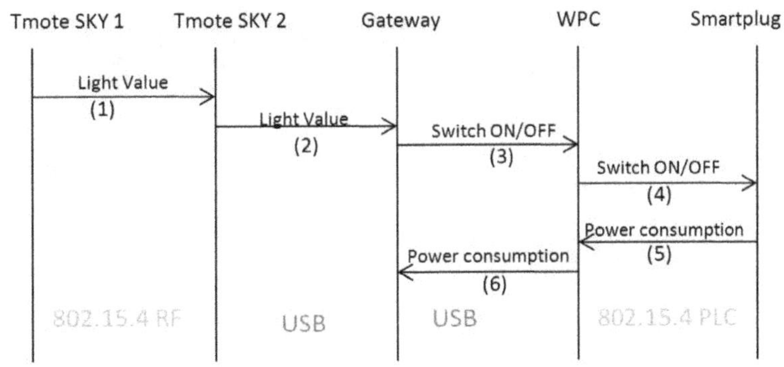

the data transmitted for debugging and checking as well as a user interface to provide feedback results and also control the experiment.

The first light sensor value read is 108 which is above the switch off value threshold, which was set to 80, so the bulb switch command is OFF. The gateway transmitted the command to switch off the bulb and the switch state was set to OFF. When we hand seal the light sensor on the first Tmote SKY, its light value falls to 69 which is under the switch threshold value (i.e., set to 70). The gateway transmitted a switch ON command to the Smartplug and the switch state came up. The different frames exchanged during these actions are schematically represented in Figure 11.

7. INTEROPERABILITY

We describe here how we implement a communication stack following the previously described characteristics on our PLC testbed, and show a real interoperability testbed in the next section.

PLC Node

Our stack is running under Contiki (Dunkels, Gronvall, & Voigt, 2004) in its 2.4 version[9]. We use the uIP IPv6 stack from Contiki which implements a full IPv6 stack. We have created a specific PLC platform in Contiki 2.4. Because our transceiver is roughly a RF transceiver emulation, the major part of the adaptation on Contiki is in the MAC layer and the uIP stack doesn't require any modifications. Our implementation of RPL is the result of a collaborative work with SICS. We have implemented the HC06 draft[10] of 6LoWPAN and the version 5 of the RPL draft[11]. To perform our tests, we used the UDP sender and UDP client example available in the Contiki OS.

PLC Gateway

There are 2 different softwares for this gateway. The first one creates an Ethernet emulation over US and is derived from the RZUSBSTICK platform available in Contiki The second one is used as a sniffer application and has been developed in the 15dot4 tools project[12]. Though this project is related to classic 15.4 RF nodes, the RF transceiver emulation of our specific PLC transceiver allows us to use this software over PLC.

PLC Networking

We have first created a RPL network on our PLC platform. We used a node with an Ethernet emulation as a sniffer with the Wireshark software. When we first plugged in a PLC node, it declared itself as a root because it got no response to its DIS RPL control message. When we connected other nodes, they received a DIO response from

Figure 12. Interoperability test setup

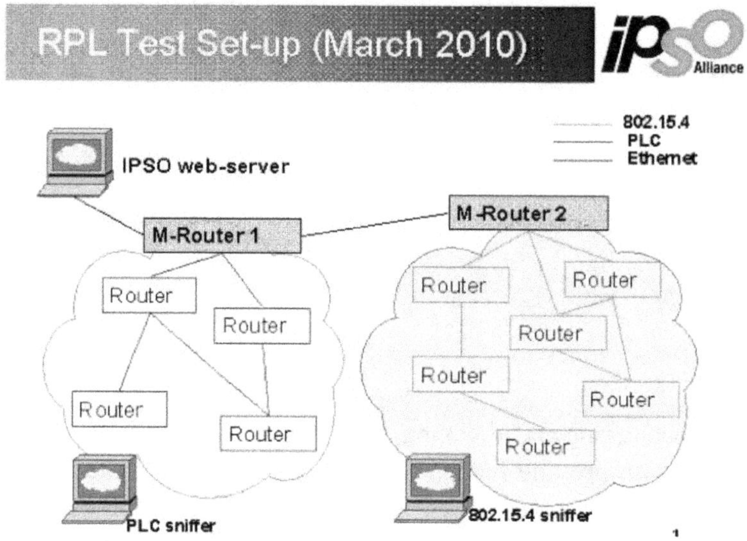

the root to their DIS request and declared themselves as leaf attached to the DAG. Through the serial interface on each node, we can send some commands and forces a node to drop the packet sent by another node. With these commands, we can block the communication between a node and the root, to detach this node from the root. When the node sends its DIS, it will only receive DIO from nodes and will associate to a node. As a result, this node will associate to the DAG with a higher rank than other nodes.

This is a simple way for manually controlling the topology of the network and reproduce unreachability that can appear in real experiments. When the DAG is settled, we successfully checked the trickle mechanism on the DIO sending, with the doubling of the time after each expiration.

Interoperability Testbed

The interoperability experiment was performed during an IPSO interoperability event[13] in March 2010. This was the first interoperability testbed between different RPL implementations. Eight companies took part in this event and five of them have tested their RPL implementations. We were the only one on PLC; the four others were using 802.15.4 RF nodes. These 5 companies used the same stack architecture, based on 802.15.4, IPv6 with 6LoWPAN (HC06 draft) and RPL (05 draft), but each had its own implementation, differing in the OS and the devices used.

As we can see in Figure 12, the network formed was using three different media: Low-power PLC, RF (802.15.4) and Ethernet. There are two clouds of nodes bridged together with two routers (M-Router 1 and M-Router 2) linked with an ethernet link. The RF cloud (yellow in the figure 12) is composed of RF nodes from different companies. Each company has its own implementation of the communication stack; all based on the 802.15.4/6LoWPAN/IPv6/RPL protocols.

A Atmel RZUSBSTICK is plugged in a computer to be used as an RF sniffer. It plots all the RF frames transmitted on a wireshark interface. This RF cloud is connected to the M-Router 2 via another Atmel RZUSBSTICK, plugged in the USB interface of the router and implementing the RPL protocol. It constitutes the RF interface of the M-Router 2. The PLC cloud (White on the Figure 12) has the same architecture. The PLC nodes (from a single company) are using the same communication stack with their own implementation. This PLC cloud is connected to the M-

Router 2 via the PLC gateway we described, plugged in the USB interface of the router. It is based on the same architecture as the Atmel RZUSBSTICK. Another PLC gateway is plugged in a computer and used to sniff all the packets transmitted on PLC, exactly the same way as for RF. All the nodes from these two clouds are routers, within the meaning of RPL, as they are able to route packets. A web server from IPSO has been connected on the M-Router 1, to show the connectivity outside of the PAN.

The M-router 1 was set as the Dag root. Each router or node connected was then associated to the DAG with a classic DIS/DIO exchange. As a result, the M-router 2 associated as a node, and our PLC nodes associated to this router. The RF nodes attached directly to the M-router 1. The Dag root has aaaaa::1 address and all the nodes (RF and PLC) have a unique global address based on their EUI-64 with a aaaa::/64 prefix.

After DAG settings, we were able to ping the node from each participating company with their global addresses from our PLC nodes. The other participants were also able to ping our node with the use of their global addresses. We had also set up one of our node as a DAG root, so that the M-router 2 and one of our node associated the DAG as leaves. This simple topology allowed us to make a ping from a PLC node to another PLC node through an ethernet link, and simulate the connection between two PLC network through a classic ethernet link.

We also successfully performed a global repair, where the DAG root increments its instance ID and every node from the DAG have to adopt this new instance ID. This testbed showed that our stack was able to easily take part of a heterogeneous network, in a LLNs context with the use of open standard protocols.

8. CONCLUSION

The IEEE 802.15.4 standard implements many primitives which provide a very complete service in a highly constrained network such as LR-WPAN. In this paper, we explored the similarities between LR-WPAN and PLC-PAN and showed a simple adaptation layer of the IEEE 802.15.4 standard over our PLC testbed.

With our adaptation, a large part of the IEEE 802.15.4 stack is implemented in order to control the WPCTMtransceiver. The MAC/PHY primitives implemented with this adaptation provide a comprehensive IEEE 802.15.4 service over PLC which opens this medium to the IP world thanks to 6LoWPAN. To our knowledge, it is the first implementation of IEEE 802.15.4/IPv6 networking over low power PLC.

Because our work relies on protocols that are currently drafts and not RFCs (6LoWPAN HC06 and RPL- 05), trivial future work will include updates from these drafts until they become becomes RFCs, especially the RPL draft. This draft still needs further revisions for improvements in the DAO mechanism and the security part. As described in the draft, RPL only defined a set core of functionalities and is not related to any particular technology. This is a strong point for the flexibility, but this also means that it leaves a part of the work to the implementation. The first experiment presented in this paper stands as a first interoperability testbed that is very hopeful, but it needs further work to be optimized on our network. Work still needs to be done, to tune the RPL protocol in a way that fulfills more restrictive requirements, without overpassing our node capacity. The OF mechanism is also a very appreciable mechanism in RPL that we need to consider in our future work, to create a specific adaptation related to our PLC network. The results of the first interoperability event with RPL that we presented brings a good feedback about the capabilities of this protocol, at least on small networks. We now need further experimentation to push the capabilities of each implementation with more complex functionalities. The next testbed will certainly include more companies and more diversity, which will enhance the interoperability aspect. Another part of our future work will be to test our stack on real applications, to measure the efficiency of the implementation, in comparison with the non-standardized protocol currently deployed. Finally, we are plannig to implement our stack and the specific WPCTMtransceiver in a simulator, to focus on specific points that could improve the stack. We will test the performance of the stack with a homogeneous/heterogeneous dense network, and measure how far we can go on the scalability side with our current implementation.

With this work, it is now possible to create an heterogeneous PLC/RF network in a LLNs context with the use of open standard protocol. More than dealing with coexistence, we know enable cooperation between these media, which is a major improvement. We have presented all the protocols of the communication stack from the physical to the network layer, including routing, to achieve the implementation of a communication stack over PLC for multi-physical layer IPv6 Networking. We have justified the need for these protocols for our LR-PAN context on PLC. Our experiments during the interoperability event has shown that our stack was able to interoperate with RF 802.15.4 nodes, with a bridge at the IP layer, creating a multi physical medium low power personal area network.

REFERENCES

Abeillé, J., Durvy, M., Hui, J., & Dawson-Haggerty, S. (2008, November). *Lightweight ipv6 stacks for smart objects: the experience of three independent and interoperable implementations* (White Paper No. 2). IPSO.

Bertoni, H. L. (1999). *Radio Propagation for Modern Wireless Systems*. Upper Saddle River, NJ: Prentice Hall.

Chauvenet, C. (2010). Bernard Tourancheau, and Denis Genon-Catalot. 802.15.4, a mac layer solution for PLC. In *Proceedings of the AICCSA'10*. Washington, DC: IEEE.

Dunkels, A., Grönvall, B., & Voigt, T. (2004). Contiki - a lightweight and flexible operating system for tiny networked sensors. In *Proceedings of the EmNetS*. Washington, DC: IEEE.

Dunkels, A., Voigt, T., & Alonso, J. (2004, January). Making tcp/ip viable for wireless sensor networks. In *Proceedings of the First European Workshop on Wireless Sensor Networks (EWSN 2004)*, Berlin.

Ha, J. Y., Jeon, J., Lee, K., Heo, J., Kim, N., Kim, S. M., et al. (2007). Design and implementation of convergence sub-layer for a heterogeneous home network. In *Proceedings of the IEEE International Symposium on Power Line Communication (ISPLC '07)* (pp. 252-256).

Hui, J. W. (2008, September). *An Extended Internet Architecture for Low-Power Wireless Networks - Design and Implementation*. Unpublished doctoral dissertation, EECS Department, University of California, Berkeley, CA.

Hui, J. W., & Culler, D. E. (2008). Ip is dead, long live ip for wireless sensor networks. In *Proceedings of the 6th ACM conference on Embedded network sensor systems (SenSys'08)*. New York: ACM.

IEEE. (2006). *IEEE Standard 802.15.4*. Washington, DC: IEEE.

IEEE. (n.d.). *Standard for broadband over power line networks: Medium access control and physical layer specifications*. Washington, DC: IEEE.

Levis, P., Brewer, E., Culler, D., Gay, D., Madden, S., & Patel, N. (2008). The emergence of a networking primitive in wireless sensor networks. *Communications of the ACM, 51*(7). doi:10.1145/1364782.1364804

Levis, P., Tavakoli, A., & Dawson-Haggerty, S. (2009, April). *Overview of existing routing protocols for low power and lossy networks*.

Montenegro, G., Kushalnagar, N., Hui, J., & Culler, D. (2007, September). *Transmission of ipv6 packets over ieee 802.15.4 networks*. Washington, DC: IEEE.

Pengfei, L., & Jiakun, L. (2009). Application of communication and remote control in PLC based on zigbee technology. In *Proceedings of the International Conference on Computational Intelligence and Security*.

Powerline Alliance HomePlug. (n.d.). *Homeplug av white paper*. Retrieved from http://www.homeplug.org

Rhee, I., Warrier, A., Aia, M., Min, J., & Sichitiu, M. L. (2008). Z-mac: a hybrid mac for wireless sensor networks. *IEEE/ACM Trans. Netw., 16*(3).

Saad, L. B., & Tourancheau, B. (2009). Multiple mobile sinks positioning in wireless sensor networks for buildings. In *Proceedings of the 2009 Third International Conference on Sensor Technologies and Applications (SENSORCOMM '09)* (pp. 264-270). Washington, DC: IEEE Computer Society.

Shih, E., Cho, S.-H., Ickes, N., Min, R., Sinha, A., Wang, A., & Chan-drakasan, A. (2001). Physical layer driven protocol and algorithm design for energy-efficient wireless sensor networks. In *Proceedings of the 7th annual international conference on Mobile computing and networking (MobiCom '01)* (pp. 272-287). New York: ACM.

van Dam, T., & Langendoen, K. (2003). An adaptive energy-efficient mac protocol for wireless sensor networks. In *Proceedings of the 1st international conference on Embedded networked sensor systems (SenSys '03)* (pp. 171-180). New York: ACM.

Watteco White Paper. (2008, December). Retrieved from http://www.watteco.com

Yamada, K., Furumura, T., Kimura, K., Kaneyama, T., Yoshida, K., Kojima, M., et al. (2007). Dual communication system using wired and wireless in home-network. In *Proceedings of the 11th international conference, KES2007 and XVII Italian workshop on neural networks conference on Knowledge-based intelligent information and engineering systems (KES'07/WIRN'07)* (pp. 469-476). Berlin: Springer Verlag.

Yang, C.-f., & Geng, R.-y. (2008). Home network using complementary communication system. In *Proceedings of the 2008 International Seminar on Business and Information Management (ISBIM '08)* (pp. 71-74). Washington, DC: IEEE Computer Society.

Ye, W., Heidemann, J., & Estrin, D. (2004). Medium access control with coordinated adaptive sleeping for wireless sensor networks. *IEEE/ACM Trans. Netw., 12*(3).

ENDNOTES

1. 1http://www.internetworldstats.com/
2. http://www.watteco.com
3. The WPC™transceiver designed by Watteco (http://www.watteco.com)
4. http://tools.ietf.org/wg/roll
5. http://tools.ietf.org/wg/roll/draft-ietf-roll-rpl
6. http://tools.ietf.org/wg/roll/draft-ietf-roll-building-routing-reqs/
7. http://www.xbow.com
8. http://www.wireshark.org
9. http http://www.sics.se/contiki/
10. http://tools.ietf.org/html/draft-ietf-6lowpan-hc-06
11. http://tools.ietf.org/html/draft-ietf-roll-rpl-05
12. http://sourceforge.net/projects/dot4-tools/ 13 http://www.ipso-alliance.org

This work was previously published in International Journal of Business Data Communications and Networking, Volume 6, Issue 4, edited by Debashis Saha and Varadharajan Sridhar, pp. 1-20, copyright 2010 by IGI Publishing (an imprint of IGI Global).

Chapter 13
A Cache Replacement Policy for Location Dependent Data in Mobile Environments

Mary Magdalene Jane F.
Dr. N. G. P. Institute of Technology, India

R. Nadarajan
PSG College of Technology, India

Maytham Safar
Kuwait University, Kuwait

ABSTRACT

Data caching at mobile clients is an efficient mechanism to enhance data accessibility and reduce access cost. The issues of cache replacement for location-dependent data have been studied under a geometric location model, and a new cache replacement policy has been proposed by taking into account the spatial and temporal factors.

Cache hit ratio is employed as the primary performance metric where the higher the cache hit ratio, the higher the local data availability, lower uplink and downlink costs and less battery consumption. The experimental results show that the proposed replacement scheme is effective when the user preference is considered and the policy significantly outperforms the conventional replacement policies.

1. INTRODUCTION

Advances in wireless networking and the proliferation of portable devices have engendered a new computing scenario, called mobile computing, in which users with power-limited mobile devices can access database information systems located at the static network while they move. Lot of research has gone into improving the functionality, computational power, size and power constraints of mobile devices such as personal digital assistants, cell phones etc. Also, there has been considerable work on improving the communication capabilities through increase in bandwidth and

DOI: 10.4018/978-1-61350-477-2.ch013

development of protocols that are fault tolerant and resilient to the error prone wireless media.

In the last few decades, LBS have created lot of hype where the geographic data helps in providing location based information. Wireless communication helps in providing the location of the service requester i.e. the user. Hence to exploit the benefits of location-based services these two technologies should go hand in hand, which requires integration. The integration of these two technologies should also help in a time critical situations where quick and efficient response could help the user. Though users can access information anytime, anywhere, even while on the move, the performance of a wireless link will always be several orders lower when compared to a wired network.

While technology has been rapidly advancing, various limitations of wireless communication and mobile devices like limited bandwidth, client power and intermittent connectivity remain primary challenges in design and implementation of location based applications.

Client side data caching is considered as one of the important techniques to relieve bandwidth constraint imposed on wireless mobile systems. The importance of data caching has been observed early and could be seen in works (Acharya, 1995) and (Barbara, 1999). Data caching is a mechanism where copies of remote data can be kept in the local memory of mobile devices to substantially reduce data retrievals from the original server. This not only reduces the uplink and downlink bandwidth consumption but also improves the latency in data access. In a majority of mobile devices like laptops, palmtops and cellular phones, wireless communication is one of the major sources of energy consumption that reduces battery life. Thus, caching frequently accessed data in mobile devices can potentially minimize communication and hence conserve battery power. In data-caching method, data is cached at the mobile client. When the client receives a query, it first searches its cache. If there is a valid copy in the cache, it returns an answer immediately. If not, the client attempts to obtain the data item from the server through on-demand access or broadcast. Client data caching is particularly important in mobile environments. The following features of caching make it attractive to all participants, including location dependent service providers, users and network managers.

- Caching reduces network bandwidth usage and load on server as a result improves the system throughput.
- Caching reduces user-perceived delays and improves data access performance, since a portion of the queries could be satisfied locally.
- Caching improves data availability in circumstances where the clients are weakly connected or totally disconnected.
- Caching reduces energy consumption since wireless communication not required for cache hits.

Figure 1 depicts a generic architecture of the wireless data dissemination systems studied. We assume that the system assumes point-to-point access for data dissemination. A cache management mechanism exists in the client. Whenever an application issues a query, the local cache manager first checks whether the desired data item is in the cache. If it is cache hit, the cache manager still needs to validate the spatial consistency of the cached item with the master copy at the server.

If the item is valid, it is returned to the application immediately. When a cache miss occurs, the cache manager sends a query request to the server. When the result arrives, the cache manager returns it to the user and retains a copy in the cache. The issue of cache replacement arises when the free cache space is not enough to accommodate a data item to be cached. Replacement schemes decide which data items are to be evicted from the cache to make room for incoming items. There are three main issues involved in the design of caches for mobile devices

Figure 1. Generic system architecture

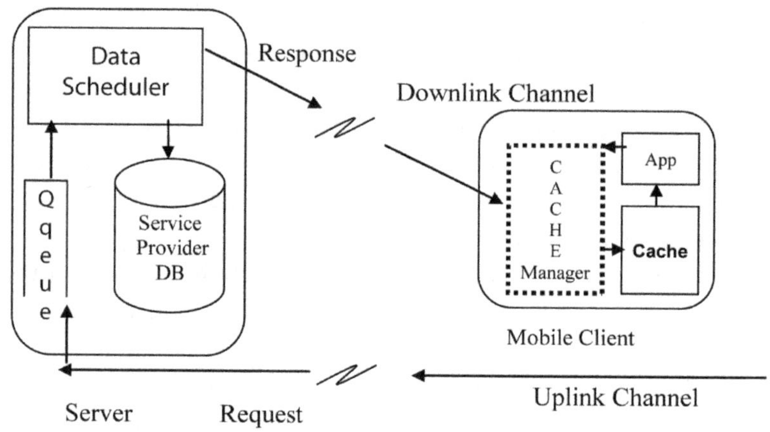

1.1 Cache Replacement

It is the process of selecting a set of victim data items to discard from the cache in order to accommodate new incoming items. A cache replacement policy determines the items to be evicted. Balamash et al. (2004) and Zheng et al. (2002) have studied the effect of replacement mechanisms and have proven that there is an improvement in overall system performance.

1.2 Cache Prefetching

It is the process of preloading data into the client cache in anticipation of future accesses. Burcea et al. (2004) and Kirchner et al. (2004) have made studies on prefetching where a cache prefetching policy or a hoarding policy automatically preloads data into cache for possible future requests. The main purpose of prefetching is to reduce the cache-miss costs.

1.3 Cache Invalidation

A cache consistency scheme or an invalidation scheme maintains consistency between data cached at the client and those at the server. Works of Xu et al. (1999 and 2003) throw light on cache invalidation.

Caching in mobile clients has to consider certain factors which are not present in traditional cache management systems. Since a mobile client may go to disconnected states either voluntarily or involuntarily and move freely between cells, a good caching mechanism should reduce the communication between the server and client. Also, location dependent data takes different values at various locations which bring in the need for good invalidation techniques. These factors along with user perceived latencies make the design of client cache management a challenge in mobile environments.

2. LOCATION DEPENDENT DATA

Location dependent data is the data whose value is determined by the location to which it is related. A location dependent query is a query that is processed on location dependent data and whose result depends on the location criteria explicitly or implicitly specified.

Spatial or location dependent queries are probably one of the most challenging issues in location-based services. Gupta et al. (2005), Marsit et al. (2005) and Seydim et al. (2001) treat a traditional query, called non location-related query, as a query whose answer does not depend on locations or retrieve location data.

Zheng et al. (2002) observe that a location dependent query is a query whose answer depends on the locations of objects where location is an attribute that determines whether the object is part of the answer or not. Location-Dependent Queries according to Jung et al. (2002) are a fundamental building block of LBS. Location-dependent queries are classified according to their purpose. Two major classes of location-dependent queries are explained below.

2.1 Location Dependent Query

Suppose a traveler issues a query to find out information about hotels in the middle of his journey, the response to this query depends on the geographical location of the client. At one place the answer might be a hotel A and at another place the response might be a hotel B. The semantics of a set of data and their values are tightly coupled with a particular location. The answer to a query depends on the geographical location where the query originates.

A Location Dependent Query denoted by Q_i ($1 <= i <= n$) is defined as a tuple < O_i, L_x:L_y, F_i > where

- O_i is the object type being queried,
- (L_x:L_y) is the location of the mobile client issuing the query,
- F_i denotes the filter condition of the query.

For example, <restaurant, 30: 42, restaurant type = French > represents the query "Show me the French restaurant nearest to my current location which is (30, 42)".

2.2 Nearest Neighbor (NN) Queries

According to Tao et al. (2005) NN queries retrieve the object of a certain class which is the closest to a certain object or location (E.g. Given my current location, where is the nearest train station). Chon et al. (2003) and Kolahdouzan et al. (2004)observe that when k objects must be retrieved instead of just the nearest one, they are called kNN queries. Nearest Neighbor queries are also distinguished as static and dynamic queries. Static NN queries are executed against a static database whereas dynaic NN queries deal with target objects that can move. In works of Benetis et al. (2006) and Wu et al. (2008) certain queries related to kNN queries called reverse kNN queries have been studied which retrieve objects that have a specified object or location among their k nearest neighbours.

2.3 Range Queries

Range Queries retrieve the objects located within a certain range or region. Range queries have been studied in the works of Trajcevski et al. (2004), Xu et al. (2003) and Yu etal (2006)

Example: Find all hotels within 5 miles from my current location. Range queries can be static range queries or dynamic range queries, depending on whether the interesting region is fixed or moving. Range queries are also called window queries when the range is a rectangular window as stated by Tao et al. (2003).

3. LOCATION - DEPENDENT CACHE REPLACEMENT POLICIES

Several location-aware cache replacement policies have been proposed for Location dependent Information Services (LDIS) over the past few years and the most predominant ones are explained in this section. In summary, the existing cache replacement polices can be classified into three groups.

Temporal-based cache replacement strategies such as LRU (Least Recently Used) with respect to web caching were studied by Balamash (2004). Policies like LFU (Least Frequently Used) and LRU-K proposed by O'Neil & O'Neil (1993) have also been studied widely in the past. Lam et al. (2000) proposed a policy called invalid-LRU where they observe that the validity of the current values of data items depend on the temporal constraints and they may become out-dated and useless with the passage of time. The policy selects the invalid items in cache for replacement. In case all the items are valid and there is no invalid item in cache, replacement is done using the original LRU principle.

The cache replacement issues for wireless data dissemination were studied in the Broadcast Disk (Bdisk) project by Acharya et al(1995) in which the cached data item with the minimum value of p/x was selected for cache replacement, where p is the item's access probability and x is its broadcast frequency. Their simulation based study showed that this strategy could significantly improve the access latency over the traditional LRU and LFU policies.

The first group of policies mentioned above is based on the assumption that clients' access patterns exhibit temporal locality that is recently or frequently accessed objects are likely to be accessed again in the near future. The next group of policies considers the farthest item for eviction.

Jung et al. (2002)designed the Manhattan Distance-based cache replacement policy to support location dependent queries in urban environments. Cache replacement decisions are made based on the distance between a client's current location and the location of each cached data object. Objects with the highest Manhattan distance from the client's current location are chosen for replacement. While the Manhattan based policy accounts for the geographical adjacency, it is still limited because it ignores the temporal access locality of mobile clients when making cache replacement decisions.

Ren and Dunham (2000) proposed a policy called FAR (Furthest Away Replacement) which uses the current location and movement direction of mobile clients to make cache replacement decisions. Cached objects are grouped into two sets. Data objects in the out direction set are always evicted first before those in the indirection set. Objects in each set are evicted in order based on their distance from the client. FAR considers only the impact of client's current location and client movement and neglects the temporal properties of clients' access pattern. It is also ineffective when mobile clients change direction frequently as objects will often be moved between the in-direction and the out-direction sets. Both the groups discussed above are unsuitable for supporting location dependent services because they do not take into account the combination of spatial and temporal factors. Few works where the cost function comprises of a temporal score and a spatial score are listed below.

PA (Probability Area) and PAID (Probability Area Inverse Distance) proposed by Zheng et al.(2002) are both cost-based replacement policies. When cache replacement takes place, a cost is computed for each cached object and the object with the lowest cost is replaced. The cost function in PA calculates the product of the access probabilities of data objects and the size of their valid scope areas. PA suffers from a further deficiency in that objects that are close to the client are often replaced before objects that are far away, simply because their valid scope area is smaller. PAID extends this by also taking into account the inverse of the distance between the client and the valid scope of the object concerned. While simulation results show that PAID outperforms existing schemes (such as LRU and FAR), the temporal property of data objects is represented by only one parameter, the access probability.

Mobility-Aware Replacement Scheme (MARS) proposed by Lai et al. (2004) behaves like PAID where it considers data updates. It takes into account factors like client access rate, access probability, update probability and client location in order to improve the effectiveness of onboard caching for mobile clients. The Weighted Predicted Region based Cache Replacement Policy proposed by Ajey kumar (2002) is an extension to PAID which uses a weighted data distance of an item along with the probability of data access to find the replacement score.

PAID is the pioneer and the most predominant cache replacement policy in literature which applies spatial and temporal factors for client side data caching in mobile environments. But it does not age the time period between last access of an item and the time of replacement. It is noted that mobile queries experience a popularity drift where the item loses its popularity after the user has used up the corresponding service. Some of such stale items may never be accessed again because the user has exhausted the service. PAID retains popular items with short period of high frequency and large valid scope area in the cache for a long time. As a result PAID might retain stale items in cache preventing fresher items to occupy this space.

4. PROPOSED POLICY

A replacement score is associated with every object in cache. When a new data object needs to be added to the cache and if there is insufficient cache space, the object with the lowest replacement score is removed. Calculation of replacement score takes into account both temporal and spatial properties of information access and the replacement score for data value j of item i is calculated as follows.

$$\text{Replacement Score} = \frac{AF_i * A(VS_{i,j})}{D(VS_{i,j})^k * A_i}$$

Where k = 1 when user is outside the VoronoiK = -1 when user is on or inside the Voronoi.

The temporal parameters are calculated per item and not per data value. AF_i denotes the access frequency for data item i and is calculated using the exponential aging method. Exponential age gives priority to the recent queries and the parameter alpha reflects the weight given to recent queries. The initial value is set to zero and is updated for every request to the data item i. $D(VS_{i,j})$ is a measure of the entry or exit possibility based on the location of the user. A_i represents the age of the data item i and not the data value. AF_i is calculated for every access to the data item i and A_i is computed during the time of replacement. The last time stamp to every data item is already maintained in the cache and inclusion of age does not cause any storage overhead.

The spatial parameters are calculated per data value in cache. It is assumed that an item is more likely to be accessed in future if the possibility of reentering the valid scope is high. The priority concerning data distance is as follows:

- User is inside a valid scope and is far from exiting it.
- User is inside a valid scope and is close to exiting it.
- User is outside the valid scope and is close to entering it.
- User is outside the valid scope and is far from entering it.

Based on these assumptions, the values for data distance $D(VS_{i,j})^k$ vary where k takes the value 1 when user is on or inside the valid scope and k takes the value -1 when user is outside the valid scope. The area of a valid scope plays a key role in eviction as suggested by Zheng (2002). A large valid scope implies that the user would remain there for a longer time thereby increasing the possibility of more queries. If the valid scope is small, the user would get another service when he exits the small Voronoi cell and also the possibility of repeated queries is less. $A(VS_{i,j})$ represents the area of the valid scope of the item. Since the approximated end points of the valid scope (Voronoi polygon) are already stored in the cache for invalidation, the policy does not need any additional parameter to be stored in cache.

The access pattern over different items follows a Zipf distribution with skewness parameter theta. When theta is set to 0, the access pattern is uniform; with increasing theta, the access pattern becomes more skewed. To answer a query, the client first checks its local cache. If the data value for the requested item with respect to the current location is available, the query is satisfied locally. Otherwise, the client submits the query and its current location to the server and retrieves the data from the server.

5. SIMULATION MODEL

This section describes the simulation model used to evaluate the performance of the proposed location-dependent replacement policy. The simulator was built using Java. It consists of a system execution model, server execution model and a client execution model.

The performance of RAAR with existing cache replacement policy PAID with respect to change in mean Query Interval (QI) where Query interval is the time interval between two successive queries. As the query interval increases, cache hit ratio decreases, because the client has low probability to remain in the same valid scope from where the previous query was issued. RAAR shows better performance than PAID for both for

Figure 2. Scope distribution – A (Uniform Voronoi with 120 points)

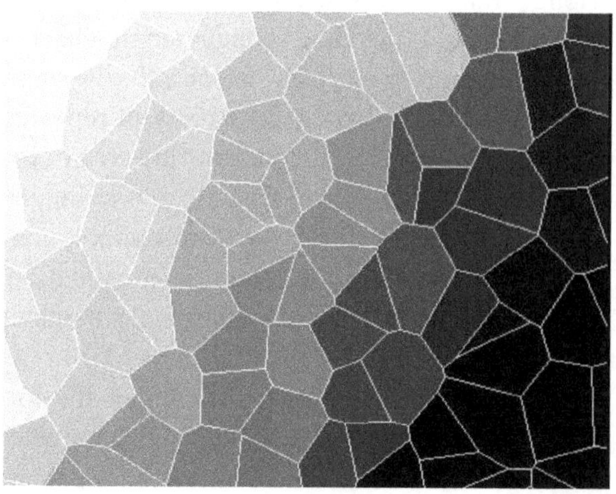

Figure 3. Scope distribution – B (Fuel stations in Turkey)

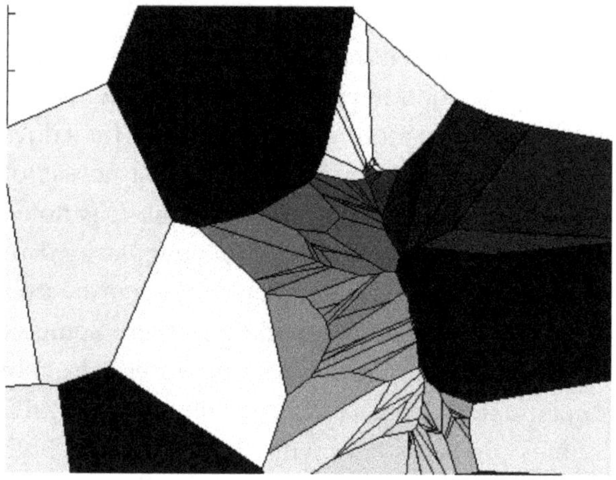

skewed and uniform distribution. Testing was done with small item sets which were active for a given period of time. Since the proposed policy ages the time since the last access, the stale items were evicted and room was created for fresher items. The average improvement perceived was 39%

When the query interval is increased both the policies perform poorly as the client tends to move more between successive query intervals. When query intervals are short, lots of queries are generated within every moving Interval and PAID might retain stale items in the cache whereas RAAR retains hot items in cache which contributes to the increased hit ratio.

The effect of Moving interval (MI) on replacement policies where MI is the time interval during which the client moves with a constant velocity. Longer MI implies that the client does not frequently change its velocity and hence, less random is the client's movement. For small MI, the randomness in client movement is more as compared to longer MI. The new policy performs better than all policies for both small and large MI. Average improvement of the new policy is 38% over the next best policy PAID .

As the MI is increased the performances of both the policies decrease. Longer MI implies that client has a higher possibility of leaving certain regions. Consequently, the cached data are less likely to be reused for subsequent queries, which lead to a worse performance. The new policy calculates data distance in terms of exit and entry probabilities and also retains the hot items where it shows better performance than PAID for long moving intervals.

The cache hit ratio of replacement policies increases with increase in cache size. This is because the cache can hold large number of data items which increases the probability of getting a cache hit. Moreover, replacement occurs less frequent as it happens with low size cache. The new policy consistently outperforms the existing policies from small size cache to large size cache. Average improvement of RAAR over PAID is 38%.

Testing has been done with small item sets which were active for a given period of time. Since the proposed policy ages the time since the last access, the stale items were evicted and room was created for fresher items

In this section, the performance of the new cache replacement policy is tested on access patterns where a popularity drift exists and once popular documents become quickly cold. One approach to create a drift is to intentionally "shut off" any item which has been accessed a fairly large number of times, pretending that these items are never accessed again. This simulates the scenario in which once popular documents quickly become cold. The user was forced to request a set of items for a length of time and these items were shut off and a new set of items were introduced.

The performance of the new policy RAAR is compared with PAID by varying the Query Interval, Moving Interval and Cache Size. The results show an increase in cache hits in both the policies. RAAR shows an increase in performance over normal Zip-f data. This could be explained as follows. Testing has been done with small item sets which were active for a given period of time. Since the proposed policy ages the time since the last access, the stale items were evicted and room was created for fresher items.

6. CONCLUSION

In this chapter, a new cache replacement policy has been proposed that evicts the stale items from the cache considering factors like re-entry possibility, area of valid scope, access frequency and recency of an item. The distinction between entry and exit possibilities and age of the data item play an important role in improving the system performance. A series of experiments were conducted to evaluate the proposed cache replacement strategy. The experimental results indicate that the proposed policy shows a substantial performance improvement over the next best policy PAID.

REFERENCES

Acharya, S., Alonso, R., Franklin, M., & Zdonik, S. (1995). Broadcast disks: Data management for asymmetric communications environments. *Proceedings of ACM SIGMOD Conference on Management of Data*, (pp. 199-210).

Ajey, K., Misra, M., & Sarje, A. K. (2007). A weighted cache replacement policy for location dependent data in mobile environments. *SAC, 07*, 920–924.

Balamash, A., & Krunz, M. (2004). An overview of Web caching replacement algorithms. *IEEE Communications Surveys and Tutorials, 6*, 44–56. doi:10.1109/COMST.2004.5342239

Barbara, D. (1999). Mobile computing and databases: A survey. *IEEE Transactions on Knowledge and Data Engineering, 11*, 108–117. doi:10.1109/69.755619

Benetis, R., Jensen, S., Karciauskas, G., & Saltenis, S. (2006). Nearest and reverse nearest neighbor queries for moving objects. *The VLDB Journal, 15*, 229–249. doi:10.1007/s00778-005-0166-4

Burcea, I., Jacobsen, H., De Lara, E., Muthusamy, V., & Petrovic, M. (2004). Disconnected operation in publish/subscribe middleware. *Proceedings of the International Conference on Mobile Data Management (MDM'04),* (pp. 39–50). IEEE Computer Society.

Chon, H. D., Agrawal, D., & Abbadi, A. E. (2003). Range and KNN query processing for moving objects in grid model. *Mobile Networks and Applications, 8*, 401–412. doi:10.1023/A:1024535730539

Gupta, M., Manghui, T., Khan, L., & Bastani, F. (2005). A study of the model and algorithms for handling location dependent continuous queries. *Knowledge and Information Systems, 8*, 414–437. doi:10.1007/s10115-005-0196-7

Huang, X., & Jensen, C. S. (2004). Towards a streams-based framework for defining location-based queries. *Proceedings of Spatio-Temporal Database Management Workshop,* (pp. 73-80).

Jung, I., You, Y., Lee, J., & Kim, K. (2002). Broadcasting and caching policies for location-dependent queries in urban areas. *Proceedings of the International Workshop on Mobile Commerce,* (pp. 54–60).

Kirchner, H., Krummenacher, R., Edwards, D., & Risse, T. (2004). A location-aware pre-fetching mechanism. *Fourth International Network Conference,* (pp. 453-460).

Kolahdouzan, M., & Shahabi, C. (2004). Continuous k-nearest neighbor queries in spatial network databases. *Proceedings of the Workshop on Spatio-Temporal Database Management,* (pp. 33-40).

Lai, K., Tari, Z., & Bertok, P. (2004). Location–aware cache replacement for mobile environments. *Proceedings of the IEEE Globecom, 04*, 3441–3447.

Lam, K.-Y., Chan, E., & Chun-Hung Yuen, J. (2000). Approaches for broadcasting temporal data in mobile computing systems. *Journal of Systems and Software, 51*(3), 175–189. doi:10.1016/S0164-1212(99)00122-3

Marsit, N., Hameurlain, A., Mammeri, Z., & Morvan, F. (2005). Query processing in mobile environments: A survey and open problems. *Proceedings of the International Conference on Distributed Frameworks for Multimedia Applications,* (pp. 150-157).

O'Neil, E., & O'Neil, P. (1993). The LRU-k page replacement algorithm for database disk buffering. *Proceedings of the ACM SIGMOD,* (pp. 296-306).

Ren, Q., & Dunham, M. (2000). Using semantic caching to manage location dependent data in mobile computing. *Proceedings of the International Conference on Mobile Computing and Networking,* (pp. 210-221).

Seydim, A. Y., Dunham, M. H., & Kumar, V. (2001). Location-dependent query processing. *Proceedings of the Second International Workshop on Data Engineering on Mobile and Wireless Access,* (pp. 47-53).

Tao, Y., Papadias, D., & Shen, Q. (2002). Continuous nearest neighbor search. *Proceedings of the International Conference on Very Large Data Bases* (VLDB'02), (pp. 287-298).

Tao, Y., Sun, J., & Papadias, D. (2003). Selectivity estimation for predictive spatio-temporal queries. *Proceedings of the International Conference on Data Engineering* (ICDE'03), (pp. 417-428).

Trajcevski, G., Wolfson, O., Hinrichs, K., & Chamberlain, S. (2004). Managing uncertainty in moving objects databases. *ACM Transactions on Database Systems, 29*, 463–507. doi:10.1145/1016028.1016030

Wu, W., Yang, F., Chan, C. Y., & Tan, K.-L. (2008). Continuous reverse k-nearest-neighbor monitoring. *Proceedings of the International Conference on Mobile Data Management,* (pp. 132-139).

Xu, B., & Wolfson, O. (2003). Time-series prediction with applications to traffic and moving objects databases. *Proceedings of the International Workshop on Data Engineering for Wireless and Mobile Access*, (pp. 56-60).

Xu, J., Tang, X., & Lee, D. L. (2003). Performance analysis of location-dependent cache invalidation schemes for mobile environments. *IEEE Transactions on Knowledge and Data Engineering, 15*, 474–488. doi:10.1109/TKDE.2003.1185846

Xu, J., Tang, X., Lee, D. L., & Hu, Q. (1999). Cache coherency in location-dependent information services for mobile environment. *Proceedings of the International Conference on Mobile Data Access*, (pp. 182-193).

Yu, B., & Kim, S. H. (2006). Interpolating and using most likely trajectories in moving objects databases. *Proceedings of the International Conference on Database and Expert Systems Applications* (DEXA'06), (pp. 718-727).

Zheng, B., Xu, J., & Lee, D. L. (2002). Cache invalidation and replacement strategies for location-dependent data in mobile environments. *IEEE Transactions on Computers, 51*, 1141–1153. doi:10.1109/TC.2002.1039841

This work was previously published in International Journal of Business Data Communications and Networking, Volume 6, Issue 3, edited by Debashis Saha and Varadharajan Sridhar, pp. 31-48, copyright 2010 by IGI Publishing (an imprint of IGI Global).

Chapter 14

Process Scheduling Using Task Duplication in Heterogeneous Distributed Systems

Jeffrey Elcock
University of the West Indies, Barbados

Pranay Chaudhuri
Heritage Institute of Technology, India

ABSTRACT

Task scheduling in heterogeneous parallel and distributed computing environments continues to be one of the most challenging problems. In this chapter, the authors investigate the Heterogeneous Earliest Finish Time (HEFT) algorithm, along with its alternative scheduling policies for the task prioritising phases, and the Critical Path on a Processor (CPOP) for scheduling tasks on a heterogeneous multi-processor system. It is shown that, by combining the HEFT algorithm selection policy with the task duplication strategy, it is possible to further reduce the schedule length produced by both HEFT and CPOP. The process scheduling algorithm presented in this chapter compares favourably with other algorithms that use a similar strategy. The proposed algorithm has a time complexity of $O(|V|^2(p + d))$*, where* $|V|$ *represents the number of tasks, p represents the number of processors, and d the maximum in-degree of tasks.*

1. INTRODUCTION

Process scheduling in a multiprocessor environment refers to the assignment of tasks to different processors, satisfying a given set of constraints, so that the schedule length or overall task completion time is minimized. A process (or an application) consists a set of interdependent tasks and may be represented by a directed acyclic graph (DAG), $G = (V, E)$, where the set of nodes V, represents the set of tasks, and the set of edges E, represents the communication between tasks. If $(n_i, n_j) \in E$, then n_i is called the immediate

DOI: 10.4018/978-1-61350-477-2.ch014

predecessor of n_j, and n_j is called the immediate successor of n_i. Given a task n_a, its set of immediate predecessors, denoted by $ipred(n_a)$, is defined as $\{n_j | (n_j, n_a) \in E\}$. If n_a has two or more immediate predecessors then n_a is referred to as a join task. The immediate successors of n_a is denoted by $isucc(n_a)$ and is defined as $\{n_j | (n_a, n_j) \in E\}$. $D(n_a)$ is the set of all the descendants of task n_a, and is defined as $\bigcup_{n_j \in isucc(n_a)} (\{n_j\} \cup D(n_j))$. It is assumed that there is one entry task n_α and one exit task n_ω for the DAG.

In both homogeneous and heterogeneous environments the process scheduling problem is NP-complete (Garey & Johnson, 1979; Ullman, 1975; Graham, Lawler, Lenstra, & Kan, 1979; Kasahara & Narita, 1984). In homogeneous environments several heuristics were proposed using both list scheduling without task duplication (Wu & Gajski, 1990; Ahmad & Kwok, 1996) and list scheduling using task duplication (Ahmad & Kwok, 1998; Bansal, Kumar & Singh, 2003; Chaudhuri & Elcock, 2005; Park & Choe, 2002). In list scheduling, tasks are assigned a priority and based on that priority a task is scheduled. In task duplication, some tasks are deliberately scheduled on more than one processor to reduce inter-processor communication.

A collection of a diverse set of resources interconnected, in most cases, with a high speed network defines as a heterogeneous/grid environment. A similar list scheduling approach has also been taken in such an environment. The Heterogeneous Earliest Finish Time (HEFT) algorithm along with its alternative scheduling policies, and the Critical Path on a Processor (CPOP) algorithm use this approach (Topcuoglu, Hariri & Wu 2002). The HEFT algorithm and its variations, firstly assigns a weight to each node based on the average computational cost, and to each edge, based on the average communication cost. Secondly, they assign a rank, using a rank function which takes into account the assigned weights, to each node. Thirdly, the nodes are prioritized on the basis of their rank value. A task is then scheduled to a processor, which returns the earliest completion time. The CPOP algorithm, on the other hand, also assigns a weight to a task based on a rank function, but it first selects the critical path tasks, defined as those tasks with the highest, but equal, weights. The processor which minimizes the cumulative computation costs of the critical path tasks is then selected and identified as the critical-path processor. Only critical path tasks are scheduled on the critical-path processor while the other tasks are scheduled on any of the remaining processors which return the earliest completion time of the task.

The Dynamic Critical Path Duplication (DCPD) algorithm (Liu, Li, Lai & Wu, 2006), is a duplication-based algorithm which uses a dynamic list-based approach for scheduling in a heterogeneous/grid environment. The DCPD algorithm determines the critical path of the DAG and dynamically selects the next node for scheduling. At each scheduling step the DCPD algorithm identifies the next task for scheduling as the dominant task (*D*tsk). To identify the *D*tsk, the DCPD algorithm first identifies the ready list – tasks where their immediate predecessors have already been scheduled. Each task n_a in the ready list is then assigned a value based on three components. The first component is *b-level*(n_a) – the length of the longest path from task n_a to the exit node. The second component is *u-level*(n_a) – the shortest average length from the entry node to n_a; and thirdly, the average computational cost of task n_a. The dominant task, *D*tsk, is the task with the maximal value of *b-level*(n_a) added to the average computational cost of n_a minus *u-level*(n_a). The DCPD algorithm then identifies the dominant processor, *Dpro*, as the one which allows the *Dtsk* to complete the earliest, utilizing idle slots whenever possible during the identification process. The *Dtsk* is then scheduled on the *Dpro*.

An algorithm called the Heterogeneous Dynamic Critical Path Duplication (HDCPD), which is also a duplication-based strategy for scheduling tasks on a set of heterogeneous processors, has been presented in (Fei, Xiaoli, Changjun, & Rong, 2009). Its strategy is to identify the dynamic critical path (DCP) at each scheduling step with the aim of scheduling these nodes so as to reduce waiting time for them. The HDCPD calculates the earliest start time (EST), the latest start time (LST) and the favorite processor (F_{pr}) for each node at each scheduling step. The F_{pr} for a node is the processor which offers the minimum runtime for that node. The HDCPD algorithm first determines the nodes of the initial critical path of the DAG. These nodes are identified as the Critical Path (CP) nodes and are all scheduled on the same processor during the scheduling process. The processor (F_{cp}) which minimizes the sum run time of the CP nodes is the one selected for scheduling the CP nodes. At each scheduling step the DCP is determined and the node with minimum difference of LST and EST is scheduled to the processor which gives the minimum of the earliest completion time plus the EST of its most important successor node – the node where its completion time is estimated to be the latest.

In this paper, we have combined the idea of task duplication with the ranking concept used in the HEFT algorithm and developed an algorithm which outperforms the HEFT algorithm, the six (6) alternative policies for the task prioritizing phase of the HEFT algorithm, the CPOP algorithm and the duplication-based algorithms; the DCPCD and the HDCPD. The rest of the paper is organized as follows. Section 2 introduces the basis of the algorithm. The proposed algorithm and its complexity are provided in Section 3. In section 4, we have presented an illustrative example. Section 5 provides experimental results and compares these results with the performances of the DCPD algorithm and the HEFT algorithm and its variations. Finally, Section 6 concludes the paper.

2. BASIS OF THE ALGORITHM

The computing environment assumed in this paper consists of a set P of p interconnected heterogeneous processors. For each task $n_i \in V$, $\tau(n_{i,k})$ denotes the computational cost of task n_i on processor p_k. $\tau(n_{i,k})$ values are stored, in our model, as a |V| x |P| computational matrix C. The average computational cost is therefore defined as:

$$\overline{\tau(n_i)} = \sum_{k=1}^{p} \tau(n_{i,k}) / p \quad (1)$$

The data transfer rates are stored in a |P| x |P| matrix H. Each processor has a start-up cost, and is represented in a |P| dimensional array S. An edge from n_i to n_j; where $n_i, n_j \in V$, represents the precedence relationship between these two tasks. The communication cost, c_{n_i,n_j}, where task n_i is scheduled on processor p_m and task n_j is scheduled on processor p_l is represented as follows:

$$c_{n_i,n_j} = S_m + \frac{DATA_{n_i,n_j}}{H_{l,m}} \quad (2)$$

where S_m is the startup cost on processor m; $DATA_{n_i,n_j}$ is the amount of data to be transferred from task n_i to task n_j; and $H_{l,m}$ is the data transfer rate between processor l and processor m. Note that $c_{n_i,n_j} = 0$ if $l = m$. If we let $\overline{S}$ be the average start-up communication cost, and $\overline{H}$ be the average transfer rate among processors, our average communication cost becomes

$$\overline{c_{n_i,n_j}} = \overline{S} + \frac{DATA_{n_i,n_j}}{\overline{H}} \tag{3}$$

The HEFT algorithm (Topcuoglu, Hariri & Wu 2002), which has taken a two phased approach to its scheduling, works in the following manner. In the first phase, the aim is to assign to each node in the DAG a weight, which is derived from its average computation cost. Similarly, each edge is assigned a weight based on its average commutation cost. The DAG is then traversed upward, assigning a rank value to each node. The upward rank of a task is defined as follows:

$$rnk_{up}(n_i) = \overline{\tau(n_i)} + \max_{n_j \in\ isucc(n_i)} (\overline{c_{n_i,n_j}} + rnk_{up}(n_j)) \tag{4}$$

The upward rank is computed recursively starting with the exit task. The rank of task $n_\omega = \overline{\tau(n_\omega)}$. For all other tasks, the rank value of each task n_i is calculated as follows. For each immediate successor n_j, its rank is added with the average communication cost c_{n_i,n_j}. The immediate successor with the largest value is selected. Task n_i is then assigned a rank value which is equal to the largest value just selected, plus its average computation cost.

The second phase is the processor selection phase. The HEFT algorithm has an insertion base policy which takes into account the possibility of inserting a task into an idle time slot. When this policy is not used, the earliest start time of task n_i on processor p_k is defined as follows

$$EST(n_i, p_k) = \max\ \{CT(p_k),\ \max_{n_r \in ipred(n_i)} (ECT(n_r, p_k) + c_{n_r,n_i})\ \} \tag{5}$$

where $CT(p_k)$, the completion time of processor p_k, is the completion time of the last task scheduled on p_k. In the insertion based policy, task n_i is not necessarily the last task scheduled on the processor. The aim of this policy is that once all the input data for n_i is received, following the existing precedence constraints, the task n_i is scheduled at the first accommodating idle time slot. This time slot could between two adjacent tasks already scheduled or as the last task on the processor. The earliest completion time of task n_i becomes

$$ECT(n_i, p_k) = \tau(n_i, p_k) + EST(n_i, p_k) \tag{6}$$

The proposed algorithm, Heterogeneous Earliest Finish Time using Task Duplication (HEFT-TD), which is similar to HEFT algorithm with respect to its selection policy, also uses a two-phase approach. However, in the second phase, when a processor, which gives the earliest completion time, is selected to assign task n_i, the idle time slots with the immediate predecessor(s) of task n_i are utilized, where ever possible, to further reduce the start time. After selecting the processor, we sort in decending order the data arrival time of those immediate predecessor(s) of task n_i which are not scheduled on the selected processor. The immediate predecessor with the largest data arrival time is then duplicated at its earliest start time on the selected processor. If this insertion results in a reduction in the start time of task n_i, then we proceed with the insertion of the next immediate predecessor. This process is continued until there are no more immediate predecessors or the start time of n_i cannot be reduced any further.

3. THE ALGORITHM

A detailed description of the algorithm is given.

Complexity Analysis

In the HEFT-TD algorithm, lines 1 through 4 require $O(|E|+|V|)$ time. The rest of the algorithm consists of a while loop. The body of the while-loop is made up of a for-loop, a repeat statement and a few assignment statements. The for-loop is executed p (the number of processors) times, and for each execution of the body of the for-loop, it takes $O(|V|)$ time. Therefore lines 7-9 has a time complexity of $O(p|V|)$. Lines 10 and 12 are both assignment statements and each requires $O(1)$ time. Let d be the maximum number of immediate predecessors of a task. Line 11 sorts, in descending order, the immediate predecessors based on the data arrival time and therefore has a time complexity of $O(d \log d)$. The body of the repeat statement (lines 13 through 19) duplicates all or some of the immediate predecessors in the selected processor, which has at most $|V|$ tasks loaded. Since the repeat statement is repeated d times, it has a complexity of $O(d|V|)$. Combining the complexities of all the steps, the overall complexity of the body of the while loop is $O(p|V|+d|V|)$ time. Since the while loop is executed $|V|$ times, the overall complexity of the algorithm HEFT-TD is $O(|V|(p|V|+d|V|))$, that is, $O(|V|^2(p+d))$.

Comparatively, the complexity of the DCPD algorithm is $O((|V|+|E|)|V|p)$ which is no better

Algorithm HEFT-TD $(V, E, P, H, C, S, \tau, c)$

```
1. For each task compute its mean computation cost
2. For each edge compute its mean communication cost
3. Compute the rnk_up for each task
4. sortLIST ← Sort the tasks in descending order of rnk_up
5. while (|sortLIST |> 0) do
6. select first task n_j, from sortLIST
7. for each processor p_i do
8.     compute ECT(n_j, p_i) using the insertion-base scheduling policy
9. od
10.    p_min ← processor p_k which minimizes the ECT of task n_j
11.    sortIMP ← Sort in descending order the data arrival time of ipred(n_i)
w.r.t p_min
12.    NewStartTime(n_j, p_min) = EST(n_j, p_min)
13.    repeat
14.            StartTime = NewStartTime(n_j, p_min)
15.            select first task t_i from sortIMP list
16.            insert t_i into p_min
17.            compute NewStartTime(n_j, p_min)
18.            sortIMP ← sortIMP - { t_i}
19.    until (|sortIMP |= 0 or NewStartTime(n_j, p_min) >= StartTime)
20.    Assign task n_j to p_min at min (StartTime, NewStartTime(n_j, p_min))
21.    sortLIST ← sortLIST - {n_j}
22. od
```

the overall runtime by scheduling these nodes in other processors is never explored.

The metrics used for the performance evaluation are Speedup, Schedule Length Ratio (SLR), Number of Occurrences of Better Quality of Schedule, and Running Time of the Algorithms, which are all explained in the following section.

5.1 Comparative Metrics

1. *Speedup:* The ratio between the sequential time and the parallel execution time of a process is defined as the speedup. The sequential time is calculated by adding, sequentially, the computational cost of each task in the graph. This is done for each processor and then the smallest value is used. The parallel execution time is the completion time of the graph, which is also referred to as the Scheduled Length (SL). Therefore

$$Speedup = \frac{\min_{p_k \in P}\{\sum_{n_i \in V} \tau(n_{i,k})\}}{SL}$$

2. *Schedule Length Ratio:* The Schedule Length (SL) is the main performance measure of a scheduling algorithm. In our experiment, a large set of task graphs with varying properties is used and therefore it becomes necessary to normalize the schedule length to the lower bound. This is called the Schedule Length Ratio (SLR). The SLR is defined as follows:

$$SLR = \frac{SL}{\sum_{n_i \in CP_{MIN}} \min_{p_k \in P}\{\tau(n_{i,k})\}}$$

   The denominator is the summation of the minimum computation costs of the tasks on the CP_{MIN} (minimum Critical Path). The CP_{MIN} is derived by first setting each task (n_i) to its minimum computational cost and calculating the length of the Critical Path (|CP|) using these values.

3. *Number of Occurrences of Better Quality of Schedules:* In our experiments while comparing two algorithms, we report the number of times an algorithm performs better than, worse than or equally with the other algorithms out of a large number of simulations with different parameters for the task graph.
4. *Running Time of the Algorithm:* The length of time an algorithm takes to schedule all tasks of the DAG is referred to as the running time of the algorithm.

5.2 Randomly Generated DAGS

In our experiment, the following input parameters are used:

- Number of tasks in the DAG (V).
- Outdgree of a node (Out_{degree}).
- As in Topcuoglu et al. (Topcuoglu, Hariri & Wu 2002), the shape parameter of the graph (α).
- Communication to computation ratio (*CCR*). It is the ratio between the average communication cost and the average computation cost.
- As in Topcuoglu et al. (Topcuoglu, Hariri & Wu 2002), the range percentage of computation costs on processors (β), is the heterogeneity factor for processors. If the percentage value is high, the difference in computation cost across the processors is significant. On the other hand, if this value is low then the computation cost of the task across processors is almost equal. For each task n_i, the average computation cost $\overline{\tau(n_i)}$ is selected randomly from a uniform distribution with the range $0,\ 2 - \overline{\Gamma(n)_{DAG}}$, where $\overline{\Gamma(n)_{DAG}}$ is the average computation cost of the given graph. For each task

n_i and processor p_k, the computation cost is randomly set so as to satisfy the following:

$$\overline{\tau(n_{i,k})} \times \left(1-\frac{\beta}{2}\right) \leq \tau(n_{i,k}) \leq \overline{\tau(n_{i,k})} \times \left(1+\frac{\beta}{2}\right)$$

For each experiment, the values discussed above, are assigned from the sets given below. Note that, in the set of Out_{degree}, the last value is the value of the number of nodes in the graph.

- SetOfNodes (V) = {20, 40, 60, 80, 100}
- Set of CCR (CCR) = {0.1, 0.5, 1.0, 5.0, 10.0}
- Set of Alpha (α) = {0.5, 1.0, 2.0}
- Set of OutDegree (Out_{degree}) = {1, 2, 3, 4, 5, V}
- Set of Beta (β) = {0.1, 0.25, 0.5, 0.75, 1.0}

With these combinations we created 2,250 different DAGs. Twenty-five (25) random DAGs were generated for each DAG type giving total of just over 56,000 DAGs for our experiment. With several input parameters and selecting each parameter from a large set ensured that the generation of diverse DAGs with different characteristics and thereby preventing any possible bias towards any one particular type of algorithm.

5.3 Performance Results

In this section, we present a performance comparison of the algorithms with respect to the different task graph features. In our first experiment, we look at the SLR performance. It was observed that the average SLR value for the HEFT-TD algorithm, as shown in Figure 2, was better than that of all others. It was better than the HEFT algorithm by 12 percent, the DCPD algorithm by 4 percent, the HDCPD algorithm by 5 percent, the CPOP algorithm by 13 percent, and approximately, 10 percent better than the average performance of the A-alternatives: HEFT-A1, HEFT-A2 and HEFT-A3. It also performed 21 percent better than the average performances of the B-alternatives: HEFT-B1, HEFT-B2 and HEFT-B3.

In Figure 3, for the average speedup, our experiment shows that the HEFT-TD algorithm outperforms all others. It is 8 percent better than the HEFT algorithm, 3 percent better than the DCPD and the HDCPD; and 8 percent better than the CPOP algorithm. It is 5 percent better than the A-alternatives and 16 percent better than the B-alternatives. For the average running time shown

Figure 2. Average SLR

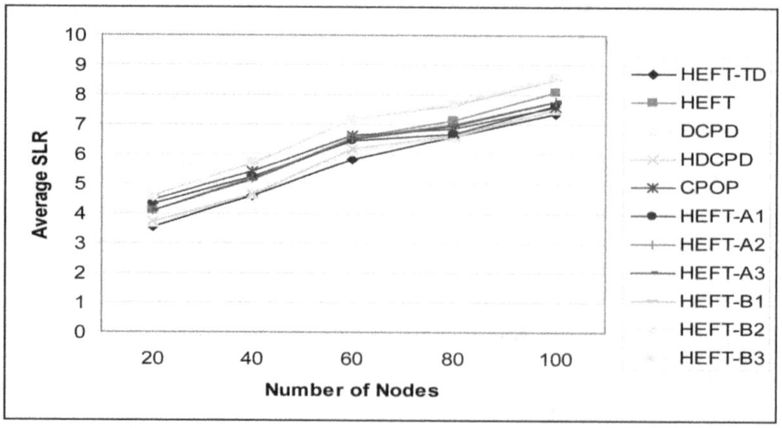

Figure 3. Average speedup

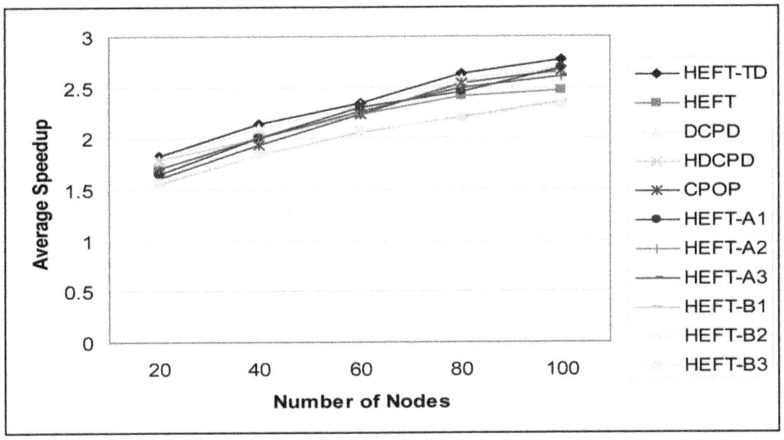

Figure 4. Average running time

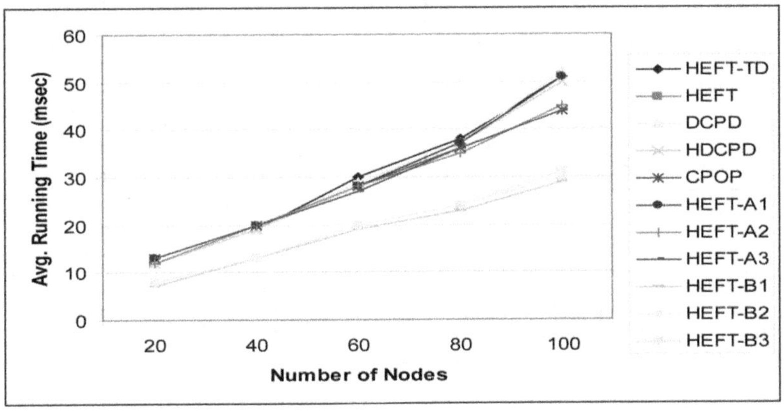

in Figure 4, HEFT-TD performed on par with all others for task graphs in which the number of nodes is less than 60 nodes. When the number of nodes is 60 or more, the HEFT-TD is slower as well as the DCPD algorithm. Of all the algorithms, the B alternatives, on an average, produce the best running times.

When the CCR values are compared, the HEFT-TD algorithm performed the best. The HDCPD performed better, and the DCPD performed evenly when compared to the HEFT-TD algorithm when the CCR values are very small (CCR < 0.5) and the number of nodes were 20. Also for these small CCR values, the A-alternatives, the HDCPD and the CPOP algorithms show a slightly better performance than both the HEFT-TD algorithm and the DCPD algorithm when the number of nodes is 40 or more. On an average, the HDCPD algorithm, the B-alternatives as well as the CPOP algorithm showed their best performances when the CCR < 5. When the comparisons are made with respect to α, the shape parameter of the graph, again the HEFT-TD algorithm performs better than the others. When $\alpha = 2.0$ (short task graphs with high parallelism), the CPOP algorithm, the HDCPD and the HEFT-A1 algo-

Table 2. Pair-wise comparison of the scheduling algorithms

		HEFT-TD	HEFT	DCPD	HDCPD	CPOP	HEFT-A1	HEFT-A2	HEFT-A3	HEFT-B1	HEFT-B2	HEFT-B3
HEFT-TD	better	*	39828	30652	33178	42796	38508	40475	40452	48643	49192	49126
	=		8025	4723	3751	396	1173	3106	2141	570	536	558
	worse		7397	20875	19431	13058	16569	12669	13657	6137	6522	5566
HEFT	better	7397	*	16012	18057	35318	27828	20226	23494	41141	41003	41126
	=	8025		2561	2843	848	2783	18423	10882	1215	2598	1922
	worse	39828		37677	35347	20084	25639	17601	21874	13894	12649	13202
DCPD	better	20875	37677	*	31632	41429	36855	37903	37940	46962	47418	47400
	=	4723	2561		3446	397	1012	2194	1772	485	542	523
	worse	30652	16012		21172	14424	18383	16153	16538	8803	8290	8327
HD-CPD	better	19431	35347	21172	*	39948	33801	35247	35703	46257	46931	47010
	=	3751	2843	3446		538	1206	2314	1983	557	677	729
	worse	33178	18057	31632		15764	21243	18689	18554	9436	8642	8511
CPOP	better	13058	20084	14424	15764	*	14176	19739	19302	33504	33774	33696
	=	396	848	397	538		5210	969	1038	506	538	548
	worse	42796	35318	41429	39948		36864	35542	35910	22240	21938	22006
HEFT-A1	better	16569	25639	18383	21243	36864	*	26036	25795	38957	39183	39084
	=	1173	2783	1012	1206	5210		3301	3534	603	617	670
	worse	38508	27828	36855	33801	14176		26913	26921	16690	16450	16496
HEFT-A2	better	12669	17601	16153	18689	35542	26913	*	17250	39918	39537	39727
	=	3106	18423	2194	2314	969	3301		21301	1171	2527	1869
	worse	40475	20226	37903	35247	19739	26036		17699	15161	14186	14654
HEFT-A3	better	13657	21874	16538	18554	35910	26921	17699	*	40126	39982	40070
	=	2141	10882	1772	1983	1038	3534	21301		1107	1811	1443
	worse	40452	23494	37940	35703	19302	25795	17250		15017	14457	14737
HEFT-B1	better	6137	13894	8803	9436	22240	16690	15161	15017	*	25795	24156
	=	570	1215	485	557	506	603	1171	1107		5410	8278
	worse	48643	41141	46962	46257	33504	38957	39918	40126		25045	23816
HEFT-B2	better	6522	12649	8290	8642	21938	16450	14186	14457	25045	*	13798
	=	536	2598	542	677	538	617	2527	1811	5410		28166
	worse	49192	41003	47418	46931	33774	39183	39537	39982	25795		14286
HEFT-B3	better	5566	13202	8327	8511	22006	16496	14654	14737	23816	14286	*
	=	558	1922	523	729	548	670	1869	1443	8278	28166	
	worse	49126	41126	47400	47010	33696	39084	39727	40070	24156	13798	

rithm perform better than the DCPD algorithm and the proposed algorithm by 12 percent for task graphs with nodes greater than 60.

The final comparison which is made is shown in Table 2. We compared each algorithm with the other remaining algorithms noting the number of times each produced better, worse or equal schedule lengths. An algorithm in each row is compared to each algorithm noted in the column of the table. Our proposed algorithm, as depicted in the first row, clearly outperforms all others when one considers the "better than" entries associated with it. Overall, the rate at which the proposed algorithm produces better schedule lengths was 57 percent better when compared to the HEFT algorithm, 17 percent when compared to the DCPD algorithm, 26 percent when compared to the HDCPD algorithm, 53 percent when compared to the CPOP algorithm, 46 percent to the average of the A-alternatives, and 76 percent to the average of the B-alternatives.

6. CONCLUSION

In this paper, we presented an algorithm termed as HEFT-TD, based on task duplication, for scheduling tasks in a heterogeneous multiprocessor environment. The performance of the HEFT-TD algorithm was compared to the HEFT algorithm, six (6) alternative policies for the task prioritizing phase of the HEFT algorithm, the DCPD algorithm, the HDCPD algorithm and the CPOP algorithm. It may be noted that in the DCPD algorithm (Liu, Li, Lai & Wu, 2006) and the HDCPD algorithm (Fei, Xiaoli, Changjun, & Rong, 2009), task duplication is used. Our experiment showed that the proposed algorithm significantly outperformed the other algorithms when the metrics average schedule length, speedup, and frequency at which the schedule length is better, were used. Using the metric average running time, the performance of the proposed algorithm, is comparable to the others.

REFERENCES

Ahmad, I., & Kwok, Y. K. (1996). Dynamic critical-path scheduling: An effective technique for allocating tasks graphs to multiprocessors. *IEEE Transactions on Parallel and Distributed Systems*, *7*(5), 560–521.

Ahmad, I., & Kwok, Y. K. (1998). On exploiting task duplication in parallel program scheduling. *IEEE Transactions on Parallel and Distributed Systems*, *9*(9), 872–892. doi:10.1109/71.722221

Bajaj, R., & Agrawal, D. P. (2004). Improving scheduling of tasks in a heterogeneous environment. *IEEE Transactions on Computers*, *15*(2), 107–118.

Bansal, S., Kumar, P., & Singh, K. (2003). An improved duplication strategy for scheduling precedence constrained graphs in multiprocessor systems. *IEEE Transactions on Parallel and Distributed Systems*, *14*(6), 533–544. doi:10.1109/TPDS.2003.1206502

Chaudhuri, P., & Elcock, J. (2005). Scheduling on multiprocessors using task duplication. In D. Tutsch (Ed.), *Proceedings of the 2005 Design, Analysis and Simulation of Distributed Systems (DASD 2005)* (pp. 28-35). San Diego, CA: SCS.

Fei, Y., Xiaoli, D., Changjun, J., & Rong, D. (2009). Directed acyclic task graph scheduling for heterogeneous computing systems by dynamic critical path duplication algorithm. *Journal of Algorithms & Computational Technology*, *3*(2), 247–270. doi:10.1260/174830109787914001

Garey, M. R., & Johnson, D. S. (1979). *Computers and intractability: A guide to the theory of NP-completeness* (pp. 238–241). W.H. Freeman.

Graham, R. L., Lawler, E. L., Lenstra, J. K., & Kan, A. H. G. R. (1979). Optimization and approximation in deterministic sequencing and scheduling: A survey. *Annals of Discrete Mathematics*, *5*, 287–326. doi:10.1016/S0167-5060(08)70356-X

Hagras, T., & Janecek, J. (2005). A high performance, low complexity algorithm for compile-time task scheduling in heterogeneous systems. *Parallel Computing*, *31*(7), 653–670. doi:10.1016/j.parco.2005.04.002

Kasahara, H., & Narita, S. (1984). practical multiprocessor scheduling algorithms for efficient parallel processing. *IEEE Transactions on Computers*, *33*(11), 1023–1029. doi:10.1109/TC.1984.1676376

Liu, C.-H., Li, C.-F., Lai, K.-C., & Wu, C.-C. (2006). Dynamic critical path duplication task scheduling algorithm for distributed heterogeneous computing systems. In *Proceedings of the 12th International Conference on Parallel and Distributed Systems (ICPADS'06)*, (pp. 365-374). Washington, DC: IEEE Computer Society.

Park, C.-L., & Choe, T.-Y. (2002). An optimal scheduling algorithm base on task duplication. *IEEE Transactions on Computers*, *51*(4), 444–448. doi:10.1109/12.995454

Topcuoglu, H., Hariri, S., & Wu, M. (2002). Performance-effective and low-complexity task scheduling for heterogeneous computing. *IEEE Transactions on Parallel and Distributed Systems*, *13*(3), 260–274. doi:10.1109/71.993206

Ullman, J. D. (1975). NP-complete scheduling problems. *Journal of Computer and System Sciences*, *10*(3), 384–393. doi:10.1016/S0022-0000(75)80008-0

Wu, M., & Gajski, D. (1990). Hypertool: A programming aid for message passing systems. *IEEE Transactions on Parallel and Distributed Systems*, *1*(3), 384–343. doi:10.1109/71.80160

This work was previously published in International Journal of Business Data Communications and Networking, Volume 6, Issue 1, edited by Debashis Saha and Varadharajan Sridhar, pp. 58-69, copyright 2010 by IGI Publishing (an imprint of IGI Global).

Section 4
Network Economy:
New Business Models

Chapter 15
Using Real Options Theory to Evaluate Strategic Investment Options for Mobile Content Delivery:
A Case Study

Divakaran Liginlal
Carnegie Mellon University

Lara Khansa
Virginia Polytechnic Institute

Stella C. Chia
City University of Hong Kong

ABSTRACT

With a rich fare of localized content, but limited regional media outlet channels, mobile content generates new business opportunities for Media News, a small media company with considerable growth potential. Two business models are considered: partnering with wireless service providers and strategic alliances with mobile content syndicators. First, the models are evaluated based on their resource requirements, market share acquisition, revenue generation, and nature, scope and control of content and bandwidth. Then, real options analysis is used to value Media News' managerial flexibility in responding to uncertainty in investment choices specific to the media industry. The modeling approach, analytical methods, and decision support tools employed in this paper serve as exemplar for engineering managers involved in strategic investment decisions, especially in emerging areas such as mobile commerce.

DOI: 10.4018/978-1-61350-477-2.ch015

INTRODUCTION

The growth of the mobile content market has been fueled by factors such as increasing subscriber growth, maturing network technologies, and mushrooming content types. Market research leader, Juniper Research, predicts that the mobile content market will reach $17.5 Billion in 2012 worldwide from $9 billion in 2007 (http://juniper-research.com/). Media companies worldwide are capitalizing on this market growth by investing in mobile content services. In this article, we present systematic methods to value investments in the media industry through a real-life example of *Media News*, a growing media company. *Media News* established an image of a lifestyle, business, and sports magazine. It covers multifarious topics, including home and garden, food and dining, arts, and special surveys, such as an annual feature about the best doctors in town. *Media News'* monthly circulation is about 20,000 and its readership is estimated at about 69,700, primarily drawn from the upper-middle class. About 85 percent of its income is from advertising while the remainder is derived from subscription and newsstand sales. *Media News* has approximately a 10 percent share of the local market as of late 2008, which is estimated to be about 15 percent of households. At the time of this study, the magazine's 17 employees included two editors, and administrative and advertising support personnel. Freelancers hired on contract produced most of the articles and pictures. *Media News* is an emerging company with substantial growth prospects. To enhance its market base and provide more real-time content, *Media News* entered into an alliance with two TV channels, MidWest-TV and BPN 14, and a web service provider, Channel 4000. *Media News'* web site has its own customized search engine and presents a variety of timely news from business to college sports. *Media News* also broadcasts its content on its TV partners' web sites, in addition to exhibiting links to its partners' real-time headlines on its own web site. *Media News'* goals behind these partnerships are to disseminate its content via multiple channels and increase circulation, thus guaranteeing a continuous source of revenues.

The traditional net present value (NPV) approach to valuing investments consists of selecting projects that yield expected returns in excess of the return required in financial markets from assets of comparable risk. The NPV approach implicitly assumes pre-commitment to future plans and considers investment decisions as "now or never" propositions. Further, van Putten and Macmillan (2004) argued that employing traditional DCF analysis to value investment decisions of growing companies produces inaccurate results, especially when the DCF value of the project is barely positive or slightly negative. Investment decisions should therefore be based on an expanded NPV criterion, which incorporates the flexibility value along with the direct NPV of expected cash flows from an immediate investment. Real options analysis (ROA) has been particularly used to value investments of media firms who are increasingly facing pressure to adapt quickly to the dynamics of the industry. Dimpfel et al. (2002) argued that ROA is particularly successful in emphasizing the high degrees of uncertainty and irreversibility peculiar to the media industry and, if used correctly, can contribute to the reactive and proactive management of flexibility in media companies. The authors argued that media companies exhibit a high degree of irreversibility due to their firm-specific, low reselling value and high operating expenses attached to the costs of infrastructure and marketing. *Media News*, being a small media company with considerable growth potential amid uncertainty, is clearly a good candidate for applying real options theory. The analytical methods employed in this paper provide useful insights to engineering managers who are constantly faced with the dilemma of which projects to invest in and which ones to abandon. The illustrations developed demonstrate how managerial decisions may be facilitated with the aid of low cost real options tools.

The remainder of this paper is organized as follows. In the next section, we evaluate two strategic models for *Media News* based on a qualitative cost-benefit analysis. We then undergo a quantitative analysis based on ROA. Using sensitivity analyses, we show that ignoring the valuation of these options can lead to the wrong decision. Finally, we provide a foundation to generalize the results of investment in the technology sector, specifically in the areas of mobile gaming and commerce.

MOBILE BUSINESS MODELS FOR *MEDIA NEWS*

Growing firms' alliances are most valuable when built with large firms that possess leading-edge technological resources. *Media News*, being a young and dynamic company, is likely to gain considerable consumer endorsement for its products and services through a strategic alliance with established service providers. Such strategic alliances fall under the category of co-specialization, thus providing the ability to exploit the synergies resulting from combining *Media News'* resources with those of its potential partners. *Media News* is also making connections with other businesses around the region by displaying their ads on its website. We propose two strategic alliances for *Media News*- one with wireless service providers (WSPs) and the other with mobile content syndicators.

Model 1: Partnering with WSPs

In this model, *Media News* hosts the entire content and actively recruits subscribers. The WSP provides the necessary technical infrastructure for content delivery to the subscriber. A direct relationship is established between the content provider and its subscribers. We apply the schematics from Weill and Vitale (2001) to depict this business model in Figure 1. *Media News* is responsible for content provision, direct interaction with subscribers and advertisers, establishment of customer relationships, and market campaign design while the WSP takes care of the technology infrastructure. A distinct feature of wireless delivery is customized information. This customization is *Media News'* sole responsibility. Reuters (http://mobile.reuters.com) was among the first media companies to experiment with such a strategic alliance to deliver financial markets' news and data to WAP-enabled phones and other handheld devices. Ericsson supplied the technology, terminals, and wireless data communication expertise, while Reuters provided news and data and introduced customers from the financial services sectors.

From a technological perspective, the wireless application protocol (WAP) provides the basis for a micro browser built into mobile devices. The mobile profile markup language (XHTMLMP) provides the source for new content, while the wireless markup language (WML) supports legacy WAP content. Markup languages facilitate the presentation layer services for wireless devices while capitalizing on their unique advantages. The accompanying client-side scripting language, WMLScript, provides for additional intelligence and presentation control. WAP gateways transform data from web sites into suitable formats for display in mobile devices. Subscribers pay for wireless service and content. The subscription fee and advertising income that *Media News* receives may also be shared with the service provider.

Model 2: Partnering with Content Syndicators

Model 2, which we present in Figure 2, allows mobile users to receive information from a variety of content sources through web-to-wireless service providers. Most news organizations adopting this model are presumed to have already set up their own web sites. *Media News* has no direct business links with wireless subscribers. The web-to-wireless providers recruit and interact with

Figure 1. Partnering with WSPs

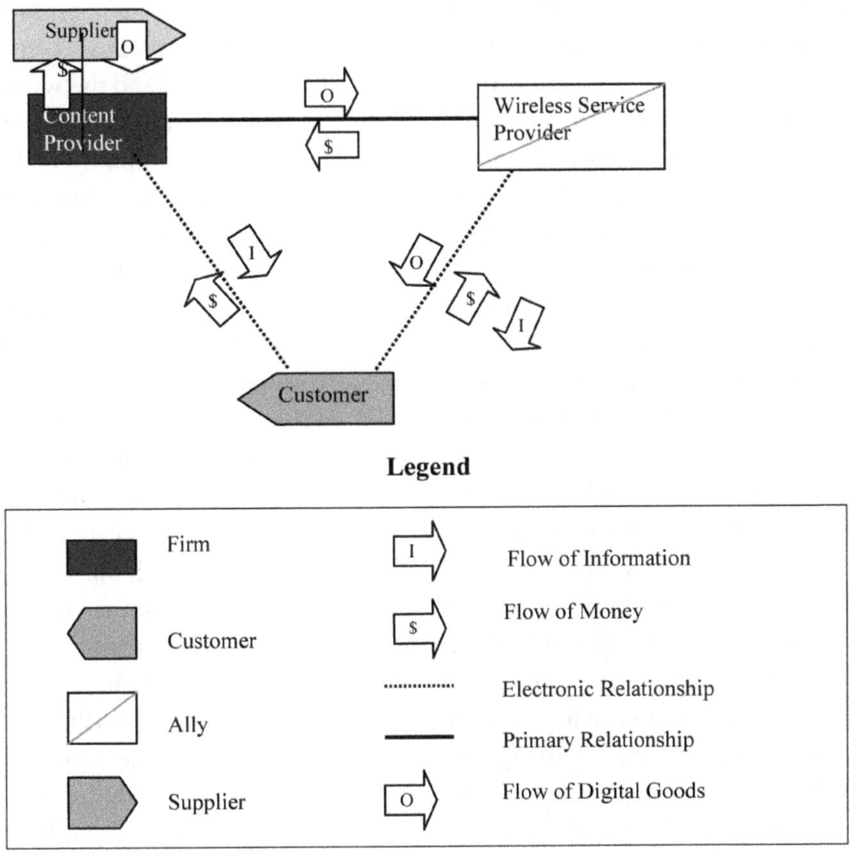

Figure 2. Partnering with content syndicators

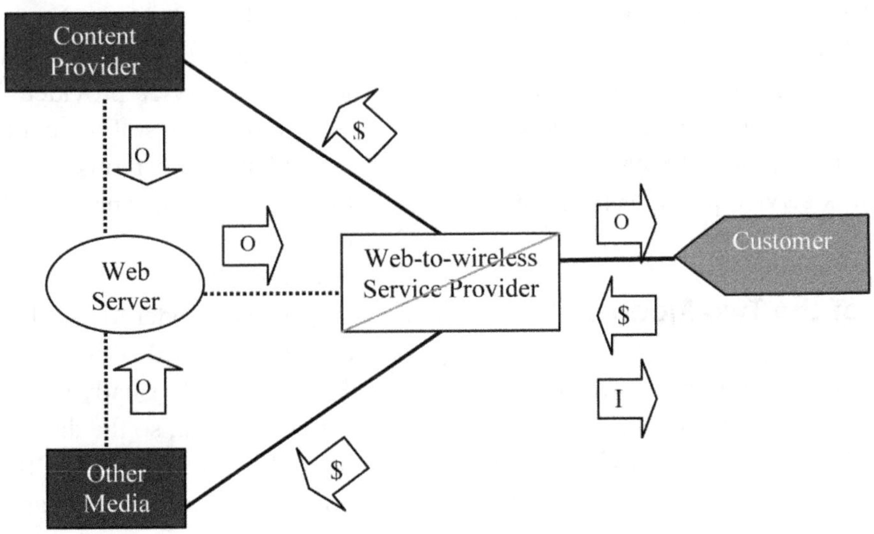

subscribers as well as advertisers. They also have to solicit subscribers using the content from the media with which they have a partnership. The content might be reorganized and integrated into a similar format by the service providers before they get customized and transmitted to the subscribers' handheld devices. AvantGo (http://avantgo.com) serves as the classic example of this model. AvantGo's content providers include MSNBC, CNET, GolfChannel, Wired Online, and The New York Times. It is estimated that more than 1,000 channels are partnering with AvantGo. The varying content types are delivered and automatically updated every time mobile users synchronize their handheld devices.

Web clipping is an example of a delivery technology for this model. The more modern version is the use of RSS (Really Simple Syndication), an XML (Extensible Markup Language) technology for distributing links or syndicating content on your website that you like others to use. In a sample web clipping scenario, a web server, in response to a query from a Palm query application (PQA), serves an appropriately formatted page. First, the query is passed via a web clipping proxy, which translates it into an HTTP query to the web server. The web server, in turn, sends back a web clipping. The page might be either static or dynamic, and the proxy compresses the HTTP response and sends it to the client. This model allows large web text or complicated graphics to be transmitted from web-to-mobile devices. *Media News* sells content to the web-to-wireless service provider. Subscribers pay the web-to-wireless service provider for access to that content.

Comparison of the Two Models

The dynamic nature of the business models associated with mobile content services poses considerable challenges to content providers such as *Media News*.

1. *Market share:* Readers, subject to service availability, may receive content even in remote areas. Under Model 1, there is a potential increase in customers for both *Media News* and the wireless provider. The wireless communication media environment is highly competitive. Users may switch providers if they find that offered content is not timely or that other companies offer richer content at lower rates. To lower this risk, *Media News* needs to offer attractive content to satisfy subscribers' interests, while keeping its fares competitive compared to those of other wireless content providers. *Media News* can reach wireless subscribers such as executives and sales and delivery personnel in addition to its regular subscribers. Under Model 2, the web-to-wireless service provider transmits content, compiled from all the media channels it partners with, to the subscribers it recruits. The greater the number of subscribers, the more readers *Media News* can potentially reach.
2. *Revenues:* Model 1 is likely to provide *Media News* with higher revenues through advertising fees and wireless subscriptions. However, specific revenue and profit-sharing agreements need to be worked out between *Media News* and the service provider. Under Model 2, *Media News* does not deal with wireless subscribers, since that is the job of the service provider. *Media News'* revenues from selling content are only dependent on the terms of its contract with the web-to-wireless provider. It doesn't vary proportionally to the number of subscribers or advertisers.
3. *Bandwidth:* Under Model 1, the amount of content delivered is limited by bandwidth. The bandwidth usually provided by WSPs is not that high, so the data transmitted are limited to small chunks. It takes a long time to transmit long paragraphs and high-resolution images, necessitating the use of

tiny displays. *Media News* needs to tailor its content accordingly. Under Model 2, the service provider optimizes its technology for all partnering media companies but hardly customizes its infrastructure for any specific partner. On the other hand, the web clipping infrastructure, underlying the second alliance model, supports rich displays with larger real estate space and multiple modes of data interaction.

4. *Control:* Model 1 gives *Media News* complete control over delivered content. Span of control extends over (a) presentation: editing and designing the layout of wireless content, including content either covered by *Media News* or purchased from other media; (b) source of content: the content is either entirely from *Media News* or from both *Media News* and its partnering media provider, in which case *Media News* gets to decide which content gets delivered by wireless; and (c) advertising: the content and layout of wireless ads are controlled by *Media News* and matched with news content to enhance ads' effects.
5. *Nature and scope of content:Media News* has an alliance with Channel 4000 and two TV stations allowing it to deliver online content. The content provided by these partners, such as real-time news and weather forecasts, compensates for the magazine's limited real-time coverage. However, *Media News'* partners' news coverage is itself regional and limited to the U.S. Midwest. Model 2 looks meaningful therefore in this regard. Besides web-to-wireless service providers, some wireless application service providers (ASPs) also favor localized information. Webraska (http://webraska.com) is one example. It provides location-specific information to mobile carriers and portals for services such as yellow pages and directions; local events, movies, and attractions; and travel and commuting. The in-depth local investigation stories reported by *Media News,* for example, the best doctors in town and the region's best sightseeing spots, always convey valuable information.

FROM BUSINESS MODELS TO REAL OPTIONS

The survivability of the telecommunications and entertainment industry has been shown to be especially risky and uncertain (Grover & Saeed, 2003; Kim, 2007; Rhéaume & Bhabra, 2008). The case of *Media News*' partnerships is no different. Figure 3 illustrates why ROA contributes to the reactive and proactive management of flexibility for a media company like *Media News*, whose expected subscriber base is a major source of uncertainty.

Media users span a wide range of socio-demographic segments and have heterogeneous tastes and exhibit a short-spanned loyalty, if any, towards the media they subscribe to. That lack of sustained loyalty is due to many reasons:

1. Media users' preferences change rapidly, especially when it comes to media formats. Media providers better keep up with technology to retain their technology-oriented customer base.
2. No objective quality criterion exists for selecting media content; media users may be tempted to be indifferent in their consideration of one media over the other.
3. In some media segments, such as advertising-based broadcasting, there is no direct interaction between the media company and the consumer. Specifically, when it comes to its interaction with users acquired through its partnerships with Wisc-TV, UPN 14, and Channel 3000, *MM* maintains an indirect and anonymously mediated relationship.

Figure 3. Relevance of ROA in valuing Media News' investment decisions

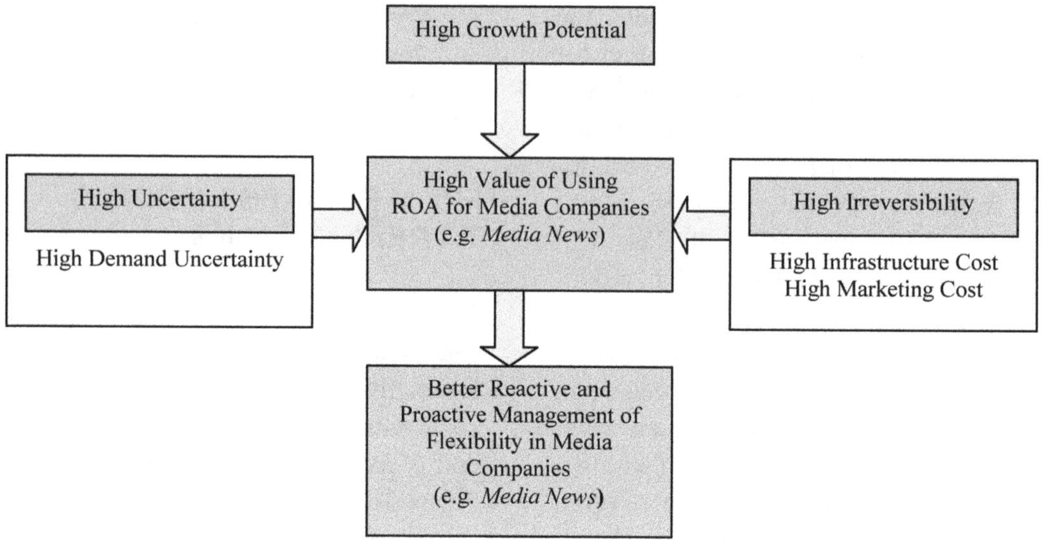

Further, once implemented, *Media News'* decisions cannot be completely reversed without incurring heavy cost burdens. Strategic flexibility in *Media News'* context is the company's ability to reformulate its strategies and reconfigure its resource allocation upon receiving up-to-date numbers. *Media News'* flexibility to abandon any one of its partnerships, for instance, is an example of defensive flexibility because it protects the company against downside losses caused by a lower than expected subscriber base, or unaccounted recurring costs.

Strategic investment decisions involve both quantitative and qualitative analyses. The numerical data used for our calculations were obtained through face-to-face interviews with *Media News'* executives, telephone interviews with marketing managers of WSPs and content syndicators, and subscription information available on media companies' websites. The structured interview sessions with the magazine executives were conducted with the aid of questionnaires, which were prepared early in advance and emailed to them. The questionnaire was designed to gather information about the magazine's strategic position on exploring wireless content services as a viable business development option. Secondly, it was designed to assess the magazine's capabilities in terms of infrastructure, human resources, customer profile, circulation, status of its existing partnerships, content-generation capabilities and data that would facilitate the determination of initial investment costs and recurring expenses. On completion of every interview session, interview transcripts were analyzed. Secondary questions and items for further clarification were then marked out for subsequent sessions. The three interview sessions conducted provided adequate information to facilitate the analysis. Secondary sources of information employed include the magazine's web site, survey data available from the media watchdog Media Audit (http://mediaaudit.com), web sites of popular WSPs such as Cingular (now AT&T), Verizon, Sprint, and NEXTEL and web-to-wireless service providers such as AvantGo, NTT DoCoMo, and EarthLink.

Background on ROA and Literature Review

A real option is the right, not the obligation to undergo an endeavor (defer, expand, contract, abandon, etc…) at an agreed upon price (exercise price) for a predetermined period (life of the option). As is the case for financial options, the value of a real option is derived from the value of the underlying risky asset (value of investment, project, or acquisition), the exercise price (amount of money invested to exercise the option), the time to maturity of the option, the volatility of the value of the underlying asset (riskiness of the underlying asset's cash flows), and the risk-free rate prevalent during the life of the option. The right to buy the underlying asset at the exercise price is a *call* option whose payoff is calculated as the difference between the underlying value and the exercise price. Alternatively, a put option is the right to sell the underlying at the exercise price with payoff at maturity equaling the difference by which the exercise price exceeds the underlying value. Options that can only be exercised upon their maturity are referred to as *European* options while those that can be exercised anytime during the lifetime of the project are entitled *American*. Real options are differentiated by the type of flexibility they represent. For example the right to defer, expand, or extend a project is an American call option, while the right to abandon a project is an American put. Various option combinations are possible by constructing portfolios of options and compounding options on options, both simultaneously and sequentially.

ROA has been amply used for the valuation of finance-related projects, retailing, auctioning, contracts, mergers & acquisitions, outsourcing, and leasing which are characterized by substantial risks (Mittendorf, 2004; Marcus & Anderson, 2006; Jiang et al., 2008; Kumar & Turnbull, 2008; Su et al., 2009). ROA has also been applied to value the flexibility embedded in operations management, facility management, manufacturing and production management, capacity planning, supply chain management, and inventory management projects (Boute et al., 2004, Kamrad & Siddique, 2004; Kamrad & Ord, 2006; West & Bengtsson, 2007; Jao et al., 2007; Zhao & Tseng, 2007; Berling, 2008; Dulluri & Raghavan, 2008; Li & Rajagopalan, 2008; Driouchi et al., 2009; Huang, 2009; Lin, 2009). In addition, ROA has been applied to value environmental projects (Guthrie & Kumareswaran, 2009; Leroux et al., 2009), power and oil & gas-related projects (Takizawa & Suzuki, 2004; Wang & Min, 2006; Tseng & Lin, 2007; Bøckman et al., 2008; Hahn & Dyer, 2008; Porchet et al., 2009), research and development projects (Pennings & Lint, 2000; Koussis et al., 2007; Kim et al., 2008a; Miller & Clarke, 2008; Wu et al., 2008c), and pharmaceutical and health-related projects (Levaggi & Moretto, 2008; Pertile, 2009), among others.

The IT literature has been particularly successful in using ROA to value IT-related investments. Many authors used ROA to derive optimal timings for IT-related investments in real world business situations (e.g. Benaroch & Kauffman, 1999, 2000; Campbell, 2002; Gordon et al., 2003; Kauffman & Li, 2005; Benaroch et al., 2006a; Benaroch et al., 2007; Ziedonis, 2007; Khansa & Liginlal, 2009). Also in IT, authors such as Kim and Sanders (2002) and Benaroch (2001, 2002) developed frameworks to assist managers in evaluating and justifying IT investment decisions. Prior research also proposed variants of ROA that are specific to peculiar types of IT investments (Benaroch et al., 2006b; Wu et al., 2008).

ROA was shown to be particularly valuable in evaluating networking and communications project due to the high level of demand uncertainty and the immensity of sunk costs. Angelou and Economides (2008) combined ROA and the more qualitative analytic hierarchy process to better value investments in information and communications technologies infrastructures. Camacho and Menezes (2009) evaluated the real options embedded in networking projects under regulated and unregulated environments. Similarly, d'Halluin et al. (2007) used ROA to determine optimal timing for network capacity investment. Kim (2006) applied real options theory to study the economic value of modularity in networks' quality of service. Kim (2007) applied ROA to evaluate IP telephony network and its savings over the total link capacity of a circuit-switched network. Kim et al. (2008b) used ROA to assess technology migration options in the wireless industry. To capture the inherent uncertainties in the Internet's Quality of Service assurances, Gupta et al. (2006) developed a utility-based options pricing approach that provides new ways for "fair risk sharing" among Internet providers and customers. In their paper, Harmantzis and Tanguturi (2007) used ROA to facilitate investment decisions related to the deferral of the expansion from 2.5G to 3G networks the expansion of a 2.5G network using Wi-Fi as an alternative technology.

Applying Real Options Analysis

In Model 1, as a sole content provider, *Media News* owns the content and the subscriber base and consequently has the power to defer partnership with the WSP until uncertainty is at least partially resolved. Alternatively, with Model 2, when purchasing content from other media providers, *Media News* possesses the option to abandon the partnership if the results do not conform to expectations.

Model 1. Analyzing the Partnership with a WSP

To establish a five-year partnership with a WSP, as determined from the direct interviews held with *Media News'* managers, an initial seed investment of $35,000 is required for hardware, software and installation costs. However, starting in year #1, an annual recurring cost of $140,000, covering maintenance costs and two additional employees' salaries, is needed for the five-year projection period. Revenues are assumed, based on *Media News'* projections, to grow at 50 percent, 40 percent, 20 percent and 20 percent in years 2, 3, 4, and 5, respectively. Using the traditional DCF model and a market risk-adjusted discount rate of 20%, year #1's break-even number of subscribers was estimated at 5,497. Monthly subscription fees, as used in the calculations, were taken off *Media News'* web site and confirmed from our interviews.

Option to defer: Traditional DCF analysis of a project quantifies its static value and makes the implicit assumption that the project decision must be made immediately, ruling out the possibility of deferring the investment until uncertainty is resolved. When it is possible to delay the investment decision without losing potential benefits, an option to defer exists (Dixit & Pyndick, 1994). Many authors have studied the dynamics of deferring investment in technology until uncertainty is partially or totally resolved. For example, Kauffman and Li (2005) analyzed the investment timing strategy for a firm that is considering adopting one of two incompatible and competing technologies. The option to defer an investment can be modeled as a call option (Dixit & Pindyck, 1994). Using the Black-Scholes pricing formulae for a call option, the defer option is computed as in Equation 1.

$$C_0 = S \times \mathrm{N}(d_1) - Ke^{-r_f T} \times \mathrm{N}(d_2) \quad (1)$$

C_0 is the value of a European option at time $t = 0$, r_f is the risk-free interest rate, S is the price of the underlying asset, σ is the volatility or risk of the underlying asset, K is the strike price of the call, i.e. the price at which the underlying asset can be purchased if the option is exercised, T is the time to maturity, and N (.) is the cumulative distribution function for the standard normal distribution. d_1 is given by

$$d_1 = \frac{\ln(S / K) + (r_f + \frac{\tilde{A}^2}{2})T}{\sigma\sqrt{T}} \quad (2)$$

and

$$d_2 = d_1 - \sigma\sqrt{T} \quad (3)$$

The details of what is imputed for each of Equation 1's parameters are shown in Appendix A (Figure 7). We used Excel 2007, one of the most ubiquitous decision analysis applications, to automate the valuation of real options and assists managers in quantifying the embedded strategic value for a range of proposed or existing investment scenarios. We used a myriad of volatility estimates pertinent to specific industries. A volatility of 20 percent was obtained using the Crystal Ball tool (Mun, 2002) on *Media News'* expected future cash flow (Appendix A, Table A.2). Excel was then used to evaluate whether *Media News* should proceed with the partnership immediately or delay until uncertainty is resolved. It is assumed that *Media News* is not likely to face any serious competition in the next five years and first-mover advantage is also precluded. Starting with 5,500 subscribers in year #1, the recommendation is not to proceed with the partnership, considering that embarking on the partnership immediately is less valuable than the option to delay it. Appendix A depicts the resulting output assuming a 10 percent cost outlay due to delaying the project.

Model 2. Analyzing the Partnership with Content Syndicators

Instead of partnering with a WSP, *Media News* can partner with other media providers. In that case, we assume, based on our interviews with *Media News'* managers, that an additional $10,000 of seed money is required initially to establish the partnership. However, the partnership will allow *Media News* to save the equivalent of one employee's salary. As before, revenues are assumed to grow at 50 percent, 40 percent, 20 percent and 20 percent in years 2, 3, 4, and 5, respectively. Using the traditional DCF model and a market risk-adjusted discount rate of 20 percent, year #1's break-even number of subscribers is 4,531.

Option to abandon: Traditional DCF analysis does not capture management's flexibility to abandon its partnership with content syndicators, if it proves unsuccessful. An abandonment option gives the right to get out of a risky situation at any time by selling it for a fixed price (the striking price). Since at any time during the five years of partnership, *Media News'* management can, after reviewing the progress of the alliance, decide to terminate the partnership with the media provider by paying some penalty. *Media News'* right to abandon its partnership is formally equivalent to an American put option on a stock. Since an American put option can be exercised at any time up to the maturity date of the option, a binomial option pricing model needs to be used (Geske & Johnson, 1984). The major assumption of option pricing models is that the underlying asset's process w(t) defined on $[0, \infty)$ is a continuous-time Brownian motion stochastic process with drift parameter μ and variance parameter σ^2. The binomial lattice approximates the process $\ln S(t)$ with a discrete process by fixing a time step Δt and n intervals on $[0, T]$, such that $\Delta t = T/n$. A binomial lattice uses two multiplicative values u (up) $= \exp[\sigma(\Delta t)^{1/2}]$ and d (down) $= 1/u = \exp[-\sigma(\Delta t)^{1/2}]$, with $u > 1 > d$ and their respective probabilities q and $1 - q$ respectively.

The abandonment option binomial tree is shown in Appendix B (Figure 8). We have derived a volatility of 45 percent using the Crystal Ball tool on *Media News'* expected future cash flow. The output of the Crystal Ball tool, along with the abandonment option binomial tree obtained using John Hull's DerivaGem tool (http://www.rotman.utoronto.ca/~hull/software/) are shown in Appendix B (Table B.2). $u = \exp[\sigma(\Delta t)^{1/2}]$ and $d = 1/u = \exp[-\sigma(\Delta t)^{1/2}]$ so with 1 period and a volatility of 45%, $\Delta t = 1$ and $\sigma = .45$, so $u = 1.5683$and $d = 0.6376$. The risk-free interest factor is $R = \exp[r_f \Delta t] = \exp[.05] = 1.0513$ (continuous compounding). The risk-neutral probabilities are computed as $q = 0.4445$ and $1\text{-}q = 1 - 0.4445 = 0.5555$. Discrete compounding is suitable for certain flows (e.g., yearly payments at stated rate), but for lattice work, especially with time periods less than a year, continuous compounding works better. Starting in the last time period, the value with the put option is the maximum of the striking price and the value without abandonment. If *Media News* does not abandon the partnership, it gets the risk-neutral expectation of the possible successor nodes, discounted back at the risk-free rate, using continuous compounding for discounting. If *Media News* abandons the partnership, it gets the striking price K. Based on the conducted interviews, the salvage value if the partnership is abandoned is $30,000. Therefore, the value of the put is maximum$\{30,000,\ V\}$. This reasoning applies at every node that is not in the last time period (See Appendix B for full lattice).

Assuming 4,520 subscribers in year #1, traditional DCF analysis gives a negative NPV, indicating that the partnership is not feasible because the present value of expected future inflows amounts to $44,097.46 and is below the initial investment of $45,000. Using the binomial approach to perform ROA, the extended NPV amounts to $49,395.10; this exceeds the initial investment of $45,000 indicating that the partnership is feasible. The difference of $5,297.64 between the extended NPV and the base NPV is the value of the abandonment option. This example illustrates how the traditional NPV approach, in the presence of managerial options, undervalues the value of the partnership and could lead to a non-optimal decision.

NUMERICAL ANALYSIS AND DISCUSSION

The analytical studies that we conducted thus far are not sufficient to make a final decision. With regard to *Media News*' partnership with a WSP, immediate partnership could potentially generate revenues and create barriers to entry for other competitors. However, given the relatively high break-even number of subscribers and *Media News*' lack of national coverage, high uncertainty surrounds year #1's projected demand for *Media News'* services, as discussed earlier. At first glance, partnering with content syndicators seems to be a strategically sound proposition in that it allows *Media News* to expand its services to include national coverage, which could in turn further develop its subscriber base.

Initial DCF analyses results, shown in Tables A.1 and B.1 in Appendices A and B respectively, reveal that the NPV of partnering with a WSP is positive at $264.82, while that of partnering with content syndicators is negative at -$902.54, using the same discount rate of 20 percent. Therefore, using simple DCF analysis, *Media News'* decision would be to choose Model 1. Using ROA, however, the value of the defer option embedded in Model 1 is $1,972.9, while the value of the option to abandon, embedded in Model 2 is $5,297.6. By adding the values of the options to the NPV obtained using DCF, we get an extended NPV of $2,237.72 for Model 1 and $4,451.06 for Model 2, as shown in Figure 4. Therefore using DCF analysis, without accounting for embedded options, could have led *Media News* to select a sub-optimal model.

Sensitivity Analyses

While various sensitivity analyses are possible with the scalable decision model we have designed, we have chosen to limit the scope of our analysis to the following two scenarios.

1. *Strike price:* To compute the defer option embedded in the partnership with the WSPs we have used a strike price of $35,000. On the other hand, we have used a strike price of $30,000 to value the abandonment option embedded in the partnership with content syndicators. We relaxed these assumptions by varying the strike price from $20,000 to $40,000. The results in Figure 5 reveal that the break-even strike price, for which *Media News* would be indifferent between both options, is between $32,000 and $33,000.

Figure 4. NPV analysis with and without ROA- Model 2

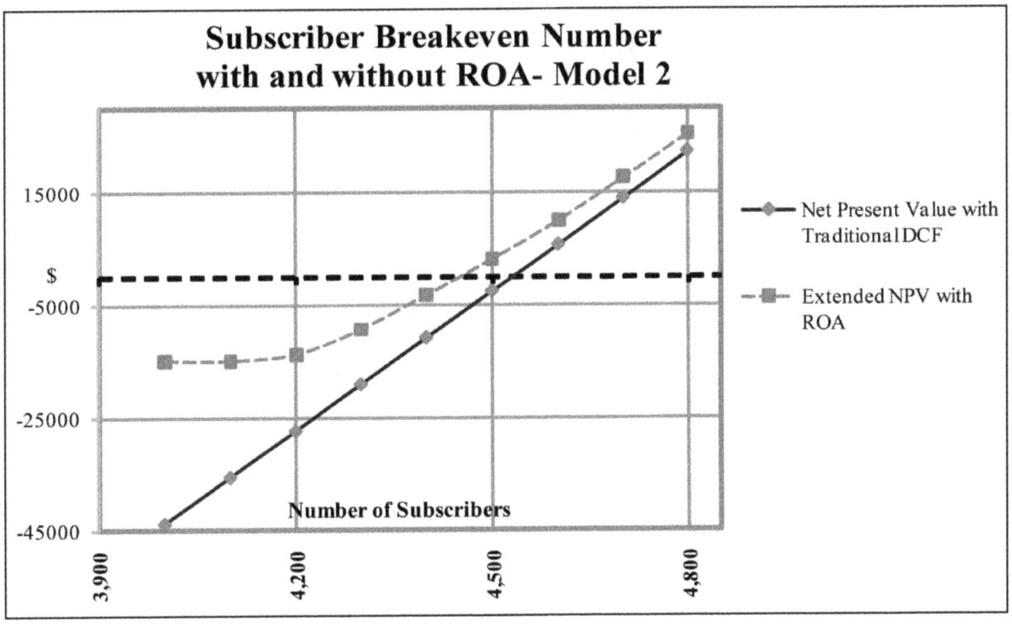

Up until this strike price, *Media News* is better off partnering with WSPs. For a strike price greater than $33,000, *Media News* is better off choosing the partnership with content syndicators. Recall that the strike price, in the case of the defer option, is the initial seed investment, which is required for hardware, software and installation costs. If *Media News* decides to delay the partnership, it would lose this seed investment. On the other hand, in the case of the partnership with content syndicators, the strike price corresponds to the penalty that *Media News* might have to pay to abandon the partnership.

2. *Volatility of the underlying cash flows:* Using the data that we collected from our interviews with *Media News*' management, we derived a volatility of 20% from the cash flow of the partnership with WSPs, and a volatility of 45% from that of the partnership with content syndicators. These initial assumptions indicate that the partnership with content syndicators is more risky, but can lead higher returns on the upside. We relaxed these initial assumptions and ran another sensitivity analyses by varying the volatility, or risk, of the underlying cash flows for a fixed strike price of $35,000. The results in Figure 6 show that for a strike price of $35,000, *Media News* is better off partnering with content syndicators for a volatility value higher than 28%.

Limitations of Real Options Analysis

Despite its usefulness, the ROA methodology has some limitations. First, misestimated volatilities can lead to significantly erroneous results in option valuation. To address this limitation, we substantiated our numerical analysis with data collected during our interviews with *Media News*' management and used this data to derive the volatilities. Second, although quite realistic to model the prices of stocks or other financial instruments, the assumption of geometric Brownian motion may be unrealistic to model the price of real assets.

Figure 5. Sensitivity analysis upon varying the strike price

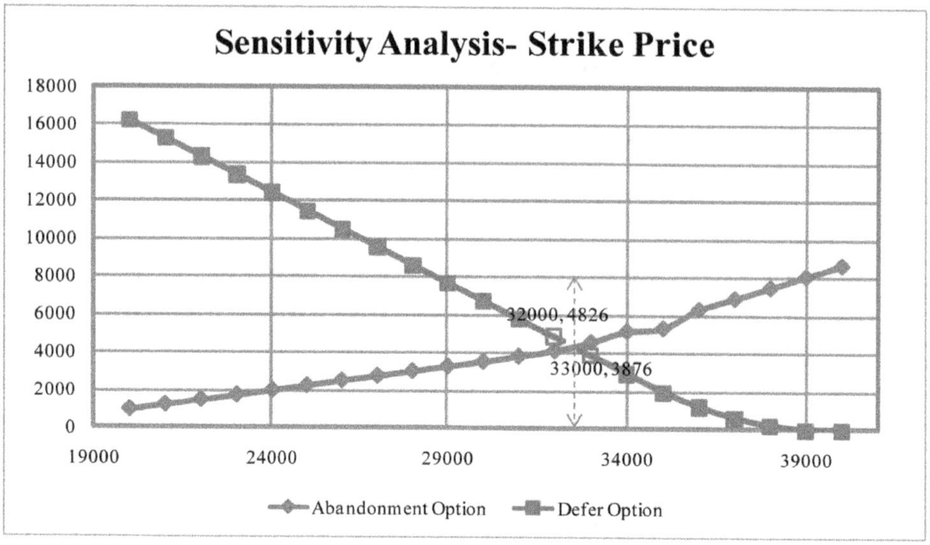

Figure 6. Sensitivity analysis upon varying cash flow volatility

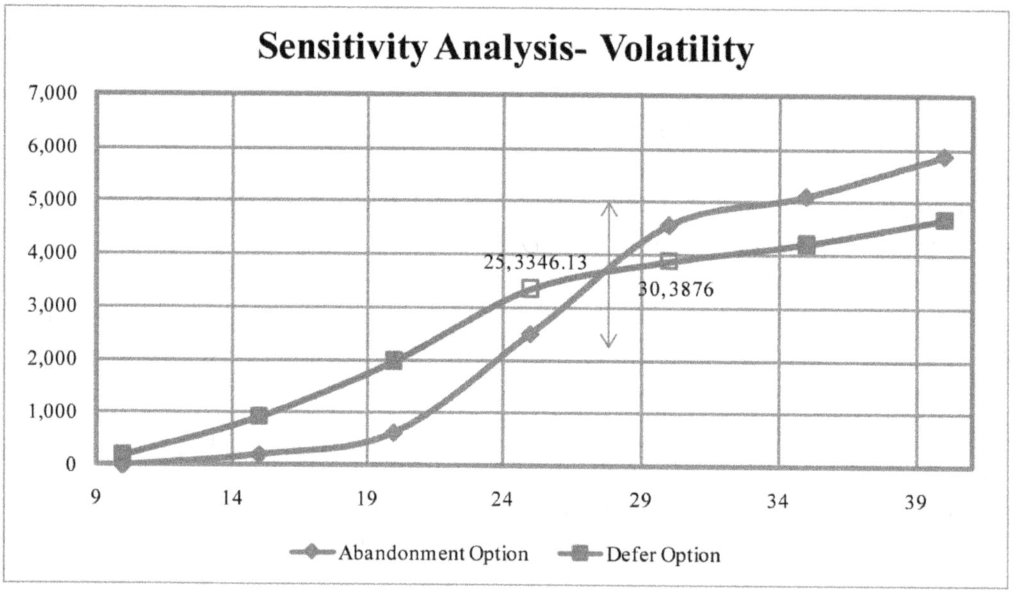

This assumption of the price of the underlying asset being random is violated if the market is a monopoly or an oligopoly as the small number of market players may influence the price of the underlying asset. In the case of *Media News*, the presence of many players dampens this limitation.

CONCLUSION

Due to limited budget provisions, local news publishers have a hard time catching up with advances in technology, and they often get relegated to outlier status. We addressed the feasibility of two business models for *Media News* to expand into the wireless market. For *Media News*, which already has an alliance with TV and wired web, wireless delivery completes the picture of print, broadcast, online, and wireless presentation. The face-to-face interviews with executives of the magazine brought out the practitioners' vision of technology and the difficulties encountered by publishers in the new era. Based on the nature of the partnership, local news publishers and WSPs may share the income in both models. The study also identified specific roles and responsibilities for each party. Wireless service companies contribute the technical component, and the news publishers contribute a plethora of rich content, such as stories and news coverage.

No matter which model the magazine prefers to adopt, it is likely that both types of partnership require content generation in a format suitable for publishing over the web. To generate and manage its own web content, *Media News* needs to recruit editors and journalists who possess the appropriate technical background. Steve Outing, president of Planetary News (http://www.planetarynews.com/), a media research firm, refers to such people as "editorial scientists" or "content engineers." An editorial scientist is someone who has been trained in computer-assisted reporting, and a content engineer is the person who is able to develop and maintain content database structures and create content presentation. Web and wireless presence alter not only a journalist's job and required skills but also the format in which the content is displayed. Media content for wireless could ordinarily be divided into two parts. One is news content, and the other is advertising. For wireless, the news content requires the following: (a) ruthless editing, (b) filtering and selecting the best content, (c) adding explicit headlines, and (d) categorization. On the other hand, wireless advertising presentation is less conclusive. Although customization and personalization are highly recommended for wireless ads, one also needs to adhere to the standards and recommendations of such bodies as the Wireless Advertising Association (http://mmaglobal.com/main). It is interesting to note that mobile content has revolutionized real-time news broadcasting. For example, Web-enabled cell phones will allow ordinary citizens to shoot and send news photos and video to the local newspaper or TV station—or publish them to their own personal Web sites—also called mob-logging. This has now become a rage in countries such as Japan where the wireless infrastructure is more advanced than in the United States. Thus, real-time news, not just print, but also videos are possible to be captured by reporters/bloggers and delivered to mobile customers' cell phones.

Traditionally, managerial decision making tends to focus on short term profits instead of aiming for more flexibility, which if exploited in a timely manner, can bring in more profits. We showed that using traditional cost-benefit analyses, such as DCF, can lead to suboptimal decisions, especially in the media industry where uncertainty and irreversibility are high. We have, instead, proposed that ROA can be used to value technology investments under uncertainty, in the particular the case of growing media firms and demonstrated how *Media News* can choose between its two expansion options using real options theory. We urge managers to exercise the options embedded within their projects. Barnett (2005) argued that real options are only valuable if management exert the necessary attention to exploit flexibility inherent in projects and extract their full theoretical value. Tiwana et al. (2007) also showed how important it is for managers not to neglect embedded real options when valuing projects. On the other hand, managerial over-commitment to projects embedding real options could also be detrimental. Tiwana et al. (2006) illustrated how real options can sometimes lead to escalation of commitment on the part of software project managers and recommended recognizing the value of timely project termination when needed.

The results of this article are easily extensible to other areas of mobile commerce both from both, the investment and technological perspectives (content engineering, for instance). Our ongoing work consists of modeling the options embedded in the partnerships in the presence of competing standards in the media industry where "winners take all" and first-mover advantage is of considerable importance. It is important for media companies to invest in and set the dominant market standard before consumers are locked in, as the costs of switching to another technology are prohibitively high for media firms. The state of competition between various technologies would determine a threshold, beyond which it would be profitable to defer investment until competition uncertainty is resolved and one of the technologies becomes the standard.

ACKNOWLEDGMENT

The authors wish to thank the management of the company featured in this article and two graduate students for assistance with conducting face-to-face interviews with the company's managers and telephone interviews with marketing managers of WSPs and content syndicators as part of data collection.

REFERENCES

Angelou, G. N., & Economides, A. A. (2008). A decision analysis framework for prioritizing a portfolio of ICT infrastructure projects. *IEEE Transactions on Engineering Management, 55*(3), 479–495. doi:10.1109/TEM.2008.922649

Barnett, M. L. (2005). Paying attention to real options. *R & D Management, 35*(1), 61–72. doi:10.1111/j.1467-9310.2005.00372.x

Benaroch, M. (2001). Option-based management of technology investment risk. *IEEE Transactions on Engineering Management, 48*(4), 428–444. doi:10.1109/17.969422

Benaroch, M. (2002). Managing information technology investment risk: A real options perspective. *Journal of Management Information Systems, 19*(2), 43–84.

Benaroch, M., Jeffery, M., Kauffman, R. J., & Shah, S. (2007). Option-based risk management: A field study of sequential information technology investment decisions. *Journal of Management Information Systems, 24*(2), 103–140. doi:10.2753/MIS0742-1222240205

Benaroch, M., & Kauffman, R. J. (1999). A case for using real options pricing analysis to evaluate information technology project investments. *Information Systems Research, 10*(1), 70–86. doi:10.1287/isre.10.1.70

Benaroch, M., & Kauffman, R. J. (2000). Justifying electronic banking network expansion using real options analysis. *Management Information Systems Quarterly, 24*(2), 197–225. doi:10.2307/3250936

Benaroch, M., Lichtenstein, Y., & Robinson, K. (2006a). Real options in information technology risk management: An empirical validation of risk-option relationships. *Management Information Systems Quarterly, 30*(4), 827–864.

Benaroch, M., Shah, S., & Jeffery, M. (2006b). On the valuation of multistage information technology investments embedding nested real options. *Journal of Management Information Systems, 23*(1), 239–261. doi:10.2753/MIS0742-1222230108

Berling, P. (2008). Real options valuation principle in the multi-period base-stock problem. *Omega-International Journal of Management Science, 36*(6), 1086–1095. doi:10.1016/j.omega.2006.05.007

Bøckman, T., Fleten, S. E., Juliussen, E., Langhammer, H. J., & Revdal, I. (2008). Investment timing and optimal capacity choice for small hydropower projects. *European Journal of Operational Research, 190*(1), 255–267. doi:10.1016/j.ejor.2007.05.044

Boute, R., Demeulemeester, E., & Herroelen, W. (2004). A real options approach to project management. *International Journal of Production Research, 42*(9), 1715–1725. doi:10.1080/00207540310001639946

Camacho, F. T., & Menezes, F. M. (2009). Access pricing and investment: a real options approach. *Journal of Regulatory Economics, 36*(2), 107–126. doi:10.1007/s11149-009-9088-1

Campbell, J. A. (2002). Real options analysis of the timing of IS investment decisions. *Information & Management, 39*(5), 337–344. doi:10.1016/S0378-7206(01)00101-X

d'Halluin, Y., Forsyth, P. A., & Vetzal, K. R. (2007). Wireless network capacity management: A real options approach. *European Journal of Operational Research, 176*(1), 584–609. doi:10.1016/j.ejor.2005.06.055

Dimpfel, M., Habann, F., & Algesheimer, R. (2002). Real options theory, flexibility and the media industry. A conceptual outline for potential application areas. *International Journal of Media Management, 4*(4), 261–272.

Dixit, A., & Pindyck, R. (1994). *Investment under uncertainty*. Princeton, NJ: Princeton University Press.

Driouchi, T., Leseure, M., & Bennett, D. (2009). A robustness framework for monitoring real options under uncertainty. *Omega-International Journal of Management Science, 37*(3), 698–710. doi:10.1016/j.omega.2007.09.004

Dulluri, S., & Raghavan, N. R. S. (2008). Collaboration in tool development and capacity investments in high technology manufacturing networks. *European Journal of Operational Research, 187*(3), 962–977. doi:10.1016/j.ejor.2006.06.059

Grover, V., & Saeed, K. (2003). The telecommunication industry revisited the changing pattern of partnerships. *Communications of the ACM, 46*(7), 119–125. doi:10.1145/792704.792709

Guthrie, G., & Kumareswaran, D. (2009). Carbon subsidies, taxes and optimal forest management. *Environmental and Resource Economics, 43*(2), 275–293. doi:10.1007/s10640-008-9238-4

Hahn, W. J., & Dyer, J. S. (2008). Discrete time modeling of mean-reverting stochastic processes for real option valuation. *European Journal of Operational Research, 184*(2), 534–548. doi:10.1016/j.ejor.2006.11.015

Harmantzis, F. C., & Tanguturi, V. P. (2007). Investment decisions in the wireless industry applying real options. *Telecommunications Policy, 31*(2), 107–123. doi:10.1016/j.telpol.2006.02.005

Huang, M. G. (2009). Real options approach-based demand forecasting method for a range of products with highly volatile and correlated demand. *European Journal of Operational Research, 198*(3), 867–877. doi:10.1016/j.ejor.2008.10.002

Jao, Y. Y., Du, J., & Jiao, J. X. (2007). A financial model of flexible manufacturing systems planning under uncertainty: identification, valuation and applications of real options. *International Journal of Production Research, 45*(6), 1389–1404. doi:10.1080/10298430600677479

Jiang, B., Yao, T., & Feng, B. C. (2008). Valuate outsourcing contracts from vendors' perspective: A real options approach. *Decision Sciences, 39*(3), 383–405. doi:10.1111/j.1540-5915.2008.00197.x

Kamrad, B., & Ord, K. (2006). Market risk and process uncertainty in production operations. *Naval Research Logistics, 53*(7), 627–640. doi:10.1002/nav.20178

Kamrad, B., & Siddique, A. (2004). Supply contracts, profit sharing, switching, and reaction options. *Management Science, 50*(1), 64–82. doi:10.1287/mnsc.1030.0157

Kauffman, R. J., & Li, X. T. (2005). Technology competition and optimal investment timing: A real options perspective. *IEEE Transactions on Engineering Management, 52*(1), 15–29. doi:10.1109/TEM.2004.839962

Khansa, L., & Liginlal, D. (2009). Valuing the flexibility of investing in security process innovations. *European Journal of Operational Research, 192*(1), 216–235. doi:10.1016/j.ejor.2007.08.039

Kim, H. J. (2006). The economic value of modularity in the IPQoS network: The real options approach (ROA). *International Journal of Business Data Communications and Networking, 2*(1), 15–31.

Kim, H. J. (2007). Suitability of IP telephony in the public switched telephone network (PSTN): A case study. *International Journal of Business Data Communications and Networking, 3*(3), 39–56.

Kim, H. J., Weiss, M., & Morel, B. (2008b). Real options and technology management: Assessing technology migration options in the wireless industry. *Telematics and Informatics, 26*(2), 180–192. doi:10.1016/j.tele.2008.05.004

Kim, K. H., Hwang, S. T., Oh, H. S., & Lee, D. J. (2008a). The impact of investment lags on investment decision. *European Journal of Operational Research, 190*(3), 696–707. doi:10.1016/j.ejor.2007.07.020

Kim, Y. J., & Sanders, G. L. (2002). Strategic actions in information technology investment based on real option theory. *Decision Support Systems, 33*(1), 1–11.

Koussis, N., Martzoukos, S. H., & Trigeorgis, L. (2007). Real R&D options with time-to-learn and learning-by-doing. *Annals of Operations Research, 151*(1), 29–55. doi:10.1007/s10479-006-0127-3

Kumar, P., & Turnbull, S. M. (2008). Optimal patenting and licensing of financial innovations. *Management Science, 54*(12), 2012–2023. doi:10.1287/mnsc.1080.0898

Leroux, A. D., Martin, V. L., & Goeschl, T. (2009). Optimal conservation, extinction debt, and the augmented quasi-option value. *Journal of Environmental Economics and Management, 58*(1), 43–57. doi:10.1016/j.jeem.2008.10.002

Levaggi, R., & Moretto, M. (2008). Investment in hospital care technology under different purchasing rules: A real option approach. *Bulletin of Economic Research, 60*(2), 159–181. doi:10.1111/j.0307-3378.2008.00274.x

Li, G., & Rajagopalan, S. (2008). Process improvement, learning, and real options. *Production and Operations Management, 17*(1), 61–74. doi:10.3401/poms.1070.0008

Lin, T. T. (2009). Applying the maximum NPV rule with discounted/growth factors to a flexible production scale model. *European Journal of Operational Research, 196*(2), 628–634. doi:10.1016/j.ejor.2008.04.004

Marcus, B., & Anderson, C. K. (2006). Online low-price guarantees - A real options analysis. *Operations Research, 54*(6), 1041–1050. doi:10.1287/opre.1060.0333

Miller, B., & Clarke, J. P. (2008). Strategic guidance in the development of new aircraft programs: a practical real options approach. *IEEE Transactions on Engineering Management, 55*(4), 566–578. doi:10.1109/TEM.2008.927817

Mittendorf, B. (2004). Information revelation, incentives, and the value of a real option. *Management Science, 50*(12), 1638–1645. doi:10.1287/mnsc.1040.0260

Mun, J. (2002). *Real options analysis, tools and techniques for valuing strategic investments and decisions*. New York: Wiley Finance Publishers.

Pennings, E., & Lint, O. (2000). Market entry, phased rollout or abandonment? A real option approach. *European Journal of Operational Research, 124*(1), 125–138. doi:10.1016/S0377-2217(99)00121-6

Pertile, P. (2009). An extension of the real option approach to the evaluation of health care technologies: the case of positron emission tomography. *International Journal of Health Care Finance and Economics, 9*(3), 317–332. doi:10.1007/s10754-008-9053-z

Porchet, A., Touzi, N., & Warin, X. (2009). Valuation of power plants by utility indifference and numerical computation. *Mathematical Methods of Operations Research*, *70*(1), 47–75. doi:10.1007/s00186-008-0231-z

Rhéaume, L., & Bhabra, H. S. (2008). Value creation in information-based industries through convergence: A study of US mergers and acquisitions between 1993 and 2005. *Information & Management*, *45*(5), 304–311. doi:10.1016/j.im.2008.03.002

Su, N., Akkiraju, R., Nayak, N., & Goodwin, R. (2009). Shared services transformation: conceptualization and valuation from the perspective of real options. *Decision Sciences*, *40*(3), 381–402. doi:10.1111/j.1540-5915.2009.00243.x

Takizawa, S., & Suzuki, A. (2004). Analysis of the decision to invest for constructing a nuclear power plant under regulation of electricity price. *Decision Support Systems*, *37*(3), 449–456. doi:10.1016/S0167-9236(03)00045-9

Tiwana, A., Keil, M., & Fichman, R. G. (2006). Information systems project continuation in escalation situations: A real options model. *Decision Sciences*, *37*(3), 357–391. doi:10.1111/j.1540-5414.2006.00131.x

Tiwana, A., Wang, J. J., Keil, M., & Ahluwalia, P. (2007). The bounded rationality bias in managerial valuation of real options: Theory and evidence from IT projects. *Decision Sciences*, *38*(1), 157–181. doi:10.1111/j.1540-5915.2007.00152.x

Tseng, C. L., & Lin, K. Y. (2007). A framework using two-factor price lattices for generation asset valuation. *Operations Research*, *55*(2), 234–251. doi:10.1287/opre.1060.0355

van Putten, A. B., & MacMillan, I. C. (2004). Making real options really work. *Harvard Business Review*, *82*(12), 134.

Wang, C. H., & Min, K. J. (2006). Electric power generation planning for interrelated projects: A real options approach. *IEEE Transactions on Engineering Management*, *53*(2), 312–322. doi:10.1109/TEM.2006.872249

Weill, P., & Vitale, M. (2001). *From place to space: Migrating to e-business models*. Cambridge, MA: Harvard Business School Press.

West, B. M., & Bengtsson, J. (2007). Aggregate production process design in global manufacturing using a real options approach. *International Journal of Production Research*, *45*(8), 1745–1762. doi:10.1080/00207540600677583

Wu, L. C., Ong, C. S., & Hsu, Y. W. (2008). Active ERP implementation management: A real options perspective. *Journal of Systems and Software*, *81*(6), 1039–1050. doi:10.1016/j.jss.2007.10.004

Wu, L. C., Ong, C. S., & Hsu, Y. W. (2008c). Knowledge-based organization evaluation. *Decision Support Systems*, *45*(3), 541–549. doi:10.1016/j.dss.2007.06.013

Zhao, T., & Tseng, C. L. (2007). Flexible facility interior layout: a real options approach. *The Journal of the Operational Research Society*, *58*(6), 729–739. doi:10.1057/palgrave.jors.2602189

Ziedonis, A. A. (2007). Real options in technology licensing. *Management Science*, *53*(10), 1618–1633. doi:10.1287/mnsc.1070.0705

APPENDIX A: ANALYSIS OF MODEL TO PARTNER WITH THE WSP

Table A.1. Cash flow for the partnership with the WSP

Parameters	**Years**					
	0	**1**	**2**	**3**	**4**	**5**
WACC	20%					
Growth Rate			50%	40%	20%	20%
Initial Investment	-$35,000					
Recurring Cost		-$140,000.00	-$140,000.00	-$140,000.00	-$140,000.00	-$140,000.00
Revenues		$82,225.00	$123,337.50	$172,672.50	$207,207.00	$248,648.40
	($35,000.00)	($57,775.00)	($16,662.50)	$32,672.50	$67,207.00	$108,648.40
NPV	$264.82					
Present Value (Income)	$35,264.82					
Number of Sub-scribers	5,500.00					

* WACC is the weighted average cost of capital

See Figure 7.

Figure 7. Defer option embedded in the partnership with the WSP decision

OUTPUTS:

d1	-0.3186	The PV of taking on the project now (265) is less than the option to delay (1973), therefore the project should NOT be undertaken now.
N(d1)	0.3750	
d2	-0.7658	There is a 22.2% to 37.5% likelihood that this project will become viable before expiration (as indicated by Nd1 and Nd2).
N(d2)	0.2219	
VALUE	**$1,972.9**	**The value of the option to delay MM's Partnership with WSP.**

PARTIALS:

? Delta	0.375	Linear exposure to changes in the project cash flows.
? Gamma	0.000	Parabolic exposure to changes in the project cash flows.
? Theta	-900.479	Linear exposure to changes in the project's time to expiration
? Vega	29902.067	Linear exposure to changes in the project's volatility or 'risk'.
? Rho	30243.720	Linear exposure to changes in the risk-free rate.

Table A.2. Volatility of the cash flow for the partnership with the WSP

Crystal Ball Report		
The following simulation is based onMun's (2002)Logarithmic Present Value Approach. *Media News*' expected cash flow is discounted to time [PV_0] and again to time 1[PV_1]. The natural logarithm of the ratio PV_1/PV_0 is then used as a forecast variable to perform a Monte Carlo simulation on *Media News*' future expected cash flow. The standard deviation of the forecast variable's distribution is the volatility estimate we used in our real options analysis. To get the standard deviation in percentage terms we divided the obtained output's standard deviation by the output's mean.		
Forecast of LN[PV_1/PV_0]		
	Summary:	
	Display Range is from 2.64 to7.51 Entire Range is from -0.96 to 7.51 After 1,260 Trials, the Std. Error of the Mean is 0.03	
	Statistics:	Value
	Trials	1260
	Mean	5.46
	Median	5.68
	Mode	---
	Standard Deviation	1.08
	Variance	1.16
	Skewness	-1.29
	Kurtosis	5.59
	Coefficient of Variability	0.20
	Range Minimum	-0.96
	Range Maximum	7.51
	Range Width	8.46
	Mean Standard Error	0.03

APPENDIX B: ANALYSIS OF MODEL TO PARTNER WITH CONTENT SYNDICATORS

Table B.1. Cash flow for the partnership with content syndicators

Parameters	Years					
	0	1	2	3	4	5
WACC	20%					
Growth Rate			50%	40%	20%	20%
Initial Investment	-$45,000					
Recurring Cost		-$110,000.00	-$110,000.00	-$110,000.00	-$110,000.00	-$110,000.00
Revenues		$67,574.00	$101,361.00	$141,905.40	$170,286.48	$204,343.78
	($45,000.00)	($42,426.00)	($8,639.00)	$31,905.40	$60,286.48	$94,343.78
NPV	($902.54)					
Present Value (Income)	$44,097.46					
Number of subscribers	4,520.00					

* WACC is the weighted average cost of capital

See Figure 8.

Figure 8. Binomial model for the partnership with content syndicators

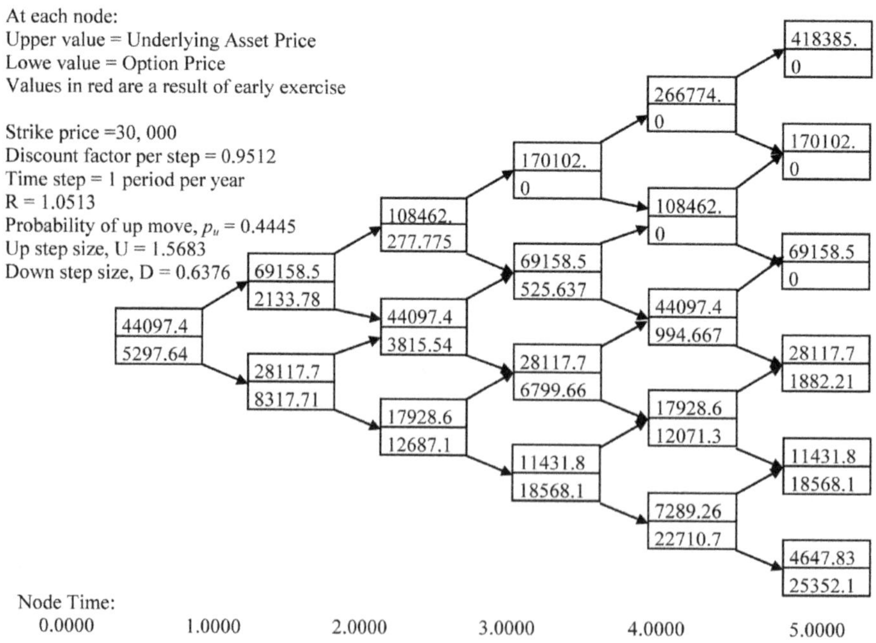

Table B.2. Volatility of the cash flow for the partnership with partnership with content syndicators

Crystal Ball Report		
The following simulation is based onMun's (2002)Logarithmic Present Value Approach. *Media News*' expected cash flow is discounted to time [PV_0] and again to time 1[PV_1]. The natural logarithm of the ratio PV_1/PV_0 is then used as a forecast variable to perform a Monte Carlo simulation on *Media News*' future expected cash flow. The standard deviation of the forecast variable's distribution is the volatility estimate we used in our real options analysis. To get the standard deviation in percentage terms we divided the obtained output's standard deviation by the output's mean.		
Forecast of LN[PV_1/PV_0]		
	Summary:	
	Display Range is from -0.18 to 2.43 Entire Range is from -5.59 to 2.65 After 36,911 Trials, the Std. Error of the Mean is 0.00	
	Statistics:	Value
	Trials	36911
	Mean	1.13
	Median	1.19
	Mode	---
	Standard Deviation	0.51
	Variance	0.26
	Skewness	-1.51
	Kurtosis	10.73
	Coefficient of Variability	0.45
	Range Minimum	-5.59
	Range Maximum	2.65
	Range Width	8.24
	Mean Standard Error	0.00

This work was previously published in International Journal of Business Data Communications and Networking, Volume 6, Issue 1, edited by Debashis Saha and Varadharajan Sridhar, pp. 17-37, copyright 2010 by IGI Publishing (an imprint of IGI Global).

Chapter 16
Valuation of Alternative Business Models in Information, Communication and Media Markets:
Convergence, Ubiquity and Pervasiveness

Álvaro Nascimento
Universidade Católica Portuguesa, Portugal

Fernando Santos
Universidade Católica Portuguesa, Portugal

ABSTRACT

Information, Communication and Media Markets (ICMM) are in a process of tremendous change. IP-based technologies and services such as broadband and triple/quadruple-play are realities that have the ability to enforce the convergence of so far parallel industries, that is, Telephony, Internet, and Broadcast. In this paper, the authors derive a scenario where a telecom operator aims to design, develop, and validate a global B3G (Beyond 3rd Generation) framework to support secure, personalized, and pervasive telecommunications services built on heterogeneous network and service infrastructures. The authors rethink the telecommunications architecture and business models to enable easy, seamless, and pervasive access to content and services, while supporting user preferences and context. This proposal involves significant changes in current industry business models. A value chain approach allows the identification of different scenarios, where the firm faces several options from which to choose. The authors investigate operators' decision process and evaluates the project based on its flexibility, using a real option approach.

DOI: 10.4018/978-1-61350-477-2.ch016

INTRODUCTION

Telephony, Internet and Broadcast had been, for a long time, distinct industries, providing differentiated services though, sometimes, supported on the same access technology. Technological evolution along with liberalization developments – especially in telephony and broadcast industries – enabled industries to diversify their service portfolios in a way that currently these industries can no longer be seen as independent. That is, technological development allowed convergence as economies of scale and scope encouraged companies to adopt broad strategies that aimed at a larger market. Telephony, Internet and Broadcast are segments of what we (hereafter) designate Information, Communication and Media (ICM) industry.

Even though cross segment competition meant convergence – which suggests homogeneous service offer – it still did not mean perfect homogeneity. ICM industry is still characterized by the diversity of agents evolved and by service and access heterogeneity (a very good reference on convergence is provided by Saxtoft, 2008)

Along with convergence, complexity and atomicity are increasingly regarded as important features of the ICM industry. Once again, technological evolution had the ability to lower entry barriers and reinforce innovation. At the same time, liberalization process enforced the separation between access and service markets. In this context, because entering is possible both in access and service markets users are given the possibility to choose service and access providers among an increasing number of alternatives. Access / Service partition enabled potential entrants to operate in service (access) layer without a need to operate in access (service) layer. By lowering investment requirements and incumbents' economies of scale, the number of agents operating in access and service layers increased considerably, which reduces sunk costs and, hence, stimulates entry.

We propose stylized overview of ICMM value chain as follows.

Figure 1 shows, on the one hand, a clear separation between service and access layers; and, on

Figure 1. Current ICCM value chain – stylized overview

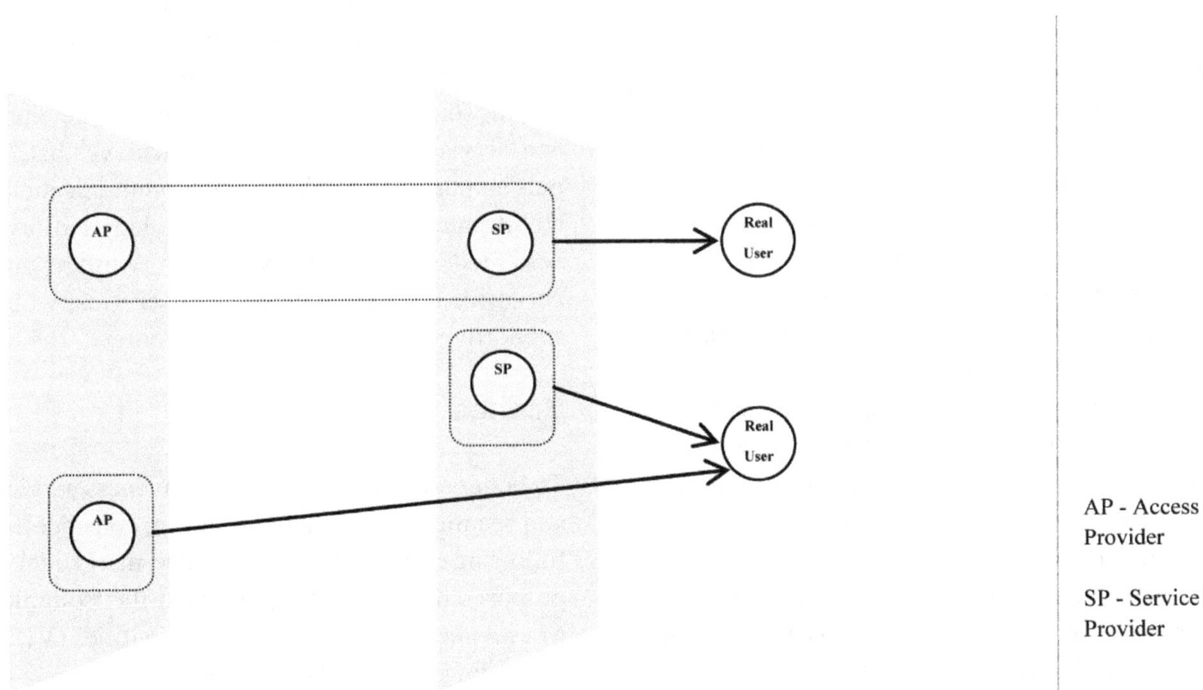

the other hand, that both access and service layers are accessible to the user. This ultimately means that users are given the possibility to choose both access and service providers.

The current trend, on the demand side, is characterized by users' demand for ubiquity. That is, the ability to purchase whatever service in a certain access technology as well as the ability to purchase a certain service in whatever access typology. Changes in users' life-styles are also forcing changes on the supply of ICMM. Contemporary life-style is increasingly flexible and mobile, which means that along with ubiquity, users are also demanding mobility. Ubiquitous and mobile service provision, anyhow and anywhere, ultimately implies multiple opportunities and threats for suppliers – market players. On the supply side, on the one hand, operators need to diversify their service portfolio and available access infrastructures (as users demand for anyhow service availability); on the other hand, operators need to geographically spread their access infrastructure availability (as users demand for anywhere service availability).

There is no simple answer for the way market players will deal with these threats and opportunities. First, their inherent characteristics are heterogeneous, but also because environmental realities – e.g., regulatory framework – vary along relevant markets.

Unambiguous is the fact that this convergence trend – allowed by technological innovation, economies of scope, but also economies of scale – is redrawing ICMM's value chain. Even though operators ultimately aim to diversify their service and access portfolios, there is high uncertain regarded the way to achieve those goals – internal development, acquisition, outsourcing, and so on.

According to the identified market trends, this paper applies real options methodology to an operator driven project/solution – denominated Daidalos – which aims to support secure, personalized and pervasive telecommunications services, built on heterogeneous network and service infrastructures. Our goal is to assess the true value of this operator driven solution by identifying the value of flexibility (the value of all embedded real options). To pursue that, and supported exclusively on public data, we focus our analysis on a stylized representation of Portuguese ICMM, by considering a relevant market composed by 5.000.000 households.

The structure of the paper is as follows: in the next section we introduce Daidalos solution and identify it's inherent flexibility as a way to justify our valuation methodology. Next, we define the model, presenting the basic conceptual differences between Net Present Value (NPV) and Real Option Approach (ROA). Next, using realistic assumptions and parameters, we calibrate the model and present the results for four types of market players – Fixed, Mobile, Internet Service Provider/Internet Access Provider (ISP/IAP) and Broadcast. Finally, we summarize the results and present the conclusions.

DAIDALOS

Daidalos is a scenario and an operator driven European funded project, which aims to design, develop and validate a global blueprint B3G (Beyond 3rd Generation) framework to support secure, personalized and pervasive telecommunications services, built on heterogeneous network and service infrastructures. To achieve that, it puts emphasis on rethinking telecommunications architecture and fundamental technology and business models issues to enable easy, seamless and pervasive access to content and services, while supporting user preferences and context

Scenes

To better understand what pervasiveness service and seamless access actually mean, consider the following example. Note, in advance, that Daidalos proposes that each real player (in the example, Maria) may have several virtual identities (VID) depending on Maria's preferences.

University

U1: Arrive at University: Maria is a new doctor giving classes at the university. On route to the university she is actively viewing her list of friends on her PDA via WLAN hotspot. Upon arrival at the university Maria is switched to the university network (based on her preferences). Maria then selects her Maria_teacher VID to be bootstrapped using the WLAN facilities of the university .She continues to view her list of friends on her PDA using UMTS, as the university WLAN network is not available to her Maria_freetime VID.

U2: Enter Classroom: Once inside the classroom, her presence information is updated automatically from free to busy. The VoIP application automatically switches to silent mode (through learning).

U3: Dean Calls Maria: Before the lecture there is a call from the Dean who wants to ask Maria something. The call is presented but no ringtone is heard (learning). Maria accepts his call manually.

U4: Start Diploma Course: Maria has initialized the big terminal with Maria_teacher VID. Maria selects the diploma course to start. The multicast session of the started diploma course is displayed on the big terminal on her desk based on indoor locationing giving her location as context information (like on all other terminals of her colleagues).

U5: Diploma Course Displayed on PDA: She stands up from her desk, and carries her PDA with her. The diploma course starts being displayed in her PDA.

U6: High Priority Call Reduces Student Media Quality: Maria pauses the active diploma course and calls an expert in the field on her PDA for a pre-arranged video conference. The performance of the applications (media quality) of the students is decreased because of the higher priority of Maria's outgoing call.

U7: Transfer Video Stream to Big Terminal: She wants the students to be aware of the call content so she transfers the video stream of the video call to the big terminal, while continuing to play the audio content from her PDA.

U8: Recover Video to PDA: After a while, Maria, using her PDA, recovers the video of the call from the big terminal to her PDA, says goodbye to the expert ends the call and restarts the multicast diploma course session.

Road

R1: Calendar Message: The class finishes earlier than expected and the husband Mark, who was waiting outside, receives a message from the University informing him that Maria has finished (Mark is subscribed to Maria's shared calendar, and it triggers the event) triggered by Maria switch of the VID to Maria_FREETIME.

R2: Car Authentication: Once together they both get into the car and Maria is the driver. Because of this, the Traffic service provider messages are now sent to the passenger's terminal (Mark), rather than distracting the driver.

R3: Handover sequence: While Maria is driving, her husband start using some broadcast application. Several network handovers occur between two operators and between two technologies within the same operator which are all triggered based on the user preferences of Maria and her husband and the contracts with the operators (roaming between two operators included). Suddenly, both are switched from the network operator to the WLAN network and the quality is heavily decreasing. They have no idea why.

Accident

A1: The highway becomes more and more congested. Maria starts the navigation application on the CAR PC, using Maria_Freetime VID. She uses the screen in the car to find a new route.

A2: Suddenly there is a danger warning message on the CAR PC screen, broadcasted about an accident in front of them. Another broadcast message automatically pops up looking for doctors in Maria's laptop. Maria, the doctor, responds to the message looking for doctors, using her laptop with WLAN in ad-hoc mode and with VID Maria_doctor.

A3: The police car stops right next to Maria's vehicle. Maria is now allowed to follow the police car to the accident passing all the cars in the queue. Maria, using her laptop, after being authenticated and authorized, thanks to federation, using Maria_DOCTOR VID, has now prioritized channel for emergency services. The police network domain is federated with Maria's operator and therefore allows her now to connect and communicate in the police network with "emergency priority".

A4: Using Maria_DOCTOR VID, Maria starts an application, showing the presence of online specialists, ready to assist her questions.

A5: Maria jumps out of the car (carrying her laptop) to the injured person and is now able to start a high-priority video call with an emergency centre to get some extra professional help while looking after the injured person.

In the next section we describe Daidalos project (concepts and roles) and investigate its implication in ICMM value chain.

Concepts, Roles and Value Chain

Daidalos aims to define a common framework that will allow the direct integration of all technologies by defining a set of five key concepts or 5 DKCs: MARQS (Mobility Management, AAA [Authentication, Authorisation and Accounting], Resource Management, QoS and Security), supporting functional integration for end-to-end services across heterogeneous technologies; VID (Virtual Identity), which separates the user from a device, thereby enables flexibility as well as privacy and personalization; USP (Ubiquitous and Seamless Pervasiveness), enabling pervasiveness across personal and embedded devices, and allowing adaptation to changing contexts, movement and user requests; SIB (Seamless Integration of Broadcast), which integrates broadcast at both the technology level, such as DVB-S/T-H, and at the services level, such TV, carouseles and datacast; and, Federation, allowing network operators and service providers to offer and receive services, and players to enter and leave the field in a dynamic business environment.

For all concepts to actually work as defined, Daidalos architecture proposes a set of new roles, stretching the set of roles embedded in the current ICMM's value chain. A brief description of them follows.

- **Aggregated Service Provider (AgSPr):** This entity is a central concept in Daidalos Value System. It provides a billable (set of) service(s) to its customers, by binding together all the required functionalities and components. Services may range from basic connectivity (e.g. mobile web) to ubiquitous pervasive value added services. This entity combines all service and network elements / capabilities (from network layer), selects contents and applications (from value layer) and proposes a unique offer to the end-user, thus providing one-stop shopping and/or billing, always relying on its brand name and customer preference / loyalty. It includes content / value added service quality monitoring. AgSPrs can establish agreements with other service providers, to expand their service offerings in terms of coverage / richness / value. Settlement with other parties can be pre-arranged or constructed in a dynamic manner, following service discovery principles.

- **Pervasive (PSPr):** This entity provides all the functionalities required for adding pervasiveness to a service, i.e., corresponds to the Pervasive Service Platform in Daidalos architecture. Its principle business is provision of enabling services related to pervasive support, i.e. personalisation, context management, event management, etc.
- **Aggregated Access Provider (AgAPr):** This entity is also a central concept in Daidalos Value System. It creates an aggregation of network accesses, thus providing a single network to the AgSPr, catering for all related functionalities, but mainly focusing on roaming and mobility support between aggregated networks. This entity is also responsible for network SLAs to the service providers utilizing the aggregated network infrastructure. Its' role is similar to that of a Virtual Operator, i.e., leasing the actual infrastructure used for service delivery. A set of enabling functions such as end-to-end QoS and content adaptation management, mobility support, resource discovery and security / privacy at network level are part of this entity's functionality.
- **Identity Provider (IPr):** According to Daidalos definition, the Identity Provider role corresponds to an entity, which provides and manages virtual identities, making sure that confidentiality and privacy is guaranteed.
- **Billing Provider (BPr):** Provides billing services to collect revenues from end-users / subscribers. This role includes the provision of the billing functions and the collection mechanism for bills. This role is embedded in the Service Provider (SPr) role.

All three analyzed markets (Telephony, Broadcast and Internet) are currently converging, and all support increasingly the user needs for mobility, and interactivity. In this context Daidalos may possibly bring some promising value, by enhancing and enabling future applications and features. As an operator driven project, Daidalos ultimately suggests a significant change in the ICMM value chain.

Many possibilities arise. Each operator is given the opportunity to play all roles or a part of them or even just one. From vertical integrated case to fully fragmented case there are numerous possibilities. Figure 2 provides a stylized overview of the Daidalos value chain. For simplicity, hereafter we consider that IPr, PSPr and BPr are played by AgSPr.

The figure shows the case where two end-users subscribe for service provision (as Daidalos suggests) to the AgSPr. One of the users has a single virtual identity and the other has two virtual identities. The AgSPr is not able to correlate virtual and real identities, which means, in practice, that AgSPr is supplying 3 users with distinct preferences. According to their revealed preferences and context, the AgSPr optimizes the service provision and access technology on which the service is provided. The service provision is accomplished by the aggregation agreements with several VASP and the access technology is optimized by the aggregation agreements between AgAPr and several AP. A link must be established between AgAPr and the AgSPr, and in limit, they may be the same operator – vertical integrated case.

From the operator point of view, the argument is simple. If customers demand pervasive and ubiquitous service, it constitutes an opportunity for operators, though one that he cannot arrest without aggregating accesses and services. Even

Figure 2. Current ICCM value chain – stylized overview

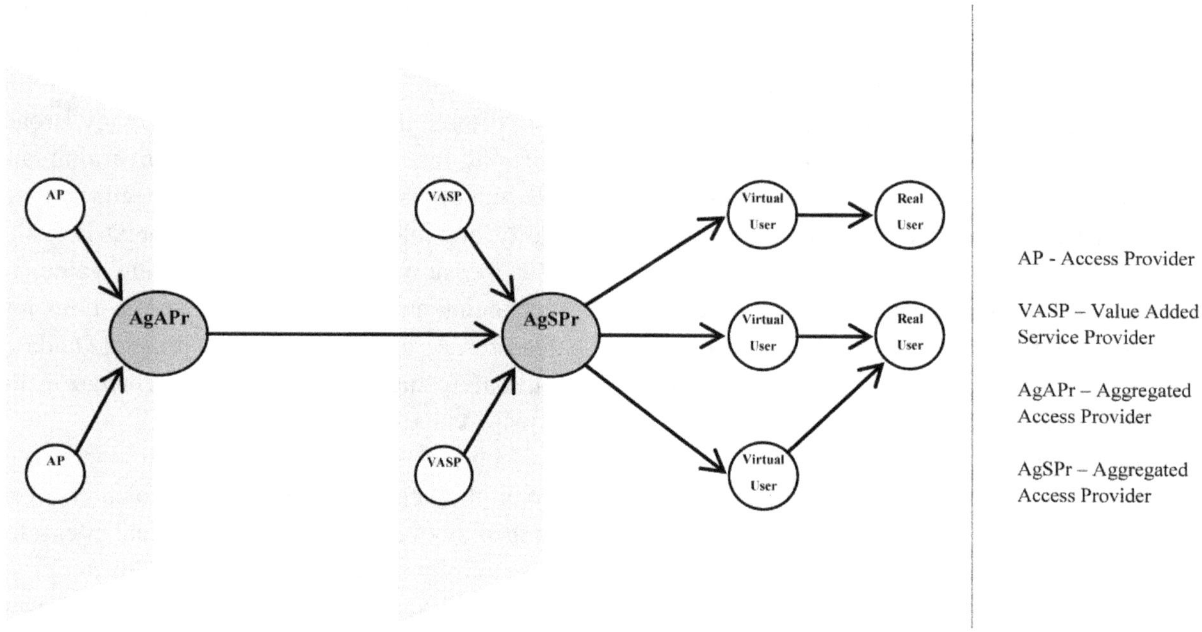

though one may find it possible for an operator to develop a portfolio of services and accesses wide enough to attend this demand, it would certainly be a second best solution when compared with aggregation.

For instance, high-speed broadband combined with pervasiveness enabled through Daidalos can contribute to increase user experience significantly. Furthermore investing in high-speed broadband is a very expensive issue with no guarantee concerning possible return on investment (ROI), therefore Daidalos concepts may also possibly help minimize necessary investment (e.g., through network initiated handover, load balancing) in this context.

From the social welfare point of view, the argument is rather valid. Even if operators' inherent characteristics were homogeneous, it is intuitive that neither M&A nor internal development would be optimal solutions for every single operator, as the former may entail competition weakening and the later would lead to unnecessary asset replication (e.g., suboptimal use of installed capacity) – both inefficient from the social welfare point of view.

We investigate the value of the Daidalos project for each of the most important types of the ICMM's operators – Fixed, Mobile, Broadcast and ISP/IAP. We investigate each case independently, assuming that operators are only willing to supply pervasive and ubiquitous services supported in the proposed Daidalos value chain, if they play AgSPr. In this case, each operator faces two alternative business models: vertical integrated Daidalos operator – playing AgSPr and AgAPr; and, vertical disintegrated Daidalos operator – playing AgSPr and outsourcing AgAPr. We further consider that IPr, BPr and PSPr roles are played by the AgSPr.

In other words, considering each player independently, we investigate whether operators are willing to supply pervasive and ubiquitous services, and if so, if they would opt by aggregating access or outsource it.

Figure 3. ICCM segmentation

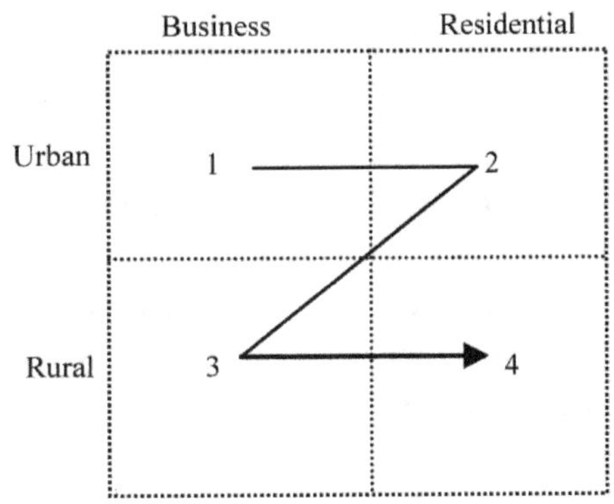

Flexibility

Considering all other things being equal, in an uncertain environment, the value of a project that allows managers to change the course of action when new information becomes available is always higher than the value of a project that does not allow an active management throughout the lifetime of the project. Which means that flexibility has value and, hence, valuing a risky project requires encountering all managerial flexibilities. As the risky nature of Daidalos solution is unsuspicious, we further identify the embedded sources of flexibility.

First of all, as in many other cases in the ICMM, marketing of an innovative service is incremental. That is, operators have incentives to test the service in a given market segment and, depending on its success, extend their offers to other market segments.

Figure 3 proposes an ICCM segmentation. On the one hand, the market may be regionally segmented by urban and rural areas. Population density is higher in urban areas than it is in rural areas and this has high impact in expected return on capital investment (especially in access layer). On the other hand, the market may be segmented by business and residential users. Figure 3 also proposes a direction of the incremental investment, justifiable by the following: operators tend to market business customers first, as they are less numerous and tend to drive demand for QoE (quality of experience) and QoS (quality of service). Operators also tend urban areas first, as urban market is one of higher value when compared to rural market – which is essentially due to the fact that scale and scope economies are most common in these areas.

In our case, however, because fixed cost is only related to access assets and capacity constraints are not considered, operators are expected to market business and residential users at the same time – say, time 0. That is, once the investment is made, there is no reason not to market residential users. Thus, there is an expansion option embedded in Daidalos solution – the

Figure 4. ICCM segmentation – Daidalos

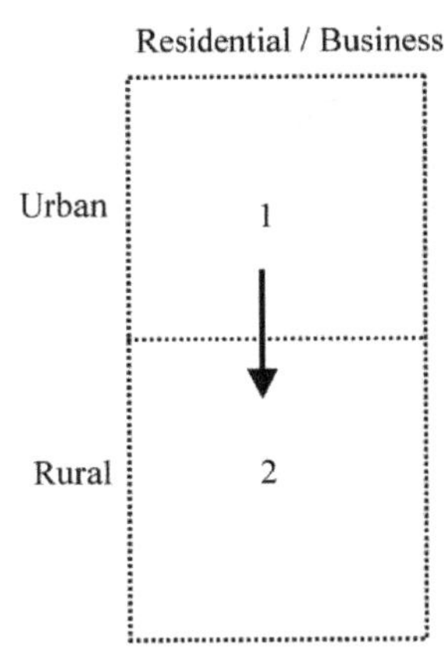

option to expand service/access provision to the rural areas.

Along with the expansion option, we find other two sources of flexibility: abandonment and shrinking options. The reasoning is as follows. Since we are considering that operators are bearing in mind investing in urban areas in time 0, if demand goes well, operators may have incentives to expand their operations to less profitable markets (rural market). Also, if demand goes bad and expansion still did not take place, operators might have incentives to abandon the project and if demand goes bad and expansion already took place, operators might have incentives to abandon rural market or even to abandon both markets.

Figure 5 proposes a decision tree to clarify the idea that Daidalos solution may be seen as an incremental investment opportunity and, hence, offers flexibility to managers.

Based on ICCM segmentation, the figure proposes a decision tree associated to the project. As one may notice, in case of success in market 1, operators have the opportunity to extend their offer to market 2. Notice, however, that we are considering here that if the project is not successful in urban areas it won't be in rural areas. We furthermore consider that operators invest in rural areas only when the project is successful in urban areas.

Till now we found three sources of flexibility – expand, shrink and abandon – associated to Daidalos solution. Note that none of them are dependent on the business model – vertical integrated Daidalos operator or vertical disintegrated Daidalos operator – adopted by market players. So far, no discussion was introduced related to that. However, once we consider that each business model constitutes in itself a solution (mutually exclusive alternatives), one might find it reason-

Figure 5. Decision tree – ill-defined case

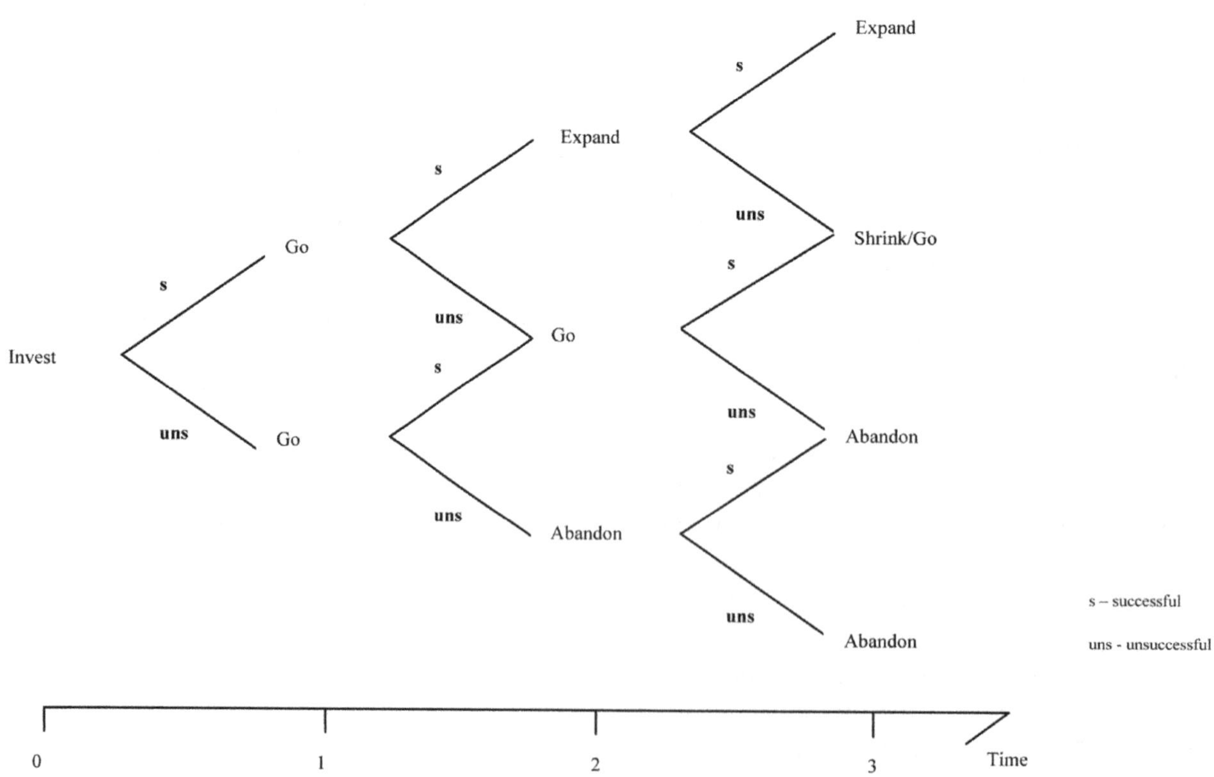

able to consider that there is always the possibility for an operator to migrate from one scenario to another – switching option.

That is, more flexibility may be found once we attend that, operators are given the possibility to play AgAPr or outsource it, both in market 1 and in market 2. Furthermore, notice that, as initial investments in urban areas are independent from those in rural areas, there are four possible scenarios from which operators may choose for expansion (mutually exclusive alternatives): scenario AA, vertical integrated Daidalos operator in markets 1 and 2; scenario BB, vertical disintegrated Daidalos operator in markets 1 and 2; scenario AB, vertical integrated Daidalos operator in market 1 and vertical disintegrated Daidalos operator in market 2; and, scenario BA, vertical disintegrated Daidalos operator in market 1 and vertical integrated Daidalos operator in market 2.

We explore the differences between the scenarios as a way to sustain that these are, in fact, alternative solutions from which market players may choose, along time, as new information is revealed.

Start by considering that, at time 0, an operator decides to aggregate accesses in market 1. In this case, it will have an initial expenditure associated to investment in ubiquity enabling switching devices as well as optimization platforms. Revenues will come from direct sales but also from AgAPr services provided to other operators that are willing to enable Daidalos in a vertical disintegrated way (if an operator decides not to aggregate accesses in market 1, it will have considerable less expenditure related to ubiquity enabling switching devices as well as optimization platforms. Though, we assume that there will be a monthly fee paid to AgAPr).

Now consider that, at time 1, information about recent demand evolution is revealed and that management decides to expand to market 2, again in a vertical integrated way – scenario AA. So, it is reasonable to consider that, in order to aggregate accesses in market 2, an additional investment will be needed due to geographic coverage.

Additionally consider that, at time 2, the demand evolution changes considerably, in a way that the value of scenario AB became greater than the value of scenario AA. In these conditions, the operator will abandon scenario AA and migrates to scenario AB, by (somehow) abandoning its AgAPr assets in rural areas (ubiquity enabling switching devices) and outsources AgAPr in rural areas. Again, the reasoning is as follows: consider scenarios AA and AB and think as if the ultimate driver of success is the demand evolution. The fact that demand conditions are similar in the two scenarios along with the fact that cost structure is different, condition success on scenarios. Hence, it is reasonable to consider that there is always the possibility for an operator to migrate from scenario AA to scenario AB and from scenario AB to AA. Moreover, as there are four scenarios, we have 10 more possible migration scenarios.

Figure 6 shows the cost structure for scenarios AA and BB – when the scenario chosen in time 1 is the same chosen in time 0 – and AB and BA – when the scenario chosen in time 1 is not the same chosen in time 0. Notice that this is a generic illustration and thus cost axis is not comparable.

Being found the sources of flexibility embedded in Daidalos solution – expansion, abandonment, shrinking and switching options – in the next section we define a valuation model able to capture uncertainty and the ability for management to exercise options along the lifetime of the project.

MODEL DEFINITION

For a long time the NPV has been the most common methodology used by firms to evaluate investments. However, the traditional NPV analysis has been widely questioned due to the fact that it ignores the value of flexibility associated to real asset investment by not capturing uncertainty –

Figure 6. Cost structure associated to scenarios AA, BB, AB and BA – generic illustration

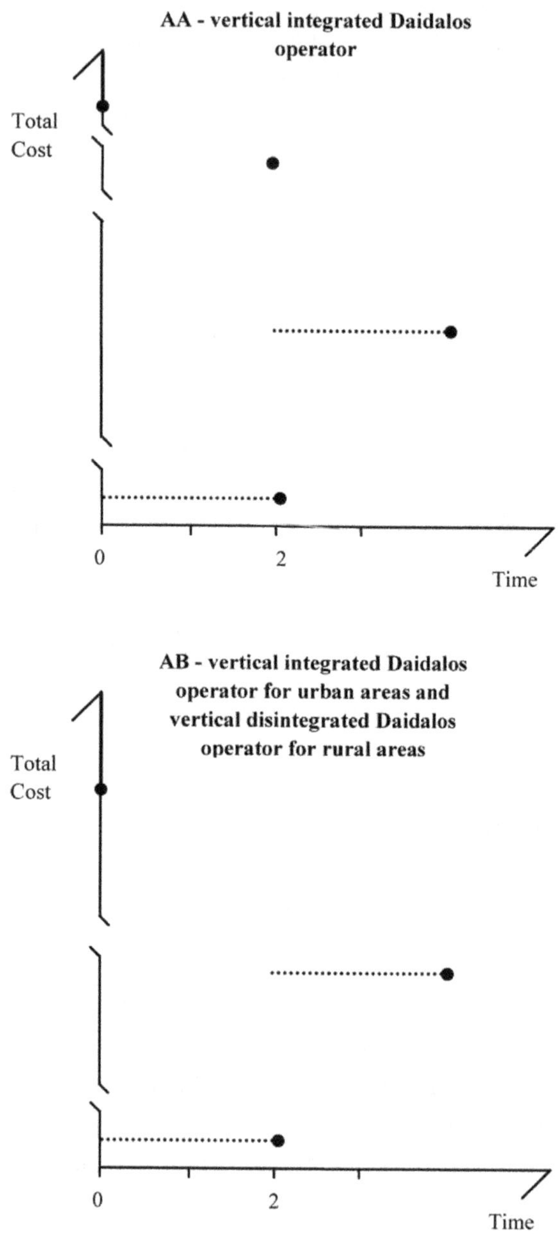

see Myers (1984), Kester (1984), Brennan and Schwartz (1985a, 1985b), McDonald and Siegel (1986), Pindyck (1991) and Trigeorgis (1993). Even though it does not concur with NPV in terms of applicability, Decision Tree Analysis (DTA) is able to incorporate flexible management, though either it gets extremely hard to model, either it violates the law of one price, by assuming, for simplicity, constant discount rate – see Trigeorgis (1996) and Copeland and Antikarov (2001). Copeland and Antikarov (2001) also state "ROA corrects both deficiencies. It models flexibility in decision making, and because it forms replicating portfolios based on the law of one price, it eliminates arbitrage opportunities and correctly prices the project with flexibility" (p. 112).

Due to non-negligible uncertainty and learning associated to Daidalos solution, we investigate the value of the Daidalos project for each of the ICMM's operators – Fixed, Mobile, Broadcast and ISP/IAP, using ROA. Even though a profound discussion on *pros* and *cons* of each real asset valuation models and approaches is out of the scope of this paper, we account that the validity of the "risk neutral" assumption used in real option pricing is dependent on the existence of one of the following, as presented by Wang and Halal (2010): "the 'portofolio matching' or 'twin portofolio', i.e. the underlying asset distribution can be 'perfectly' mimicked by a financial security; "the complete market theory, which includes the real asset investment into the financial asset market. Therefore, any real asset can be replicated 'from an equivalent portfolio of traded securities'. Once the market of financial assets and real assets is a complete one, i.e., a unique market price of risk applies to both financial assets and real assets, 'no arbitrage' assumption and risk neutral approach are valid in pricing any asset within this complete market." (p. 15) – for literature review and deep discussion on advantages and disadvantages of each real asset valuation models – NPV, DTA and ROA – see Wang and Halal (2010).

Despite the considerable consensus about the usefulness of ROA to capture the value of flexibility, there is no consensus about the most adequate approach. One approach may be the adequate to solve problem x, but may not be to solve problem y. This means that a practitioner should find the appropriate approach to solve a given problem. Table 1 (adapted from Borison, A., 2003) summarizes the major analytical approaches to real options.

As the differences rely on their underlying assumptions and mechanics but also, obviously, on their applicability, we will consider the Marketed Asset Disclaimer (MAD) approach to pursue with our analysis. More precisely, the one proposed by Copeland and Antikarov (2001): "If you are trying to value real options on a research and development program, or the right to shut down and reopen a General Motors automobile assembly plant, where do you find a twin security? Instead of searching in financial markets,

Table 1. Summary of major analytical approaches to real options analysis

Approach	Nature of Capital Markets for Underlying Assets	Data Source	Proponents
Classic	Complete with respect to all corporate investments	Capital market	Amram & Kulatilaka (1999) Brennan & Schwartz (1985) Trigeorgis & Manson (1987) Copeland, Koller & Murrin (1994)
Subjective	Complete with respect to all corporate investments	Subjective judgment	Howell et al. (2001) Luehrman (1997)
Marketed Asset Disclaimer	Absent with respect to all corporate investments	Subjective judgment	Copeland & Antikarov (2001) Trigeorgis (1999) Brealey & Myers (2000)
Revised Classic	Complete with respect to market-dominated corporate investments; absent with private-dominated corporate investments	Capital market for market-dominated investments; subjective judgment for private-dominated investments	Dixit & Pindyck (1994) Amram & Kulatilaka (2000)
Integrated	Complete with respect to market risks of corporate investments; absent with private risks of corporate investments	Capital market for market risks; subjective judgment for private risks	Smith & Nau (1995) Smith & McCardle (1998) Constantinides (1978) Luenberger (1998)

we recommend that you use the present value of the project itself, without flexibility, as the underlying risky asset – the twin security. What is better correlated with the project than the project itself? We are willing to make the assumption that the present value of the cash flows of the project without flexibility (i.e. the traditional NPV) is the best unbiased estimate of the market value of the project were it a traded asset. We call this assumption the Marketed Asset Disclaimer (MAD) … Note that the MAD makes assumptions no stronger than those used to estimate the project NPV in the first place" (pp. 94-95).

The valuation traditionally embeds four steps, but since managerial flexibilities were already identified in the previous section, in the next section we calibrate the model and present the major findings following the remaining three steps: compute base case present value without flexibility using DCF valuation model - NPV; model the uncertainty using event trees; and, calculate real option value (ROA).

CALIBRATION AND RESULTS

We evaluate the Daidalos project taking as a starting point a stylized representation of Portuguese ICMM. As stated before, to investigate the value of the Daidalos project for each of the ICMM's operators – Fixed, Mobile, Broadcast and ISP/IAP – we will use ROA as flexibility is embedded in the project.

To do so we (i) estimate the value of the project without flexibility – using NPV – and consider it as a base case to (ii) model the project uncertainty, which is an input to (iii) investigate the value of each scenario (AA, AB, BA, and BB) with flexibility – using ROA. Some conclusions are addressed in each stage of the valuation, and results are tested along the way.

NPV: Base Case

The value of Daidalos project without flexibility is the expected free cash flows of the project discounted at a risk-adjusted rate – in this case the WACC – as stated in the following equation.

$$NPV = -I_0 + \sum_{t=1}^{n} \frac{E(FCF_t)}{(1+WACC)^t} \qquad (1)$$

The construction of our base case followed a simple argument. Operators face urban market as a 'primary' market for the provision of Daidalos. They may provide it as a vertical integrated Daidalos Operator (scenarios AA and AB) or as a vertical disintegrated Daidalos Operator (scenarios BA and BB). We find it hard to associate flexibility to this specific project – provide Daidalos in urban areas – in a way that we find DCF an appropriate tool to estimate its value.

Even though some subjectiveness is associated to the inputs used to compute NPV, caution was taken to construct a base case as closer to the Portuguese market as possible. Overall, Portuguese operators' gross margins where checked. Table 2 and Table 3 synthesize the set of assumptions used to compute the NPV.

The project has 10 years length. In the first year demand will be of about 1% of the relevant market (5.000.000 households) and will grow approximately 33% a year (continuous growth rate) to 20% of the relevant market. Prices are expected to decline along the life of the project. Revenues from AgAPr are expected to be of about 8 EUR per user (10% of price per unit) and will be steady during the 10 years. Costs per unit will be of about 50 EUR in the first year and are expected to decline (though less than price). Initial investment is 500.000.000 EUR for fixed operators and will depreciate 10% a year. For other operators the initial investment is higher: 10%, 5% and 30% for mobile, broadcast and ISP/IAP, respectively. The differences are due to know-how,

Table 2. Set of assumptions for base case construction – scenarios AA and AB

Input	Unit	Fixed	Mobile	Broadcast	ISP/IAP
Relevant market	Nr	5.000.000	5.000.000	5.000.000	5.000.000
Maturity	Years	10	10	10	10
Quantity sold year 1	Nr	50.000	50.000	50.000	50.000
Quantity sold year 10	Nr	1.000.000	1.000.000	1.000.000	1.000.000
Implicit annual growth rate quantity	%	33	33	33	33
Price year 1	EUR	80	80	80	80
Price year 10	EUR	30	30	30	30
Implicit annual growth rate price	%	-11	-11	-11	-11
Revenue AgAPr year 1	EUR	8,00	8,00	8,00	8,00
Revenue AgAPr year 10	EUR	8,00	8,00	8,00	8,00
Implicit annual growth rate revenue AgAPr	%	0	0	0	0
Cost per unit year 1	EUR	50,00	55,00	55,00	57,50
Cost per unit year 10	EUR	25,00	27,50	27,50	28,75
Annual growth rate cost	%	-8	-8	-8	-8
Initial investment	EUR	500.000.000,00	550.000.000,00	525.000.000,00	650.000.000,00
Depreciation	EUR	50.000.000,00	55.000.000,00	52.500.000,00	65.000.000,00
WACC	%	10	13	16	20

Table 3. Set of assumptions for base case construction – scenarios BA and BB

Input	Unit	Fixed	Mobile	Broadcast	ISP/IAP
Relevant market	Nr	5.000.000	5.000.000	5.000.000	5.000.000
Maturity	Years	10	10	10	10
Quantity sold year 1	Nr	50.000	50.000	50.000	50.000
Quantity sold year 10	Nr	1.000.000	1.000.000	1.000.000	1.000.000
Implicit annual growth rate quantity	%	33	33	33	33
Price year 1	EUR	80	80	80	80
Price year 10	EUR	30	30	30	30
Implicit annual growth rate price	%	-11	-11	-11	-11
Cost AgAPr year 1	EUR	-8,00	-8,00	-8,00	-8,00
Cost AgAPr year 10	EUR	-8,00	-8,00	-8,00	-8,00
Implicit annual growth rate revenue AgAPr	%	0	0	0	0
Cost per unit[1] year 1	EUR	50,00	55,00	55,00	57,50
Cost per unit[1] year 10	EUR	25,00	27,50	27,50	28,75
Annual growth rate cost	%	-8	-8	-8	-8
Initial investment	EUR	50.000.000,00	55.000.000,00	57.750.000,00	75.075.000,00
Depreciation	EUR	5.000.000,00	5.500.000,00	5.775.000,00	7.507.500,00
WACC	%	10	13	16	20

Table 4. Value of the project without flexibility

Scenario	Unit	Fixed	Mobile	Broadcast	ISP/IAP
AA and AB	EUR	279.137.951,99	- 98.814.809,77	- 200.184.220,14	- 440.243.391,47
BA and BB	EUR	- 145.687.100,81	- 199.521.039,78	- 140.423.557,72	- 136.571.650,64

installed capacity and diversity in the access layer. No growth rate was considered for the computation of residual value.

These two scenarios are in line with scenarios AA and AB. There are two differences: on the one hand, initial investment is only 10% of the initial investment of the other scenarios; and, because there is a need to outsource AgAPr, what constitutes a revenue from AgAPr services in scenarios AA and AB here constitutes a cost. Cost per unit hereafter stands for costs not related to AgAPr outsourcing.

Considering that there is no flexibility associated with the project and given the assumptions above, only Fixed operators would be willing to invest in scenarios AA and AB (vertical integrated Daidalos operator in urban areas). Note also that scenarios AA or AB are only more favorable for Fixed and Mobile operators. The others would prefer scenarios BA or BB. The results are summarized in Table 4.

Notice that, for each scenario, the differences in NPV, between operators, are essentially due to differences in WACC initial investment and cost structure. On the other hand, despite the significant differences in initial investment requirement, for each operator, the differences in NPV, between scenarios, are mainly due to the costs of outsourcing AgAPr. There might be an equilibrium price for outsourcing AgAPr, which is expected to be highly dependent on negotiation. Even though these interactive and dynamic issues are out of the scope of this article, we investigate the sensibility of NPV to the net revenue of providing AgAPr outsourcing. Figure 7 shows the results.

On the xx axis we represent the net revenue of providing AgAPr, per user, per month. Note first that the NPV is highly sensible to this particular input. Of course, in scenarios AA/AB we have strictly increasing functions and in scenarios BA and BB, we have strictly decreasing functions. Though, 8 euros appear to proxy well a

Figure 7. Sensitivity analysis of NPV to change in the net revenue of AgAPr (in euros)

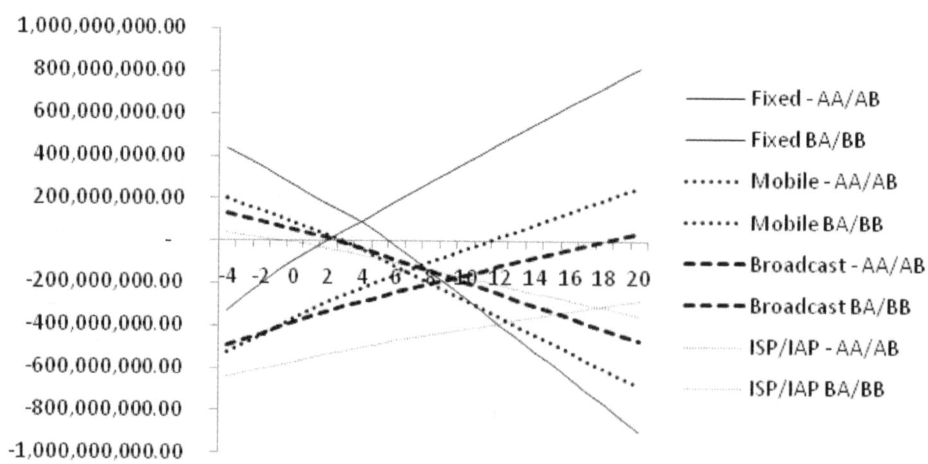

competitive price. Furthermore, as we shall notice, when flexibility is considered, the results are considerably different and this question no longer arises.

ROA: Modeling Uncertainty

The volatility needed for the binomial tree is the volatility of the rate of return. In our case, there are several sources of uncertainty. Therefore, we will use a Monte Carlo analysis to combine these uncertainties into one: the volatility of the (rate of return of the) project – σ_D. There is uncertainty about four input parameters: quantity, price, revenue/cost from AgAPr and cost per unit. We will estimate each variable's volatility by constructing 95 percent confidence intervals.

Quantity sold in year 1 – Q_1 – is 50.000 and is expected to grow on average at a constant 33 percent rate for the next 10 years. Thus, quantity sold in year 10 – Q_{10} – is expected to reach 1.000.000. Price in year 1 – P_1 – is 80 EUR and is expected to decline on average at a constant -11 percent rate for the next 10 years. Thus, price in year 10 – P_{10} – is expected to be 30 EUR. Cost (revenue) from AgAPr outsourcing in year 1 – Ag_1 – is -8 EUR (8 EUR) and is expected to stay in the same level for the next 10 years. Thus, Cost (revenue) from AgAPr outsourcing in year 10 – Ag_{10} – is -8 EUR (8 EUR). Cost per unit in year 1 – Cu_1 – is 50 EUR and is expected to decline on average at a constant -8 percent rate for the next 10 years. Thus, cost per unit in year 10 – Cu_{10} – is 25 EUR. Note that this last estimate is only valid for fixed operators as Cu_1 differs between operators.

$$Q_t = \left[Q_1 e^{tr - 2\sigma\sqrt{t}}, Q_1 e^{tr + 2\sigma\sqrt{t}} \right] \tag{2}$$

$$P_t = \left[P_1 e^{tr - 2\sigma\sqrt{t}}, P_1 e^{tr + 2\sigma\sqrt{t}} \right] \tag{3}$$

$$Ag_t = \left[Ag_1 e^{tr - 2\sigma\sqrt{t}}, Ag_1 e^{tr + 2\sigma\sqrt{t}} \right] \tag{4}$$

$$Cu_t = \left[Cu_1 e^{tr - 2\sigma\sqrt{t}}, Cu_1 e^{tr + 2\sigma\sqrt{t}} \right] \tag{5}$$

Each variable's volatility is calculated by solving one of the equations 6. For quantity and revenue/cost from AgAPr we use the first; for price and cost per unit we use the second. V_T^{upper} stands for the upper limit of the 95 percent confidence interval and V_T^{lower} stands for the lower limit of the 95 percent confidence interval. Q_{10}^{upper} is 3.750.000 units; P_{10}^{upper} is 23 EUR; Ag_{10}^{upper} is 10 EUR; and, Cu_{10}^{upper} is 19 EUR. Once again, note that this estimate for Cu_{10}^{upper} is only valid for fixed operators.

$$volatility\ estimate: \sigma_Q\ or\ \sigma_{Ag} = \frac{\ln\left(\frac{V_T^{upper}}{V_0}\right) - \sum_{i=1}^{n} r_i}{2\sqrt{T}},$$

$$\sigma_P\ or\ \sigma_{Cu} = \frac{\sum_{i=1}^{n} r_i - \ln\left(\frac{V_T^{lower}}{V_0}\right)}{2\sqrt{T}} \tag{6}$$

Volatility estimates are 22%, 5%, 4% and 5%, for σ_Q, σ_P, σ_{Ag} and σ_{Cu}, respectively. Confidence intervals are constructed for time 2 to 10 and results are shown in Figure 8.

Quantity follows a normal distribution and correlates with price. Price follows a lognormal distribution and correlates with itself. Revenue/cost from AgAPr follows a lognormal distribution and correlates with quantity. Cost per unit follows a lognormal distribution.

As stated, to pursue with the ROA valuation, these volatilities must be aggregated into a single one – the volatility of the rates of return of the project (base case). To do so, we took a three-step methodology: First, we define the assumptions in our base case (NPV) – those related to correlation between variables, autocorrelation probability distribution and volatility estimates of the four sources of uncertainty. Second, we define the forecast variable as the percentage changes in

Figure 8. Confidence intervals – inputs for Monte Carlo simulation

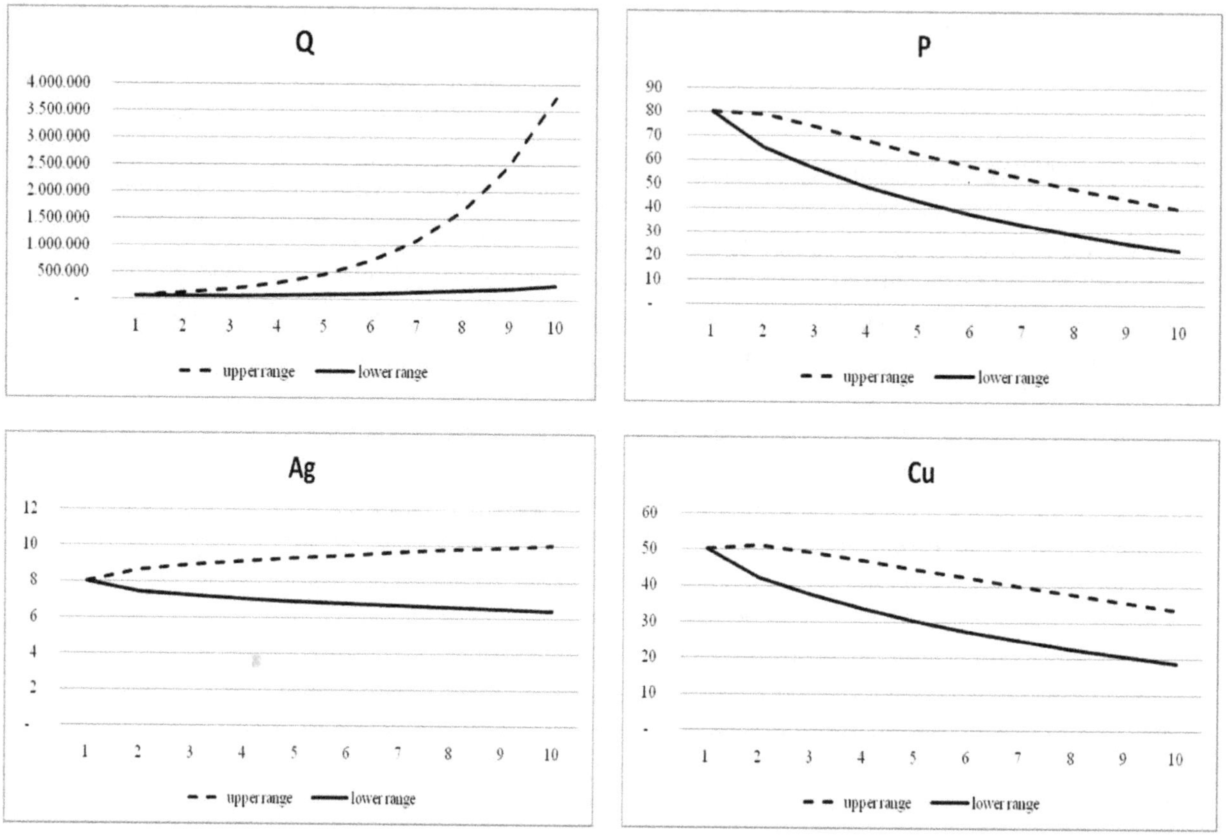

the value of the project from one time period to the next. We compute this, interacting the input variables the present value at time 1 to vary and by keeping constant the present value at time 0. Finally, run the simulation 10.000 trials.

The process was repeated for each operator both for AA/AB and BA/BB scenarios. The results are shown in Table 5.

Note that volatility of scenario AA/AB will be used to real option valuation in both AA and AB scenarios (and the same with scenario BA/BB), as the base case – underlying asset – is exactly the same.

ROA

This methodology will be repeated in two stages. In a first stage we consider that an operator has no possibility to migrate from one scenario to another. This means that we assess the value of each scenario – AA to BB – independently, con-

Table 5. Volatility of rate of return – σ_D

Scenario	Fixed	Mobile	Broadcast	ISP/IAP
AA and AB	20%	18%	16%	14%
BA and BB	23%	20%	18%	16%

Table 6. Expansion options' strikes

Scenario	Unit	Fixed	Mobile	Broadcast	ISP/IAP
AA	EUR	500,000,000.00	550,000,000.00	525,000,000.00	650,000,000.00
AB	EUR	50,000,000.00	55,000,000.00	52,500,000.00	65,000,000.00
BA	EUR	500,000,000.00	550,000,000.00	525,000,000.00	650,000,000.00
BB	EUR	50,000,000.00	55,000,000.00	57,750,000.00	75,075,000.00

Table 7. ROA – without considering switching

Scenario	Unit	Fixed	Mobile	Broadcast	ISP/IAP
AA	EUR	532,812,174	269,009,694	170,054,899	91,811,725
AB	EUR	532,663,455	269,064,243	170,054,899	91,811,725
BA	EUR	-13,939,531	-19,508,779	-14,892,602	-12,751,629
BB	EUR	-13,939,531	-19,508,779	-14,892,602	-12,751,629

sidering only expansion, abandon and shrinking options. This may appear tricky but it will turn very useful, as we will clearly notice that there are business models 'dominated' by others. Then, in the second stage, we investigate the value of the 'dominant' business models considering the possibility of switching between each other.

Some considerations should be addressed related to the options considered in the first stage. Abandonment and shrinking options have no strike and can be exercised multiple times. That is, there is no exercise price and once they are exercised, there is always the possibility to re-enter the market. Table 6 synthesizes the characteristics of the expansion options, for each of the four scenarios.

The exercise price for expanding in a disintegrated way is sensibly 10% of the exercise price for expanding in an integrated way – which is similar to the initial investment in urban areas. Besides from this, we model this expansion by increasing cash flows 25% or 5% - for integrated and disintegrated case, respectively. That is, whenever the decision is to be in the urban areas providing Daidalos vertically integrated, the current cash flows are 25 percent larger than it would be in urban areas exclusively (for disintegrated case, analogously).

Despite NPV approach suggested that only fixed operator would be willing to provide Daidalos, when considering its' inherent flexibility the results turn out to be considerably different. Note that all operators are willing to be Daidalos providers. More, it turns out that the disintegrated business model in the urban areas is uninteresting for all of them. Table 7 summarizes the results.

Even though disintegrated business model appears uninteresting in the urban areas, there is no clear dominating business model in rural areas – mobile operators would slightly prefer scenario AB and fixed operators would prefer AA, while broadcast and internet service/access providers would be indifferent. This result arises the question of whether or not switching between AA and AB throughout the lifetime of the project is indeed a valuable option. We will investigate this, but first we find it interesting to show how sensitive is the project's real value to the net revenue of the AgAPr outsourcing (analogously to the analysis provided for NPV) in scenarios AA and AB.

Figure 9. Sensitivity analysis of ROA to change in the net revenue of AgAPr (in euros)

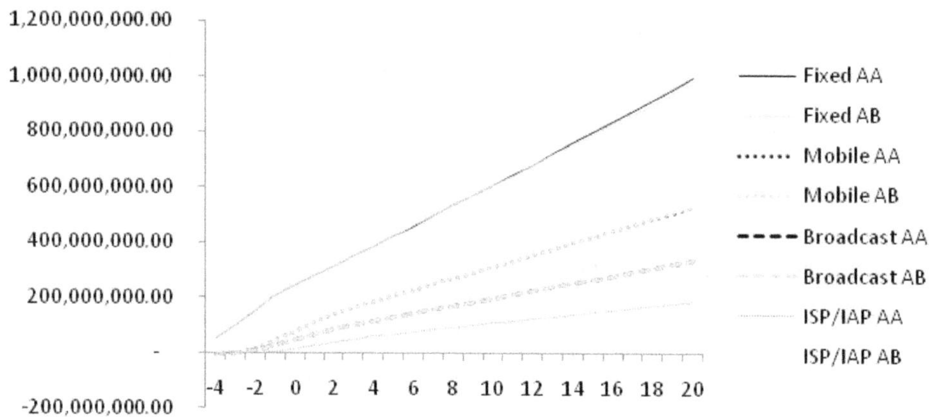

Figure 10. Sensitivity analysis of the value of flexibility to change in the net revenue of AgAPr (in euros)

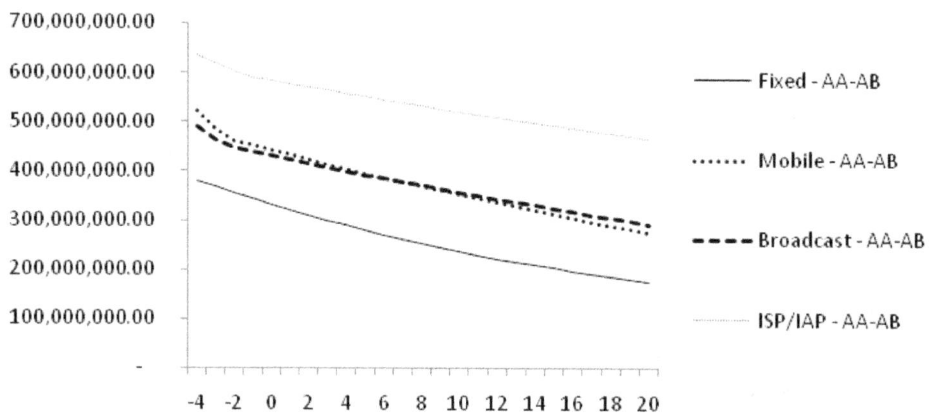

In Figure 9, on the xx axis we represent the net revenue of providing AgAPr, per user per month. ROA sensitivity is considerable. Notice however, that contrary to NPV, negative values are very uncommon. The value of flexibility is, indeed, very high. It is also interesting to analyze the extent of which the value of flexibility varies with the net revenue of providing AgAPr. Figure 10 proposes this analysis.

Again, on the xx axis we represent the net revenue of providing AgAPr, per user per month. In the yy axis we represent the difference between scenario AA with flexibility (ROA) and scenario AA/AB without flexibility (NPV). The value of flexibility is a strictly decreasing function of the net revenue of providing AgAPr. This happens because of the ability of NPV to capture the corresponding increase in net revenue.

As stated before, this analysis concludes with the introduction of the last source of flexibility – the ability of switching between vertical integrated and vertical disintegrated business models. Our previous results enable us to exclude from the analysis the switching most of the switching possibilities, as the vertical integrated business model clearly dominates the vertical disintegrated

Table 8. ROA – considering switching

Scenario	Unit	Fixed	Mobile	Broadcast	ISP/IAP
A (A/B)	EUR	534,494,163	269,064,243	170,054,899	91,811,725

Figure 11. Sensitivity analysis of the ROA to change in project volatility (in euros)

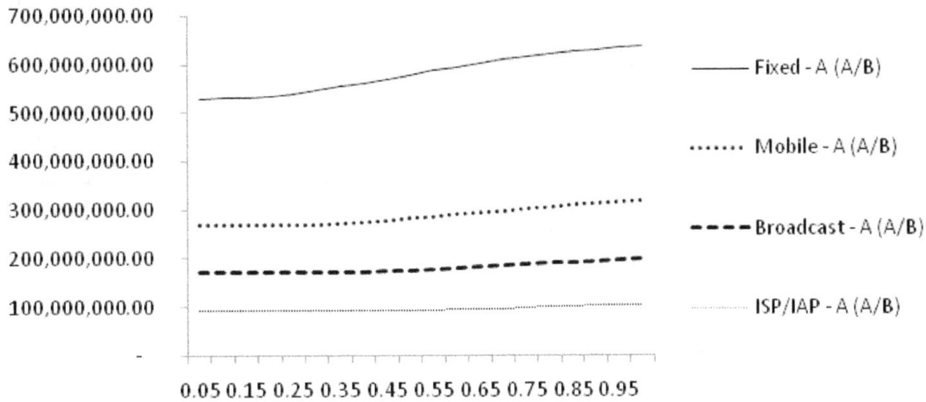

business model in the urban areas. As some previous results already indicated (though indirectly), this additional value is by far marginal. Table 8 provides the results.

Due to the fact that the switching option does not provide significant additional value, the results presented for scenarios AA and AB may be extrapolated to this particular situation.

Most of the analysis provided till now, in particular the sensitivity analysis presented, somehow lack of rigorous treatment. Note that it was not taken into account the fact that by changing an input to the base case (e.g., net revenue of AgAPr outsourcing), we were changing the base of calculating the volatility of the project – which is the central input to the ROA approach. The following result clarifies that the resolution of these imprecision wouldn't change the results considerably (Figure 11).

CONCLUSION

By analyzing current trend in the information, communication and media markets, we proposed to evaluate a project within the lines of Daidalos solution in a relevant market similar to the Portuguese ICMM. The valuation takes into account the difference between the traditional Fixed, Mobile, Broadcast and ISP/IAP operators – which constitute the set of potential providers of such pervasive and ubiquitous services (and accesses) – using realistic assumptions and parameters. We identify various sources of flexibility, which constitute managerial options associated to the project: expand, shrink, abandon and switch. The methodology used follows the Real Option Approach under the Marketed Asset Disclaimer assumption.

Primarily, we find that the value of a project with such characteristics is highly dependent on the exercise of its' inherent options and that operators will adopt a vertical integrated business model in urban areas – though, in rural areas, it is not clear which business model would be adopt by each of market player types. We also find that provision will take place in rural areas only in very favorable states of nature and that the value of the switching option is in all cases neglectible. Moreover, we find that the real value of the project is relatively insensitive to changes in the volatility of the underlying asset – project without flexibility.

REFERENCES

Alleman, J., Madden, G., & Kim, H. (2008). Real Option Methodology Applied to the ICT Sector: A Survey. *Communications & Strategies*, *70*, 27.

Amram, M., & Kulatilaka, N. (1999). *Real Options: Managing Strategic Investment in an Uncertain World*. Boston: Harvard Business School Press.

Amram, M., & Kulatilaka, N. (2000). Strategy and Shareholder Value Creation: The Real Options Frontier. *Journal of Applied Corporate Finance*, *15*(2), 15–28. doi:10.1111/j.1745-6622.2000.tb00051.x

Benaroch, M., & Kauffman, R. (1999). A Case for Using Real Options Pricing Analysis to Evaluate Information Technology Project Investments. *Information Systems Research*, *10*(1), 70–86. doi:10.1287/isre.10.1.70

Borison, A. (2003). *Real Options Analysis: Where are the Emperor's Clothes*? Paper presented at the Seventh Annual Real Option Conference.

Brealey, R., & Myers, S. (Eds.). (2000). *Principles of Corporate Finance*. Boston: Irwin McGraw-Hill.

Brennan, M. J., & Schwartz, E. S. (1985). Evaluating Natural Resource Investments. *The Journal of Business*, *58*(2), 135–157. doi:10.1086/296288

Brennan, M. J., & Schwartz, E. S. (1985). New Approach to Evaluating Natural Resource Investments. *Midland Corporate Journal of Finance*, *3*(1), 37–47.

Constantinides, G. M. (1978). Market Risk Adjustment in Project Valuation. *The Journal of Finance*, *33*, 603–616. doi:10.2307/2326572

Copeland, T., & Antikarov, V. (Eds.). (2001). *Real Options: A Practitioner's Guide*. New York: Texere.

Copeland, T., Koller, T., & Murrin, J. (2000). *Valuation: Measuring and Managing the Value of Companies*. New York: John Wiley and Sons.

Dixit, A., & Pindyck, R. (1994). *Investment Under Uncertainty*. Princeton, NJ: Princeton University Press.

Ferreira, N., Kar, J., & Trigeorgis, L. (2009). Option Games – The key to competing in Capital-Intensive Industries. *Harvard Business Review*, 101–107.

Fichman, R. (2004). Real Options and IT Platform Adoption: Implications for Theory and Practice. *Information Systems Research*, *15*(2), 132–154. doi:10.1287/isre.1040.0021

Harmantzis, F., & Tanguturi, V. (2005). *Migration to 3G Wireless Broadband Internet and Real Options: The case of an Operator in India*. Telecommunications Policy Journal.

Harmantzis, F., & Tanguturi, V. (2007). Investment Decisions in the Wireless Industry: Applying Real Options. *Telecommunications Policy*, *31*, 107–123. doi:10.1016/j.telpol.2006.02.005

Harmantzis, F., & Trigeorgis, L. (2006). *Flexible Investment Decisions in the Telecommunications Industry: Case Applications using Real Options* (Working Paper No. 06-06). Orlando, FL: NET Institute.

Howell, S., Stark, A., Newton, D., Paxson, D., Cavus, M., Pereira, J., & Patel, K. (2001). *Real Options: Evaluating Corporate Investment Opportunities in a Dynamic World*. Upper Saddle River, NJ: Prentice Hall.

Kester, W. (1984). Today's Options for Tommorow's Growth. *Harvard Business Review*, (62): 153–160.

Kumar, R. (2002). Managing Risks in IT projects: an Option Perspective. *Information & Management*, *40*, 63–74. doi:10.1016/S0378-7206(01)00133-1

Luehrman, T. A. (1997). What's It Worth? A General Manager's Guide to Valuation. *Harvard Business Review*, 132–142.

Luenberger, D. G. (1998). *Investment Science*. New York: Oxford University Press.

McDonald, R., & Siegel, D. (1986). The Value of Waiting to Invest. *The Quarterly Journal of Economics*, *101*(4), 707–728. doi:10.2307/1884175

Myers, S. (1984). Finance Theory and Financial Strategy. *Interfaces*, *14*, 126–137. doi:10.1287/inte.14.1.126

Pindick, R. (1991). Irreversability, Uncertainty, And Investment. *Journal of Economic Literature*, *29*(3), 1110–1148.

Saxtoft, C. (2008). *Convergence – user expectations, communications enablers and business opportunities*. New York: Wiley.

Smith, J. E., & McCardle, K. F. (1998). Valuing Oil Properties: Integrating Option Pricing and Decision Analysis Approaches. *Operations Research*, *46*(2), 198–217. doi:10.1287/opre.46.2.198

Smith, J. E., & Nau, R. F. (1995). Valuing Risky Projects: Option Pricing Theory and Decision Analysis. *Management Science*, *41*(5), 795–816. doi:10.1287/mnsc.41.5.795

Trigeorgis, L. (1993). *Quantifying Sequential Option Interdependencies* (Working Paper). Boston: Boston University.

Trigeorgis, L. (1996). *Real Options: Managerial Flexibility and Strategy in Resource Allocation*. Cambridge, MA: MIT press.

Trigeorgis, L. (1999). *Real Options: A Primer* (pp. 3-33). Alphen aan den Rijn, The Netherlands: Kluwer.

Trigeorgis, L., & Mason, S. P. (1987). Valuing Managerial Flexibility. *Midland Corporate Finance Journal*, *5*(1), 14–21.

Wang, A., & Halal, W. (2010). Comparison of Real Asset Valuation Models: A Literature Review. *International Journal of Business and Management*, *5*(5).

This work was previously published in International Journal of Business Data Communications and Networking, Volume 6, Issue 2, edited by Debashis Saha and Varadharajan Sridhar, pp. 75-95, copyright 2010 by IGI Publishing (an imprint of IGI Global).

Compilation of References

Abbasi, A. A., Iqbal, J., & Malik, A. N. (2010). *A novel QoS optimization architecture for 4G networks.* International Conference on Computer Electrical and Systems Science and Engineering.

Abeillé, J., Durvy, M., Hui, J., & Dawson-Haggerty, S. (2008, November). *Lightweight ipv6 stacks for smart objects: the experience of three independent and interoperable implementations* (White Paper No. 2). IPSO.

Abid, M. A., & Belghith, A. (2009-3). *Functional enhancements of proactive routing in mobile ad hoc networks.* Third IFIP International Conference on New Technologies, Mobility and Security (IFIP NTMS 2009), Cairo, Egypt, 20-23 December, 2009.

Acharya, S., Alonso, R., Franklin, M., & Zdonik, S. (1995). Broadcast disks: Data management for asymmetric communications environments. *Proceedings of ACM SIGMOD Conference on Management of Data*, (pp. 199-210).

Achieng, R., & Chan, H. A. (2006). *A queuing theoretic scheme for a QoS-Enabled 4G wireless access*. Retrieved from http://utsescholarship.lib.uts.edu.au

Ahamed, S. I., Rahman, F., & Hoque, E. (2008). *ERAP: ECC based RFID authentication protocol.* 12th IEEE International Workshop on Future Trends of Distributed Computing Systems.

Ahmad, I., & Kwok, Y. K. (1996). Dynamic critical-path scheduling: An effective technique for allocating tasks graphs to multiprocessors. *IEEE Transactions on Parallel and Distributed Systems*, *7*(5), 560–521.

Ahmad, I., & Kwok, Y. K. (1998). On exploiting task duplication in parallel program scheduling. *IEEE Transactions on Parallel and Distributed Systems*, *9*(9), 872–892. doi:10.1109/71.722221

Ahmed, T., & Mushtaq, M. (2007). P2P Object-based adaptivE Multimedia Streaming (POEMS). *Journal of Network and Systems Management*, *15*(3), 289–310. doi:10.1007/s10922-007-9068-7

Aho, A., Hopcroft, J., & Ullman, J. (1974). *The design and analysis of computer Algorithms*. Reading, MA: Addison-Wesley.

Ajey, K., Misra, M., & Sarje, A. K. (2007). A weighted cache replacement policy for location dependent data in mobile environments. *SAC*, *07*, 920–924.

Akyildiz, I. F., Melodia, T., & Chowdhury, K. R. (2007). A Survey on wireless multimedia sensor networks. *Computer Networks*, *51*(4), 921–960. doi:10.1016/j.comnet.2006.10.002

Alhaisoni, M. (2010). Resource-awareness and trade-off optimisation in P2P video streaming. *International Journal of Advanced Media and Communication*, *4*(1). doi:10.1504/IJAMC.2010.030006

Alhaisoni, M., & Liotta, A. (2009). Characterization of signaling and traffic in Joost. *Peer-to-Peer Networking and Applications*, *2*(1), 75–83. doi:10.1007/s12083-008-0015-5

Alhaisoni, M., et al. (2009). Streaming Layered Video over P2P Networks. In *Proceedings of the 6th International Conference on Advances in Mobile Computing and Multimedia (MOMM2009)*, Kuala lumpurm, Malaysia. Washington, DC: IEEE.

Alleman, J., Madden, G., & Kim, H. (2008). Real Option Methodology Applied to the ICT Sector: A Survey. *Communications & Strategies*, *70*, 27.

Amram, M., & Kulatilaka, N. (1999). *Real Options: Managing Strategic Investment in an Uncertain World*. Boston: Harvard Business School Press.

Amram, M., & Kulatilaka, N. (2000). Strategy and Shareholder Value Creation: The Real Options Frontier. *Journal of Applied Corporate Finance*, *15*(2), 15–28. doi:10.1111/j.1745-6622.2000.tb00051.x

Andi, W. C., Yeo, C. K., & Lee, B. S. (2010). Environment-aware QoS framework for multi-interface terminal. *Journal of Computer Communications*, *33*(9), 1049–1055. .doi:10.1016/j.comcom.2010.01.010

Angelou, G. N., & Economides, A. A. (2008). A decision analysis framework for prioritizing a portfolio of ICT infrastructure projects. *IEEE Transactions on Engineering Management*, *55*(3), 479–495. doi:10.1109/TEM.2008.922649

ANSI/IEEE Standards. (1999). *Wireless LAN medium access control (MAC) and physical layer (PHY) specification*. Retrieved from http:// standards.ieee.org/ getieee802/802.11.html

Aslam, B., & Zou, C. (2009). *Distributed certificate and application architecture for VANETs*. SIGCOMM- Project CPSC 534.

Awerbuch, B., Folmes, D., Nita-Rotaru, C., & Rubens, H. (2002). An on demand secure routing protocol resilient to Byzantine failures. *Proceedings of the 3rd ACM workshop on Wireless Security* (pp.21-30).

Bajaj, R., & Agrawal, D. P. (2004). Improving scheduling of tasks in a heterogeneous environment. *IEEE Transactions on Computers*, *15*(2), 107–118.

Bakhtiari, S., Safavi-Naini, R., & Pieprzyk, J. (1995). *Cryptographic hash functions: A survey.* Technical Report 95-09.

Balamash, A., & Krunz, M. (2004). An overview of Web caching replacement algorithms. *IEEE Communications Surveys and Tutorials*, *6*, 44–56. doi:10.1109/COMST.2004.5342239

Banerjee, S. (2002). *Scalable application layer multicast*. New York: ACM.

Bansal, S., Kumar, P., & Singh, K. (2003). An improved duplication strategy for scheduling precedence constrained graphs in multiprocessor systems. *IEEE Transactions on Parallel and Distributed Systems*, *14*(6), 533–544. doi:10.1109/TPDS.2003.1206502

Barbara, D. (1999). Mobile computing and databases: A survey. *IEEE Transactions on Knowledge and Data Engineering*, *11*, 108–117. doi:10.1109/69.755619

Barnett, M. L. (2005). Paying attention to real options. *R & D Management*, *35*(1), 61–72. doi:10.1111/j.1467-9310.2005.00372.x

Basu, P., Khan, N., & Little, T. D. C. (2001). A mobility based metric for clustering in mobile ad hoc networks. *Proceedings of IEEE ICDCS 2001 Workshop on Wireless Networks and Mobile Computing*, (pp. 13-418). Phoenix, AZ.

Batina, L., Guajardo, J., Kerins, T., Mentens, N., Tuyls, P., & Verbauwhede, I. (2007). Public-key cryptography for RFID-tags. *IEEE International Conference on Pervasive Computing and Communications Workshops*, (pp. 217-222).

Belghith, A., & Abid, M. A. (2009-1). *Autonomic self tunable proactive routing in mobile ad hoc networks*. The 5th IEEE International Conference on Wireless and Mobile Computing, Networking and Communications (IEEE WiMob'09), Marrakech, Morocco, October 12-14. 2009.

Belghith, A., & Abid, M. A. (2009-2). *Dynamically self adjustable proactive routing protocols for mobile ad hoc networks.* The 34th Annual IEEE Conference on Local Area Networks (IEEE LCN'09), Zurich, Switzerlan, October 20-23.

Belghith, A., Mnaouer, A. B., & Abid, M. A. (2009). *Suitability analysis of probabilistic routing for dynamic ad-hoc networks.* The 5th IEEE International Conference on Wireless and Mobile Computing, Networking and Communications (IEEE WiMob 2009), Marrakech, Morocco, October 12-14, 2009.

Bellare, M., Shi, H., & Zhang, C. (2005). Foundations of group signatures: The case of dynamic groups, topics in cryptography. In Menezes, A. (Ed.), *CT-RSA Proceedings* (*Vol. 3376*). Lecture Notes in Computer Science Springer-Verlag.

Benaroch, M. (2001). Option-based management of technology investment risk. *IEEE Transactions on Engineering Management*, *48*(4), 428–444. doi:10.1109/17.969422

Benaroch, M. (2002). Managing information technology investment risk: A real options perspective. *Journal of Management Information Systems*, *19*(2), 43–84.

Benaroch, M., Jeffery, M., Kauffman, R. J., & Shah, S. (2007). Option-based risk management: A field study of sequential information technology investment decisions. *Journal of Management Information Systems*, *24*(2), 103–140. doi:10.2753/MIS0742-1222240205

Benaroch, M., & Kauffman, R. J. (1999). A case for using real options pricing analysis to evaluate information technology project investments. *Information Systems Research*, *10*(1), 70–86. doi:10.1287/isre.10.1.70

Benaroch, M., & Kauffman, R. J. (2000). Justifying electronic banking network expansion using real options analysis. *Management Information Systems Quarterly*, *24*(2), 197–225. doi:10.2307/3250936

Benaroch, M., Lichtenstein, Y., & Robinson, K. (2006a). Real options in information technology risk management: An empirical validation of risk-option relationships. *Management Information Systems Quarterly*, *30*(4), 827–864.

Benaroch, M., Shah, S., & Jeffery, M. (2006b). On the valuation of multistage information technology investments embedding nested real options. *Journal of Management Information Systems*, *23*(1), 239–261. doi:10.2753/MIS0742-1222230108

Benaroch, M., & Kauffman, R. (1999). A Case for Using Real Options Pricing Analysis to Evaluate Information Technology Project Investments. *Information Systems Research*, *10*(1), 70–86. doi:10.1287/isre.10.1.70

Benetis, R., Jensen, S., Karciauskas, G., & Saltenis, S. (2006). Nearest and reverse nearest neighbor queries for moving objects. *The VLDB Journal*, *15*, 229–249. doi:10.1007/s00778-005-0166-4

Beraldi, R., Querzoni, L., & Baldoni, R. (2006). A hint-based probabilistic protocol for unicast communications in MANETs. *Ad Hoc Networks*, *4*, 547-566. Elsevier, 2006.

Berling, P. (2008). Real options valuation principle in the multi-period base-stock problem. *Omega-International Journal of Management Science*, *36*(6), 1086–1095. doi:10.1016/j.omega.2006.05.007

Bertoni, H. L. (1999). *Radio Propagation for Modern Wireless Systems*. Upper Saddle River, NJ: Prentice Hall.

Bettstetter, C., Lassila, P., & Virtamo, J. (2006). Spatial node distribution of the random waypoint mobility model with applications. *IEEE Transactions on Mobile Computing*, *5*(6), 680–694. Retrieved from http://ieeexplore.ieee.org/lpdocs/epic03/ wrapper.htm?arnumber=1624340. doi:10.1109/TMC.2006.86

Bhalaji, N., Banerjee, S., & Shanmugam, A. (2008). A novel routing technique against packet dropping attack in ad hoc networks. *Journal of Computer Science*, *4*(7), 538–544. doi:10.3844/jcssp.2008.538.544

Bhalaji, N., & Shanmugam, A. (2009). Association between nodes to combat black hole attack in DSR based MANET. *International Conference on WOC '09, Wireless and Optical Communication Network 2009* (pp. 1-5).

Bhattacharjee, P. S., Saha, D., & Mukherjee, A. (2004). An approach for location area planning in a personal communication services network (PCSN). *IEEE Transactions on Wireless Communication*, *3*(4).

Blum, J., & Eskandarian, A. (2004). The Threat of Intelligent Collisions. *IT Professional*, *6*(1), 24–29. doi:10.1109/MITP.2004.1265539

Blum, J., Eskandarian, A., & Hoffman, L. (2003). Mobility management in IVC networks. *Proceedings of IEEE Intelligent Vehicles Symposium*, *9*(11), 150–155.

Bøckman, T., Fleten, S. E., Juliussen, E., Langhammer, H. J., & Revdal, I. (2008). Investment timing and optimal capacity choice for small hydropower projects. *European Journal of Operational Research*, *190*(1), 255–267. doi:10.1016/j.ejor.2007.05.044

Boneh, D., & Shacham, H. (2004). Group signatures with verifier- local revocation. In *CCS'04* (pp. 168–177). New York, NY: ACM Press.

Borison, A. (2003). *Real Options Analysis: Where are the Emperor's Clothes*? Paper presented at the Seventh Annual Real Option Conference.

Bosselaers, A., Govaerts, R., & Vandewalle, J. (1996). Fast hashing on the Pentium. *Lecture Notes in Computer Science*, 298–313. doi:10.1007/3-540-68697-5_23

Boukerche, A., de Oliveira, H. A. B. F., Nakamura, E. F., & Loureiro, A. A. F. (2007). Towards an integrated solution for node localization and data routing in sensor networks. *Proceedings 12th IEEE Symposium on Computers and Communications* (ISCC '07), July 2007.

Boute, R., Demeulemeester, E., & Herroelen, W. (2004). A real options approach to project management. *International Journal of Production Research*, *42*(9), 1715–1725. doi:10.1080/00207540310001639946

Boyden, S., Mahanti, A., & Williamson, C. (2005). Characterizing the behaviour of real video streams. In *SCS Symposium on Performance Evaluation of Computer and Telecommunication Systems* (pp. 783-791). Philadelphia, PA: SCS.

Brakmo, L., & Peterson, L. (1995). TCP Vegas: End to End Congestion Avoidance on a Global Internet. *IEEE Journal on Selected Areas in Communications*, *13*(8), 1465–1480. doi:10.1109/49.464716

Braun, M., Hess, E., & Meyer, B. (2008). Using elliptic curves on RFID tags. *IJCSNS International Journal of Computer Science and Network Security, 8*(2).

Brealey, R., & Myers, S. (Eds.). (2000). *Principles of Corporate Finance*. Boston: Irwin McGraw-Hill.

Brennan, M. J., & Schwartz, E. S. (1985). Evaluating Natural Resource Investments. *The Journal of Business*, *58*(2), 135–157. doi:10.1086/296288

Brennan, M. J., & Schwartz, E. S. (1985). New Approach to Evaluating Natural Resource Investments. *Midland Corporate Journal of Finance*, *3*(1), 37–47.

Buchegger, S., & Le Boudec, J. (2002). Performance analysis of the confident protocol. *Proceedings of the 3rd ACM Symposium on Mobile Ad Hoc Networking & Computing* (pp. 226-236).

Burcea, I., Jacobsen, H., De Lara, E., Muthusamy, V., & Petrovic, M. (2004). Disconnected operation in publish/subscribe middleware. *Proceedings of the International Conference on Mobile Data Management (MDM'04),* (pp. 39–50). IEEE Computer Society.

Buttyán, L., Holczer, T., & Vajda, I. (2006). Optimal key-trees for tree-based private authentication. *Workshop on Privacy Enhancing Technologies - PET2006.* Cambridge, United Kingdom.

Cablelabs. (2005). *Data-over-cable service interface specification DOCSIS 2.0.* Radio Frequency Interface Specification CM-SP-RFIv2.0-I08-050408, 8th release, Cablelabs.

Calandriello, G., Papadimitratos, P., Hubaux, J.-P., & Lioy, A. (2007). Efficient and robust pseudonymous authentication in VANET. In *Proceedings of the Fourth ACM International Workshop on Vehicular Ad Hoc Networks*, Canada.

Camacho, F. T., & Menezes, F. M. (2009). Access pricing and investment: a real options approach. *Journal of Regulatory Economics*, *36*(2), 107–126. doi:10.1007/s11149-009-9088-1

Camp, T., Boleng, J., & Davies, V. (2002). A survey of mobility models for ad hoc network research. *Wireless Communication and Mobile Computing (WCMC): Special ISSUE on Mobile Ad Hoc Networking: Research. Trends and Applications*, *2*(5), 483–502.

Campbell, J. A. (2002). Real options analysis of the timing of IS investment decisions. *Information & Management*, *39*(5), 337–344. doi:10.1016/S0378-7206(01)00101-X

Capkun, S., Hamdi, M., & Hubaux, J. P. (2001). GPS-free positioning in mobile ad-hoc networks. *Proceedings of the 34th Hawaii International Conference on System Sciences* (HICSS), 2001.

Chang, L. H., Wang, C. C., & Lee, T. H. (2010). A handover mechanism using IEEE 802.21 in heterogeneous wireless networks. *Communications in Computer and Information Science*, *77*, 95–108. .doi:10.1007/978-3-642-13405-0_10

Chang, B., Lin, S., & Liang, Y. (2006). Minimizing roaming overheads for vertical handoff in heterogeneous wireless mobile networks. *International Conference on Wireless Communications and Mobile Computing* (pp. 957-962).

Chaudhuri, P., & Elcock, J. (2005). Scheduling on multiprocessors using task duplication. In D. Tutsch (Ed.), *Proceedings of the 2005 Design, Analysis and Simulation of Distributed Systems (DASD 2005)* (pp. 28-35). San Diego, CA: SCS.

Chauvenet, C. (2010). Bernard Tourancheau, and Denis Genon-Catalot. 802.15.4, a mac layer solution for PLC. In *Proceedings of the AICCSA'10*. Washington, DC: IEEE.

Chawathe, Y. (2000). *Scattercast: An architecture for Internet broadcast distribution as an infrastructure service*. Doctoral dissertation, University of California at Berkley.

Cherkasova, L., & Gupta, M. (2004). Analysis of enterprise media server workload: Access patterns, locality, content evolution and rates of change. [TON]. *IEEE/ACM Transactions on Networking, 12*(5), 781–794. doi:10.1109/TNET.2004.836125

Choi, Y., & Choi, S. (2007, May). Service charge and energy-aware vertical handoff in integrated IEEE 802.16e/802.11 networks. *26th IEEE International Conference on Computer Communications IEEE INFOCOM 2007* (pp. 589-597).

Chon, H. D., Agrawal, D., & Abbadi, A. E. (2003). Range and KNN query processing for moving objects in grid model. *Mobile Networks and Applications, 8*, 401–412. doi:10.1023/A:1024535730539

Chung, W. H. (2004). Probabilistic analysis of routes on mobile ad hoc networks. *IEEE Communications Letters, 8*(8), 506–508. doi:10.1109/LCOMM.2004.833841

Clausen, T., & Jacquet, P. (2003). *Optimized link state routing protocol (OLSR). Request for Comment 3626*. MANET Working Group.

Clausen, T., Jacquet, P., & Viennot, L. (2002). *Comparative study of routing protocols for mobile ad hoc networks*. The First Annual Mediterranean Ad HOC Networking Workshop, September 2002.

Constantinides, G. M. (1978). Market Risk Adjustment in Project Valuation. *The Journal of Finance, 33*, 603–616. doi:10.2307/2326572

Copeland, T., & Antikarov, V. (Eds.). (2001). *Real Options: A Practitioner's Guide*. New York: Texere.

Copeland, T., Koller, T., & Murrin, J. (2000). *Valuation: Measuring and Managing the Value of Companies*. New York: John Wiley and Sons.

Costa, C., Cunha, I., Borges, A., Ramos, C., Rocha, M., Almeida, J., & Ribeiro-Neto, B. (2004). *Analyzing client interactivity in streaming media.* In *13th International Conference on World Wide Web* (pp. 534-543). New York, NY: ACM.

d'Halluin, Y., Forsyth, P. A., & Vetzal, K. R. (2007). Wireless network capacity management: A real options approach. *European Journal of Operational Research, 176*(1), 584–609. doi:10.1016/j.ejor.2005.06.055

Dai, L. (2007). *Maximizing Throughput in Layered Peer-to-Peer Streaming*. New York: ICC.

Demirkol, I., Ersoy, C., Caglayan, M. U., & Delic, H. (2004, May). Location area planning and cell to switch assignment in cellular networks using simulated annealing. *IEEE Transactions on Wireless Communications, 3*(3), 880–890. doi:10.1109/TWC.2004.827767

Deng, H., Li, W., & Agarwal, D. P. (2002). Routing security in wireless ad hoc networks. [University of Cincinnati.]. *IEEE Communications Magazine, 40*(10), 70–75. doi:10.1109/MCOM.2002.1039859

Deshpande, H., Bawa, M., & García-Molina, H. (2001). *Streaming live media over a peer-to-peer network (Tech. Rep.)*. Stanford University.

Desmedt, Y. (1994). Threshold cryptography. *European Transactions on Telecommunications, 5*(4), 449–457. doi:10.1002/ett.4460050407

Desmedt, Y., & Frankel, Y. (1990). Threshold crypto system. *Advances in Cryptology-Crypto'89, 9th Annual International Cryptology Conference, 435.* (pp. 307-315). Springer.

Di Crescenzo, G., Zhang, T., & Pietrowicz, S. (2007). Anonymity Notions for Public Key Infrastructure in Vehicular Networks. In *Proceedings of the IEEE International Conference on Mobile Ad hoc and Sensor Systems*.

Dimitriou, T. (2005). A lightweight RFID protocol to protect against traceability and cloning attacks. *Proceedings of the First International Conference on Security and Privacy for Emerging Areas in Communications Networks - in SECURECOMM'05,* (pp. 59-66). Washington, DC: IEEE Computer Society.

Dimitriou, T. (2006). *A secure and efficient RFID protocol that could make big brother (partially) obsolete*. International Conference on Pervasive Computing and Communications – PerCom 2006.

Dimpfel, M., Habann, F., & Algesheimer, R. (2002). Real options theory, flexibility and the media industry. A conceptual outline for potential application areas. *International Journal of Media Management*, *4*(4), 261–272.

Din, D. R., & Tseng, S. S. (2002). Genetic algorithm for solving dual-homing cell assignment problem of the two-level wireless ATM network. *Computer Communications*, *25*(17), 1536–1547. doi:10.1016/S0140-3664(02)00086-5

Diot, C. (2000). Deployment issues for the IP multicast service and architecture. *IEEE Network*, *14*(1), 78–88. doi:10.1109/65.819174

Dixit, A., & Pindyck, R. (1994). *Investment Under Uncertainty*. Princeton, NJ: Princeton University Press.

Dixit, A., & Pindyck, R. (1994). *Investment under uncertainty*. Princeton, NJ: Princeton University Press.

Driouchi, T., Leseure, M., & Bennett, D. (2009). A robustness framework for monitoring real options under uncertainty. *Omega-International Journal of Management Science*, *37*(3), 698–710. doi:10.1016/j.omega.2007.09.004

Dube, R., Rais, C. D., Wang, K. Y., & Tripathi, S. K. (1997). Signal stability-based adaptive routing (SSA) for ad hoc mobile networks. *IEEE Personal Communications Magazine*, *4*(1), 36–45. doi:10.1109/98.575990

Dubois-Ferriere, H., Grossglauser, M., & Vetterli, M. (2003). Age matters: Efficient route discovery in mobile ad hoc networks using encounter ages. *Proc. of ACM MobiHoc03*, 1-3 Annapolis, MD, USA, 2003.

Dulluri, S., & Raghavan, N. R. S. (2008). Collaboration in tool development and capacity investments in high technology manufacturing networks. *European Journal of Operational Research*, *187*(3), 962–977. doi:10.1016/j.ejor.2006.06.059

Dunkels, A., Grönvall, B., & Voigt, T. (2004). Contiki - a lightweight and flexible operating system for tiny networked sensors. In *Proceedings of the EmNetS*. Washington, DC: IEEE.

Dunkels, A., Voigt, T., Ritter, H., & Alonso, J. (2004, June). Distributed TCP Caching for Wireless Sensor Networks. In *Proceedings of the 3rd Annual Mediterranean Ad Hoc Networking Workshop*, Turkey.

Dunkels, A., Voigt, T., & Alonso, J. (2004, January). Making tcp/ip viable for wireless sensor networks. In *Proceedings of the First European Workshop on Wireless Sensor Networks (EWSN 2004)*, Berlin.

Endrayanto, A. I., Van Der Berg, H., & Boucherie, R. J. (2005). An analytical model for CDMA downlink rate optimization taking into account uplink coverage restrictions. *International Journal on Performance Evaluation*, *59*(2-3), 225–246. .doi:10.1016/j.peva.2004.07.004

Fei, Y., Xiaoli, D., Changjun, J., & Rong, D. (2009). Directed acyclic task graph scheduling for heterogeneous computing systems by dynamic critical path duplication algorithm. *Journal of Algorithms & Computational Technology*, *3*(2), 247–270. doi:10.1260/174830109787914001

Ferreira, N., Kar, J., & Trigeorgis, L. (2009). Option Games – The key to competing in Capital-Intensive Industries. *Harvard Business Review*, 101–107.

Fichman, R. (2004). Real Options and IT Platform Adoption: Implications for Theory and Practice. *Information Systems Research*, *15*(2), 132–154. doi:10.1287/isre.1040.0021

Fonseca, E., & Festag, A. (2006). *A survey of existing approaches for secure ad hoc routing and their applicability to VANETS*. NEC Technical Report NLE-PR-2006-19, NEC Network Laboratories.

Gannaro, M., Jarecki, S., Karawczyk, H., & Rabin, T. (1996). Robust threshold DSS signatures. [Springer.]. *Advances in Cryptology-Eurococrypt*, *96*, 354–371.

Garey, M. R., & Johnson, D. S. (1979). *Computers and intractability: A guide to the theory of NP-completeness* (pp. 238–241). W.H. Freeman.

Gelabert, X., Pérez-Romero, J., Sallent, O., Agustí, R., & Casadevall, F. (2008). *Radio resource management in heterogeneous networks*. Heterogeneous Wireless Access Networks - Architectures and Protocols.

George, A., Rashad, S., Kumar, A., & Kantardzic, M. (2008). *A predictive handoff scheme for multi-hop heterogeneous wireless networks*. 3rd International ICST Conference on Communication System Software and Middleware. doi:10.1109/COMSWA.2008.4554472

Gerlach, M., Festag, A., Leinmuller, T., Goldacker, G., & Harsch, C. (2005). *Security architecture for vehicular communications*. In WIT2005, Hamburg, Germany.

Ghosh, D., & Lott, C. (2007). Uplink-downlink imbalance in wireless cellular networks. *IEEE International Conference on Communications*, (pp. 4275 – 428). doi: 10.1109/ICC.2007.705

Giordano, S., & Stojmenovic, I. (2003). *Position-based ad hoc routes in ad hoc networks* (pp. 287–300). Boca Raton, FL: CRC Press, Inc.

Girault, M. (1992). Self-certified public keys. *Proceedings of Eurocrypt'91* [Springer-Verlag.]. *LNCS*, *547*, 490–497.

Girault, M., Juniot, L., & Robshaw, M. J. B. (2007). *The feasibility of on-the-tag public key cryptography*. Workshop on RFID Security, RFIDSec'07.

Giupponi, L., Agust, R., Pérez-Romero, J., & Sallent, O. (2006). A framework for JRRM with resource reservation and multiservice provisioning in heterogeneous networks. *Mobile Networks and Applications*, *11*(6), 825–846. .doi:10.1007/s11036-006-0052-3

Gódor, G., Antal, M., & Imre, S. (2008). *Mutual authentication protocol for low computational capacity RFID systems. IEEE Globecom 2008 CCNS*. New Orleans, LA, USA: IEEE.

Gódor, G., & Imre, S. (2010). Simple lightweight authentication protocol: security and performance considerations. *International Journal of Business Data Communications and Networking*, *6*(3), 66–94. doi:10.4018/jbdcn.2010070104

Gódor, G., & Antal, M. (2008). *Improved lightweight mutual authentication protocol for RFID systems*. 13th IFIP International Conference on Personal Wireless Communications. Toulouse, France: Springer.

Gódor, G., & Imre, S. (2009). *Performance analysis of the simple lightweight authentication protocol*. The 7th International Conference on Advances in Mobile Computing and Multimedia – MoMM2009, Kuala Lumpur, Malaysia, 14-16 December 2009.

Gollan, L., & Meinel, C. (2002). Digital signatures for automobiles. In *Proceedings of Systemics, Cybernetics and Informatics* (p. 02). SCI.

Gopal, D., & Abdulkadir, T. (2006). Self-similarity and Internet performance. *Journal of Research and Practice in Information Technology*, *38*(2), 171–180.

Gordejuela-Sanchez, F., Juttner, A., & Zhang, J. (2009). A multiobjective optimization framework for IEEE 802.16e network design and performance Analysis. *IEEE Journal on Selected Areas in Communications*, *2*(27), 202–216. .doi:10.1109/JSAC.2009.090211

Gordijn, J., Yu, E., & Van der Raadt, B. (2006). e-service design using i* and e3 value modeling. *IEEE Software*, *23*(3), 26–33. doi:10.1109/MS.2006.71

GPP TS 23.002. (2011). *Technical specification group services and systems aspects: Network architecture.*

GPP TS 25.401. (2011). *Technical specification group radio access network: UTRAN overall description.*

GPP TS 25.413. (2011). *Technical specification group radio access network: UTRAN Iu interface radio access network application part (RANAP) signaling.*

GPP TS 25.415. (2011). *Technical specification group radio access network: UTRAN Iu interface user plane protocols.*

GPP TS 29.060. (2011). *Technical specification group core network, general packet radio service (GPRS): GPRS tunnelling protocol (GTP) across the Gn and Gp interface.*

GPP TS 29.414. (2011). *Technical specification group core network and terminals: Core network Nb data transport and transport signaling.*

GPP TS 29.415. (2011). *Technical specification group core network and terminals: Core network Nb interface user plane protocols.* Britvic, V. (2004). Steps in UMTS network design. *Proceedings of the 12th IEEE Mediterranean Electro-technical Conference,* vol. 2 (pp. 461-464).

Graham, R. L., Lawler, E. L., Lenstra, J. K., & Kan, A. H. G. R. (1979). Optimization and approximation in deterministic sequencing and scheduling: A survey. *Annals of Discrete Mathematics, 5,* 287–326. doi:10.1016/S0167-5060(08)70356-X

Grover, V., & Saeed, K. (2003). The telecommunication industry revisited the changing pattern of partnerships. *Communications of the ACM, 46*(7), 119–125. doi:10.1145/792704.792709

Gupta, M., Manghui, T., Khan, L., & Bastani, F. (2005). A study of the model and algorithms for handling location dependent continuous queries. *Knowledge and Information Systems, 8,* 414–437. doi:10.1007/s10115-005-0196-7

Guthrie, G., & Kumareswaran, D. (2009). Carbon subsidies, taxes and optimal forest management. *Environmental and Resource Economics, 43*(2), 275–293. doi:10.1007/s10640-008-9238-4

Ha, J. Y., Jeon, J., Lee, K., Heo, J., Kim, N., Kim, S. M., et al. (2007). Design and implementation of convergence sub-layer for a heterogeneous home network. In *Proceedings of the IEEE International Symposium on Power Line Communication (ISPLC '07)* (pp. 252-256).

Hagras, T., & Janecek, J. (2005). A high performance, low complexity algorithm for compile-time task scheduling in heterogeneous systems. *Parallel Computing, 31*(7), 653–670. doi:10.1016/j.parco.2005.04.002

Hahn, W. J., & Dyer, J. S. (2008). Discrete time modeling of mean-reverting stochastic processes for real option valuation. *European Journal of Operational Research, 184*(2), 534–548. doi:10.1016/j.ejor.2006.11.015

Harmantzis, F. C., & Tanguturi, V. P. (2007). Investment decisions in the wireless industry applying real options. *Telecommunications Policy, 31*(2), 107–123. doi:10.1016/j.telpol.2006.02.005

Harmantzis, F., & Tanguturi, V. (2005). *Migration to 3G Wireless Broadband Internet and Real Options: The case of an Operator in India.* Telecommunications Policy Journal.

Harmantzis, F., & Tanguturi, V. (2007). Investment Decisions in the Wireless Industry: Applying Real Options. *Telecommunications Policy, 31,* 107–123. doi:10.1016/j.telpol.2006.02.005

Harmantzis, F., & Trigeorgis, L. (2006). *Flexible Investment Decisions in the Telecommunications Industry: Case Applications using Real Options* (Working Paper No. 06-06). Orlando, FL: NET Institute.

Harmatos, J. (2002) Planning of UMTS core networks. *Personal, Indoor and Mobile Radio Communications. The 13th IEEE International Symposium, 2*(15-18), 740–744.

Harmatos, J., Jüttner, A., & Szentesi, Á. (1999). *Cost-based UMTS transport network topology optimisation.* Paper presented at the International Conference on Computer Communications, Tokyo, Japan.

He, T., Stankovic, J. A., Lu, C., & Abdelzaher, T. (2003, May). SPEED: A Stateless Protocol for Real-Time Communication in Sensor Networks. In *Proceedings of 23rd International Conference on Distributed Computing Systems (ICDCS),* USA (p. 46).

Hefeeda, M. (2003). *PROMISE: peer-to-peer media streaming using CollectCast.* New York: ACM.

Hefeeda, M., & Bhargava, B. (2003). On-Demand Media Streaming Over the Internet. In *Proceedings of the Ninth IEEE Workshop on Future Trends of Distributed Computing Systems.* Washington, DC: IEEE Computer Society.

Hei, H., Liang, C., Liang, J., Liu, Y., & Ross, K. W. (2007). A measurement study of large-scale P2P IPTV system. *IEEE Transactions on Multimedia, 9*(8), 1672–1687. doi:10.1109/TMM.2007.907451

Howell, S., Stark, A., Newton, D., Paxson, D., Cavus, M., Pereira, J., & Patel, K. (2001). *Real Options: Evaluating Corporate Investment Opportunities in a Dynamic World.* Upper Saddle River, NJ: Prentice Hall.

Hu, Y. C., Perrig, A., & Johnson, D. B. (2002). Ariadne: A secure on- demand routing protocol for ad hoc networks. *Proceedings of the 8th ACM International Conference of Mobile Computing and Networking* (pp. 12-23).

Huang, M. G. (2009). Real options approach-based demand forecasting method for a range of products with highly volatile and correlated demand. *European Journal of Operational Research, 198*(3), 867–877. doi:10.1016/j.ejor.2008.10.002

Huang, C. J., Chuang, Y. T., Guan, C. T., Yang, D. X., & Chen, Y. J. (2007). A self-adaptive bandwidth reservation scheme for 4G mobile WiMAX. *IEEE International Workshop on Cross Layer Design,* (pp. 55–59).

Huang, X., & Jensen, C. S. (2004). Towards a streams-based framework for defining location-based queries. *Proceedings of Spatio-Temporal Database Management Workshop,* (pp. 73-80).

Hubaux, J. P., Capkun, S., & Luo, J. (2004). The security and privacy of Smart vehicles. *IEEE Security and Privacy Magazine, 3*, 49–55.

Hubaux, J. P., Buttyan, L., & Capkun, S. (2001). The quest for security in mobile ad hoc networks. *Proceedings of the 2nd ACM International Symposium on Mobile Ad hoc Networking and Computing* (pp.146-155).

Hui, J. W. (2008, September). *An Extended Internet Architecture for Low-Power Wireless Networks - Design and Implementation.* Unpublished doctoral dissertation, EECS Department, University of California, Berkeley, CA.

Hui, J. W., & Culler, D. E. (2008). Ip is dead, long live ip for wireless sensor networks. In *Proceedings of the 6th ACM conference on Embedded network sensor systems (SenSys '08)*. New York: ACM.

Hull, B., Jamieson, K., & Balakrishnan, H. (2004, November). Mitigating Congestion in Wireless Sensor Networks. In *Proceedings of the 2nd ACM Conference on Embedded Networked Sensor Systems (ACM SenSys)*, USA (pp. 134-147).

Hyytia, E., Lassila, P., & Virtamo, J. (2006). Spatial node distribution of the random waypoint mobility model with applications. *IEEE Transactions on Mobile Computing, 5*(6), 680–694. doi:10.1109/TMC.2006.86

Iancu, B., Dadarlat, V., & Peculea, A. (2010). A novel end-to-end QoS framework over heterogeneous networks - An architectural approach. *In ICAI'10 Proceedings of the 11th WSEAS International Conference on Automation & Information, 51*(2).

IEEE. (2006). *IEEE Standard 802.15.4*. Washington, DC: IEEE.

IEEE. (n.d.). *Standard for broadband over power line networks: Medium access control and physical layer specifications*. Washington, DC: IEEE.

IEEE. *802.15 WPANTM Task Group*. (n.d.). Retrieved August 2008, from http://ieee802.org/15/index.html

IEEE. *P802.11, Main General Info Page*. (n.d.). Retrieved August 2008, from www.ieee802.org/11/main.html

IETF RFC 2225. (1999). *Classical IP and ARP over ATM.*

IETF RFC 761. (1980). *User data protocol.*

IETF RFC 791. (1981). *Internet protocol.*

ITU-T I.363.2. (1997). *B-ISDN ATM adaptation layer specification: Type 2 AAL Series I: Integrated services digital network - Overall network aspects and functions - Protocol layer requirements.*

ITU-T I.363.5. (1995). *B-ISDN ATM adaptation layer specification: Type 5 AAL - Series I: Integrated services digital network overall network aspects and functions - Protocol layer requirements.*

ITU-T I.366.2. (2007). *AAL type 2 service specific convergence sub-layer for narrow-band services.*

ITU-T Q1901. (2001). *Series Q: Switching and signalling: Specifications of signaling related to bearer independent call control (BICC), bearer independent call control protocol corrigendum 1.*

ITU-T Q1902.1. (2001). *Series Q: Switching and signalling: Specifications of signaling related to bearer independent call control (BICC), bearer independent call control protocol (capability set 2): Functional description.*

ITU-T Q1902.2. (2001). *Series Q: Switching and signalling: Specifications of signaling related to bearer independent call control (BICC), bearer independent call control protocol (capability set 2): General functions of messages and parameters.*

ITU-T Q1902.3. (2001). *Series Q: Switching and signalling: Specifications of signaling related to bearer independent call control (BICC), bearer independent call control protocol (capability set 2): Formats and codes.*

ITU-T Q1902.4. (2001). *Series Q: Switching and signalling: Specifications of signaling related to bearer independent call control (BICC), bearer independent call control protocol (capability set 2): Basic call procedures.*

ITU-T Q1902.5. (2001). *Series Q: Switching and signalling: Specifications of signaling related to bearer independent call control (BICC), bearer independent call control protocol (capability set 2): Extensions to the application transport mechanism in the context of the bearer independent call control.*

ITU-T. (2002). *Series H: Audiovisual and multimedia systems, infrastructure of audiovisual services- Communication procedures. H.248.1: Gateway control protocol.*

Iyer, Y. G., Gandham, S., & Venkatesan, S. (2005, October). STCP: a Generic Transport Layer Protocol for Wireless Sensor Networks. In *Proceedings of the 14th IEEE International Conference on Computer Communications and Networks (ICCCN)*, USA (pp. 449-454).

Jamaa, S. B., Altman, Z., Picard, J. M., & Fourestie, B. (2004). Multi-objective strategies for automatic cell planning of UMTS networks. *59th IEEE Vehicular Technology Conference (VTC), vol. 4* (pp. 2420-2424).

Jane, Y., Peter, H. J., & Chong. (2005). A survey of clustering schemes for mobile ad hoc networks. *IEEE Communications Survey & Tutorial, 7*(1).

Jao, Y. Y., Du, J., & Jiao, J. X. (2007). A financial model of flexible manufacturing systems planning under uncertainty: identification, valuation and applications of real options. *International Journal of Production Research, 45*(6), 1389–1404. doi:10.1080/10298430600677479

Jiang, B., Yao, T., & Feng, B. C. (2008). Valuate outsourcing contracts from vendors' perspective: A real options approach. *Decision Sciences, 39*(3), 383–405. doi:10.1111/j.1540-5915.2008.00197.x

Jin, S., & Bestavros, A. (2001). GISMO: A generator of Internet streaming media objects and workloads. *ACM SIGMETRICTS Performance Evaluation Review, 29*(3), 2–10. doi:10.1145/507553.507554

Johnson, D. B., Maltz, D. A., & Hu, Y.-C. (2004). *The dynamic source routing protocol for mobile ad hoc networks* (DSR). Internet draft, IETF MANET Working Group 2004.

Johnson, D., Maltz, D. A., Fu, Y. C., & Jetcheva, J. G. (2002). *The dynamic source routing protocol for mobile ad hoc networks* (DSR) (pp 1-25). Retrieved from http://www.monarch.cs.rice.edu/ monarch-papers/dsr-chapter00.pdf

Juels, A. (2006). RFID security and privacy: A research survey. *IEEE Journal on Selected Areas in Communications, 24*(2), 381–394. doi:10.1109/JSAC.2005.861395

Juels, A. (2003). *The blocker tag: Selective blocking of RFID tags for consumer privacy.* 8th ACM Conference on Computer and Communications Security.

Jung, I., You, Y., Lee, J., & Kim, K. (2002). Broadcasting and caching policies for location-dependent queries in urban areas. *Proceedings of the International Workshop on Mobile Commerce*, (pp. 54–60).

Junqi, W., Zhengbing, H., Jiuming, Y., & Jun, S. (2007). The degradation of bandwidth in QoS scheme for 4G mobile networks. *IEEE International Conference on Wireless Communications, Networking and Mobile Computing,* (pp. 2020 – 2024).

Just, M., Eranakis, E., & Wan, T. (2004). Resisting malicious packet dropping wireless ad hoc networks. Ad-Hoc. *Mobile and Wireless Networks, 2865*(2003), 151-163.

Juttner, A., Orban, A., & Fiala, Z. (2005). Two new algorithms for UMTS access network topology design. *European Journal of Operational Research, 164*, 456–474. doi:10.1016/j.ejor.2003.11.027

Kamrad, B., & Ord, K. (2006). Market risk and process uncertainty in production operations. *Naval Research Logistics, 53*(7), 627–640. doi:10.1002/nav.20178

Kamrad, B., & Siddique, A. (2004). Supply contracts, profit sharing, switching, and reaction options. *Management Science, 50*(1), 64–82. doi:10.1287/mnsc.1030.0157

Karthikeyan, N. K., & Narayanasamy, P. (2008). A novel mobility based bandwidth reduction algorithm in cellular mobile networks. *International Journal of Computer Science and Network Security, 8*(5), 62–69.

Kasahara, H., & Narita, S. (1984). practical multiprocessor scheduling algorithms for efficient parallel processing. *IEEE Transactions on Computers, 33*(11), 1023–1029. doi:10.1109/TC.1984.1676376

Katz, R. H., & Brewer, E. A. (1996, January). The case for wireless overlay networks. *SPIE Multimedia Computing and Networking, 2667*, 77–88.

Kauffman, R. J., & Li, X. T. (2005). Technology competition and optimal investment timing: A real options perspective. *IEEE Transactions on Engineering Management, 52*(1), 15–29. doi:10.1109/TEM.2004.839962

Kazaa. (n.d.). *Home Page Kazaa*. Retrieved from www.kazaa.com

Kester, W. (1984). Today's Options for Tommorow's Growth. *Harvard Business Review*, (62): 153–160.

Khansa, L., & Liginlal, D. (2009). Valuing the flexibility of investing in security process innovations. *European Journal of Operational Research, 192*(1), 216–235. doi:10.1016/j.ejor.2007.08.039

Kim, S. W., Kim, B. S., & Fang, Y. (2005). Downlink and uplink resource allocation in IEEE 802.11 Wireless LANs. *IEEE Transactions on Vehicular Technology, 54*, 320–327. doi:10.1109/TVT.2004.838887

Kim, H. J. (2006). The economic value of modularity in the IPQoS network: The real options approach (ROA). *International Journal of Business Data Communications and Networking, 2*(1), 15–31.

Kim, H. J. (2007). Suitability of IP telephony in the public switched telephone network (PSTN): A case study. *International Journal of Business Data Communications and Networking, 3*(3), 39–56.

Kim, H. J., Weiss, M., & Morel, B. (2008b). Real options and technology management: Assessing technology migration options in the wireless industry. *Telematics and Informatics, 26*(2), 180–192. doi:10.1016/j.tele.2008.05.004

Kim, K. H., Hwang, S. T., Oh, H. S., & Lee, D. J. (2008a). The impact of investment lags on investment decision. *European Journal of Operational Research, 190*(3), 696–707. doi:10.1016/j.ejor.2007.07.020

Kim, Y. J., & Sanders, G. L. (2002). Strategic actions in information technology investment based on real option theory. *Decision Support Systems, 33*(1), 1–11.

Kim, T. W., & Kim, H. G. (2011). A study on the delay of streaming packets during IP handover between heterogeneous networks. *13th International Conference on Advanced Communication Technology*, (pp. 1275–1278).

Kirchner, H., Krummenacher, R., Edwards, D., & Risse, T. (2004). A location-aware pre-fetching mechanism. *Fourth International Network Conference*, (pp. 453-460).

Kleinrock, L. (1975). Queueing systems, volume 1: Theory. In Wiley Interscience (Ed.), *Queueing systems*.

Knuth, D. (1969). The *art of computer programming*: *Vol. 2. Semi numerical algorithms*. Reading, MA: Addison-Wesley.

Kolahdouzan, M., & Shahabi, C. (2004). Continuous k-nearest neighbor queries in spatial network databases. *Proceedings of the Workshop on Spatio-Temporal Database Management*, (pp. 33-40).

Konstantopoulos, C., Gavalas, D., & Pantziou, E. G. (2008). Clustering in mobile ad hoc networks through neighborhood stability-based mobility prediction. [Elsevier.]. *Computer Networks, 52*(9), 1797–1824. doi:10.1016/j.comnet.2008.01.018

Kou, M., & Zhen, Y. (2009). *Dynamic uplink/downlink resource allocation for TDD OFDMA access network.* International Conference on Communications and Mobile Computing. doi 10.1109/CMC.2009.191

Koussis, N., Martzoukos, S. H., & Trigeorgis, L. (2007). Real R&D options with time-to-learn and learning-by-doing. *Annals of Operations Research, 151*(1), 29–55. doi:10.1007/s10479-006-0127-3

Krishnamurthy, V., Fowler, K., & Sazonov, E. (2008). The Effect of Time Synchronization of Wireless Sensors on the Modal Analysis of Structures. *International Journal of Smart Materials Systems and Structures, 17*(11), 1–13.

Kuang, T., & Williamson, C. (2002). A measurement study of RealMedia streaming traffic. In *SPIE ITCOM* (pp. 68–79). Boston, MA: SPIE. doi:10.1117/12.473407

Kumar, P., & Turnbull, S. M. (2008). Optimal patenting and licensing of financial innovations. *Management Science, 54*(12), 2012–2023. doi:10.1287/mnsc.1080.0898

Kumar, R. (2002). Managing Risks in IT projects: an Option Perspective. *Information & Management, 40*, 63–74. doi:10.1016/S0378-7206(01)00133-1

Kunz, A., Gonzalez, M., Tcaciuc, S., Ruland, C., & Rapp, V. (2005). Analysis of the QoS requirements under radio access network planning aspects for GPRS/EDGE and UMTS. *Wireless Networks, Communications and Mobile Computing, 2005 International Conference, vol. 1* (pp. 386-391).

Lai, K., Tari, Z., & Bertok, P. (2004). Location–aware cache replacement for mobile environments. *Proceedings of the IEEE Globecom, 04*, 3441–3447.

Lam, K.-Y., Chan, E., & Chun-Hung Yuen, J. (2000). Approaches for broadcasting temporal data in mobile computing systems. *Journal of Systems and Software, 51*(3), 175–189. doi:10.1016/S0164-1212(99)00122-3

Lan, K. C., & Heidemann, J. (2001). *Structural modeling of realaudio traffic* (Tech. Rep. ISI-TR-544). USC/Information Sciences Institute.

Law, A. M., & Kelton, W. D. (2000). *Simulation modeling and analysis* (3rd ed.). Boston, MA: McGraw-Hill Higher Education.

Layuan, L., Chunlin, L., & Peiyan, Y. (2007). Performance evaluation and simulations of routing protocols in ad hoc networks. *International Journal of Computer Communications, 30*(8), 1890–1898. doi:10.1016/j.comcom.2007.02.015

Le, T. V., Burmester, M., & Medeiros, B. D. (2007). *Forward-secure RFID authentication and key exchange*. Cryptology ePrint Archive, Report 2007/051.

Lee, S., Sriram, K., Kim, K., Kim, Y. H., & Golmie, N. (2009, February). Vertical handoff decision algorithms for providing optimized performance in heterogeneous wireless networks. *IEEE Transactions on Vehicular Technology, 52*(2), 865–881.

Lee, T., Seo, S., Lee, U., & Song, J. (2009). A new soft handover mechanism using DCHs in 3GPP HSDPA networks. *Journal of Networks, 4*(3), 184–191. doi:10.4304/jnw.4.3.184-191

Lee, S. J., Su, W., & Gerla, M. (2001). Wireless ad hoc multicast routing with mobility prediction. *Mobile Networks and Applications, 6*(4), 351–360. doi:10.1023/A:1011478628358

Lee, I., & Guan, L. (2003). Centralized peer-to-peer streaming with layered video. In *Proceedings of the 2003 International Conference on Multimedia and Expo* (Vol. 2, pp. 513-516). Washington, DC: IEEE Computer Society.

Lee, S., Asano, T., & Kim, K. (2006). *RFID mutual authentication scheme based on synchronized secret information.* The 2006 Symposium on Cryptography and Information Security – SCIS.

Leinmueller, T., Buttyan, L., Hubaux, J. P., Kargl, F., Kroh, R., Papadimitratos, P., et al. (2006). *SEVECOM "Secure Vehicle Communication".* Paper presented at the IST Mobile and Wireless Communication Summit (MobileSummit2006), Mykonos, Greece.

Leroux, A. D., Martin, V. L., & Goeschl, T. (2009). Optimal conservation, extinction debt, and the augmented quasi-option value. *Journal of Environmental Economics and Management, 58*(1), 43–57. doi:10.1016/j.jeem.2008.10.002

Levaggi, R., & Moretto, M. (2008). Investment in hospital care technology under different purchasing rules: A real option approach. *Bulletin of Economic Research, 60*(2), 159–181. doi:10.1111/j.0307-3378.2008.00274.x

Levis, P., Brewer, E., Culler, D., Gay, D., Madden, S., & Patel, N. (2008). The emergence of a networking primitive in wireless sensor networks. *Communications of the ACM, 51*(7). doi:10.1145/1364782.1364804

Levis, P., Tavakoli, A., & Dawson-Haggerty, S. (2009, April). *Overview of existing routing protocols for low power and lossy networks.*

Li, C., Huang, M., & Chu, Y. (2008). A Secure and Efficient Communication Scheme with Authenticated Key Establishment and Privacy Preserving for Vehicular Ad hoc Networks. *Science Direct Computer Communications, 31*, 2803–2814.

Li, G., & Rajagopalan, S. (2008). Process improvement, learning, and real options. *Production and Operations Management, 17*(1), 61–74. doi:10.3401/poms.1070.0008

Li, J., Blake, C., DeCouto, D. S. J., & Lee, H. I. (2001, July). Capacity of Ad Hoc wireless networks. In *Proceedings of the 7th International Conference on Mobile Computing and Networking (SIGMOBILE)*, Italy (pp. 61-69).

Lim, C. H., & Kwon, T. (2006). Strong and robust RFID authentication enabling perfect ownership transfer. *Lecture Notes in Computer Science,* Springer-Verlag, North Carolina, USA, December 2006.

Lin, C. R., & Gerla, M. (1997). Adaptive clustering for mobile wireless networks. *IEEE Journal on Selected Areas in Communications, 15*(7), 1265–1275. doi:10.1109/49.622910

Lin, T. T. (2009). Applying the maximum NPV rule with discounted/growth factors to a flexible production scale model. *European Journal of Operational Research, 196*(2), 628–634. doi:10.1016/j.ejor.2008.04.004

Lin, P., Ngo, H., Qiao, C., Wang, X., Wang, T., & Qian, D. Y. (2006). *Minimum cost wireless broadband overlay network planning*. International Symposium on a World of Wireless, Mobile and Multimedia Networks.

Liotta, A. (2002). Exploiting agent mobility for large-scale network monitoring. *IEEE Network, 16*(3), 7–15. doi:10.1109/MNET.2002.1002994

Little, T. D. C., & Agrawal, A. (2005). An Information Propagation Scheme for VANETs. In *Proceeding in 8th International IEEE Conference on Intelligent Transportation Systems (ITSC2005),* Vienna, Austria.

Liu, K., Deng, J., Varshney, P. K., & Balakrishnan, K. (2007). An acknowledgment based approach for the detection of routing misbehavior in MANETs. *IEEE Transactions on Mobile Computing, 6*(5), 536–550. doi:10.1109/TMC.2007.1036

Liu, Y. (2008). A survey on peer-to-peer video streaming systems. *Peer-to-Peer Networking and Applications, 1*(1), 18–28. doi:10.1007/s12083-007-0006-y

Liu, C.-H., Li, C.-F., Lai, K.-C., & Wu, C.-C. (2006). Dynamic critical path duplication task scheduling algorithm for distributed heterogeneous computing systems. In *Proceedings of the 12th International Conference on Parallel and Distributed Systems (ICPADS'06)*, (pp. 365-374). Washington, DC: IEEE Computer Society.

Liu, Y., et al. (2004). BLocation-aware topology matching in P2Psystems. In *Proceedings of INFOCOM.*

Lo, A., Lu, N. W., Jacobsson, M., Prasad, V., & Niemegeers, I. (2006). *Personal networks: An overlay network of wireless personal area networks and 3G networks.* 3rd Annual International Conference on Mobile and Ubiquitous Systems – Workshops.

Luehrman, T. A. (1997). What's It Worth? A General Manager's Guide to Valuation. *Harvard Business Review*, 132–142.

Luenberger, D. G. (1998). *Investment Science*. New York: Oxford University Press.

Luna, C. E., Kondi, L. P., & Katsaggelos, A. K. (2003). Maximizing user utility in video streaming applications. *IEEE Transactions on Circuits and Systems for Video Technology, 13*, 141–148. doi:10.1109/TCSVT.2002.808439

Luo, J., Fan, M., & Ye, D. (2008). Black hole attack prevention based on authentication mechanism. *Proceedings of ICCS 2008, IEEE 2008* (pp.173-177).

Luo, Z., Chan, T., & Li, J. S. (2005). A lightweight mutual authentication protocol for RFID networks. *International Conference on e-Business Engineering*, (pp. 620-625).

Ma, X., Chen, X., & Refai, H. (2009). Performance and Reliability of DSRC Vehicular Safety Communication: A formal Analysis, EURASIP. *Journal on Wireless Communications and Networking, 3*, 1687–1472.

Magagula, L. A., Chan, H. A., & Falowo, O. E. (2011). (in press). Handover approaches for seamless mobility management in next generation wireless networks. *Journal of Wireless Communication and Mobile Computing*. .doi:10.1002/wcm.1074

Magagula, L. A., Falowo, O. E., & Chan, H. A. (2010). Enhancing PMIPv6 for better handover performance among heterogeneous wireless networks in a micromobility domain. *EURASIP Journal on Wireless Communications and Networking, 2010*. .doi:10.1155/2010/274935

Magagula, L. A., Falowo, O. E., & Chan, H. A. (2010). PMIPv6-HC: Handover mechanism for reducing handover delay and packet loss in NGWN. In. *Proceedings of GLOBECOM, 2010,* 1–5. doi:.doi:10.1109/GLOCOM.2010.5684108

Magagula, L. A., Chan, H. A., & Falowo, O. E. (2010). Achieving seamless mobility through handover coordination in a network-based localized mobility managed heterogeneous environment. *21st Annual IEEE International Symposium on Personal, Indoor and Mobile Radio Communications*, (pp. 2505–2510). doi:10.1109/PIMRC.2010.5671745

Maitra, M., Mukherjee, A., & Saha, D. (2002). *A heuristic-based approach for minimization of inter-switch handoff cost with load balancing for optimal location area planning in a PCSN* (pp. 172–176). IEEE ICPWC.

Mandal, S., Saha, D., & Mahanti, A. (2004). Heuristic search techniques for cell to switch assignment in location area planning for cellular networks. *IEEE International Conference on Communications (ICC)* (pp. 4307-4311).

Mao, S., Bushmitch, D., Narayanan, S., & Panwar, S. S. (2003, October). MRTP: a Multiflow Real time Transport Protocol for ad hoc networks. In *Proceedings of the IEEE 58th Vehicular Technology Conference (VTC)*, USA (Vol. 4, pp. 2629-2634).

Maple, C., Guo, L., & Zhang, J. (2004). Parallel genetic algorithms for third generation mobile network planning. *International Conference on Parallel Computing in Electrical Engineering* (pp. 229–236).

Marcus, B., & Anderson, C. K. (2006). Online low-price guarantees - A real options analysis. *Operations Research*, *54*(6), 1041–1050. doi:10.1287/opre.1060.0333

Marfia, G., et al. (2007). P2P Streaming Systems: A Survey and Experiments. *ST Journal of Research*, 1-4.

Marsit, N., Hameurlain, A., Mammeri, Z., & Morvan, F. (2005). Query processing in mobile environments: A survey and open problems. *Proceedings of the International Conference on Distributed Frameworks for Multimedia Applications*, (pp. 150-157).

Mbarushimana Amine, C., & Shahrabi, A. (2007). *Comparative study of reactive and proactive routing protools performance in mobile ad hoc networks*. The 21st International Conference on Advanced Information Networking and Applications Workshop, AINA'07, IEEE 2007.

McDonald, A. B., & Znati, T. F. (1999). A mobility-based framework for adaptive clustering in wireless ad hoc networks. *IEEE Journal on Selected Areas in Communications*, *17*(18), 1466–1486. doi:10.1109/49.780353

McDonald, R., & Siegel, D. (1986). The Value of Waiting to Invest. *The Quarterly Journal of Economics*, *101*(4), 707–728. doi:10.2307/1884175

McLoone, M., & Robshaw, M. J. B. (2007). *Public key cryptography and RFID tags. LNCS 377/2006* (pp. 372–384). Berlin, Germany: Springer.

Melendi, D., Pañeda, X. G., Vilas, M., García, R., & García, V. (2007). A methodology for the deployment of live audio and video services. In *2nd International Conference on Signal Processing and Multimedia Applications* (pp. 15-21). Barcelona, Spain: INSTICC Press.

Melendi, D., Vilas, M., García, R., Pañeda, X. G., & García, V. (2006). Characterization of a real Internet radio service. In *32nd EuroMicro Conference on Software Engineering and Advanced Applications* (pp. 356-363). Dubrovnik, Croatia: IEEE.

Mena, A., & Heidemann, J. (2000). An empirical study of real audio traffic. In *19th IEEE International Conference on Computer Communications* (pp. 1001-1010). Tel-Aviv, Israel: IEEE.

Merchant, A., & Sengupta, B. (1995). Assignment of cells to switches in PCS networks. *IEEE/ACM Transactions on Networking*, *3*(5), 521–526. doi:10.1109/90.469954

Milanovic, N., Malek, M., Davidson, A., & Milutinovic, V. M. (2004). Routing and security in mobile ad hoc networks. *IEEE Computer*, *37*(2), 61–65. doi:10.1109/MC.2004.1266297

Miller, B., & Clarke, J. P. (2008). Strategic guidance in the development of new aircraft programs: a practical real options approach. *IEEE Transactions on Engineering Management*, *55*(4), 566–578. doi:10.1109/TEM.2008.927817

Miller, V. (1986). *Use of elliptic curves in cryptography*. International Cryptology Conference.

Miranda, H., & Rodrigues, L. (2002). Preventing selfishness in open mobile ad hoc networks. *Proceedings of Seventh CaberNet Radicals Workshop*.

Mittendorf, B. (2004). Information revelation, incentives, and the value of a real option. *Management Science*, *50*(12), 1638–1645. doi:10.1287/mnsc.1040.0260

Molnar, D., & Wagner, D. (2005). *A scalable, delegatable pseudonym protocol enabling ownership transfer of RFID Tags*. Springer.

Molnar, D., & Wagner, D. (2004). Privacy and security in library RFID: Issues, practices, and architectures. *Conference on Computer and Communications Security - ACM CCS*, (pp. 210-219). Washington, DC, USA.

Montenegro, G., Kushalnagar, N., Hui, J., & Culler, D. (2007, September). *Transmission of ipv6 packets over ieee 802.15.4 networks*. Washington, DC: IEEE.

Mun, J. (2002). *Real options analysis, tools and techniques for valuing strategic investments and decisions*. New York: Wiley Finance Publishers.

Munoz, J., & Syracuse, N. (2002). Proceedings of The 53 Internet Engineering Task Force Networks. *Wireless Networks*, *9*, 443–460.

Mushtaq, M., & Ahmed, T. (2006). *Adaptive Packet Video Streaming Over* (p. 2P). Networks Using Active Measurements.

Myers, S. (1984). Finance Theory and Financial Strategy. *Interfaces*, *14*, 126–137. doi:10.1287/inte.14.1.126

Nagpal, R., Shrobe, H., & Bachrach, J. (2003). Organizing a global coordinate system from local information on an ad hoc sensor network. *Proceedings of the Second International Workshop Information Processing in Sensor Networks* (IPSN), 2003.

Nasser, N., Hasswa, A., & Hassanein, H. (2006, October). Handovers in fourth generation heterogeneous networks. *IEEE Communications Magazine*, *44*(10), 96–103. doi:10.1109/MCOM.2006.1710420

Neruda, M., & Bestak, R. (2008). Evolution of 3GPP core network. *Systems. Signals and Image Processing, IWSSIP*, *2008*, 25–28.

Neubauer, T., & Toeltsch, M. (2005). UMTS radio network planning-maximizing return on investment. *Elektrotechnik und Informationstechnik*, *122*(3), 108–113.

Nguyen, T., & Zakhor, A. (2002). Distributed video streaming over Internet. In *Multimedia Computing and Networking* (pp. 186–195). California: SPIE/ACM.

Nguyen, T., & Zakhor, A. (n.d.). Distributed video streaming over Internet. *Citeseer.*

Niculescu, D., & Nath, B. (2003). DV based positioning in ad hoc networks. [Springer Netherlands.]. *Telecommunication Systems*, *22*(1-4), 267–280. doi:10.1023/A:1023403323460

Niyato, D., & Hossain, E. (2005). Call admission control for QoS provisioning in 4G wireless networks: Issues and approaches. *IEEE Network*, *19*(5), 5–11. doi:10.1109/MNET.2005.1509946

Nohara, Y., Inoue, S., Baba, K., & Yasuura, H. (2006). Quantitative evaluation of unlinkable ID matching schemes. *Workshop on Privacy in the Electronic Society - WPES*, (pp. 55-60). Alexandria, Virginia, USA.

Nohl, K., & Evans, D. (2006). Quantifying information leakage in tree-based hash protocols. *Conference on Information and Communications Security – ICICS'06*, (pp. 228-237). Raleigh, North Carolina, USA.

Novatnack, J., Greenwald, L., & Arora, H. (2005). Evaluating ad hoc routing protocols with respect to quality of service. *Wireless and Mobile Computing Networking and Communications WiMob*, *05*, 2005.

O'Neil, E., & O'Neil, P. (1993). The LRU-k page replacement algorithm for database disk buffering. *Proceedings of the ACM SIGMOD*, (pp. 296-306).

Ohkubo, M., Suzuki, K., & Kinoshita, S. (2003). *Cryptographic approach to "privacy-friendly" tags.* RFID Privacy Workshop. MA, USA: MIT. OMNeT++. (n.d.). *Community site.* Retrieved from http://www.omnetpp.org

Ouyang, Y., & Fallah, M. H. (2009). Evolving core networks from GSM to UMTS R4 version. *International Journal of Mobile Network Design and Innovation*, *3*(2), 93–102. doi:10.1504/IJMNDI.2009.030841

Ouyang, Y., & Fallah, M. H. (2009). A study of throughput for Iu-CS and Iu-PS interface in UMTS core network. *Performance Computing and Communications Conference (IPCCC), IEEE 28th International* (pp. 349-353).

Ouyang, Y., & Fallah, M. H. (2010). A study of throughput for Nb, Mc and Nc interface in UMTS core network. *Performance Computing and Communications Conference (IPCCC), 28th IEEE International* (pp.371-375).

Padmanabhan, V. N., Wang, H. J., Chou, P. A., & Sripanidkulchai, K. (2002). Distributing streaming media content using cooperative networking. In *12th International Workshop on Network and Operating Systems Support for Digital Audio and Video* (pp. 177-186). Miami, FL: ACM.

Pañeda, X. G., Melendi, D., García, R., Vilas, M., & García, V. (2007). A methodology for performance, content analysis, and coniguration of audio/video-on-demand services. *International Journal of Business Data Communications and Networking, 3*(4), 17–46. doi:10.4018/jbdcn.2007100102

Papadimitratos, P., & Haas, Z. J. (2002). Securing the Internet routing infrastructure. *IEEE Transactions on Communications, 10*(40), 60–68.

Papadimitratos, A., Buttyan, L., & Holczer, T., GWoch, E., Freudiger, J., Raya, M., Ma, Z., Kargl, F., Kung, A., & Hubaux, J. P. (2008). Secure Vehicular Communication Systems: Design and Architecture. *IEEE Communications Magazine*, 100–109. doi:10.1109/MCOM.2008.4689252

Papadimitratos, P., & Haas, Z. J. (2002). *Secure routing for mobile ad hoc networks*. SCS Communication Networks and Distributed Systems Modeling and Simulation Conference (CNDS 2002), San Antonio, TX.

Papadimitratos, P., Gligor, V., & Hubaux, J. P. (2006). Securing Vehicular Communications Assumptions, Requirements, and Principles. In *Proceedings of the Workshop on Embedded Security in Cars (ESCAR).*

Park, C.-L., & Choe, T.-Y. (2002). An optimal scheduling algorithm base on task duplication. *IEEE Transactions on Computers, 51*(4), 444–448. doi:10.1109/12.995454

Park, S., Zou, C. C., & Turgut, D. (2009). Reliable traffic information propagation in vehicular ad hoc networks. In Beyah, R., McNair, J., & Corbett, C. (Eds.), *Security in ad-hoc and sensor networks* (pp. 261–291). World Scientific Press. doi:10.1142/9789814271097_0009

Park, S. J., Vedantham, R., Sivakumar, R., & Akyildiz, I. F. (2004, May). A scalable approach for reliable downstream data delivery in wireless sensor networks. In *Proceedings of the 5th ACM International Symposium on Mobile Ad Hoc Networking and Computing (ACM Mobihoc)*, Japan (pp. 78-89).

Parno, B., & Perrig, A. (2005). Challenges in securing vehicular networks. In *Proceedings of ACM Workshop on Hot Topics in Networks* (HotNets-IV), College Park, MD, USA.

Patcha, A., & Mishra, A. (2003). Collaborative security architecture for black hole attack prevention in mobile ad hoc networks. *Proceedings of RAWCON, 03*, 75–78.

Pengfei, L., & Jiakun, L. (2009). Application of communication and remote control in PLC based on zigbee technology. In *Proceedings of the International Conference on Computational Intelligence and Security.*

Pennings, E., & Lint, O. (2000). Market entry, phased rollout or abandonment? A real option approach. *European Journal of Operational Research, 124*(1), 125–138. doi:10.1016/S0377-2217(99)00121-6

Peris-Lopez, P., Hernandez-Castro J. C., Estevez-Tapiador J. M., & Ribagorda, A. (2006). *A minimalist mutual-authentication protocol for low-cost RFID tags.*

Perkins, C. E., & Bhagwat, P. (1999). *Highly dynamic destination-sequenced distance-vector routing (DSDV) for mobile computers.* ACM SIGCOMM'94 Conference on Communications Architectures, Protocols and Applications, 1999.

Perkins, C. E., Royer, E. M., & Chakeres, I. E. (2003). *Ad hoc on demand distance vector routing protocol.* Internet draft, MANET Working Group 2003.

Perkins, C. E., Royer, E. B., & Das, S. (2003). *Ad hoc On-Demand Distance Vector (AODV) Routing (IETF RFC 3561).*

Pertile, P. (2009). An extension of the real option approach to the evaluation of health care technologies: the case of positron emission tomography. *International Journal of Health Care Finance and Economics, 9*(3), 317–332. doi:10.1007/s10754-008-9053-z

Petty, K. (1994). *Freeway service patrol (FSP) 1.1: The analysis software for the FSP Project. California PATH Research Report, UCB-ITS-PRR-95-20*. Berkeley: Institute of Transportation Studies, University of California.

Pierre, S., & Houeto, F. (2002). Assigning cells to switches in cellular mobile networks using taboo search. *IEEE Transaction on Systems, Man and Cybernetics. Part B, 32*(3), 351–356.

Pindick, R. (1991). Irreversability, Uncertainty, And Investment. *Journal of Economic Literature, 29*(3), 1110–1148.

Plobi, K., Nowey, T., & Mletzko, C. (2006). Towards a security architecture for vehicular ad hoc networks. availability, reliability and security. *Proceedings of the First International Conference on Availability, Reliability and Security*, April 20-22, ARES 2006 (pp. 374-381).

Pontes, A. B., Silva, D. D. P., Jailton, J. J., Rodrigues, O. J., & Dias, K. L. (2008, October). Handover management in integrated WLAN and mobile WiMAX networks. *IEEE Wireless Communications, 15*(5), 86–95. doi:10.1109/MWC.2008.4653137

Porchet, A., Touzi, N., & Warin, X. (2009). Valuation of power plants by utility indifference and numerical computation. *Mathematical Methods of Operations Research, 70*(1), 47–75. doi:10.1007/s00186-008-0231-z

Poschmann, A., Robshaw, M., Vater, F., & Paar, C. (2009). *Lightweight cryptography and RFID: Tackling the hidden overheads*. 7th International Conference on Information Security and Cryptology 2009 – ICISC 2009, Seoul, Korea.

Potdar, V., Sharif, A., & Chang, E. (2009, May). Wireless Sensor Networks: A Survey. In *Proceedings of the 23rd International Conference on Advanced Information Networking and Applications Workshops (WAINA)*, UK (pp. 636-641).

Poupard, G., & Stern, J. (1998). Security analysis of a practical "on the fly" authentication and signature generation. *Proc. of Eurocrypt '98* [Springer-Verlag.]. *LNCS, 1403*, 422–436.

Powerline Alliance HomePlug. (n.d.). *Homeplug av white paper*. Retrieved from http://www.homeplug.org

Qiu, D., & Srikant, R. (2004). Modeling and performance analysis of BitTorrent-like peer-to-peer networks. *ACM SIGCOMM Computer Communication Review, 34*(4), 367–378. doi:10.1145/1030194.1015508

Quintero, A., & Pierre, S. (2002, November). A memetic algorithm for assigning cells to switches in cellular mobile networks. *IEEE Communications Letters, 6*(11), 484–486. doi:10.1109/LCOMM.2002.805515

Quintero, A., & Pierre, S. (2003a, September). Assigning cells to switches in cellular mobile networks: A comparative study. *Computer Communications, 26*(9), 950–960. doi:10.1016/S0140-3664(02)00224-4

Quintero, A., & Pierre, S. (2003b, February). Evolutionary approach to optimize the assignment of cells to switches in personal communications networks. *Computer Communications, 9*(2), 927–938. doi:10.1016/S0140-3664(02)00238-4

Rachedi, A., & Benslimane, A. (2010). A Secure and Resistant Architecture against Attacks for Mobile Ad Hoc Networks. *Journal of Security and Communication Network, 3*, 150–166. doi:10.1002/sec.116

Ragusa, C. (2005). An adaptive clustering approach for the management of dynamic systems. *IEEE Journal on Selected Areas in Communications, 23*(12), 2223–2235. doi:10.1109/JSAC.2005.857203

Rahman, A. A., & Hailes, S. (1997). A distributed trust model. *Proceedings of the Workshop on New Security Paradigms*, (pp. 48-60).

Ramaswamy, S., Fu, H., Sreekantaradhya, M., Dixon, J., & Nygard, K. (2003). Prevention of *cooperative black hole attack in wireless ad hoc networks* (pp. 1-7). Retrieved from www.cs.ndsu.nodak.edu/~nygard/ research/ BlackHoleMANET.pdf

Ramesh, B. H. S., Gowrishankar, G., & Satyanarayana, P. S. (2010). Call admission control for next generation wireless networks using higher order Markov model. *Computer and Information Science Journal, 3*(1), 192–196.

Rappaport, T. S. (1995). *Wireless Communications: Principles and Practice*. Upper Saddle River, NJ: Prentice Hall.

Raya, M., Paradimitratos, P., & Hubaux, J. P. (2006). *Securing Vehicular Communications*. IEEE Wireless Communications.

Raya, M., Aziz, A., & Hubaux, J. P. (2006). *Efficient Secure Aggregation in VANETs*. In *Proceedings of VANET06*, Los Angeles, CA.

RealNetworks. (2002). *Helix universal server administration guide*. USA: RealNetworks.

Rejaie, R., & Ortega, A. (2003). PALS: peer-to-peer adaptive layered streaming. In *Proceedings of the 13th international workshop on Network and operating systems support for digital audio and video*, Monterey, CA (pp. 153-161). New York: ACM.

Ren, Q., & Dunham, M. (2000). Using semantic caching to manage location dependent data in mobile computing. *Proceedings of the International Conference on Mobile Computing and Networking*, (pp. 210-221).

Rerney, L. M., & Nilsson, M. (2001). Investigating the energy consumption of a wireless network interface in an ad hoc networking environment. *Proceedings of IEEE INFOCOM'01*.

Rhéaume, L., & Bhabra, H. S. (2008). Value creation in information-based industries through convergence: A study of US mergers and acquisitions between 1993 and 2005. *Information & Management*, *45*(5), 304–311. doi:10.1016/j.im.2008.03.002

Rhee, I., Warrier, A., Aia, M., Min, J., & Sichitiu, M. L. (2008). Z-mac: a hybrid mac for wireless sensor networks. *IEEE/ACM Trans. Netw.*, *16*(3).

Robshaw, M. (2008). *cryptoGPS: Anti-counterfeiting solution for RFID tags*. Orange Labs. Tech. Rep.

Rogaway, P., & Shrimpton, T. (2004). Cryptographic hash-function basics: Definitions, implications, and separations for preimage resistance, second-preimage resistance, and collision resistance. *Lecture Notes in Computer Science*, 371–388. doi:10.1007/978-3-540-25937-4_24

Roth, M., & Wicker, S. (2003). Termite: Emergent ad-hoc networking. *Proceedings of the Second Mediterranean Workshop on Ad-Hoc Networks*, Mehdia, Tunisia, 2003.

Saad, C., Benslimane, A., & Konig, J.-C. (2008). AT-dist: A distributed method for localization with high accuracy in sensor networks. *International Journal Studia Informatica Universalis, Special Issue on Wireless Ad Hoc and Sensor Networks, 6*(1).

Saad, L. B., & Tourancheau, B. (2009). Multiple mobile sinks positioning in wireless sensor networks for buildings. In *Proceedings of the 2009 Third International Conference on Sensor Technologies and Applications (SENSORCOMM '09)* (pp. 264-270). Washington, DC: IEEE Computer Society.

Salhani, M. (2008). *Modélisation et Simulation des Réseaux Mobiles de 4ème Génération Université de Toulouse* (Doctoral dissertation, Toulouse University)

Samaan, N., Benmammar, B., Krief, F., & Karmouch, A. (2005). *Prediction-based advanced resource reservation in mobile environments* (pp. 1411–1414). Electrical and Computer Engineering.

Sampigethaya, K., Huang, L., Li, M., Poovendran, R., Mastsuura, K., & Sezaki, K. (2005). Providing Location Privacy for VANET. In *Proceedings of Embedded Security in Cars (ESCAR)*. CARAVAN.

Sampigethaya, K., Huang, L., Li, M., Poovendran, R., Matsuura, K., & Sezaki, K. (2005). CARAVAN: Providing location privacy for VANET. In *Proceedings of the Workshop on Embedded Security in Cars* (ESCAR)'05.

Sankarasubramaniam, Y., Akan, O. B., & Akyildiz, I. F. (2003, June). ESRT: Event-to-Sink Reliable Transport in Wireless Sensor Networks. In *Proceedings of the 4th ACM International Symposium on Mobile Ad Hoc Networking and Computing (ACM Mobihoc)*, USA (pp. 177-188).

Sarma, S., Weis, S., & Engels, D. (2003). Radio-frequency identification: Security risks and challenges. *Cryptobytes. RSA Laboratories*, *6*, 2–9.

Saxtoft, C. (2008). *Convergence – user expectations, communications enablers and business opportunities*. New York: Wiley.

Schuba, L., Krsul, V., Kuhn, G., Spafford, H., & Sunda, A. (1997). *Analysis of a denial of service attack on TCP.* IEEE Symposium on Security and Privacy 1997.

Schulzrinne, H., Rao, A., & Lanpier, R. (1998). *Real time streaming protocol (RTSP). RFC2326*. USA: The Internet Society.

Schwarz, H., et al. (2006, October). Overview of the scalable H. 264/MPEG4-AVC extension. In *Proceedings of the Conference on Image Processing*, Atlanta. Citeseer.

Sergio Marti, T. J. Giuli, Lai, K., & Baker, M. (2000). Mitigating routing misbehavior in mobile ad hoc networks. *Proceedings of The 6th Annual International Conference on Mobile computing and Networking*, (pp. 255–265).

Seydim, A. Y., Dunham, M. H., & Kumar, V. (2001). Location-dependent query processing. *Proceedings of the Second International Workshop on Data Engineering on Mobile and Wireless Access*, (pp. 47-53).

Shafiee, K., Attar, A., & Leung, V. C. M. (2011, March). Optimal distributed vertical handoff strategies in vehicular heterogeneous networks. *IEEE Journal on Selected Areas in Communications*, *29*(3), 534–544. doi:10.1109/JSAC.2011.110304

Shalak, R., Sandrasegaran, K., Agbinya, J., & Subenthiran, S. (2004). UMTS core network planning model and comparison of vendor product performance. *Advanced Communication Technology, the 6th International Conference, vol. 2* (pp. 685- 689).

Shamir, A. (1979). How to share secret. *ACM Transactions on Communications*, *22*(11), 612–613. doi:10.1145/359168.359176

Sharif, A., Ahmad, R. F., & Potdar, V. (2010, April). Adaptive Channel Coding and Modulation Scheme Selection for Achieving High Throughput in Wireless Networks. In *Proceedings of the 6th International Symposium on Frontiers of Information Systems and Network Applications (FINA 2010)*, WA.

Sharif, A., Potdar, V., & Chang, E. (2009, May). Wireless Multimedia Sensor Network Technology: A Survey. In *Proceedings of the 7th IEEE International Conference on Industrial Informatics (INDIN)*, UK (pp. 606-613).

Sharif, A., Potdar, V., & Rathnayaka, A. J. D. (2009, December). Performance Evaluation of different Transport layer protocols on the IEEE 802.11 and IEEE 802.15.4 MAC/PHY layers for WSN. In *Proceedings of the 7th International Conference on Advances in Mobile Computing & Multimedia (MoMM2009)*, Malaysia (pp. 300-306).

Sharif, A., Potdar, V., & Rathnayaka, A. J. D. (2010, April). Prioritizing Information for Achieving QoS Control in WSN. In *Proceedings of the International Conference on Advanced Information Networking and Applications (AINA)*, WA.

Shen, W., & Zeng, Q. (2006). *A novel decision strategy of vertical handoff in overlay wireless networks*. Fifth IEEE International Symposium on Network Computing and Applications (NCA'06).

Shen, Y., et al. (2005). Streaming layered encoded video using peers. In *Proceedings of the ICME.*

Shih, E., Cho, S.-H., Ickes, N., Min, R., Sinha, A., Wang, A., & Chan-drakasan, A. (2001). Physical layer driven protocol and algorithm design for energy-efficient wireless sensor networks. In *Proceedings of the 7th annual international conference on Mobile computing and networking (MobiCom '01)* (pp. 272-287). New York: ACM.

Shyu, S. J., Lin, B. M. T., & Hsiao, T. (2006). Ant colony optimization for the cell assignment problem in PCS networks. *Computers & Operations Research*, *33*(6), 1713–1723. doi:10.1016/j.cor.2004.11.026

Simić, I. S. (2007, June). Evolution of mobile base station architectures. *Microwave Review, 13*(1).

Sivagurunathan, S., Subathra, P., Mohan, V., & Ramaraj, N. (2009). Authentic Vehicular Environment Using a Cluster Based Key Management. *European Journal of Scientific Research*, *36*(2), 299–307.

Smith, J. E., & McCardle, K. F. (1998). Valuing Oil Properties: Integrating Option Pricing and Decision Analysis Approaches. *Operations Research*, *46*(2), 198–217. doi:10.1287/opre.46.2.198

Smith, J. E., & Nau, R. F. (1995). Valuing Risky Projects: Option Pricing Theory and Decision Analysis. *Management Science*, *41*(5), 795–816. doi:10.1287/mnsc.41.5.795

Smith, C., & Collins, D. (2007). *3G wireless networks* (2nd ed.).

Sridhar, K. N., & Chan, M. C. (2005). Stability and hop-count based approach for route computation in MANET. In *Proceedings of ICCCN05.*

Sripanidkulchai, K., Ganjam, A., Maggs, B., & Zhang, H. (2004). The feasibility of supporting large-scale live streaming applications with dynamic application end-points. *ACM SIGCOMM Computer Communication Review*, *34*(4), 107–120. doi:10.1145/1030194.1015480

Sripanidkulchai, K., Maggs, B., & Zhang, H. (2004). An analysis of live streaming workloads on the Internet. In *ACM SIGCOMM Internet Measurement Conference* (pp. 41-54). Sicily, Italy: ACM.

Stamatiou, P. (2007). *Privacy implications of RFID tags.* Retrieved April 9, 2010, from http://paulstamatiou.com/ wp-content/ uploads/2007/11/pstam_rfid_paper.pdf

Stann, F., & Heideman, J. (2003, October). RMST: Reliable Data Transport in Sensor Networks. In *Proceedings of the 1st IEEE International Workshop on Sensor Network Protocols and Applications (SNPA)*, USA (pp. 102-113).

Stemm, M., & Katz, R. H. (1998). Vertical handoffs in wireless overlay networks. *Mobile Networks and Applications, 3*(4), 335–350. doi:10.1023/A:1019197320544

Su, G. M., Han, Z., Wu, M., & Ray Liu, K. J. (2007). Joint uplink and downlink optimization for real-time multiuser video streaming over WLANs. *IEEE Journal of Selected Topics in Signal Processing, 1*(2), 280–294. .doi:10.1109/JSTSP.2007.901518

Su, N., Akkiraju, R., Nayak, N., & Goodwin, R. (2009). Shared services transformation: conceptualization and valuation from the perspective of real options. *Decision Sciences, 40*(3), 381–402. doi:10.1111/j.1540-5915.2009.00243.x

Suzuki, S. C., & Nakada, K. (1991). *An authentication technique based on distributed security management for the global mobility network.* Symposium on Security and Privacy. IEEE Computer Society Press.

Takenaka, T., Mineno, H., Tokunaga, Y., Miyauchi, N., & Mizuno, T. (2007). Performance evaluation of optimized link state routing-based localization. *2007 IEEE International Performance, Computing, and Communications Conference*, (pp. 150-159).

Takizawa, S., & Suzuki, A. (2004). Analysis of the decision to invest for constructing a nuclear power plant under regulation of electricity price. *Decision Support Systems, 37*(3), 449–456. doi:10.1016/S0167-9236(03)00045-9

Taleb, T., Fadlullah, Z., Benslimane, A., & Letaief, K. B. (2009). *Towards an Effective Risk-conscious and Collaborative Vehicular Collision Avoidance System.* IEEE Transactions on Vehicular Technology.

Tamea, G., Vegni, A. M., Inzerilli, T., & Cusani, R. (2009, September). A probability based vertical handover approach to prevent ping-pong Effect. *6th International Symposium on Wireless Communication Systems,* 2009 (pp. 181-185).

Tamilselvan, L., & Sankaranarayanan, V. (2008). Preventive black hole attack in MANET. *Journal of Networks, 3*(5), 13–20. doi:10.4304/jnw.3.5.13-20

Tao, Y., Papadias, D., & Shen, Q. (2002). Continuous nearest neighbor search. *Proceedings of the International Conference on Very Large Data Bases* (VLDB'02), (pp. 287-298).

Tao, Y., Sun, J., & Papadias, D. (2003). Selectivity estimation for predictive spatio-temporal queries. *Proceedings of the International Conference on Data Engineering* (ICDE'03), (pp. 417-428).

TCP. (n.d.). Retrieved August 2008, from http://www.ibiblio.org/pub/ docs/rfc/rfc793.txt

Technologies, O. P. N. E. T. (2009). *OPNET modeler: Making networks and applications perform.* Retrieved December, 2009, from http://www.opnet.com

The Network Simulator — NS-2. (n.d.). Retrieved July 2008, from http://www.isi.edu/ nsnam/ns/index.html

Thompson, D., Chaudhry, N., & Thompson, W. (2006). *RFID security threat model. Proceedings Acxiom Laboratory for Applied Research (ALAR)*. Conf.

Tiwana, A., Keil, M., & Fichman, R. G. (2006). Information systems project continuation in escalation situations: A real options model. *Decision Sciences, 37*(3), 357–391. doi:10.1111/j.1540-5414.2006.00131.x

Tiwana, A., Wang, J. J., Keil, M., & Ahluwalia, P. (2007). The bounded rationality bias in managerial valuation of real options: Theory and evidence from IT projects. *Decision Sciences, 38*(1), 157–181. doi:10.1111/j.1540-5915.2007.00152.x

Todini, A., Baiocchi, A., Merliny, S., Messinaz, D., Morettix, M., & Scanavino, B. … Veronesiy, D. (2006). Cross-layer design of packet scheduling and resource allocation algorithms for 4G cellular systems. *Personal Multimedia Communications,* (pp. 100–105).

Toh, C.-K. (1997). Associativity-based routing for ad hoc mobile networks. *International Journal on Wireless Personal Communications*, *4*(2), 103–139. doi:10.1023/A:1008812928561

Toiruul, B., & Lee, K. (2006). An advanced mutual-authentication algorithm using AES for RFID systems. *IJCSNS International Journal of Computer Science and Network Security*, *6*(9).

Topcuoglu, H., Hariri, S., & Wu, M. (2002). Performance-effective and low-complexity task scheduling for heterogeneous computing. *IEEE Transactions on Parallel and Distributed Systems*, *13*(3), 260–274. doi:10.1109/71.993206

Trajcevski, G., Wolfson, O., Hinrichs, K., & Chamberlain, S. (2004). Managing uncertainty in moving objects databases. *ACM Transactions on Database Systems*, *29*, 463–507. doi:10.1145/1016028.1016030

Trigeorgis, L. (1996). *Real Options: Managerial Flexibility and Strategy in Resource Allocation*. Cambridge, MA: MIT press.

Trigeorgis, L., & Mason, S. P. (1987). Valuing Managerial Flexibility. *Midland Corporate Finance Journal*, *5*(1), 14–21.

Trigeorgis, L. (1993). *Quantifying Sequential Option Interdependencies* (Working Paper). Boston: Boston University.

Trigeorgis, L. (1999). *Real Options: A Primer* (pp. 3-33). Alphen aan den Rijn, The Netherlands: Kluwer.

Tseng, C. L., & Lin, K. Y. (2007). A framework using two-factor price lattices for generation asset valuation. *Operations Research*, *55*(2), 234–251. doi:10.1287/opre.1060.0355

Tseng, Y., Li, Y. F., & Chang, Y. (2003). On route life time in multi-hop mobile ad hoc networks. *IEEE TOC*, *2*(4).

Tsudik, G. (2006). YA-TRAP: Yet another trivial RFID authentication protocol. *Fourth Annual IEEE International Conference Pervasive Computing and Communications Workshops, 2006. PerCom Workshops 2006*, *6*(9B).

UDP. (n.d.). Retrieved August 2008, from http://tools.ietf.org/html/rfc768

Ullman, J. D. (1975). NP-complete scheduling problems. *Journal of Computer and System Sciences*, *10*(3), 384–393. doi:10.1016/S0022-0000(75)80008-0

van Dam, T., & Langendoen, K. (2003). An adaptive energy-efficient mac protocol for wireless sensor networks. In *Proceedings of the 1st international conference on Embedded networked sensor systems (SenSys '03)* (pp. 171-180). New York: ACM.

Van der Merwe, J., Sen, S., & Kalmanek, C. (2002). *Streaming video traffic: Characterization and network impact*. In 7th International Workshop on Web Content Caching and Distribution, Boulder, Colorado.

Van der Raadt, B., Gardin, T., & Yu, E. (2005). Exploring Web services from a business value perspective. In *13th IEEE International Conference on Requirements Engineering* (pp. 53-62). Paris, France: IEEE.

van Putten, A. B., & MacMillan, I. C. (2004). Making real options really work. *Harvard Business Review*, *82*(12), 134.

Veloso, E., Almeida, V., Meira, W., Bestavros, A., & Jin, S. (2002). A hierarchical characterization of a live streaming media workload. In *2nd ACM SIGCOMM Internet Measurement Workshop 2002* (pp. 117-130). Marseille, France: ACM.

Wan, C. Y., Campbell, A. T., & Krishnamurthy, L. (2002, September). PSFQ: A Reliable Transport Protocol for Wireless Sensor Networks. In *Proceedings of ACM International Workshop on WSN and Applications (WSNA)*, USA (pp. 1-11).

Wan, C. Y., Eisenman, S. B., & Campbell, A. T. (2003, October). CODA: Congestion detection and avoidance in sensor networks. In *Proceedings of the 1st ACM Conference on Embedded Networked Sensor Systems (ACM SenSys)*, USA (pp. 266-279).

Wang, N., Huang, Y., & Chen, W. (2008). A Novel Secure Communication scheme in Vehicular Ad hoc Networks. *Science Direct Computer Communications*, *31*, 2827–2837.

Wang, C. H., & Min, K. J. (2006). Electric power generation planning for interrelated projects: A real options approach. *IEEE Transactions on Engineering Management*, *53*(2), 312–322. doi:10.1109/TEM.2006.872249

Wang, A., & Halal, W. (2010). Comparison of Real Asset Valuation Models: A Literature Review. *International Journal of Business and Management*, *5*(5).

Watteco White Paper. (2008, December). Retrieved from http://www.watteco.com

Weerasinghe, H., & Fu, H. (2008). Preventing cooperative black hole attacks in mobile ad hoc networks: Simulation implementation and evaluation. *International Journal of Software Engineering and Its Application*, *2*(3), 39–54.

Weill, P., & Vitale, M. (2001). *From place to space: Migrating to e-business models*. Cambridge, MA: Harvard Business School Press.

Weis, S. (2003). *Security and privacy in radio-frequency identification devices*. Massachusetts, USA: Massachusetts Institute of Technology (MIT).

Weis, S., Sarma, S., Rivest, R., & Engels, D. (2003). Security and privacy aspects of low-cost radio frequency identification systems. *International Conference on Security in Pervasive Computing - SPC 2003*, (pp. 454-469). Boppard, Germany.

West, B. M., & Bengtsson, J. (2007). Aggregate production process design in global manufacturing using a real options approach. *International Journal of Production Research*, *45*(8), 1745–1762. doi:10.1080/00207540600677583

Wiegand, T., & Sullivan, G. (2007). The H. 264/AVC video coding standard. *IEEE Signal Processing Magazine*, *24*(2), 148–153. doi:10.1109/MSP.2007.323282

Wu, Y., & Pierre, S. (2004). A new hybrid constraint-based approach for 3G network planning. *IEEE Communications Letters*, *8*(5), 277–279. doi:10.1109/LCOMM.2004.827437

Wu, Y., & Pierre, S. (2005). Optimization of 3G mobile network design using a hybrid search strategy. *Journal of Communications and Networks*, *7*(4), 471–477.

Wu, M., & Gajski, D. (1990). Hypertool: A programming aid for message passing systems. *IEEE Transactions on Parallel and Distributed Systems*, *1*(3), 384–343. doi:10.1109/71.80160

Wu, L. C., Ong, C. S., & Hsu, Y. W. (2008). Active ERP implementation management: A real options perspective. *Journal of Systems and Software*, *81*(6), 1039–1050. doi:10.1016/j.jss.2007.10.004

Wu, L. C., Ong, C. S., & Hsu, Y. W. (2008c). Knowledge-based organization evaluation. *Decision Support Systems*, *45*(3), 541–549. doi:10.1016/j.dss.2007.06.013

Wu, W., Yang, F., Chan, C. Y., & Tan, K.-L. (2008). Continuous reverse k-nearest-neighbor monitoring. *Proceedings of the International Conference on Mobile Data Management*, (pp. 132-139).

Wu, Y., & Pierre, S. (2003). Optimization of access network design in 3G networks. *IEEE Canadian Conference on Electrical and Computer Engineering (CCECE)* (pp. 781–784).

Xiao, X., et al. (2008). On optimal scheduling for layered video streaming in heterogeneous peer-to-peer networks. In *Proceedings of the 16th ACM international conference on Multimedia*, Vancouver, British Columbia, Canada (pp. 785-788). New York: ACM.

Xiaofei, L., Jin, H., Liu, T., Ni, L. M., & Deng, D. (2006). AnySee: Peer-to-peer live streaming. In *25th IEEE International Conference on Computer Communications* (pp. 1-10). Barcelona, Spain: IEEE.

Xu, J., Tang, X., & Lee, D. L. (2003). Performance analysis of location-dependent cache invalidation schemes for mobile environments. *IEEE Transactions on Knowledge and Data Engineering*, *15*, 474–488. doi:10.1109/TKDE.2003.1185846

Xu, B., & Wolfson, O. (2003). Time-series prediction with applications to traffic and moving objects databases. *Proceedings of the International Workshop on Data Engineering for Wireless and Mobile Access*, (pp. 56-60).

Xu, D., Hefeeda, M., Hambrusch, S., & Bhargava, B. (2002). On peer-to-peer media streaming. In *22nd International Conference on Distributed Computing Systems* (pp. 363-371). Vienna, Austria: IEEE.

Xu, J., Tang, X., Lee, D. L., & Hu, Q. (1999). Cache coherency in location-dependent information services for mobile environment. *Proceedings of the International Conference on Mobile Data Access*, (pp. 182-193).

Xu, Z., et al. (2003). Building topology-aware overlays using global soft-state. In *Proceedings of ICDCS*. Citeseer.

Yamada, K., Furumura, T., Kimura, K., Kaneyama, T., Yoshida, K., Kojima, M., et al. (2007). Dual communication system using wired and wireless in home-network. In *Proceedings of the 11th international conference, KES2007 and XVII Italian workshop on neural networks conference on Knowledge-based intelligent information and engineering systems (KES'07/WIRN'07)* (pp. 469-476). Berlin: Springer Verlag.

Yang, H., Shu, J., Meng, X., & Lu, S. (2006). SCAN: Self organized network layer security in mobile ad hoc networks. *IEEE Journal on Selected Areas in Communications, 24*(2), 261–273. doi:10.1109/JSAC.2005.861384

Yang, H., Luo, H., Ye, F., Lu, S., & Zhang, L. (2004). Security in mobile ad-hoc networks- Challenges and solutions. *IEEE Transactions on Wireless Communications, 11*(1).

Yang, C.-f., & Geng, R.-y. (2008). Home network using complementary communication system. In *Proceedings of the 2008 International Seminar on Business and Information Management (ISBIM '08)* (pp. 71-74). Washington, DC: IEEE Computer Society.

Ye, W., Heidemann, J., & Estrin, D. (2004). Medium access control with coordinated adaptive sleeping for wireless sensor networks. *IEEE/ACM Trans. Netw., 12*(3).

Yoon, J., Liu, M., & Noble, B. (2003). Random waypoint considered harmful. IEEE INFOCOM 2003 Twenty-Second Annual Joint Conference of the IEEE Computer and Communications Societies (pp. 1312-1321). IEEE. Retrieved from http://ieeexplore.ieee.org/lpdocs/epic03/wrapper.htm?arnumber=1208967

Yoshimura, Y., & Masugi, M. (2004). A QoS monitoring method for video streaming service based on presentation timeline detection at user clients. In *10th IEEE Asia-Pacific Conference on Communications, vol. 2* (pp. 681-685). Beijing, China: IEEE.

Yu, B., & Kim, S. H. (2006). Interpolating and using most likely trajectories in moving objects databases. *Proceedings of the International Conference on Database and Expert Systems Applications* (DEXA'06), (pp. 718-727).

Yu, D., Li, H., & Gruber, I. (2003). Path availability in ad hoc network. *Proceedings of the 10th International Conference on Telecommunications* (ICT03), volume 1, (pp. 383–387).

Yu, M. J., Choi, S. G., Kim, H. S., & Cho, K. S. (2011). An improved scheme for reducing handover latency in heterogeneous networks. *13th International Conference on Advanced Communication Technology*, (pp. 1534–1539).

Yuan, X. (1999). On the extended Bellman-Ford algorithm to solve two-constrained quality of service routing problems. *Proceedings of The Eighth International Conference on Computer Communications and Networks* (IC3N'99), Boston, USA, 1999.

Yun, M., Kim, T., Zhou, Y., Arora, A., Gomes, J., & Choi, H. A. (2009). Uplink resource management design in multi-access wireless networks. In Liu, B. (Eds.), *Wireless algorithms, systems, and applications* (pp. 499–508). doi:10.1007/978-3-642-03417-6_49

Zhai, H., Chen, X., & Fang, Y. (2005). How well can the IEEE 802.11 wireless LAN support quality of service? *Wireless Communications, 4*(6), 825–846. doi10.1.1.60.9241

Zhang, H., & Dong, Y.-N. (2007). A novel path stability computation model for wireless ad hoc networks. *IEEE Signal Processing Letters, 14*(12). doi:10.1109/LSP.2007.907994

Zhang, H., Arora, A., Choi, Y. R., & Gouda, M. G. (2005, May). Reliable Bursty Convergecast in Wireless Sensor Networks. In *Proceedings of the 6th ACM International Symposium on Mobile Ad Hoc Networking and Computing (ACM Mobihoc)*, USA (pp. 266-276).

Zhao, T., & Tseng, C. L. (2007). Flexible facility interior layout: a real options approach. *The Journal of the Operational Research Society, 58*(6), 729–739. doi:10.1057/palgrave.jors.2602189

Zhao, B., et al. (2002). Locality-aware mechanisms for large-scale networks. In *Proceedings of the Workshop Future Directions Distrib. Comput. (FuDiCo)*. Citeseer.

Zhao, M., & Zhou, J. (2009). Cooperative black hole attack prevention for mobile ad hoc networks. 2009 *International Symposium on Information Engineering and Electronic Commerce*, (pp.26-30).

Zheng, B., Xu, J., & Lee, D. L. (2002). Cache invalidation and replacement strategies for location-dependent data in mobile environments. *IEEE Transactions on Computers*, *51*, 1141–1153. doi:10.1109/TC.2002.1039841

Zhou, L., & Hass, Z. J. (1999). Securing ad hoc networks. *IEEE Network Magazine*, *13*(6), 41–50.

Zhou, T., Choudhury, R., Ning, P., & Chakrabarty, K. (2007). Privacy Preserving Detection of Sybil Attacks in Vehicular Ad hoc Networks. In *Proceedings of the Fourth Annual International Conference on Mobile and Ubiquitous Systems: Networking and Services (MobiQuitous)*.

Zhou, X., Yang, X., Wei, P., & Hu, Y. (2007). Dynamic group signature with forward security and its application. In *Proceedings of the Sixth International Conference on Grid and Cooperative Computing* (pp. 473-480).

Zhuang, W., Gan, Y., Loh, K., & Chua, K. (2003, November). Policy-based QoS management architecture in an integrated UMTS and WLAN environment. *IEEE Communications Magazine*, *41*(11), 118–125. doi:10.1109/MCOM.2003.1244931

Ziedonis, A. A. (2007). Real options in technology licensing. *Management Science*, *53*(10), 1618–1633. doi:10.1287/mnsc.1070.0705

Zygmunt, J., Haas, M. R., & Pearlman, P. S. (2002). *The interzone routing protocol (IERP) for ad hoc networks*. Internet Draft, work in progress.

About the Contributors

Debashis Saha is currently a full professor with the MIS Group, Indian Institute of Management (IIM)-Calcutta (Kolkata, India). Previously, he was with CSE Department at Jadavpur University (Kolkata, India). He received his BE (Hons) degree from Jadavpur University (Kolkata, India), and the MTech and PhD degrees from the Indian Institute of Technology (IIT)-Kharagpur (India), all in electronics and telecommunications engineering. His research interests include pervasive communication and computing, network operations, management and security, wireless networking and mobile computing, ICT for development, and network economics. He has supervised thirteen doctoral theses, published about 250 research papers in various conferences and journals, and directed four funded research projects on networking. He is a member of the organizing/program committee of several international conferences, and is a regular reviewer of several international journals. He has co-authored several book chapters, a monograph, and five books including Networking Infrastructure for Pervasive Computing: Enabling Technologies and Systems (Norwell, MA: Kluwer, 2002) and Location Management and Routing in Mobile Wireless Networks (Boston, MA: Artech House, 2003). Dr. Saha is the recipient of the prestigious career award for Young Teachers from AICTE, Government of India, and is a SERC Visiting Fellow with the Department of Science and Technology (DST), Government of India. He is a Fellow of West Bengal Academy of Science and Technology (WAST), Senior Life Member of Computer Society of India, Senior Member of IEEE, member of ACM, member of AIS, and member of the International Federation of Information Processing Working Group's 6.8 and 6.10. He was the founding Chair of Calcutta Chapter of IEEE Communications Society (2003-2008) which won the 'Best Chapter of the World' award in 2008.

Varadharajan Sridhar is a Research Fellow at Sasken Communication Technologies, Bangalore (India) wherein he looks after strategic research initiatives, knowledge management, and academic partnerships. Prior to Sasken, he was a professor of information management and Dean (Research and Consulting) at the Management Development Institute (India). He had also taught at Ohio University and American University in the US; University of Auckland in New Zealand, and at the Indian Institute of Management (Lucknow, India). He received his BE from the University of Madras (India), Post Graduate Diploma in industrial engineering from the National Institute for Training in Industrial Engineering (Mumbai, India), and PhD in MIS from the University of Iowa (USA). Dr. Sridhar's primary research interests are in the area of telecommunication management and policy and global software development. He has published many research articles, business cases, and chapters in edited books in his area of research. The book co-edited by him titled "Recent Advances in Broadband Integrated Network Operations and Services Management" was published by IGI Global in 2011. His latest book titled "The Telecom Revolution in India: Technology, Regulation and Policy" has been published by the Oxford University Press India in October 2011. He has been a member of the Government of India committees in telecom and IT. Dr. Sridhar contributes regularly in leading Indian business newspapers on telecom policy related issues. He is the Co-Editor of International Journal of Business Data Communications and Networking and is on the editorial board of the Journal of Global Information Management. He is a member of Association for Computing Machinery and Association of Information Systems.

* * *

Mohamed Amine Abid is currently a PhD student at the National School of Computer Sciences (Ecole Nationale des Sciences de l'Informatique, ENSI), University of Manouba, Tunisia. He is a current member of the HANA Research Group within the laboratory CRISTAL at ENSI. He received his main engineering diploma and his Master of Science both in Computer Science from the same institution (ENSI) in June 2008 and July 2009, respectively. He is currently serving as a teaching assistant at the National School of Computer Sciences, University of Manouba. His research interests focus on Networking. Particularly, he works on mobile ad hoc networks, autonomic management, performance evaluation, and protocol specification and optimization. Being at the final stage of his PhD, his research work has been subject of scientific publications in several international conferences (LCN, WiMob, NTMS, WMNC, GIIS…) and journals (IJBDCN, IJPUC…).

Majed Alhaisoni earned his BSc in Computer Science from Qassim University, Saudi Arabia (2005), and his MSc in Computer and Information Networks from the University of Essex, UK (2007). He is now a PhD candidate at the School of Computer Science and Electronic Engineering, University of Essex, UK. His research interests encompass video coding techniques, peer-to-peer networking and applications, and distributed computing. His PhD is in the area of P2P streaming.

Abdelfettah Belghith is currently a Professor of Computer Science at the National School of Computer Sciences (Ecole Nationale des Sciences de l'Informatique, ENSI), University of Manouba, Tunisia and since 1992. He received in 1979 his main Engineering Diploma in Computer Sciences from the Faculty of Sciences of Tunis, University of Tunis II, his Master of Science and his Ph.D. both in Computer Science from the University of California at Los Angeles (UCLA) respectively in 1982 and 1987. He is currently the Director of ENSI and the Director of the Doctoral School (Graduate School) STICODE of the University of Manouba, the responsible of the Network and Distributed Systems Master degree and the head of the HANA Research Group within the Laboratory CRISTAL at ENSI. His research and teaching interests focus on Advanced Heterogeneous Networking and Applications. Particularly, he works on ad hoc networks, sensor networks, mesh networks, WiMax, and LTE, inter vehicular communications, management and optimization of hand off mechanisms, QoS management in wired and wireless networks, performance evaluation, and protocol optimization and specification. He is the author of more than 190 journal articles and conference papers. He is currently serving on several reputed international conference Technical Program Committees (IEEE Globecom, IEEE ICC, IEEE LCN, IEEE WiMob,…). He organized and served as the technical Committee Chair and the organizing committee chair of many international and regional conferences. He has been involved in several national and international collaborative research projects with both academia and industry. He has built strong academic and research cooperation with several international institutions with whom several Master's and PhDs have been conducted under joint supervisions. As the head of the HANA Research Group, he supervised over sixty Master and PhD theses on advanced networking and networking applications.

Noureddine Boudriga is an internationally known scientist/academic. He received his Ph.D. in Algebraic Topology from the University of Paris (France) and his Ph.D. in Computer Science from the University of Tunis (Tunisia). He is currently a Professor of Telecommunications at the University of Carthage, Tunisia, and the director of the Communication Networks and Security Research Laboratory (CNAS). Dr. Boudriga is the recipient of the Tunisian Presidential Award in Science and Research (2004). He has served as the General Director and Founder of the Tunisian National Digital Certification Agency. He has been very actively involved in research and has authored or co-authored many chapters and books, including more than 250 published journal and conference papers. Prof. Boudriga is the President of the Tunisia Scientific Telecommunications Society.

Sergio Cabrero is a telecommunication Engineer from the University of Oviedo and a Ph.D. student. He is also an assistant professor in the Department of Computer Science at the University of Oviedo. His current research interests are in telecommunication networks, digital TV, multimedia services, and mobile ad hoc networks.

Pranay Chaudhuri received his B.Sc. (Physics) and B.Tech. (Electronics) from Calcutta University. He obtained his M.E. and Ph.D., both in Computer Science & Engineering, from Jadavpur University, Calcutta. At present he holds the position of Principal at Heritage Institute of Technology, Calcutta, India. A former Professor of Computer Science and Head, Department of Computer Science, Mathematics and Physics, University of the West Indies, Cave Hill Campus, Barbados, Professor Chaudhuri, has more than 28 years of teaching, research and administrative experience in premier institutions/universities across the globe. Professor Chaudhuri's research interests include parallel and distributed algorithms, self-stabilization, and graph theory. He has published extensively in leading international journals and is the author of a book titled, *Parallel Algorithms: Design and Analysis* (Prentice-Hall, 1992). Professor Chaudhuri is the recipient of several national and international awards for his research contributions. He is also the recipient of the *Vice-Chancellor's Award for Excellence 2007* at the University of the West Indies for all-round excellent performance in research and service to the University community.

Stella C. Chia is associate professor in the Department of Media and Communication at the City University of Hong Kong. She received her PhD from the University of Wisconsin-Madison. Dr. Chia's research interests include media effects and technology. Her research work has won top-faculty paper award in the Mass Communication Division at International Communication Association.

Jeffrey Elcock received his B.Sc (Mathematics and Computer Science) from the University of the West Indies, Cave Hill, Campus, Barbados; and M.Sc. (Computation) from Oxford University. He obtained his PhD in Computer Science from the University of the West Indies, Cave Hill, Campus, Barbados. He has held senior positions in both the banking and credit card industries. He is currently a Lecturer at the University of the West Indies, Cave Hill, Campus, Barbados. His research interest includes algorithms and complexity, distributed systems, and Grid Computing.

M. Hosein Fallah is an Associate Professor of Technology Management at Stevens Institute of Technology in New Jersey. His research interest is in the area of innovation management with a focus on the telecommunications industry. Prior to joining Stevens, he was the Director of Network Planning and Systems Engineering at Bell Laboratories. He has over 30 years of experience in the areas of systems engineering, product/service realisation, software engineering, project management, and R&D effectiveness. He holds a BS in Engineering from AIT and MS and PhD in Applied Science from the University of Delaware.

Roberto García has a Ph.D. degree from the University of Oviedo and a telecommunication engineering degree from The Technical University of Madrid. He is an Associate Professor in the Department of Computer Science and Engineering, University of Oviedo, and a former Associate Professor of the Electronics Department at the University of Alcalá, Madrid. His current research interests are in telecommunication networks and services applied to the performance analysis, modeling, and simulation of telecommunication systems and multimedia services. He is also taking part in several research projects on the national and European levels.

Victor Garcia is Telecommunications Engineer from the Technical University of Madrid and Ph.D. from the University of Oviedo. Nowadays, he is Associate Professor with the Department of Computer Science of the University of Oviedo. His current research interests are in the area of multimedia systems and services, in content distribution networks, digital TV services, mobile ad hoc networks, wireless networks, performance analysis, modeling, and simulation of telecommunication systems. He is also taking part in several research projects on the national and European levels.

Mohammed Ghanbari (M'78-SM'97-F01) is a Professor of Video Networking at the University of Essex, UK. He is best known for the pioneering work on layered video coding, now is known as SNR scalability in the standard video codecs, which earned him the Fellowship of IEEE in 2001. His research field includes a wide range of video networking issues appearing in 11 international patents,480 technical papers and 5 books, where his book on Video coding: an introduction to standard codecs, received the Rayleigh prize as the best book of the year 2000 by IET.

Győző Gódor received his M.Sc. degree in Electrical Engineering from Budapest University of Technology and Economics (BUTE), Budapest, Hungary, in 2003. He is currently a Ph.D. candidate, and he is an Assistant Professor at the Department of Telecommunications of BUTE. He is a member of Mobile Communications and Computing Laboratory (MC^2L), an associate member of the IEEE and IEEE Communications Society. His research interests include mobile and wireless network security and reliability, security issues and lightweight authentication protocols of heterogenous and sensor networks, elliptic curve cryptography based authentication protocols for small computational capacity environment, and QoS support and medium access methods for wireless LANs.

Sándor Imre was born in Budapest in 1969. He received the M.Sc. degree in Electrical Engineering from the Budapest University of Technology (BME) in 1993. Next he started his PhD. studies at BME and obtained Dr. Univ. degree in 1996, Ph.D. degree in 1999 and DSc degree from the Hungarian Academy of Sciences in 2007. Currently he is carrying his activities as Professor and Head of Dept. of Telecommunications. He is secretary of Telecommunication Systems Committee of the Hungarian Academy of Sciences. He participates the Editorial Board of two journals: *Infocommunications Journal* and *Hungarian Telecommunications*. He was invited to join the Mobile Innovation Centre as R&D Director in 2005. His research interest includes mobile and wireless systems quantum computing and communications. Especially he has contributions on different wireless access technologies, mobility protocols, security and privacy, reconfigurable systems, quantum computing based algorithms, and protocols.

Lara Khansa is assistant professor in the Department of Business Information Technology at Virginia Polytechnic Institute and State University. She received the PhD in MIS, MS in computer engineering, and MBA in finance and investment banking from the University of Wisconsin, Madison. Her primary research interests include the economics of information security and regulatory economics, and their implications for the future of the IT industry landscape. Dr. Khansa has published in EJOR, CACM, and Computers & Security, among others. She is a member of the Association for Information Systems (AIS), the IEEE, and the Beta Gamma Sigma National Honor Society.

Divakaran Liginlal (Lal) is an associate teaching professor of Information Systems at Carnegie Mellon University in Qatar. Lal holds a BS in telecommunication engineering from CET, a MS in computer science from IISc, and a PhD in MIS from the University of Arizona. Before joining CMU, he taught at the University of Wisconsin-Madison. His research in information security and decision support systems has been published in CACM, IEEE-TKDE, IEEE-SMC, EJOR, Computers & Security, DSS, and Fuzzy Sets & Systems. Lal's teaching and research have been supported by organizations such as Microsoft Corporation, Hewlett Packard and CISCO.

Antonio Liotta holds the Chair of Communication Network Protocols at the Eindhoven University of Technology (The Netherlands) where he leads the Autonomic Networks team. He was previously a Reader at the University of Essex (UK), where he led the Pervasive Services team. Antonio has over 100 publications to his credit. He is a Fellow of the U.K. Higher Education Academy and serves the Peer Review College of the U.K. Engineering and Physical Sciences Research Council, and the Advisory Board of Editors of the Journal of Network and System Management (Springer) and of the International Journal of Network Management (Wiley).

Mary Magdalene Jane F. is currently Professor at the Computer Applications department at Dr.N.G.P. Institute of Technology affiliated to Anna University, India. She received her Ph.D. degree from Anna University, Chennai, India. She has over one twenty book chapters and conference/journal articles. Research interests include mobile computing, pervasive computing and spatial databases. She has been a reviewer for conferences/journals. She is a member of ACM, CSI (Computer Society of India), and ISTE (Indian Society for Technical Education).

Madhubanti Maitra received her B.E (Hons), M.E, and Ph.D. degrees from Jadavpur University (Kolkata, India) all in Electrical Engineering. Presently, she is associated with Jadavpur University as an Associate Professor with Electrical Engineering Department. To her credit, she is author of several journal papers, conference papers, and book chapters. Her research interests include mobile computing, control system engineering, digital signal processing, et cetera. She is a member of IEEE and for a long period of time, she has been acting as an executive committee member of Calcutta Chapter of IEEE Communication Society.

Swarup Mandal received his Bachelor of Engineering degree from Jadavpur University (India), Post Graduate Degree in Management from Indian Institute of Management Calcutta (IIMC), and Doctoral degree from IIMC. He has worked for 19 years both in industry and academics after his graduation. He published a number of research papers in reputed journals and conferences in the area of technology and management. He is currently Chairman of IEEE ComSoc Calcutta and also convener of EST OWG of WG21 of ISO/ IEC/SC7 responsible for SAM standards.

David Melendi is a computer science Engineer from the University of Oviedo and received his Ph.D. from the University of Oviedo. Nowadays, he is an Associate Professor in the Department of Computer Science at the University of Oviedo and a member of the SYMM Working Group of the W3C. His current research interests are in multimedia systems and services, content distribution networks, digital TV services, mobile ad- hoc networks, performance analysis, and modeling and simulation of telecommunication systems. He is also taking part in several research projects on the national and European levels.

R. Nadarajan is Professor and Head of the Department of Mathematics and Computer Applications, PSG College of Technology, India. He obtained his PhD degree from Annamalai University, India. He is the author of over 30 journal papers, as well as numerous proceeding papers and technical reports. He has been a reviewer for many journals. His research interests include object computing, data mining, and database technologies. He is the Chairman, Computer Society of India Coimbatore chapter, and a member of Member of OR Society of India and ISTE (Indian Society for Technical Education).

Álvaro Nascimento is the dean of the Faculty of Economics and Management at the Portuguese Catholic University, in Porto, Portugal. He is consultant to financial and telecommunication companies on regulatory and competition issues; and he has research and publications in the subjects of regulation of financial markets and telecommunications. Álvaro holds a PhD in Finance from Cass Business School (UK) and a MSc in International Trade and Finance from Lancaster University (UK). He was a visiting research fellow at the London Business School in the Regulation Initiative. He was the lead economist on competition issues for some of the largest M&A operations in Portugal and advisor to the financial services industry. He has conducted research for the Portuguese Competition Authority and the Health Regulatory Authority. He has presented to a professional audience on topics such as asset pricing, financial risks and regulation of the financial services and telecommunications industries.

Ye Ouyang is a Senior Member of Technical Staff in VPIsystems. His research interest is in the emerging telecommunication network technologies, services, strategies, and policies, focused on mobile networks convergence, transition, and network performance analysis. Dr. Ouyang has intensive experience in planning and implementation of 2-4G networks. He was actively involved in more than 20 nationwide mobile networks projects, in particular planning the 1st GSM for a country in North Africa and the 1st UMTS core network for a country in South Asia. He holds a PhD in Telecommunications from Stevens Institute of Technology, an MS in System Engineering Man agement from Tufts University, and an ME and a BE in Control Engineering and Information Engineering from Southeast University.

Xabiel G. Pañeda is a Computer Science Engineer with a Ph.D. from the University of Oviedo. Nowadays, he is an Associate Professor in the Department of Computer Science at the University of

Oviedo and a member of the SYMM Working Group of the W3C. His current research interests are in multimedia systems and services, content distribution networks, digital interactive TV services, and mobile ad hoc networks. He is also taking part in several research projects on the national and European levels.

Ayan Paul received his PGDBA from the Indian Institute of Technology, Kharagpur. Currently, he is pursuing his Ph.D. at Jadavpur University (Kolkata, India). His research interests include wireless networking and mobile computing, network economics, network operations, and management. He is a member of IEEE and an executive committee member of Calcutta Chapter of IEEE Communication Society. He has over eight years of experience in the telecom industry in various technologies.

Vidyasagar Potdar holds the position of Head of Planning, Policy and Business Development at Digital Ecosystems & Business Intelligence Institute at Curtin University, Australia. Dr. Potdar's main area of research is Wireless Sensor Networks. He heads the Wireless Sensor Networks Research Lab http://wsn.debii.curtin.edu.au, which undertakes research in the areas of wireless hardware development, communication protocol development, advanced cross layer protocol design etc. He has attracted more than $450,000 in research funding to solve the problem of traffic congestion in Perth Metropolitan Areas using Wireless Sensor Networks. He is currently supervising 8 full time PhD candidates and has published more than 85 scientific research articles in refereed journals, peer reviewed conference papers, book chapters etc. most of his publications are in top Tier 1 conferences and journals including IEEE and ACM. He can be reached via http://debii.curtin.edu.au/~vidy

A.J.D. Rathnayaka graduated with B.Sc(Hons) in Electronics and Telecommunication Engineering from University of Moratuwa, Sri Lanka. Currently she is working as a Training Research Associate at WSN lab, DEBII, Curtin University of Technology in the field of Wireless Sensor Networks.

Samir K Sadhukhan received M. Sc. degree in Applied Mathematics from Calcutta University, India in the year 1978, M.Tech. degree in Computer Science from Indian Statistical Institute, India in the year 1983 and MBA degree from Jadavpur University, India in the year 1996. He is currently working as SSA in the Indian Institute of Management Calcutta. His research interests are in Mobile and Wireless network planning and algorithms in heuristic search.

Maytham Safar is currently an Associate Professor at the Computer Engineering Department at Kuwait University. He received his Ph.D. degree in Computer Science from the University of Southern California in 2000. He has over one hundred books, book chapters, and conference/journal articles. Current research interests include social networks, complex networks, location based services, and geographic information systems. He is a Senior Member of IEEE since 2008, and the First Kuwaiti to become an ACM Senior member in 2009. He is also a member of IEEE Standards Association, IEEE Computer Society, IEEE Geoscience & Remote Sensing Society, IADIS, @WAS, SDIWC and INSNA. He served on over one hundred conference committees as a reviewer and/or a scientific program committee member. He also served as a member of steering committee, organizing committee, and publicity committee of over twenty conferences. He served as a reviewer for over twenty journals. He also served as a member on the editorial board and committees on over ten journals. He established the first complex networks research group (Synergy) at Kuwait University, http://synergy.ku.edu.kw, 2010.

Fernando Santos holds a MSc in Finance from the Universidade Católica Portuguesa, where he was a teaching assistant. He is currently in the Postgraduate Program in Economics at Universidad Carlos III.

Atif Sharif holds a B.Sc(Hons) in Electrical Engineering from the University of Engineering and Technology Taxila, Pakistan, and M.Sc in Electrical Engineering from Center of Advanced Studies in Engineering (CASE. Currently, he is a Ph.D research scholar at Curtin University of Technology, Australia. He was a senior member of DSP-Embedded group in AERO (Oct, 2001-Jan, 2008). He also worked with number of professional leaders like Cadence, Xilinx, Altera, Mentor Graphics on various cutting edge technologies involving onechip solution. In 2000- 2001 he worked with a team of Communications Enabling Technologies and successfully design and implemented Multiprotocol Label Switching (MPLS) Domain, Programmed an emulator of fastest WAN routing and forwarding technology, MPLS.

S. Sivagurunathan is an Assistant Professor in Computer Science and Applications at Gandhigram Rural Institute – Deemed University, Gandhigram, Dindigul (India). He received his B.Sc., degree in Physics and Post Graduation in Computer Applications from the Madurai Kamaraj University (India) and PhD from Anna University, Chennai (India). He has twelve years of teaching experience. His primary research interests are in the area of wireless networks, storage area networks, and network security. He has published many research articles and chapters in text books in his area of research. He is a life member of Computer Society of India.

P. Subathra is an Assistant Professor in the Department of Computer Science and Engineering, Thiagarajar College of Engineering, Madurai, Tamil Nadu, India. She has a BE (Electrical and Electronic Engineering) and ME (Computer Science and Engineering) degrees from Madurai Kamaraj University, Madurai, Tamil Nadu, India. She did her Ph.D. in Anna University, Chennai. She has eight years of teaching experience. She has published about five articles in the proceedings of national and international conferences and journals. Her current research interests include communication networks, mobile ad-hoc networks, and network security. She is a life member is ISTE, Cryptography Society of India, and Computer Society of India.

Sihem Trabelsi received her Master's degree in Information and Communication Technologies from the High School of Communications of Tunis (Sup'Com) in 2006. She is currently working toward the Ph.D. degree at Sup'Com in the Communications Networks and Security (CNAS) Research Laboratory, and she holds the position of head of the Internet services unit at the National Telecommunications Commission. Her research interests include heterogeneous wireless access networks, mobility and resource management, and QoS optimization.

Index

D

E

F

G

H

I

K

L

M

N

O

P

Q

R

S

T

U

V

W

X

www.ingramcontent.com/pod-product-compliance
Lightning Source LLC
LaVergne TN
LVHW062140200926
848484LV00007B/1839